Ω Perspectives in Mathematical Logic

Ω-Group:
R. O. Gandy H. Hermes A. Levy G. H. Müller
G. E. Sacks D. S. Scott

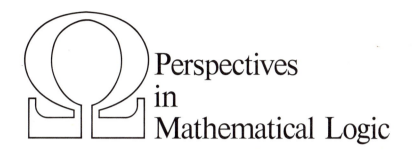

Jon Barwise

Admissible Sets and Structures

An Approach to Definability Theory

Springer-Verlag
Berlin Heidelberg New York 1975

JON BARWISE
Department of Mathematics, University of Wisconsin
Madison, WI 53706 / USA
and
U.C.L.A., Los Angeles, CA 90024 / USA

With 22 Figures

AMS Subject Classification (1970): 02 F 27, 02 B 25, 02 H 10, 02 K 35

ISBN 3-540-07451-1 Springer-Verlag Berlin Heidelberg New York
ISBN 0-387-07451-1 Springer-Verlag New York Heidelberg Berlin

Library of Congress Cataloging in Publication Data. Barwise, Jon. Admissible sets and structures. (Perspectives in mathematical logic). Bibliography: p. Includes index. 1. Admissible sets. 2. Definability theory. I. Title. QA9.B29. 511'.3. 75-33102.

Typesetting and printing: Zechnersche Buchdruckerei, Speyer. Bookbinding: Konrad Triltsch, Würzburg.

To my mother
and the memory of my father

Preface to the Series

On Perspectives. *Mathematical logic arose from a concern with the nature and the limits of rational or mathematical thought, and from a desire to systematise the modes of its expression. The pioneering investigations were diverse and largely autonomous. As time passed, and more particularly in the last two decades, interconnections between different lines of research and links with other branches of mathematics proliferated. The subject is now both rich and varied. It is the aim of the series to provide, as it were, maps or guides to this complex terrain. We shall not aim at encyclopaedic coverage; nor do we wish to prescribe, like Euclid, a definitive version of the elements of the subject. We are not committed to any particular philosophical programme. Nevertheless we have tried by critical discussion to ensure that each book represents a coherent line of thought; and that, by developing certain* themes, *it will be of greater interest than a mere assemblage of results and techniques.*

The books in the series differ in level: some are introductory some highly specialised. They also differ in scope: some offer a wide view of an area, others present a single line of thought. Each book is, at its own level, reasonably self-contained. Although no book depends on another as prerequisite, we have encouraged authors to fit their book in with other planned volumes, sometimes deliberately seeking coverage of the same material from different points of view. We have tried to attain a reasonable degree of uniformity of notation and arrangement. However, the books in the series are written by individual authors, not by the group. Plans for books are discussed and argued about at length. Later, encouragement is given and revisions suggested. But it is the authors who do the work; if, as we hope, the series proves of value, the credit will be theirs.

History of the Ω-Group. *During 1968 the idea of an integrated series of monographs on mathematical logic was first mooted. Various discussions led to a meeting at Oberwolfach in the spring of 1969. Here the founding members of the group (R. O. Gandy, A. Levy, G. H. Müller, G. Sacks, D. S. Scott) discussed the project in earnest and decided to go ahead with it. Professor F. K. Schmidt and Professor Hans Hermes gave us encouragement and support. Later Hans Hermes joined the group. To begin with all was fluid. How ambitious should we be? Should we write the books ourselves? How long would it take? Plans for authorless books were promoted, savaged and scrapped. Gradually there emerged a form and a method. At the end of an infinite discussion we found our name, and that of the series. We established our centre in Heidelberg. We agreed to meet twice a year together with authors, consultants and assistants, generally in Oberwolfach. We soon found the value of collaboration: on the one hand the permanence of the founding group gave coherence to the over-all plans; on the other hand the stimulus of new contributors kept the project alive and flexible. Above all, we found how intensive discussion could modify the authors ideas and our own. Often the battle ended with a detailed plan for a better book which the author was keen to write and which would indeed contribute a perspective.*

Acknowledgements. *The confidence and support of Professor Martin Barner of the Mathematisches Forschungsinstitut at Oberwolfach and of Dr. Klaus Peters of Springer-Verlag made possible the first meeting and the preparation of a provisional plan. Encouraged by the Deutsche Forschungsgemeinschaft and the Heidelberger Akademie der Wissenschaften we submitted this plan to the Stiftung Volkswagenwerk where Dipl. Ing. Penschuck vetted our proposal; after careful investigation he became our adviser and advocate. We thank the Stiftung Volkswagenwerk for a generous grant (1970–73) which made our existence and our meetings possible.*

Since 1974 the work of the group has been supported by funds from the Heidelberg Academy; this was made possible by a special grant from the Kultusministerium von Baden-Württemberg (where Regierungsdirektor R. Goll was our counsellor). The success of the negotiations for this was largely due to the enthusiastic support of the former President of the Academy, Professor Wilhelm Doerr. We thank all those concerned.

Finally we thank the Oberwolfach Institute, which provides just the right atmosphere for our meetings, Drs. Ulrich Felgner and Klaus Gloede for all their help, and our indefatigable secretary Elfriede Ihrig.

Oberwolfach
September 1975

R. O. Gandy — *H. Hermes*
A. Levy — *G. H. Müller*
G. Sacks — *D. S. Scott*

Author's Preface

It is only before or after a book is written that it makes sense to talk about *the* reason for writing it. In between, reasons are as numerous as the days. Looking back, though, I can see some motives that remained more or less constant in the writing of this book and that may not be completely obvious.

I wanted to write a book that would fill what I see as an artificial gap between model theory and recursion theory.

I wanted to write a companion volume to books by two friends, H. J. Keisler's *Model Theory for Infinitary Logic* and Y.N. Moschovakis' *Elementary Induction on Abstract Structures*, without assuming material from either.

I wanted to set forth the basic facts about admissible sets and admissible ordinals in a way that would, at long last, make them available to the logic student and specialist alike. I am convinced that the tools provided by admissible sets have an important rôle to play in the future of mathematical logic in general and definability theory in particular. This book contains much of what I wish every logician knew about admissible sets. It also contains some material that every logician ought to know about admissible sets.

Several courses have grown out of my desire to write this book. I thank the students of these courses for their interest, suggestions and corrections. A rough first draft was written at Stanford during the unforgettable winter and spring of 1973. The book was completed at Heatherton, Freeland, Oxfordshire during the academic year 1973—74 while I held a research grant from the University of Wisconsin and an SRC Fellowship at Oxford. I wish to thank colleagues at these three institutions who helped to make it possible for me to write this book, particularly Professors Feferman, Gandy, Keisler and Scott. I also appreciate the continued interest expressed in these topics over the past years by Professor G. Kreisel, and the support of the Ω-Group during the preparation of this book. I would like to thank Martha Kirtley and Judy Brickner for typing and John Schlipf, Matt Kaufmann and Azriel Levy for valuable comments on an earlier version of the manuscript. I owe a lot to Dana Scott for hours spent helping prepare the final manuscript. I would also like to thank Mrs. Nora Day and the other residents of Freeland for making our visit in England such a pleasant one.

A final but large measure of thanks goes to my family: to Melanie for allowing me to use her room as a study during the coal strike; to Jon Russell for help with the corrections; but most of all to Mary Ellen for her encouragement and patience. To Mary Ellen, on this our eleventh anniversary, I promise to write at most one book every eleven years.

September 19, 1975 K.J.B.
Santa Monica

Table of Contents

Major Dependencies

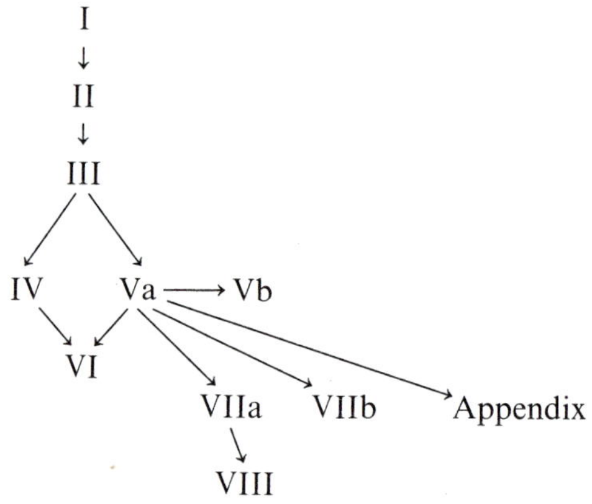

(Va denotes the first three §§ of Chapter V, similarly for VIIa.)

Introduction

Since its beginnings in the early sixties, admissible set theory has become a major source of interaction between model theory, resursion theory and set theory. In fact, for the student of admissible sets the old boundaries between fields disappear as notions merge, techniques complement one another, analogies become equivalences, and results in one field lead to results in another. This is the view of admissible sets we hope to share with the reader of this book.

Model theory, recursion theory and set theory all deal, in part, with problems of definability and set existence. *Definability theory* is (by definition) that part of mathematical logic which deals with such problems. The Craig Interpolation Theorem, Kleene's analysis of Δ_1^1 sets by means of the hyperarithmetic sets, Gödel's universe L of constructible sets and Shoenfield's Absoluteness Lemma are all major contributions to definability theory. The theory of admissible sets takes such apparently divergent results and makes them converge in a single coherent body of thought, one with ramifications for all parts of logic.

This book is written for the student who has taken a good first space year graduate course in logic. The specific material we presuppose can be summarized as follows. The student should understand the completeness, compactness and Löwenheim-Skolem theorems as well as the notion of elementary submodel. He should be familiar with the basic properties of recursive functions and recursively enumerable (hereinafter r.e.) sets. The student should have seen the development of intuitive set theory in some formal theory like ZF (Zermelo-Fraenkel set theory). His life will be more pleasant if he has some familiarity with the constructible sets before reading §§ II.5, 6 or V.4—8, but our treatment of constructible sets is self-contained.

A logical presentation of a reasonably advanced part of mathematics (which this book attempts to be) bears little relation to the historical development of that subject. This is particularly true of the theory of admissible sets with its complicated and rather sensitive history. On the other hand, a student is handicapped if he has no idea of the forces that figured in the development of his subject. Since the history of admissible sets is impossible to present here, we compromise by discussing how some of the older material fits into the current theory. We concentrate on those topics that are particularly relevant to this book. The prerequisites for understanding the introduction are rather greater than those for understanding the book itself.

Recursive ordinals and hyperarithmetic sets. In retrospect, the study of admissible ordinals began with the work of Church and Kleene on *notation systems and recursive ordinals* (Church-Kleene [1937], Church [1938], Kleene [1938].) This study began as a recursive counterpart to the classical theory of ordinals; *the least nonrecursive ordinal* ω_1^c is the recursive analogue of ω_1, the first uncountable ordinal. (Similarly for ω_2^c and ω_2, etc.) The theory of recursive ordinals had its most important application when Kleene [1955] used it in his study of the class of *hyperarithmetic sets*, the smallest reasonably closed class of sets of natural numbers which can be considered as given by the structure $\mathcal{N}=\langle\omega, +, \times\rangle$ of natural numbers. Kleene's theorem that

$$\text{hyperarithmetic} = \Delta_1^1$$

provided a construction process for the class of Δ_1^1 sets and constituted the first real breakthrough into (applied) second order logic. One of our aims is to provide a similar analysis for any structure $\mathfrak{M}$. Given $\mathfrak{M}$ we construct the smallest admissible set $\mathbb{HYP}_{\mathfrak{M}}$ above $\mathfrak{M}$ (in § II.5) and use it in the study of definability problems over $\mathfrak{M}$ (in Chapters IV and VI).

The study of hyperarithmetic sets generated a lot of discussion of the analogy between, on the one hand, the Π_1^1 and hyperarithmetic sets, and the r.e. and recursive sets on the other. These analogies became particularly striking when expressed in terms of representability in ω-logic and first order logic, by Grzegorczyk, Mostowski and Ryll-Nardzewski [1959]. The analogy had some defects, though, as the workers realized at the time. For example, the image of a hyperarithmetic function is hyperarithmetic, not just Π_1^1 as the analogy would suggest.

Kreisel [1961] analyzed this situation and discovered that the correct analogy is between Π_1^1 and hyperarithmetic on the one hand and r.e. and *finite* (not recursive) on the other. He went on to develop a recursion theory on the hyperarithmetic sets via a notation system. (He also proved the Kreisel Compactness Theorem for ω-logic: If a Π_1^1 theory T of second order arithmetic is inconsistent in ω-logic, then some hyperarithmetic subset $T_0\subseteq T$ is inconsistent in ω-logic.) This theory was expanded in the *metarecursion theory* of Kreisel-Sacks [1965]. Here one sees how to develop, by means of an ordinal notation system, an attractive recursion theory on ω_1^c such that for $X\subseteq\omega$:

X is Π_1^1 iff X is ω_1^c-r.e.,

X is Δ_1^1 iff X is ω_1^c-finite.

In § IV.3 we generalize this, by means of $\mathbb{HYP}_{\mathfrak{M}}$, to show that for any countable structure $\mathfrak{M}$ and any relation R on $\mathfrak{M}$:

R is Π_1^1 on $\mathfrak{M}$ iff R is $\mathbb{HYP}_{\mathfrak{M}}$-r.e.,

R is Δ_1^1 on $\mathfrak{M}$ iff R is $\mathbb{HYP}_{\mathfrak{M}}$-finite,

thus providing a construction process for the Δ_1^1 relations over any countable structure $\mathfrak{M}$ whatsoever. The use of notation systems then allows us to transfer results from $\mathbb{HYP}_{\mathfrak{M}}$ to $\mathfrak{M}$ itself (see §§ V.5 and VI.5).

Constructible sets. The other single most important line of development leading to admissible sets also goes back to the late thirties. It began with the introduction by Gödel [1939] of the class L of constructible sets, in order to provide a model of set theory satisfying the axiom of choice and generalized continuum hypothesis (GCH).

Takeuti [1960, 1961] discovered that one could develop L by means of a recursion theory on the class Ord of all ordinals. He showed that Gödel's proof of the GCH in L corresponds to the following recursion theoretic stability: If κ is an uncountable cardinal and if $F\colon \mathrm{Ord} \to \mathrm{Ord}$ is ordinal-recursive then $F(\beta)<\kappa$ for all $\beta<\kappa$. In modern terminology, every uncountable cardinal is *stable*. Takeuti's definition of the ordinal-recursive functions was by means of schemata, Tagué [1964] provided an equivalent definition by means of an equation calculus obtained by adjoining an infinitistic rule to Kleene's equation calculus for ordinary recursion theory.

Admissible ordinals and admissible sets. The notion of admissible ordinal can be viewed as a common generalization of metarecursion theory and Takeuti's recursion theory on Ord. Kripke [1964] introduced admissible ordinals by means of an equation calculus. Platek [1965] gave an independent equivalent definition using schemata and another by means of machines as follows. Let α be an ordinal. Imagine an idealized computer capable of performing computations involving less than α steps. A function F computed by such a machine is called *α-recursive*. The ordinal α is said to be admissible if, for every α-recursive function F, whenever $\beta<\alpha$ and $F(\beta)$ is defined then $F(\beta)<\alpha$, that is, the initial segment determined by α is closed under F.

The first admissible ordinal is ω. An ordinal like $\omega+\omega$ cannot be admissible since, for $\alpha>\omega$, the equation

$$F(\beta)=\sup_{\gamma<\beta}(\omega+\gamma)$$

defines an α-recursive function and $F(\omega)=\omega+\omega$. The second admissible ordinal is, in fact, ω_1^c and the ω_1^c-recursion theory of Kripke and Platek agrees with that from metarecursion theory (see §§ IV.3 and V.5). The theorem of Takeuti mentioned above implies that every uncountable cardinal is admissible. The important advance made possible by the definition of admissible ordinal is that it allows one to study recursion theory on important ordinals (like ω_1^c) which are not cardinals.

Takeuti's work had shown that recursion theory on Ord amounts to definability theory on L. Analogously, the Kripke-Platek theory on an admissible ordinal α has a definability version on $L(\alpha)$, the sets constructible before stage α. It is this second approach which is most useful and is the one followed here. It leads us to consider *admissible sets*, sets $\mathbb{A}$ which, like $L(\alpha)$ for α admissible, satisfy *closure conditions* which insure a reasonable definability theory on $\mathbb{A}$. These principles are formalized in a first order set theory KP. In order to study general definability, though, not just definability theory in transitive sets, we must strengthen the general theory weakening KP to a new theory KPU. But this is taken up in detail at once in Chapter I.

Infinitary Logic. There is just one other idea that needs to be introduced here, that of infinitary logic. The model theory of $\mathsf{L}_{\omega\omega}$, the usual first order predicate calculus, consists largely of global results, results which have to do with all models of some first order theory. These results have little to say about any one particular structure since only finite structures can be characterized up to isomorphism by a theory of $\mathsf{L}_{\omega\omega}$. Recursion theory, on the other hand, is a local theory about the single structure $\mathcal{N}$. If we are to have a global theory with non-trivial local consequences, we must extend the model theory of $\mathsf{L}_{\omega\omega}$ to stronger logics, logics which can characterize structures and properties not characterizable in $\mathsf{L}_{\omega\omega}$. For the study of admissible sets the appropriate logics turn out to be admissible fragments $\mathsf{L}_{\mathbb{A}}$ of $\mathsf{L}_{\infty\omega}$, as developed in Barwise [1967, 1969a, b]. The countable case is studied in Chapter III; the uncountable case, in Chapters VII and VIII.

Some material not covered in this book. This book is a perspective on admissible sets, not a definitive treatment. It is far bigger and contains somewhat less material than we foresaw when we began writing. In particular, the following topics, all highly relevant to definability theory, are either omitted or slighted:

recursion theory in higher types,
Spector classes,
non-monotonic inductive definitions,
relative recursion theory on admissible ordinals,
forcing on admissible sets,
forcing and infinitary compactness arguments.

It is planned that some of these topics will be treated in other books in this series.

A note to the casual reader. There is one bit of notation that might be confusing to the casual reader of this book. We use $\mathfrak{A}$ for arbitrary models of the theory KPU or, more generally, for arbitrary structures for the language $\mathsf{L}^* = \mathsf{L}(\in, \ldots)$ in which KPU is formulated. We switch to the notation $\mathbb{A}$ when our structure is well founded. In 99.44% of the uses $\mathbb{A}$ will denote an admissible set.

Part A

The Basic Theory

"Logic is logic. That's all I say."

Oliver Wendel Holmes
The Deacon's Masterpiece

Chapter I
Admissible Set Theory

Admissible sets are the intended models of a certain first order theory. In this chapter we discuss the theory itself and show how to develop a significant part of intuitive set theory within it.

1. The Role of Urelements

Our approach to admissible sets is unorthodox in several respects, the most obvious being that we allow admissible sets to contain urelements. Bluntly put, we consider admissible sets which are built up out of the stuff of mathematics, not just the sets built up from the empty set. To make this a little clearer, and to see why it is an obvious step to take, we begin by reviewing the development of ZF, Zermelo-Fraenkel set theory, as it is correctly presented (in, for example, § 9.1 of Shoenfield [1967]).

The fundamental tenet of set theory is that, given a collection M of mathematical objects, subcollections are themselves perfectly reasonable mathematical objects, as are collections of these new objects, and so on. We begin with a collection M of objects called *urelements* (sometimes called *points*, *atoms* or *individuals*, depending principally on our subject), which we think of as being given outright. The objects in M might be real numbers, elements of some group or even physical objects. We construct sets out of the objects of M in stages. At each stage α we are allowed to form sets out of urelements and the sets formed at earlier stages. An object is a *set on* M just in case it is formed at some stage in this construction; the collection of all sets on M is denoted by $\mathbb{V}_M$.

Now it turns out, and it must have been a surprising discovery, that if we allow strong enough principles of construction at each stage α, and if we assume that there are enough stages, then urelements become superfluous. All ordinary mathematical objects occur, up to isomorphism, in $\mathbb{V}$, *i.e.* in $\mathbb{V}_M$ where M is the empty collection. It is consistent with this that the extensionality axiom of ZF explicitly rule out the existence of objects which are not sets; the combination of the power set and replacement axioms is so strong as to make urelements unnecessary.

Set theory, as formalized in ZF, provides an extremely powerful and elegant way to organize existing mathematics. It is not without its drawbacks, nevertheless. While it is too weak to decide some questions (like the continuum hypothesis) which seem meaningful (even important), it is in some ways too strong. Some examples:

(1) The most obvious advantage of the axiomatic method is lost since ZF has so few recognizable models in which to interpret its theorems.

(2) Important distinctions on the nature of the sets asserted to exist are completely lost.

(3) The principle of parsimony, of established value throughout the mathematical ages, is violated at every turn.

(4) Large parts of mathematical practice are distorted by the demand that all mathematical objects be *realized as* sets (as opposed to being *isomorphic to* sets). If these objections are not too clear, they should becomes so as we investigate the theory of admissible sets. At any rate these considerations, and others familiar to anyone versed in generalized recursion theory, eventually dictate the study of set theories weaker than ZF, weaker in the principles of set existence which they attempt to formalize. The theory we have in mind here, of course, is the Kripke-Platek theory KP for admissible sets.

It is at this point that one is tempted to make a simplifying mistake. We have first thrown out urelements from ZF because ZF is so strong. When we then weaken ZF to KP we must remember to reexamine the justification for banning the urelements. Doing so, we discover that the justification has completely disappeared. In this book we readmit urelements by "weakening" KP to a theory KPU. The original KP will be equivalent to the theory

$$\text{KPU} + \text{"there are no urelements".}$$

This approach has many advantages. The chief is that it allows us to form, for any structure $\mathfrak{M} = \langle M, R_1 \ldots R_k \rangle$ a particularly important admissible set $\mathbb{HYP}_{\mathfrak{M}}$ above $\mathfrak{M}$, one which is of great use in the study of definability over $\mathfrak{M}$. The approach has no disadvantages since we can always restrict attention to the special case where there are no urelements.

1.1—1.4 Examples

1.1. The point made in (1) above becomes clearer when we recall that if ZF is consistent, so is

$$\text{ZF} + \text{"There is no transitive model of ZF".}$$

(Prove this without using Gödel's Incompleteness Theorems!)

1.2. The observation in (2) is illustrated by considering, for example, an arbitrary abelian group $\mathfrak{G} = \langle G, + \rangle$. Consider the following subgroups of G:

$$pG = \{px \mid x \in G\},$$

$$T = \{x \mid nx = 0 \text{ for some natural number } n > 0\} = \text{the torsion subgroup of } G,$$

$$D = \bigcup \{H \mid H \text{ is a divisible subgroup of } G\} = \text{the divisible part of } G.$$

While these definitions are clearly increasing in logical complexity, there is no distinction to be made between them from ZF's point of view. We will return to this example in Chapter IV.

1.3. As an example of the way one is tempted to violate the principle of parsimony when working in ZF, one need only look in the average text on set theory. There you will find the power set axiom (a very strong axiom from our point of view) used to verify a simple fact like the existence of $a \times b$.

1.4. The point made in (4) above is illustrated by considering the real line. While we know how to construct something isomorphic to the real line in ZF (either by Cauchy sequences or by Dedekind cuts), in practise the mathematician is not interested in the details of this construction. For example, he would never think of worrying about what the elements of $\sqrt{2}$ happen to be.

1.5 Notes. The notes at the end of sections are used to collect historical remarks, credit for theorems (when possible) and various remarks which might otherwise have gone into footnotes.

In the early days of set theory, certainly in the work of Zermelo, urelements were an integral part of the subject. The rehabilitation of urelements in the context of admissible set theory is such a simple idea that it would be silly to assign credit for it to any one person. Probably everyone who has thought at all about infinitary logic and admissible sets has had a similar idea.

Karp [1968] suggests the study of nontransitive admissible sets. Kreisel [1971] points out that "the principal gap in the existing model theoretic [generalized recursion theory] ... is its preoccupation with *sets* (that is sets built up from the empty set by some cumulative operation ...); not even sets of individuals are treated." Barwise [1974] contains the first published treatment of admissible sets with urelements. This book grew out of that paper, to some extent. It is worth remembering that the defense of urelements given in § 1 would have been unnecessary not too long ago. Perhaps it will be equally pointless sometime in the future.

2. *The Axioms of* KPU

Let L be a first order language with equality, some relation, function and constant symbols and let $\mathfrak{M} = \langle M, \text{---} \rangle$ be a structure for this language L. We wish to form admissible sets which have M as a collection of urelements; these admissible sets are the intended models of a theory KPU which we begin to develop in this section.

The theory KPU is formulated in a language $\mathsf{L}^* = \mathsf{L}(\in, \ldots)$ which extends L by adding a membership symbol $\in$ and, possibly other function, relation and constant symbols. Rather than describe L^* precisely, we describe its class of structures, leaving it to the reader to formalize L^* in a way that suits his tastes.

2.1 Definition. A structure $\mathfrak{A}_{\mathfrak{M}} = (\mathfrak{M}; A, E, \ldots)$ for L^* consists of

(i) a structure $\mathfrak{M} = \langle M, \text{---}\rangle$ for the language L, where $M = \emptyset$ is kept open as a possibility (the members of M are the *urelements* of $\mathfrak{A}_{\mathfrak{M}}$);

(ii) a nonempty set A disjoint from M (the members of A are the *sets* of $\mathfrak{A}_{\mathfrak{M}}$);

(iii) a relation $E \subseteq (M \cup A) \times A$ (which interprets the *membership* symbol $\in$);

(iv) other functions, relations and constants on $M \cup A$ to interpret any other symbols in $\mathsf{L}(\in, \ldots)$ (that is the symbols in the list indicated by the three dots).

The equality symbol of L^* is always interpreted as the usual equality relation.

We use variables of L^* subject to the following conventions: Given a structure $\mathfrak{A}_{\mathfrak{M}} = (\mathfrak{M}; A, E, \ldots)$ for L^*,

$p, q, p_1, \ldots$	range over M (urelements),
$a, b, c, d, f, r, a_1, \ldots$	range over A (sets),
$x, y, z, \ldots$	range over $M \cup A$.

This notation gives us an easy way to assert that something holds of sets, or of urelements. For example, $\forall p\, \exists a\, \forall x\, (x \in a \leftrightarrow x = p)$ asserts that $\{p\}$ exists for any urelement p, whereas $\forall p\, \exists a\, \forall q\, (q \in a \leftrightarrow q = p)$ asserts that there is a set a whose intersection with the class of all urelements is $\{p\}$.

We sometimes use (e.g. in 2.2(iii)) u, v, w to denote any kind of variable.

The axioms of KPU are of three kinds. The axioms of extensionality and foundation concern the basic nature of sets. The axioms of pair, union and Δ_0 separation deal with the principles of set construction available to us. The most important axiom, Δ_0 collection, guarantees that there are enough stages in our construction process. In order to state the latter two axioms we need to define the notion of Δ_0 formula of $\mathsf{L}(\in, \ldots)$, of Lévy [1965].

2.2 Definition. The collection of Δ_0 *formulas* of a language $\mathsf{L}(\in, \ldots)$ is the smallest collection Y containing the atomic formulas of $\mathsf{L}(\in, \ldots)$ closed under:

(i) if φ is in Y, then so is $\neg\varphi$;

(ii) if φ, ψ are in Y, so are $(\varphi \wedge \psi)$ and $(\varphi \vee \psi)$;

(iii) if φ is in Y, then so are $\forall u \in v\, \varphi$ and $\exists u \in v\, \varphi$ for all variables u and v.

The importance of Δ_0 formulas rests in the metamathematical fact that any predicate defined by a Δ_0 formula is absolute (see 7.3), and the empirical fact (which we will verify) that many predicates occuring in nature can be defined by Δ_0 formulas (see Table 1).

2.3 Definition. The theory KPU (relative to a language $\mathsf{L}(\in, \ldots)$) consists of the universal closures of the following formulas:

Extensionality: $\forall x\, (x \in a \leftrightarrow x \in b) \rightarrow a = b$;

Foundation: $\exists x\, \varphi(x) \rightarrow \exists x [\varphi(x) \wedge \forall y \in x\, \neg\varphi(y)]$ for all formulas $\varphi(x)$ in which y does not occur free;

Pair: $\exists a\,(x \in a \wedge y \in a)$;

Union: $\exists b\,\forall y \in a\,\forall x \in y\,(x \in b)$;

Δ_0 *Separation:* $\exists b\,\forall x\,(x \in b \leftrightarrow x \in a \wedge \varphi(x))$ for all Δ_0 formulas in which b does not occur free;

Δ_0 *Collection:* $\forall x \in a\,\exists y\,\varphi(x,y) \rightarrow \exists b\,\forall x \in a\,\exists y \in b\,\varphi(x,y)$ for all Δ_0 formulas in which b does not occur free.

Note that the formulas $\varphi(x)$, $\varphi(x,y)$ used above may have other free variables.

2.4 Definition. KPU$^+$ is KPU plus the axiom:

$$\exists a\,\forall x\,[x \in a \leftrightarrow \exists p\,(x = p)],$$

which asserts that there is a set of all urelements.

2.5 Definition. KP is KPU plus the axiom:

$$\forall x\,\exists a\,(x = a),$$

which asserts that every object is a set, i.e. that there are no urelements.

2.6 A word of caution. There are some axioms built into our definition of structure for $\mathsf{L}(\in, \ldots)$. The following sentences make these conditions explicit and should be considered part of the axioms of KPU:

$\forall p\,\forall a\,(p \neq a)$ (cf. 2.1(ii));

$\exists a\,(a = a)$ (expresses $A \neq \emptyset$ in 2.1(ii));

$\forall p\,\forall x\,(x \notin p)$ (cf. 2.1(iii)).

2.7 Notes. The notions of Δ_0 and Σ_1 are due to Lévy [1965]. The axioms of KP go back to Platek (the P in KP), in particular, to Platek [1966]. He defined an admissible set A to be a transitive, nonempty set closed under TC satisfying Δ_0 separation and Σ reflection. Kripke [1964] (the K in KP) had, independently, a similar notion with Σ reflection replaced by Σ replacement. (For the models Kripke had in mind (L_α's) they are equivalent; but in general it is Σ reflection which matters.) Both of these men were influenced by Kreisel [1959] and Kreisel [1965]. See, e.g. Kreisel [1965, p. 199(b)]. (For the notion of a Σ formula, see 4.1 below.)

3. Elementary Parts of Set Theory in KPU

In this section we show how to define some of the elementary concepts of intuitive set theory in KPU. We thus want to show that certain sentences of $\mathsf{L}(\in, \ldots)$ are logical consequences of KPU. We do this here by translating these sentences into English and then giving their proofs in English, being careful to use only axioms from KPU. For example, rather than state:

$$\mathrm{KPU} \vdash \forall x \, \forall y \, \exists! a \, \forall z \, [z \in a \leftrightarrow z = x \vee z = y],$$

we state:

Given x, y, there is a unique set $a = \{x, y\}$ with only x, y as members;

and then we give an informal proof of the latter. (Given x, y, there is a b with $x, y \in b$, by Pair. By Δ_0 separation there is an a with $z \in a \leftrightarrow z \in b \wedge [z = x \text{ or } z = y]$, the part in brackets being a Δ_0 formula. By Extensionality, there can be at most one such a.) Thus all results in this section are proved in KPU.

3.1 Proposition. (i) *There is a unique set* 0 *with no elements.*
(ii) *Given* a, *there is a unique set* $b = \bigcup a$ *such that* $x \in b$ *iff* $\exists y \in a \, (x \in y)$.
(iii) *Given* a, b *there is a unique set* $c = a \cup b$ *such that* $x \in c$ *iff* $x \in a$ *or* $x \in b$.
(iv) *Given* a, b *there is a unique set* $c = a \cap b$ *such that* $x \in c$ *iff* $x \in a$ *and* $x \in b$.

Proof. These are all routine. By 2.1 (ii) there is a set b. For (i) we apply Δ_0 separation to b and the formula $x \neq x$. For (ii) use the union axiom to get a b' such that $\forall y \in a \, \forall x \in y \, (x \in b')$, and then form

$$b = \{x \in b' \mid \exists y \in a \, (x \in y)\}$$

by Δ_0 separation. For (iii), form $\bigcup \{a, b\}$. To prove (iv), let $c = \{x \in a \mid x \in b\}$, which exists by Δ_0 separation. In each case uniqueness follows from the axiom of extensionality. □

We define, as ususal, the ordered pair of x, y by

$$\langle x, y \rangle = \{\{x\}, \{x, y\}\}$$

and prove that $\langle x, y \rangle = \langle z, w \rangle$ iff $x = z$ and $y = w$.

3.2 Proposition. *For all* a, b *there is a set* $c = a \times b$, *the Cartesian product of* a *and* b, *such that*

$$c = \{\langle x, y \rangle \mid x \in a \text{ and } y \in b\}.$$

Proof. By Table 1 the predicate of a, b, u:

u is an ordered pair $\langle x, y \rangle$ with $x \in a$ and $y \in b$

is Δ_0 so we can use Δ_0 separation once we know that there is a set c with $\langle x, y \rangle \in c$ for all $x \in a$, $y \in b$. This follows from Δ_0 collection as follows. Given any $x \in a$ we first show that there is a w_x such that $\langle x, y \rangle \in w_x$ for all $y \in b$. Why? Well, given $y \in b$ there is a set $d = \langle x, y \rangle$. So, by Δ_0 collection there is a set w_x such that $\langle x,y\rangle \in w_x$ for all $y\in b$. Now, apply Δ_0 collection again. We have

$$\forall x\in a\, \exists w \underbrace{\forall y\in b\, \exists d\in w\, (d=\langle x,y\rangle)}_{\Delta_0}$$

so there is a c_1 such that for all $x\in a$, $y\in b$, $\langle x,y\rangle \in w$ for some $w\in c_1$. Thus, if $c=\bigcup c_1$, then $\langle x,y\rangle \in c$ for all $x\in a$, $y\in b$. □

The above is a good example of the principle of parsimony. In ZF, where one has the power set axiom, the set c needed in the proof can be taken to be just $P(P(a\cup b))$, but this proof does not carry over to KPU.

We can define ordered n-tuples, for $n>2$, as follows, by induction on n:

$$\langle x_1,\ldots,x_n\rangle = \langle x_1, \langle x_2,\ldots,x_n\rangle\rangle$$

and, similarly,

$$a_1\times\cdots\times a_n = a_1\times(a_2\times\cdots\times a_n)\,.$$

Thus $a_1\times\cdots\times a_n$ is the set of n-tuples $\langle x_1,\ldots,x_n\rangle$ with $x_i\in a_i$ for $i=1,\ldots,n$.

Now that we have ordered pairs, we can give the usual definitions of intuitive notions like relation, function, etc., all by Δ_0 formulas as in Table 1.

A set a is *transitive*, written Tran(a), iff

$$\forall y\in x\, \forall z\in y\, (z\in a)\,,$$

so that Tran(a) is a Δ_0 formula. Urelements are not considered transitive. Every set of urelements is transitive. The empty set 0 is transitive.

3.3 Definition. Let $\mathscr{S}(a)=a\cup\{a\}$.

3.4 Exercise. Prove (by induction) that for each n,

$$\mathrm{KPU}\vdash \forall x_1,\ldots,\forall x_n\, \exists a\, (a=\{x_1,\ldots,x_n\})\,.$$

3.5 Exercise. Show that if a is a set of transitive sets, then $\bigcup a$ is transitive. Show that if a is transitive and $b\subseteq a$, then $a\cup\{b\}$ is transitive. In particular, if a is transitive, so is $\mathscr{S}(a)$.

3.6 Definition. An *ordinal* is a transitive set a such that every member x of a is also a transitive set. Thus, we may write this definition as:

$$\mathrm{Ord}(a) \leftrightarrow \mathrm{Tran}(a) \wedge \forall x\in a\, \mathrm{Tran}(x)\,.$$

We use $\alpha,\beta,\gamma,\ldots$ to range over ordinals. We write $\alpha<\beta$ for $\alpha\in\beta$. An ordinal α is a *natural number* if for all $\beta\leqslant\alpha$, if $\beta\neq 0$ then $\beta=\mathscr{S}(\gamma)$ for some γ. We use variables $n,m,\ldots$ over natural numbers.

3.7 Exercise. We assume that the reader has some familiarity with ordinal numbers. He should verify that all the usual things are provable in KPU:

(i) 0 is an ordinal;

(ii) If α is an ordinal so is $\mathscr{S}(\alpha)$, usually written $\alpha+1$.

(iii) If $\alpha \neq \beta$ then $\alpha<\beta$ or $\beta<\alpha$. (This uses the axiom of foundation!)

(iv) For all α, $\alpha \nless \alpha$.

(v) If a is a set of ordinals, then $\bigcup a$ is an ordinal β with $\alpha \leqslant \beta$ whenever $\alpha \in a$, and $\exists \alpha \in a\,(\gamma \leqslant \alpha)$ whenever $\gamma<\beta$. (Thus β is the supremum of a, and we write $\beta=\sup(a)$.)

(vi) If $\alpha<\beta$ then $\alpha+1 \leqslant \beta$.

(vii) Every nonempty set of ordinals has a smallest element.

3.8 Definition. A set a is *finite* if there is a one-one function f with $\operatorname{dom}(f)=a$ and range some natural number n. A set a is *countable* if there is a one-one function f with domain a such that $f(x)$ is a natural number for every $x \in a$.

3.9 Exercise. (i) Show that every member of an ordinal is an ordinal.

(ii) Show that a set is an ordinal iff it is transitive and its elements are linearly ordered by $\in$.

(iii) Show that an ordinal is finite iff it is a natural number.

Table 1. Some Δ_0 Predicates

Predicate	*Abbreviation*	Δ_0 *Definition*
$x \subseteq y$		$\forall z \in x\ (z \in y)$
$a=\{y,z\}$		$y \in a \wedge z \in a \wedge \forall x \in a\ (x=y \vee x=z)$
$a=\langle y,z\rangle$		$\exists b \in a\, \exists c \in a\ (b=\{y\} \wedge c=\{y,z\} \wedge a=\{b,c\})$
$a=\langle x,y\rangle$ *for some* y	$1^{\mathrm{st}}(a)=x$	$\exists c \in a\, \exists y \in c\ (a=\langle x,y\rangle)$
$a=\langle x,y\rangle$ *for some* x	$2^{\mathrm{nd}}(a)=y$	$\exists c \in a\, \exists x \in c\ (a=\langle x,y\rangle)$
$a=\langle x,y\rangle$ *for some* x,y	"*a is an ordered pair*"	$\exists c \in a\, \exists x \in c\, \exists y \in c\ (a=\langle x,y\rangle)$
a is a relation	$\operatorname{Reln}(a)$	$\forall x \in a$ "*x is an ordered pair*"
f is a function	$\operatorname{Fun}(f)$	$\operatorname{Reln}(f) \wedge \forall a \in f\, \forall b \in f\ (1^{\mathrm{st}}a=1^{\mathrm{st}}b \rightarrow 2^{\mathrm{nd}}a=2^{\mathrm{nd}}b)$
r is a relation with domain a	$\operatorname{dom}(r)=a$	$\operatorname{Reln}(r) \wedge \forall b \in r\ (1^{\mathrm{st}}b \in a) \wedge \forall x \in a\, \exists b \in r\ (1^{\mathrm{st}}b=x)$
r is a relation with range a	$\operatorname{rng}(r)=a$	$\operatorname{Reln}(r) \wedge \forall b \in r\ (2^{\mathrm{nd}}b \in a) \wedge \forall x \in a\, \exists b \in r\ (2^{\mathrm{nd}}b=x)$
r is a relation with field a	$\operatorname{field}(r)=a$	$a=\operatorname{dom}(r) \cup \operatorname{rng}(r)$
$y=f(x)$		$\operatorname{Fun}(f) \wedge \langle x,y\rangle \in f$
$a=\bigcup b$		$\forall x \in b\, \forall y \in x\ (y \in a) \wedge \forall y \in a\, \exists x \in b\ (y \in x)$

4. Some Derivable Forms of Separation and Replacement

Our development of set theory progressed smoothly as long as the predicates involved were definable by Δ_0 formulas. With the notions of finite and countable in 3.8 we hit the first examples of predicates which cannot be so expressed.

For example, if we write either of these out they take the form

$$\exists f\, \varphi(f,a)$$

where φ is Δ_0. A formula of the form $\exists u\, \varphi(u)$, where φ is Δ_0, is called a Σ_1 *formula*. It turns out that a wide class of formulas are equivalent to Σ_1 formulas and that we can use these formulas in various forms of separation, collection and replacement.

4.1 Definition. The class of Σ *formulas* is the smallest class Y containing the Δ_0 formulas and closed under conjunction and disjunction (2.2(ii)), bounded quantification (2.2(iii)) and satisfying:

(i) if φ is in Y so is $\exists u\, \varphi$ for all variables u.

The class of Π *formulas*, on the other hand, is the smallest class Y' containing the Δ_0 formulas closed under conjunction, disjunction, bounded quantification and satisfying:

(ii) if φ is in Y' so is $\forall u\, \varphi$, for all variables u.

For example, the two formulas:

$$\forall b \in a\, [b \text{ is countable}] \quad \text{and} \quad \forall x \in a\, \exists b\, [\mathrm{Tran}(b) \wedge x \in b],$$

are Σ but not Σ_1. Clearly the negation of any Σ formula is logically equivalent to a Π formula and vice versa. As a corollary to Theorem 4.3 we will see that for every Σ formula φ, there is a Σ_1 formula φ' such that

$$\mathrm{KPU} \vdash \varphi \leftrightarrow \varphi'.$$

Given a formula φ and a variable w not appearing in φ, we write $\varphi^{(w)}$ for the result of replacing each *unbounded* quantifier in φ by a *bounded* quantifier; that is we replace:

$$\exists u \quad \text{by} \quad \exists u \in w, \quad \text{and}$$

$$\forall u \quad \text{by} \quad \forall u \in w,$$

for all variables u. Thus $\varphi^{(w)}$ is a Δ_0 formula. If φ is Δ_0 then $\varphi^{(w)} = \varphi$, since there are no unbounded quantifiers in φ. We *always* assume that w *does not* already appear in φ.

4.2 Lemma. *For each Σ formula φ the following are logically valid (i.e., true in all structures $\mathfrak{A}_{\mathfrak{M}}$):*

(i) $\varphi^{(u)} \wedge u \subseteq v \rightarrow \varphi^{(v)}$,

(ii) $\varphi^{(u)} \rightarrow \varphi$,

where $u \subseteq v$ *abbreviates the formula* $\forall x\,[x\in u \to x\in v]$. (Actually it is the universal closures of these formulas which are true in all $\mathfrak{A}_{\mathfrak{M}}$ since φ may have other free variables. We will not bother with this comment in the future.)

Proof. Both facts are proved by induction following the inductive definition 4.1 of Σ formula. Let us just prove the first, the second being similar. Fix a structure $\mathfrak{A}_{\mathfrak{M}} = (\mathfrak{M}; A, E, \ldots)$ and $x, y \in A \cup M$ so that $x \subseteq y$ is true in $\mathfrak{A}_{\mathfrak{M}}$. For Δ_0 formulas φ, we have, obviously, $\varphi = \varphi^{(x)} = \varphi^{(y)}$. Assume first that $(\varphi \wedge \psi)^{(x)}$ (i.e., assume it's true in $\mathfrak{A}_{\mathfrak{M}}$). Hence, $\varphi^{(x)}$ and $\psi^{(x)}$. By induction $\varphi^{(y)}$ and $\psi^{(y)}$, so $(\varphi \wedge \psi)^{(y)}$. Similarly for $(\varphi \vee \psi)^{(x)} \to (\varphi \vee \psi)^{(y)}$ and bounded quantifiers.

Now assume $(\exists w\, \varphi(w))^{(x)}$, so there is a $w \in x$ such that $\varphi(w)^{(x)}$. By induction $\varphi(w)^{(y)}$; and, since $x \subseteq y$, $\exists w \in y(\varphi(w)^{(y)})$; i.e., $(\exists w\, \varphi(w))^{(y)}$. □

4.3 Theorem. (The Σ Reflection Principle). *For all* Σ *formulas* φ *we have the following:*

$$\mathrm{KPU} \vdash \varphi \leftrightarrow \exists a\, \varphi^{(a)}.$$

(Here a is any set variable not occurring in φ; we will not continue to make these annoying conditions on variables explicit.) *In particular, every* Σ *formula is equivalent to a* Σ_1 *formula in* KPU.

Proof. We know from the previous lemma that $\exists a\, \varphi^{(a)} \to \varphi$ is valid, so the axioms of KPU come in only in showing $\varphi \to \exists a\, \varphi^{(a)}$. The proof is by induction on φ, the case for Δ_0 formulas being trivial. We take the three most interesting cases, leaving the other two to the reader.

Case 1. φ is $\psi \wedge \theta$. Assume that

$$\mathrm{KPU} \vdash \psi \leftrightarrow \exists a\, \psi^{(a)}, \quad \text{and}$$

$$\mathrm{KPU} \vdash \theta \leftrightarrow \exists a\, \theta^{(a)},$$

as induction hypothesis, and prove that

$$\mathrm{KPU} \vdash (\psi \wedge \theta) \to \exists a\, [\psi \wedge \theta]^{(a)}.$$

Let us work in KPU, assuming $\psi \wedge \theta$ and proving $\exists a\, [\psi^{(a)} \wedge \theta^{(a)}]$. Now there are a_1, a_2 such that $\psi^{(a_1)}, \theta^{(a_2)}$, so let $a = a_1 \cup a_2$. Then $\varphi^{(a)}$ and $\psi^{(a)}$ hold by the previous lemma.

Case 2. φ is $\forall u \in v\, \psi(u)$. Assume that

$$\mathrm{KPU} \vdash \psi \leftrightarrow \exists a\, \psi^{(a)}.$$

Again, working in KPU, assume $\forall u \in v\, \psi(u)$ and prove $\exists a\, \forall u \in v\, \psi(u)^{(a)}$. For each $u \in v$ there is a b such that $\psi(u)^{(b)}$, so by Δ_0 collection there is an a_0 such that

$\forall u \in v\, \exists b \in a_0\, \psi(u)^{(b)}$. Let $a = \bigcup a_0$. Now, for every $u \in v$, we have $\exists b \subseteq a\, \psi(u)^{(b)}$; so $\forall u \in v\, \psi(u)^{(a)}$, by the previous lemma.

Case 3. φ is $\exists u\, \psi(u)$. Assume $\psi(u) \leftrightarrow \exists b\, \psi(u)^{(b)}$ proved and suppose $\exists u\, \psi(u)$ true. We need an a such that $\exists u \in a\, \psi(u)^{(a)}$. If $\psi(u)$ holds, pick b so that $\psi(u)^{(b)}$ and let $a = b \cup \{u\}$. Then $u \in a$ and $\psi(u)^{(a)}$, by the previous lemma. ▯

In Platek's original definition of admissible set he took the Σ reflection principle as basic. It is very powerful, as we'll see below. The Δ_0 collection axiom is easier to verify in particular structures, however, and is also more like the replacement axioms with which one is familiar from ZF.

4.4 Theorem. (The Σ Collection Principle). *For every Σ formula φ the following is a theorem of* KPU: *If $\forall x \in a\, \exists y\, \varphi(x,y)$ then there is a set b such that $\forall x \in a\, \exists y \in b\, \varphi(x,y)$ and $\forall y \in b\, \exists x \in a\, \varphi(x,y)$.*

Proof. Assume that

$$\forall x \in a\, \exists y\, \varphi(x,y)\,.$$

By Σ reflection there is a set c such that

(1) $$\forall x \in a\, \exists y \in c\, \varphi^{(c)}(x,y)\,.$$

Let

(2) $$b = \{y \in c \mid \exists x \in a\, \varphi^{(c)}(x,y)\}\,,$$

by Δ_0 separation. Now since $\varphi^{(c)}(x,y) \rightarrow \varphi(x,y)$ by 4.2, (1) gives us:

$$\forall x \in a\, \exists y \in b\, \varphi(x,y)\,;$$

whereas (2) gives us:

$$\forall y \in b\, \exists x \in a\, \varphi(x,y)\,. \quad ▯$$

4.5 Theorem. (Δ Separation). *For any Σ formula $\varphi(x)$ and Π formula $\psi(x)$, the following is a theorem of* KPU: *If for all $x \in a$, $\varphi(x) \leftrightarrow \psi(x)$, then there is a set $b = \{x \in a \mid \varphi(x)\}$.*

Proof. Assume $\forall x \in a\, (\varphi(x) \leftrightarrow \psi(x))$. Then $\forall x \in a\, [\varphi(x) \vee \neg \psi(x)]$, which is equivalent to a Σ formula, so there is a c such that $\forall x \in a\, [\varphi^{(c)}(x) \vee \neg \psi^{(c)}(x)]$. Let, by Δ_0 separation, $b = \{x \in a \mid \varphi^{(c)}(x)\}$. Clearly every $x \in b$ satisfies $\varphi(x)$. If $x \in a$ and $\varphi(x)$ then $\psi(x)$, so $\psi^{(c)}(x)$ (since $\psi(x) \rightarrow \psi^{(c)}(x)$); so $\varphi^{(c)}(x)$. Thus $x \in b$. ▯

4.6 Theorem. (Σ Replacement). *For each Σ formula $\varphi(x,y)$ the following is a theorem of* KPU: *If $\forall x \in a\, \exists!\, y\, \varphi(x,y)$ then there is a function f, with* $\mathrm{dom}(f) = a$, *such that $\forall x \in a\, \varphi(x, f(x))$.*

Proof. By Σ Collection there is a set b such that $\forall x \in a\, \exists y \in b\, \varphi(x, y)$. Using Δ Separation there is an f such that

$$\begin{aligned} f &= \{\langle x, y\rangle \in a \times b \mid \varphi(x, y)\} \\ &= \{\langle x, y\rangle \in a \times b \mid \neg \exists z\, [\varphi(x, z) \wedge y \neq z]\}. \quad \Box \end{aligned}$$

The above is sometimes unsuable because of the uniqueness requirement $\exists!$ in the hypothesis. In these situations it is usually 4.7 which comes to the rescue.

4.7 Theorem. (Strong Σ Replacement). *For each Σ formula $\varphi(x, y)$ the following is a theorem of* KPU: *If $\forall x \in a\, \exists y\, \varphi(x, y)$ then there is a function f with* $\mathrm{dom}(f) = a$ *such that*

(i) $\forall x \in a\, f(x) \neq 0$;

(ii) $\forall x \in a\, \forall y \in f(x)\, \varphi(x, y)$.

Proof. By Σ Collection there is a b such that $\forall x \in a\, \exists y \in b\, \varphi(x, y)$ and $\forall y \in b\, \exists x \in a\, \varphi(x, y)$. Hence there is a w, by 4.3, such that

$$\forall x \in a\, \exists y \in b\, \varphi^{(w)}(x, y), \quad \text{and} \quad \forall y \in b\, \exists x \in a\, \varphi^{(w)}(x, y).$$

For any fixed $x \in a$ there is a unique set c_x such that

$$c_x = \{y \in b \mid \varphi^{(w)}(x, y)\}$$

by Δ_0 Separation and Extensionality; so, by Σ Replacement, there is a function f with domain a such that $f(x) = c_x$ for each $x \in a$. $\Box$

4.8—4.9 Exercises. There are a number of minor variations on the above.

4.8. For example, prove that, for each Σ formula φ,

$$\mathrm{KPU} \vdash \varphi \rightarrow \exists a\, (x_1 \in a \wedge \cdots \wedge x_n \in a \wedge \varphi^{(a)}).$$

4.9. Given a Σ formula φ let φ^{*a} denote the result of replacing some, but not necessarily all, existential quantifiers $\exists u$ by $\exists u \in a$ for some new set variable a. Show that: $\mathrm{KPU} \vdash \varphi \leftrightarrow \exists a\, \varphi^{*a}$.

5. *Adding Defined Symbols to* KPU

The introduction of defined relation and function symbols is a common practice, but it must be used with just a little care in KPU. In a theory like ZF one is able to take any formula $\varphi(x_1, \ldots, x_n)$, define a new relation symbol by

(R) $$\forall x_1 \ldots \forall x_n\, [\mathsf{R}(x_1, \ldots, x_n) \leftrightarrow \varphi(x_1, \ldots, x_n)],$$

and then use R as an atomic formula in other formulas—even in the axiom of replacement. After all, one could always go back and replace R by φ. For KPU, however, where we must pay attention to the syntactic form our axioms take, a definition like (R) would work, at first glance, only if the φ in (R) were Δ_0. We have tacitly used this form of introducing new relation symbols repeatedly in § 3. Using the principles of § 4 we may allow ourselves a bit more freedom.

5.1 Definition. Let $\varphi(x_1, \ldots, x_n)$ be a Σ formula of L^* and $\psi(x_1, \ldots, x_n)$ be a Π formula of L^* such that

$$\mathrm{KPU} \vdash \varphi \leftrightarrow \psi .$$

Let R be a new n-ary relation symbol and define R by (R) above. R is then called a Δ *relation symbol* of KPU.

To be really precise it would be the triple $\mathsf{R}, \varphi, \psi$ such that the above hold which constitute a Δ *definition of the relation symbol* R, but we do not need to be this careful. The next lemma shows that we can treat Δ relation symbols as though they were atomic formulas of L^*. Here, and elsewhere, we abbreviate $x_1, \ldots, x_k$ by $\vec{x}$.

5.2 Lemma. *Let* KPU *be formulated in* L^* *and let* R *be a* Δ *relation symbol of* KPU. *Let* KPU′ *be* KPU *as formulated in* $\mathsf{L}^*(\mathsf{R})$, *plus the defining axiom* (R) *above.*

(i) *For every formula* $\theta(x_1, \ldots, x_k, \mathsf{R})$ *of* $\mathsf{L}^*(\mathsf{R})$, *there is a formula* $\theta_0(x_1, \ldots, x_k)$ *of* L^* *such that* KPU+(R) *implies*

$$\theta(\vec{x}, \mathsf{R}) \leftrightarrow \theta_0(\vec{x}) .$$

Moreover, if θ *is a* Σ *formula of* $\mathsf{L}^*(\mathsf{R})$ *then* θ_0 *is a* Σ *formula of* L^*.

(ii) *For every* Δ_0 *formula* $\theta(x_1, \ldots, x_k, \mathsf{R})$ *of* $\mathsf{L}^*(\mathsf{R})$ *there are* Σ *and* Π *formulas* $\theta_0(x_1, \ldots, x_k)$, $\theta_1(x_1, \ldots, x_k)$ *of* L^* *such that* KPU+(R) *implies*

$$\theta(\vec{x}, \mathsf{R}) \leftrightarrow \theta_0(\vec{x}), \quad \text{and} \quad \theta(\vec{x}, \mathsf{R}) \leftrightarrow \theta_1(\vec{x}) .$$

(iii) KPU′ *is a conservative extension of* KPU. *That is, for any sentence* θ *of* L^*,

$$\mathrm{KPU}' \vdash \theta \quad \text{iff} \quad \mathrm{KPU} \vdash \theta .$$

Proof. Let us suppose that R is defined by

$$\mathsf{R}(x_1, \ldots, x_n) \leftrightarrow \varphi(x_1, \ldots, x_n) ,$$

where φ is a Σ formula, and that

$$\mathrm{KPU} \vdash \varphi(x_1, \ldots, x_n) \leftrightarrow \psi(x_1, \ldots, x_n) ,$$

where ψ is a Π formula. The first sentence in (i) is obvious since we may replace R by its definition. It is to make the second sentence of (i) true that we need R to be a Δ relation symbol of L*. Using de Morgan's laws, push all negations in θ inside as far as possible so that they only apply to atomic formulas. Now replace each positive (i.e., unnegated) occurrence of R in θ by φ, each occurrence $\neg$R by the Σ formula equivalent to $\neg\psi$. The result is called θ_0. Since

$$\mathrm{KPU}' \vdash \mathsf{R} \leftrightarrow \varphi \quad \text{and} \quad \mathrm{KPU}' \vdash \neg \mathsf{R} \leftrightarrow \neg\psi\,,$$

it is clear that

$$\mathrm{KPU}' \vdash \theta(x_1, \ldots, x_k, \mathsf{R}) \leftrightarrow \theta_0(x_1, \ldots, x_k)\,.$$

It is also clear that this transformation takes Σ formulas into Σ formulas. Note, however, that the transformation does not take Δ_0 formulas into Δ_0 formulas, but only into Σ formulas. However, since the Δ_0 formulas are closed under negation, (ii) immediately follows from (i). To prove (iii) it suffices to show that every axiom of KPU′ is turned into a theorem of KPU when R is replaced as above. For example, Δ_0 Separation of KPU′ becomes Δ Separation in KPU and Δ_0 Collection for KPU′ becomes a consequence of Σ Collection for KPU. □

Using this lemma we can clear up a point which may have been bothering the reader. One way of formalizing $\mathsf{L}^* = \mathsf{L}(\in, \ldots)$ is to make it a single sorted language with predicate symbols U for urelements and S for sets. In this way $\forall p(\ldots p \ldots)$ would stand for $\forall x(\mathsf{U}(x) \rightarrow (\ldots x \ldots))$, and $\forall a(\ldots)$ would stand for $\forall x(\mathsf{S}(x) \rightarrow (\ldots x \ldots))$, and "$x$ is an urelement" would be a Δ_0 formula, $\mathsf{U}(x)$. The other way of formalizing L* is to have a many-sorted language with the three sorts of variables

$$p, q, \ldots\,,$$
$$a, b, \ldots\,, \quad \text{and}$$
$$x, y, \ldots\,.$$

The predicate "x is an urelement" is no longer Δ_0, but it is Δ. Our definition of L* insures that

$$x \text{ is an urelement} \leftrightarrow \exists p\,(x = p), \quad \text{and}$$

$$x \text{ is not an urelement} \leftrightarrow \exists a\,(x = a)\,;$$

so the predicate and its negation are Σ_1. The lemma assures us that we can introduce a new symbol by:

$$\mathsf{U}(x) \leftrightarrow x \text{ is an urelement},$$

and use it in Δ_0 formulas without fear. Similarly we can introduce

$$\mathsf{S}(x) \leftrightarrow \exists a\,(x=a)$$
$$\leftrightarrow \neg\exists p\,(x=p)$$

for "x is a set" and treat it as a Δ_0 formula.

A predicate of intuitive set theory is said to be a Δ *predicate of* KPU if it can be defined by a Δ relation symbol. Using the above lemma we see that we may treat Δ predicates as though they were defined by atomic formulas of L^*. Furthermore, the Δ predicates are closed under $\wedge$, $\vee$, $\forall u \in v$, $\exists u \in v$. Using these observations, we see that all the predicates listed in Table 2 are indeed Δ predicates.

The introduction of defined relation symbols is a convenience, but the introduction of defined function symbols is a practical necessity (though theoretically a luxury). The conditions necessary for us to be able to do this are given in the following definition.

5.3 Definition. Let $\varphi(x_1, \ldots, x_n, y)$ be a Σ formula of L^* such that

$$\mathrm{KPU} \vdash \forall x_1, \ldots, x_n \exists! y\, \varphi(x_1, \ldots, x_n, y).$$

Let F be a new n-ary function symbol and define F by:

$$(\mathsf{F}) \qquad \forall x_1, \ldots, x_n, y[\mathsf{F}(x_1, \ldots, x_n) = y \leftrightarrow \varphi(x_1, \ldots, x_n, y)].$$

F is then called a Σ *function symbol* of KPU.

The next lemma lets us treat Σ function symbols as though they were atomic symbols of the basic language L^*.

5.4 Lemma. *Let* KPU *be formulated in* L^* *and let* F *be a* Σ *function symbol of* KPU. *Let* KPU′ *be* KPU *as formulated in* $\mathsf{L}^*(\mathsf{F})$, *plus the defining axiom* (F) *above.*

(i) *For every formula* $\theta(x_1, \ldots, x_k, \mathsf{F})$ *of* $\mathsf{L}^*(\mathsf{F})$ *there is a formula* $\theta_0(x_1, \ldots, x_k)$ *of* L^* *such that* KPU$+(\mathsf{F})$ *implies*

$$\theta(\vec{x}, \mathsf{F}) \leftrightarrow \theta_0(\vec{x}).$$

Moreover, if θ *is a* Σ *formula of* $\mathsf{L}^*(\mathsf{F})$ *then* θ_0 *is a* Σ *formula of* L^*.

(ii) *For every* Δ_0 *formula* $\theta(x_1, \ldots, x_k, \mathsf{F})$ *of* $\mathsf{L}^*(\mathsf{F})$ *there are* Σ *and* Π *formulas* $\theta_0(x_1, \ldots, x_k)$, $\theta_1(x_1, \ldots, x_k)$ *of* L^* *such that* KPU$+(\mathsf{F})$ *implies*

$$\theta(\vec{x}, \mathsf{F}) \leftrightarrow \theta_0(\vec{x}), \quad \textit{and}$$
$$\theta(\vec{x}, \mathsf{F}) \leftrightarrow \theta_1(\vec{x}).$$

(iii) KPU′ *is a conservative extension of* KPU.

Table 2. Some Δ predicates

Predicate	*Abbreviation*	*Definition*
x is an urelement	$\mathsf{U}(x)$	$\exists p\,(x=p)$ (*or* $\forall a\,(x\neq a)$)
x is a set	$\mathsf{S}(x)$	$\exists a\,(x=a)$ (*or* $\forall p\,(x\neq p)$)
x is transitive	$\mathrm{Tran}(x)$	$\mathsf{S}(x)\wedge\forall y\in x\,\forall z\in y\,(z\in x)$
x is an ordinal	$\mathrm{Ord}(x)$	$\mathrm{Tran}(x)\wedge\forall y\in x\,\mathrm{Tran}(y)$
x is a limit ordinal	$\mathrm{Lim}(x)$	$\mathrm{Ord}(x)\wedge x\neq 0\wedge\forall y\in x\,\exists z\in x\,(z=y\cup\{y\})$
x is a natural number	$\mathrm{Nat\,No}(x)$	$\mathrm{Ord}(x)\wedge\forall y\in x\neg\mathrm{Lim}(y)\wedge\neg\mathrm{Lim}(x)$
less than for ordinals	$\alpha<\beta$	$\mathrm{Ord}(\alpha)\wedge\mathrm{Ord}(\beta)\wedge\alpha\in\beta$
less than or equal	$\alpha\leqslant\beta$	$\alpha<\beta\vee\alpha=\beta$.

Proof. Note that if $\varphi(x_1,\ldots,x_n,y)$ is a Σ formula and if $\mathsf{F}(x_1,\ldots,x_n)=y$ iff $\varphi(x_1,\ldots,x_n,y)$, then we can get a Σ definition for $\mathsf{F}(x_1,\ldots,x_n)\neq y$ by

(1) $$\mathsf{F}(x_1,\ldots,x_n)\neq y \quad\text{iff}\quad \exists z[\varphi(x_1,\ldots,x_n,z)\wedge y\neq z].$$

Thus the graph of F is a Δ predicate. The only complication, then, that can occur here but not in the previous lemma, is that F may occur in θ in complicated contexts like:

$$\mathsf{F}(\mathsf{G}(x))=\mathsf{H}(y) \quad\text{and}\quad \mathsf{R}(\mathsf{F}(x),y).$$

Call a formula *simple* if F only appears in simple contexts like:

$$\mathsf{F}(x_1,\ldots,x_n)=y \quad\text{and}\quad \mathsf{F}(x_1,\ldots,x_n)\neq y.$$

Repeated uses of the equivalences below allow us to transform every formula into an equivalent simple formula in such a way that Σ formulas transform into Σ formulas:

$$\mathsf{F}(\mathsf{G}(x),x_2,\ldots,x_n))=y\leftrightarrow\exists z\,[\mathsf{G}(x)=z\wedge\mathsf{F}(z,x_2,\ldots,x_n)=y],$$

$$\mathsf{F}(\mathsf{G}(x),x_2,\ldots,x_n))\neq y\leftrightarrow\exists z\,[\mathsf{G}(x)=z\wedge\mathsf{F}(z,x_2,\ldots,x_n)\neq y],$$

$$\mathsf{F}(x_1,\ldots,x_n)=\mathsf{H}(y)\quad\leftrightarrow\exists z\,[\mathsf{H}(y)=z\wedge\mathsf{F}(x_1,\ldots,x_n)=y],$$

$$\mathsf{F}(x_1,\ldots,x_n)\neq\mathsf{H}(y)\quad\leftrightarrow\exists z\,[\mathsf{H}(y)=z\wedge\mathsf{F}(x_1,\ldots,x_n)\neq z],$$

$$\varphi(\mathsf{F}(\vec{x}),\ldots)\qquad\leftrightarrow\exists z\,[z=\mathsf{F}(\vec{x})\wedge\varphi(z,\ldots)]\quad(\varphi\text{ quantifier free}).$$

The proof now proceeds as in 5.2, replacing occurrences of $\mathsf{F}(x_1,\ldots,x_n)=y$ by $\varphi(x_1,\ldots,x_n,y)$, occurrences of $\mathsf{F}(x_1,\ldots,x_n)\neq y$ by the Σ formula in (1). □

When we use 5.1 (or 5.3) to introduce a Δ relation symbol R (or Σ function symbol F) we often abuse notation by using KPU to denote the new theory KPU′ of 5.2 (or 5.4). The lemmas insure us that we can't get into trouble with this abuse of notation.

Table 3. Some Σ operations

Operation	*Domain*	*Abbreviation*	*Σ Definition (the unique z such that)*
domain of f	all functions f	$\mathrm{dom}(f)$	see Table 1
range of f	all functions f	$\mathrm{rng}(f)$	see Table 1
the first coordinate of x	all ordered pairs x	$1^{\mathrm{st}}x$	see Table 1
the second coordinate of x	all ordered pairs x	$2^{\mathrm{nd}}x$	see Table 1
the restriction of f to a	all functions f and sets a	$f\upharpoonright a$	$z=\{x\in f \mid 1^{\mathrm{st}}x\in a\}$
the image of f restricted to a	all functions f and sets a	$f''a$	$z=\{x\in \mathrm{rng}(f) \mid \exists y\in a(f(x)=y)\}$
successor	all sets x	$\mathscr{S}(x)$	$z=x\cup\{x\}$
ordinal successor	all ordinals α	$\alpha+1$	$z=\mathscr{S}(\alpha)$
supremum	sets of ordinals	$\sup(a)$	$z=\bigcup a$

An operation of intuitive set theory is a Σ *operation of* KPU if it can be defined by a Σ function symbol of KPU. The following exercises summarize some of the ways, in addition to 5.3, we have of defining Σ operations. The most important method, though, must wait for the next section.

5.5—5.7 Exercises

5.5. Every function symbol of L^* is a Σ function symbol.

5.6. The Σ operations are closed under composition.

5.7. The Σ operations are closed under definition by cases. That is, if $\mathsf{G}_1, \ldots, \mathsf{G}_k$ are n-ary Σ operations and $\varphi_1(x_1, \ldots, x_n), \ldots, \varphi_k(x_1, \ldots, x_n)$ are Σ formulas such that

$$\mathrm{KPU} \vdash \forall x \left[\underline{\bigvee}_{i\leqslant k} \varphi_i(x_1, \ldots, x_n)\right]$$

$\underline{\bigvee}$ indicates exclusive or), then we may define a Σ operation F by:

$$\mathsf{F}(x_1, \ldots, x_n) = \begin{cases} \mathsf{G}_1(x_1, \ldots, x_n) & \text{if} \quad \varphi_1(x_1, \ldots, x_n)\,, \\ \vdots & \\ \mathsf{G}_k(x_1, \ldots, x_n) & \text{if} \quad \varphi_k(x_1, \ldots, x_n)\,. \end{cases}$$

Frequently we are interested in the value of a function symbol only for certain kinds of objects. For example, we want to define $1^{\mathrm{st}}a$ to be the first coordinate of a if a is an ordered pair, but we don't really care what $1^{\mathrm{st}}a$ means otherwise. To introduce $1^{\mathrm{st}}a$ as a function symbol then, we should, to be completely rigorous, first do something like prove: $\forall x\, \exists! y\, \varphi(x,y)$, where $\varphi(x,y)$ is:

x is an ordered pair with first coordinate y, or

x is not an ordered pair and y is the empty set,

and then define:

$$1^{\mathrm{st}}x = y \quad \text{iff} \quad \varphi(x,y)\,.$$

Similarly, we are interested in $\bigcup x$ only when x is a set. We will not bother with such details in the future, as long as it is clear that the intended domain of our new function symbol is Δ definable.

6. *Definition by Σ Recursion*

Definition by recursion is a powerful tool. It will allow us to introduce, in accordance with 5.3, operations such as ordinal addition, ordinal multiplication and the support function sp:

$$\mathrm{sp}(p)=\{p\},$$

$$\mathrm{sp}(a)=\bigcup_{x\in a}\mathrm{sp}(x),$$

which gives the set of urelements which go into the construction of a set a. Before showing how to justify such recursions we must first prove outright what is in effect a special case.

6.1 Theorem (Existence of Transitive Closure). *We can introduce a* Σ *function symbol* TC *into* KPU *so that the following becomes a theorem of* KPU: *For every* x, $\mathrm{TC}(x)$ *is a transitive set such that* $x\subseteq \mathrm{TC}(x)$; *and for any other transitive set* a, *if* $x\subseteq a$, *then* $\mathrm{TC}(x)\subseteq a$.

The axiom of foundation will be used in the proof of 6.1, in the form of *Proof by Induction over* $\in$. If one takes the contrapositive of foundation one gets the following scheme. For every formula φ the following is a theorem of KPU:

$$\forall x(\forall y\in x\,\varphi(y)\rightarrow\varphi(x))\rightarrow\forall x\,\varphi(x).$$

Thus in proving $\forall x\,\varphi(x)$, we pick an arbitrary x and prove $\varphi(x)$ using, in the proof, $\varphi(y)$ for any $y\in x$. (Of coure if x is an urelement then there are no such $y\in x$.)

Proof of 6.1. If we had the ordinal ω at our disposal (we cannot prove it exists in KPU) we could use it to define

$$\mathrm{TC}(a)=a\cup(\bigcup a)\cup(\bigcup\bigcup a)\cup\cdots.$$

This definition should be kept in mind to understand the following proof. Define $Q(x,a)$ to be:

$$x\subseteq a\wedge \mathrm{Tran}(a)\wedge\forall b(x\subseteq b\wedge \mathrm{Tran}(b)\rightarrow a\subseteq b).$$

Thus Q is defined by a Π formula and $Q(x,a)$ iff a is the smallest transitive set containing x. It is clear that $Q(x,a)\wedge Q(x,a')\rightarrow a=a'$.

Now let $P(x,a)$ be the following Σ predicate:

x is an urelement $\wedge a=0$, or

x is a set, $x\subseteq a$, $\mathrm{Tran}(a)\wedge\forall z\in a\,\exists f\,[\mathrm{Fun}(f)\wedge\mathrm{dom}(f)$

is a natural number $n+1=\{0,\ldots,n\}\wedge z=f(0)\in f(1)\in\cdots\in f(n)\in x.]$

(This can be easily formalized without writing "$\cdots$"; so there is no hidden recursion.) A simple induction on natural numbers n shows that $P(x,a)\rightarrow Q(x,a)$. In particular, $P(x,a)\wedge P(x,a')\rightarrow a=a'$.

If we can prove that for every x there is an a such that $P(x,a)$ then we will be able to define a Σ function symbol TC by

$$\mathrm{TC}(x)=a \quad \text{iff} \quad P(x,a)$$

and $\mathrm{TC}(x)$ will have the desired property of the transitive closure of x. We still need to show that $\forall x\,\exists a\,P(x,a)$. If x is an urelement, take $a=0$. Thus, we need only prove $\forall b\,\exists a\,P(b,a)$, which we do by induction on $\in$. Given b, in proving $\exists a\,P(b,a)$ we may assume

$$\forall x\in b\,\exists c\,P(x,c)$$

and hence, by the above,

$$\forall x\in b\,\exists!c\,P(x,c)\,.$$

By Σ replacement there is a function g with $\mathrm{dom}(g)=b$, such that $P(x,g(x))$ holds for all $x\in b$. Let

$$a=b\cup(\bigcup\mathrm{rng}(g))$$
$$=b\cup\textstyle\bigcup_{x\in b}g(x)\,.$$

It is clear that $b\subseteq a$ and it is not difficult to check that a is transitive. Let us verify the last clause of $P(b,a)$. Thus, let $z\in a$. If $z\in b$ then take $f=\{\langle 0,z\rangle\}$. Now assume $z\in\bigcup\mathrm{rng}(g)$, i.e. $z\in g(x)$ for some $x\in b$. But then there is an h such that $\mathrm{dom}(h)$ is an integer $n+1$, $h(0)=z$, $h(i)\in h(i+1)$ and $h(n)\in x$ since $P(x,g(x))$. Let $f=h\cup\{\langle n+1,x\rangle\}$. Then $f(0)=z\in f(1)\in f(2)\in\cdots\in f(n+1)=x\in b$ so $P(a,b)$. □

6.2 Exercise. Verify

(i) $\mathrm{TC}(p)=0$, and

(ii) $\mathrm{TC}(a)=a\cup\bigcup\{\mathrm{TC}(x)\mid x\in a\}$.

Once we have Theorem 6.4 we could use the equations in 6.2 to define TC; unfortunately we need 6.1 and 6.3 to state and prove 6.4. The following is a strengthening of the method of proof by induction over $\in$.

6.3 Theorem (Proof by Induction over TC). *For any formula $\varphi(x)$ the following is a theorem of* KPU: *If, for each x, $(\forall y \in \mathrm{TC}(x)\, \varphi(y))$ implies $\varphi(x)$, then $\forall x\, \varphi(x)$.*

Proof. We show, under the hypothesis, that $\forall x \forall y \in \mathrm{TC}(x)\, \varphi(y)$. This implies $\forall x\, \varphi(x)$, since $x \in \mathrm{TC}(\{x\})$. We may assume, by induction on $\in$, that for all $z \in x$

(1) $\qquad \forall y \in \mathrm{TC}(z)\, \varphi(y)$

in showing $\forall y \in \mathrm{TC}(x)\, \varphi(y)$. But by the hypothesis, (1) implies $\varphi(z)$ so we have $\varphi(y)$, for all $y \in x \cup \bigcup\{\mathrm{TC}(z) \mid z \in x\} = \mathrm{TC}(x)$. □

The following theorem is of central importance to all that follows.

6.4 Theorem (Definition by Σ Recursion). *Let* G *be an $n+2$-ary Σ function symbol, $n \geqslant 0$. It is possible to define a new Σ function symbol* F *so that the following is a theorem of* KPU (+ *the defining axiom* (F)): *for all $x_1, \ldots, x_n, y$,*

(i) $\qquad \mathsf{F}(x_1, \ldots, x_n, y) = \mathsf{G}(x_1, \ldots, x_n, y, \{\langle z, \mathsf{F}(x_1, \ldots, x_n, z)\rangle \mid z \in \mathrm{TC}(y)\})$.

Before turning to the rather tedious proof of 6.4, let us make some remarks on variations which follow from it. For example, we could replace 6.4(i) by:

$$\mathsf{F}(x_1, \ldots, x_n, y) = \mathsf{G}(x_1, \ldots, x_n, y, \{\langle z, \mathsf{F}(x_1, \ldots, x_n, z)\rangle \mid z \in y)\}.$$

(Let $\mathsf{G}'(\vec{x}, y, f) = \mathsf{G}(\vec{x}, y, f \upharpoonright y)$, and apply 6.4 to G'.) We could also start out with two functions G, H and define

$$\mathsf{F}(x_1, \ldots, x_n, p) = \mathsf{H}(x_1, \ldots, x_n, p),$$

$$\mathsf{F}(x_1, \ldots, x_n, a) = \mathsf{G}(x_1, \ldots, x_n, a, \{\langle z, \mathsf{F}(x_1, \ldots, x_n)\rangle : z \in \mathrm{TC}(a)\}).$$

This is the form we usually use. (Let $\mathsf{G}'(\vec{x}, y, f)$ be $\mathsf{H}(\vec{x}, y)$, if y is an urelement, otherwise $\mathsf{G}(\vec{x}, y, f)$ if y is a set. Then apply 6.4 to G'.)

Proof of 6.4. To be a little more formal, what we really want to prove about F, once we find a way of defining it, is that for all $x_1, \ldots, x_n, y$ there is an f such that

(1) $\qquad f$ *is a function* $\wedge\ \mathrm{dom}(f) = \mathrm{TC}(y)$,

(2) $\qquad \forall w \in \mathrm{dom}(f)(f(w) = \mathsf{F}(x_1, \ldots, x_n, w))$, and

(3) $\qquad \mathsf{F}(x_1, \ldots, x_n, y) = \mathsf{G}(x_1, \ldots, x_n, y, f)$.

This suggests the correct defining formula for F. Let $n=1$ to simplify notation. Let $P(x, y, z, f)$ be the Σ predicate given by:

$$f \text{ is a function} \wedge \mathrm{dom}(f)=\mathrm{TC}(y)$$
$$\wedge \forall w\in \mathrm{TC}(y)(f(w)=\mathsf{G}(x,w,f\upharpoonright \mathrm{TC}(w)))$$
$$\wedge z=\mathsf{G}(x,y,f)\,.$$

We will prove:

(4) $$\forall x\,\forall y\,\exists! z\,\exists f\,P(x,y,z,f)\,;$$

and so we can introduce a Σ function symbol F by:

(5) $$\mathsf{F}(x,y)=z\leftrightarrow \exists f\,P(x,y,z,f)\,,$$

where it is clear that the right-hand side of (5) is a Σ formula. In order to prove (4) it suffices to prove;

(6) $$P(x,y,z,f)\wedge P(x,y,z',f')\rightarrow z=z'\wedge f=f'\,,\quad \text{and}$$

(7) $$\forall y\,\exists z\,\exists f\,P(x,y,z,f)\,.$$

We prove both (6) and (7) by induction on $\mathrm{TC}(y)$. We use, in these proofs, lines (8), (9) below which are obtained by inspecting the definition of P:

(8) $$P(x,y,z,f)\rightarrow z=\mathsf{G}(x,y,f)\,;$$

(9) $$P(x,y,z,f)\wedge w\in\mathrm{TC}(y)\rightarrow P(x,w,f(w),f\upharpoonright\mathrm{TC}(w))\,.$$

We now prove (6) by induction on $\mathrm{TC}(y)$. Thus, we may assume that for $w\in\mathrm{TC}(y)$ there is at most one u and g with $P(x,w,u,g)$ and prove that $P(x,y,z,f)\wedge P(x,y,z',f')\rightarrow z=z'\wedge f=f'$. Since $z=\mathsf{G}(x,y,f)$ and $z'=\mathsf{G}(x,y,f')$, it suffices to prove $f=f'$. But f and f' are functions with common domain $\mathrm{TC}(y)$ so it suffices to show that $f(w)=f'(w)$ for all $w\in\mathrm{TC}(y)$. But by (9), $P(x,w,f(w),f\upharpoonright\mathrm{TC}(w))$ and $P(x,w,f'(w),f'\upharpoonright\mathrm{TC}(w))$; so $f(w)=f'(w)$ by the induction hypothesis. It remains to prove (7), and this is where Δ_0 Collection enters in the guise of Σ Replacement. We prove $\exists z\,\exists f\,P(x,y,z,f)$ assuming, by induction on TC, that $\forall w\in\mathrm{TC}(y)\,\exists u\,\exists g\,P(x,w,u,g)$; and hence, by (6), there is a unique u_w, g_w such that $P(x,w,u_w,g_w)$. By Σ Replacement the function

$$f=\{\langle w,u_w\rangle\mid w\in\mathrm{TC}(y)\}$$

exists. To prove (7) it suffices to prove $P(x,y,\mathsf{G}(x,y,f),f)$ and this will follow from $\forall z\in\mathrm{TC}(x)(f(z)=\mathsf{G}(x,z,f\upharpoonright\mathrm{TC}(z)))$. Since we have $P(x,z,u_z,g_z)$ we have $f(z)=u_z=\mathsf{G}(x,y,g_z)$. Thus, all we have to show is $f\upharpoonright\mathrm{TC}(z)=g_z$. For $w\in\mathrm{dom}(g_z)$ $=\mathrm{TC}(z)$, (9) implies $P(x,w,g_z(w),g_z\upharpoonright\mathrm{TC}(w))$. Thus by (6) we have $g_z(w)=u_w=f(w)$; so $g_z=f\upharpoonright\mathrm{TC}(w)$ as desired. This proves (7). Now let us introduce F by line (5) and go back to prove 6.4(i). By (5) we have

$$\mathsf{F}(x,y)=\mathsf{G}(x,y,f) \quad \text{where} \quad P(x,y,\mathsf{G}(x,y,f),f),$$

so we need only show that

$$f=\{\langle z,\mathsf{F}(x,z)\rangle \mid z\in \mathrm{TC}(y)\}.$$

For $z\in\mathrm{TC}(y)$ we have, by (9), $P(x,z,f(z),f{\upharpoonright}\mathrm{TC}(z))$ so, by (5), $\mathsf{F}(x,z)=f(z)$ as desired. □

6.5 Exercise. Prove that if two operations F_1, F_2 both satisfy 6.4(i) in place of F for all $x_1,\dots,x_n,y$ then $F_1(x_1,\dots,x_n,y)=F_2(x_1,\dots,x_n,y)$, for all $x_1,\dots,x_n,y$.

In applications of 6.4 one does not usually bother to introduce the explicit function symbols G,H first.

6.6 Corollary (Δ Predicates Defined by Recursion). *Let P,Q be Δ predicates of $n+1$, $n+2$ arguments respectively, $n\geqslant 0$. We can introduce a Δ predicate R by definition so that the following are provable in the resulting* KPU:

(i) $\mathsf{R}(x_1,\dots,x_n,p)\leftrightarrow P(x_1,\dots,x_n,p)$;

(ii) $\mathsf{R}(x_1,\dots,x_n,a)\leftrightarrow Q(x_1,\dots,x_n,a,\{b\in\mathrm{TC}(a)\mid \mathsf{R}(x_1,\dots,x_n,b)\})$.

Proof. Introduce the characteristic functions G,H of P,Q respectively. Use Σ Recursion to define the characteristic function F of R and then note that

$$\begin{aligned}\mathsf{R}(x_1,\dots,x_n,y) &\leftrightarrow \mathsf{F}(x_1,\dots,x_n,y)=1\\ &\leftrightarrow \mathsf{F}(x_1,\dots,x_n,y)\neq 0,\end{aligned}$$

so that R is shown to be Δ. □

In Table 4 we give some examples of operations defined by recursion. The reader not familiar with this type of thing should work through the following exercises.

6.7—6.9 Exercises

6.7. (The rank function). (i) Show how to make the definition of rk given in Table 4 fit into the form of Theorem 6.4.

(ii) Prove that $\mathrm{rk}(x)$ is an ordinal, $\mathrm{rk}(\alpha)=\alpha$ for ordinals α, and $\mathrm{rk}(y)<\mathrm{rk}(x)$ whenever $y\in\mathrm{TC}(x)$.

(iii) Prove that $\mathrm{rk}(a)=\{\mathrm{rk}(y)\mid y\in\mathrm{TC}(a)\}$. (This could be used to give a different recursive definition of rk.)

6.8. (The support function). (i) Show how to make the definition of sp given in Table 4 fit into the form demanded by 6.4.

(ii) Prove that $\mathrm{sp}(a)=\{x\in \mathrm{TC}(a) \mid x$ is an urelement$\}$.

6.9. (Ordinal addition). (i) Show how to make ordinal addition fit into the form of 6.4.

(ii) Prove:

$$\alpha+0=\alpha;$$

$$\alpha+(\beta+1)=(\alpha+\beta)+1; \quad \text{and}$$

$$\alpha+\lambda=\sup\{\alpha+\beta \mid \beta<\gamma\}, \quad \text{if} \quad \mathrm{Lim}(\lambda).$$

(iii) Prove:

$$\alpha+(\beta+\gamma)=(\alpha+\beta)+\gamma;$$

$$\beta\leqslant\alpha+\beta;$$

$$0<\beta\Rightarrow\alpha<\alpha+\beta;$$

$$\alpha<\beta\Rightarrow\exists!\,\gamma\leqslant\beta(\alpha+\gamma=\beta);$$

$$\beta<\gamma\Rightarrow\alpha+\beta<\alpha+\gamma;$$

$$\alpha\leqslant\beta\wedge\gamma\leqslant\delta\Rightarrow\alpha+\gamma\leqslant\beta+\delta.$$

To conclude this section we point out that, like much of axiomatic mathematics, the development of set theory in KPU is largely a matter of refining proofs from ZF. Among its rewards is the Σ recursion theorem (I.6.4). Since we end with a Σ operation symbol, the operation defined by recursion is absolute. The usual development in ZF completely looses track of this vital information. (This is relevant to the point we made in § 1, line (2).)

Table 4. Some Σ Operations Defined by Recursion

Operation	*Domain*	*Abbreviation*	*Recursive definitions*
rank function	everything	$\mathrm{rk}(x)$	$\mathrm{rk}(p)=0$ $\mathrm{rk}(a)=\sup\{\mathrm{rk}(x)+1 \mid x\in a\}$
support function	everything	$\mathrm{sp}(x)$	$\mathrm{sp}(p)=\{p\}$ $\mathrm{sp}(a)=\bigcup_{x\in a}\mathrm{sp}(x)$
ordinal addition	pairs of ordinals α,β	$\alpha+\beta$	$\alpha+\beta=\alpha\cup\sup\{(\alpha+\gamma)+1 \mid \gamma<\beta\}$
ordinal multiplication	pairs of ordinals α,β	$\alpha\beta$	$\alpha\beta=\sup\{\alpha\gamma+\alpha \mid \gamma<\beta\}$
collapsing function (cf. I.7)	pairs a, x	$\mathrm{C}_a(x)$	$\mathrm{C}_a(p)=p$ $\mathrm{C}_a(b)=\{\mathrm{C}_a(x) \mid x\in a\cap b\}$
constructible sets (cf. II.5)	pairs a,α	$\mathrm{L}(a,\alpha)$	$\mathrm{L}(a,0)=\mathrm{TC}(a)$ $\mathrm{L}(a,\alpha+1)=\mathscr{D}(\mathscr{S}(\mathrm{L}(a,\alpha)))$ $\mathrm{L}(a,\lambda)=\bigcup_{\alpha<\lambda}\mathrm{L}(a,\alpha)$, for limit λ.

7. The Collapsing Lemma

We return to the development of set theory in KPU to discuss an important operation C of two arguments; we write $\mathrm{C}_x(y)$ instead of $\mathrm{C}(x,y)$. The operation is defined in KPU using Σ Recursion by the equations:

$$\mathrm{C}_x(p)=p\,;$$

$$\mathrm{C}_x(a)=\{\mathrm{C}_x(y)\mid y\in a\cap x\}\,.$$

(This falls under the second variation on Theorem 6.4.) C will be called Mostowski's *collapsing function*. We shall compute $\mathrm{C}_x(y)$ for some specific values of x and y after we have a lemma to aid us. In this section we will only be interested $\mathrm{C}_x(y)$ for $y\in x$.

7.1 Lemma. (i) $\mathrm{C}_p(a)=0$, *for all* p,a.

(ii) *If* $a\subseteq b$ *and* a *is transitive, then* $\mathrm{C}_b(x)=x$ *for all* $x\in a$.

(iii) *For any* b *the set* $\{\mathrm{C}_b(x)\mid x\in b\}=\mathrm{C}_b(b)$ *is transitive.*

Proof. (i) is obvious. We prove (ii) by $\in$-induction. Thus, given $x\in a$ we suppose that $\mathrm{C}_b(y)=y$ for all $y\in x$. But since a is transitive, $x\subseteq a\subseteq b$, so we have

$$\begin{aligned}\mathrm{C}_b(x)&=\{\mathrm{C}_b(y)\mid y\in x\cap b\}\\&=\{\mathrm{C}_b(y)\mid y\in x\}\\&=\{y\mid y\in x\}\\&=x\,.\end{aligned}$$

To prove (iii), let $a=\{\mathrm{C}_b(x)\mid x\in b\}$. We must show that a is transitive. Let $z\in y\in a$. Thus $y=\mathrm{C}_b(x)$ for some $x\in b$, hence $z\in\{\mathrm{C}_b(x')\mid x'\in x\cap b\}$; so $z=\mathrm{C}_b(x')$ for some $x'\in b$. Hence $z\in a$. ▯

7.2 Example. Let $b=\{0,1,2,4,\{1,3,4\},\{1,4\}\}$. If we let $a=3=\{0,1,2\}$ then 7.1 (ii) applies to give:

$$\mathrm{C}_b(0)=0\,,$$

$$\mathrm{C}_b(1)=1\,,$$

$$\mathrm{C}_b(2)=2\,.$$

Let us compute $\mathrm{C}_b(4)$:

$$\begin{aligned}\mathrm{C}_b(4)&=\{\mathrm{C}_b(x)\mid x\in b,\ x\in 4\}\\&=\{\mathrm{C}_b(x)\mid x=0,1,2\}\\&=\{0,1,2\}\\&=3\,.\end{aligned}$$

Thus C_b "collapses" 4 to 3 since 3 wasn't in b. Now let us compute $C_b(\{1,3,4\})$ and $C_b(\{1,4\})$:

$$\begin{aligned} C_b(\{1,3,4\}) &= \{C_b(x) \mid x \in \{1,3,4\} \cap b\} \\ &= \{C_b(1), C_b(4)\} \\ &= \{1,3\}; \end{aligned}$$

$$\begin{aligned} C_b(\{1,4\}) &= \{C_b(x) : x \in \{1,4\} \cap b\} \\ &= \{C_b(1), C_b(4)\} \\ &= \{1,3\}. \end{aligned}$$

Thus both the sets $\{1,3,4\}$ and $\{1,4\}$ are collapsed to $\{1,3\}$, all because 3 was left out of b. Note that

$$\{C_b(x) : x \in b\} = \{0,1,2,3,\{1,3\}\},$$

which is a transitive set, just as 7.1 (iii) foretold.

7.3 Definition. For any set b let c_b denote the restriction of $C_b(\cdot)$ to b; i.e. $c_b = \{\langle x, C_b(x)\rangle : x \in b\}$, and let

$$\text{clpse}(b) = \text{rng}(c_b) = \{C_b(x) : x \in b\} = C_b(b).$$

Note that the function c_b exists (as a set) by Σ replacement and that clpse(b) is a transitive set by 7.1 (iii).

A set b is *extensional* if for every two distinct sets $a_1, a_2 \in b$ there is an $x \in b$ such that x is in one of a_1, a_2 but not both; in other symbols,

$$\forall x \in b\, (x \in a_1 \leftrightarrow x \in a_2) \rightarrow a_1 = a_2.$$

We would like to say that b is extensional if

$$\langle b, \in \cap\, b^2\rangle \models \text{"Extensionality"},$$

but we cannot do this because we have not yet defined syntax and semantics (say $\models$) in KPU. So, what we have done is simply to write this out in full.

In Example 7.2, b was not extensional because of the two sets

$$a_1 = \{1,3,4\}, \quad a_2 = \{1,4\}.$$

Any transitive set is extensional, as is any set of ordinals. The next lemma shows that any extensional set is isomorphic to a transitive set.

7.4 Theorem (The Collapsing Lemma). *If a is extensional then c_a maps a one-one onto the transitive set* clpse(a). *Furthermore, for all* $x, y \in a$

(i) $x \in y$ *iff* $c_a(x) \in c_a(y)$.

In other words, c_a is an isomorphism of $\langle a, \in \cap\, a^2 \rangle$ *onto* $\langle \text{clpse}(a), \in \cap\, \text{clpse}(a)^2 \rangle$.

Proof. We need to show c_a is one-one and that $c_a(x) \in c_a(y)$ implies $x \in y$. We prove both of these by proving $\forall x \forall y\, P(x, y)$ where $P(x, y)$ is the conjunction of:

$$x, y \in a \wedge c_a(x) = c_a(y) \rightarrow x = y\,,$$

$$x, y \in a \wedge c_a(x) \in c_a(y) \rightarrow x \in y\,, \quad \text{and}$$

$$x, y \in a \wedge c_a(y) \in c_a(x) \rightarrow y \in x\,.$$

Given an x_0 we can assume, by induction on $\in$,

(1) $\forall x \in x_0\, \forall y\, P(x, y)$

in our proof of $\forall y\, P(x_0, y)$. Given an arbitrary y_0 we can assume

(2) $\forall y \in y_0\, P(x_0, y)$

in our proof of $P(x_0, y_0)$, again using $\in$-induction. Thus, suppose $x_0, y_0 \in a$.

Case 1. $c_a(x_0) = c_a(y_0)$. Suppose $x_0 \neq y_0$. We see that both x_0, y_0 must be sets since $c_a(p) = p$. But then, since a is extensional there is a $z \in a$ with $z \in (x_0 \cup y_0) - (x_0 \cap y_0)$. Suppose $z \in x_0 - y_0$, the other possibility being similar. Then $c_a(z) \in c_a(x_0) = c_a(y_0)$ but, by (1), $P(z, y_0)$ so $z \in y_0$, a contradiction.

Case 2. $c_a(x_0) \in c_a(y_0)$. But then $c_a(x_0) = c_a(z)$, for some $z \in y_0$, but $P(x_0, z)$ by (2), so $x_0 = z$ and $x_0 \in y_0$.

Case 3. $c_a(y) \in c_a(x_0)$. Similar to Case 2. □

We hint at some of the types of applications of the collapsing lemma in the exercises.

7.5—7.8 Exercises

7.5. Show that if a is finite so is $C_b(a)$. [Hint: Use induction on natural numbers.]

7.6. Show in KPU that a set a of ordinals is finite iff clpse(a) is a natural number. This shows that the predicate "a is a finite set of ordinals" is a Δ predicate in KPU. (For contrast see the remarks in 9.1.)

7.7. Assuming intuitive set theory, or ZF, use the collapsing lemma and the Löwenheim-Skolem theorem to show that for every transitive A there is a countable transitive set B such that $\langle A,\in\rangle \equiv \langle B,\in\rangle$. ($\equiv$ denotes elementary equivalence; we use $\langle A,\in\rangle$ for $\langle A,\in \cap A^2\rangle$ when A is transitive.) Show that $\equiv$ cannot in general be replaced by $\prec$ (elementary substructure).

7.8. Let a,b be transitive sets, f an isomorphism of $\langle a,\in\rangle$ and $\langle b,\in\rangle$. Show that if $f(p)=p$ for all urelements $p\in a$ then $f(x)=x$ for all $x\in a$ and hence $a=b$.

7.9 Notes. The collapsing lemma is due to Mostowski [1949] and is one of the standard tools of the set-theorist. (See also the notes to § 9.) The value $C_b(x)$ of the collapsing function is of interest even when $x\notin b$. For example if b is countable one can use $C_b(x)$ as a kind of countable approximation to x. Using a notion of "almost all" due to Kueker and Jech, one can prove that if P is a Σ predicate and $P(x)$ holds, then $P(C_b(x))$ holds for almost all countable sets b. For more on this see Kueker [1972], Jech [1973] and Barwise [1974].

8. Persistent and Absolute Predicates

In this section we discuss the reason for the restriction to Δ_0 formulas in the axioms of separation and collection. The rationale behind this restriction rests in one of the basic notions of the subject, that of absoluteness.

Recall the discussion of $\mathbb{V}_M$ from § 1. The sets in $\mathbb{V}_M$ come in stages and separation tells us what principles are allowed in forming the sets at each stage. The content of Δ_0 Separation is that we allow ourselves to form the set $b=\{x\in a \mid \varphi(x,y)\}$ at stage α if we already have formed a and y, but only if the meaning of $\varphi(x,y)$ is completely (or absolutely) determined solely on the basis of the sets formed before stage α. In other words, when we come to a later stage β and form $\{x\in a \mid \varphi(x,y)\}$ we want to get the same set b, even though there are now more sets around which might conceivably affect the truth of $\varphi(x,y)$ by altering the range of any unbounded quantifiers in φ.

Similar considerations apply to collection. Suppose that, in the process of building $\mathbb{V}_M$, we suddenly notice that $\forall x\in a\, \exists y\, \varphi(x,y)$ is true. We want to be able to form at the next stage a set b for which $\forall x\in a\, \exists y\in b\, \varphi(x,y)$ is true, and remains true. But what if the introduction of this very set b destroyed the truth of $\varphi(x,y)$ for some $x\in a$? This can happen if φ has unbounded universal quantifiers in it. If we want this stability, we must apply collection only if $\varphi(x,y)$ cannot become false when we add new sets to our universe of set theory. That is, $\varphi(x,y)$ should persist.

The aim of this section is to extract formal consequences from these ideas.

8.1 Definition. Let $\mathfrak{A}_{\mathfrak{M}}=(\mathfrak{M}; A, E, \ldots)$ be a structure for L^*. For $a\in A$ we define

$$a_E=\{y\in M\cup A \mid yEa\}\,.$$

Note that the value of a_E in 8.1 depends on $\mathfrak{A}_{\mathfrak{M}}$, and a. The import of 8.1 is clear. Speaking very loosely, the set a_E is "the set that a believes itself to be". In the natural intended structures a_E will just be a itself.

The usual notion of substructures has an obvious generalization to L*. We say that $\mathfrak{B}_{\mathfrak{N}}$ is an *extension* of $\mathfrak{A}_{\mathfrak{M}}$, and write $\mathfrak{A}_{\mathfrak{M}} \subseteq \mathfrak{B}_{\mathfrak{N}}$ (where $\mathfrak{A}_{\mathfrak{M}} = (\mathfrak{M}; A, E, \ldots)$ and $\mathfrak{B}_{\mathfrak{N}} = (\mathfrak{N}; B, E', \ldots)$) if $\mathfrak{M} \subseteq \mathfrak{N}$ (as L-structures), if $A \subseteq B$, and if the interpretations $E, \ldots$ are just the restrictions to $M \cup A$ of the interpretations $E', \ldots$.

A moment's reflection shows that $\mathfrak{A}_{\mathfrak{M}} \subseteq \mathfrak{B}_{\mathfrak{N}}$ is not really the natural notion of extension when one is thinking of models of set theory. For suppose $a \in A$. The trouble is that a may be "schizophrenic" in its role as a set in $\mathfrak{A}_{\mathfrak{M}}$ and as a set in $\mathfrak{B}_{\mathfrak{N}}$. The relation $\mathfrak{A}_{\mathfrak{M}} \subseteq \mathfrak{B}_{\mathfrak{N}}$ guarantees that $a_E \subseteq a_{E'}$ but it does not rule out the possibility that for some $x \in B - A$, $x \in (a_{E'} - a_E)$. This is clearly a chaotic situation (since a set is supposed to be determined by its members), so we introduce a stronger notion of extension suitable for the study of set theory.

8.2 Definition. Given structures $\mathfrak{A}_{\mathfrak{M}} = (\mathfrak{M}; A, E, \ldots)$ and $\mathfrak{B}_{\mathfrak{N}} = (\mathfrak{N}; B, E', \ldots)$ for L*, we say that $\mathfrak{B}_{\mathfrak{N}}$ is an *end extension* of $\mathfrak{A}_{\mathfrak{M}}$, written either as:

$$\mathfrak{A}_{\mathfrak{M}} \subseteq_{\text{end}} \mathfrak{B}_{\mathfrak{N}} \quad \text{or} \quad \mathfrak{B}_{\mathfrak{N}} \supseteq_{\text{end}} \mathfrak{A}_{\mathfrak{M}},$$

if $\mathfrak{A}_{\mathfrak{M}} \subseteq \mathfrak{B}_{\mathfrak{N}}$ and if for each $a \in A$, $a_E = a_{E'}$. One sometimes reads, aloud, $\mathfrak{A}_{\mathfrak{M}} \subseteq_{\text{end}} \mathfrak{B}_{\mathfrak{N}}$ as "$\mathfrak{A}_{\mathfrak{M}}$ is an *initial substructure* of $\mathfrak{B}_{\mathfrak{N}}$".

8.3 Example. If A is a transitive set, $B \supseteq A$, $E = \in \cap A^2$, and $E' = \in \cap B^2$, then

$$(\mathfrak{M}; A, E) \subseteq_{\text{end}} (\mathfrak{M}; B, E') ;$$

for in both structures any $a \in A$ has $a_E = a_{E'} = a$. If A were not transitive, however, this could fail.

8.4 Lemma. *Let $\mathfrak{A}_{\mathfrak{M}}, \mathfrak{B}_{\mathfrak{N}}$ be structures for* L*, $\mathfrak{B}_{\mathfrak{N}} \supseteq_{\text{end}} \mathfrak{A}_{\mathfrak{M}}$. *If φ is a Σ formula of* L* *then for any $x_1, \ldots, x_n \in \mathfrak{A}_{\mathfrak{M}}$, $\mathfrak{A}_{\mathfrak{M}} \models \varphi[x_1, \ldots, x_n]$ implies $\mathfrak{B}_{\mathfrak{N}} \models \varphi[x_1, \ldots, x_n]$.*

Proof. This just repeats the proof of Lemma 4.2 proceeding by induction on φ. The end extension hypothesis is used to assure that $\forall x \in a$ has the same meaning in $\mathfrak{A}_{\mathfrak{M}}$ and $\mathfrak{B}_{\mathfrak{N}}$. ☐

8.5 Definition. A formula $\varphi(u_1, \ldots, u_n)$ of L* is said to be *persistent* relative to a theory T of L* if for all models $\mathfrak{A}_{\mathfrak{M}}, \mathfrak{B}_{\mathfrak{N}}$ of T with $\mathfrak{B}_{\mathfrak{N}} \supseteq_{\text{end}} \mathfrak{A}_{\mathfrak{M}}$, and all $x_1, \ldots, x_n$ in $\mathfrak{A}_{\mathfrak{M}}$:

$$\mathfrak{A}_{\mathfrak{M}} \models \varphi[x_1, \ldots, x_n] \quad \text{implies} \quad \mathfrak{B}_{\mathfrak{N}} \models \varphi[x_1, \ldots, x_n].$$

The formula φ is *absolute* relative to T if for all $\mathfrak{A}_{\mathfrak{M}}, \mathfrak{B}_{\mathfrak{N}}, x_1, \ldots, x_n$ as above:

$$\mathfrak{A}_{\mathfrak{M}} \models \varphi[x_1, \ldots, x_n] \quad \text{iff} \quad \mathfrak{B}_{\mathfrak{M}} \models \varphi[x_1, \ldots, x_n].$$

The significance of 8.5 should be clear enough. Absolute formulas don't shift their meaning on us as we move from $\mathfrak{A}_{\mathfrak{M}}$ to its end extension $\mathfrak{B}_{\mathfrak{N}}$ and back again. Absoluteness is a precious attribute.

8.6 Corollary. *All Σ formulas are persistent and all Δ_0 formulas are absolute (relative to any theory* T*).*

Proof. By 8.4 all Σ formulas are persistent, hence all Δ_0 formulas are persistent. But the Δ_0 formulas are closed under negation and φ is absolute iff φ and $\neg\varphi$ are both persistent. □

8.7 Example. Let $\mathfrak{A}_{\mathfrak{M}}$ and $\mathfrak{B}_{\mathfrak{N}}$ be models of KPU, $\mathfrak{A}_{\mathfrak{M}} \subseteq_{\text{end}} \mathfrak{B}_{\mathfrak{N}}$. We can interpret all the definitions and theorems of KPU in these two models. For example, let $a \in \mathfrak{A}_{\mathfrak{M}}$. Since $\mathrm{Ord}(x)$ is a Δ_0 formula,

$$\mathfrak{A}_{\mathfrak{M}} \models \mathrm{Ord}(a) \quad \text{iff} \quad \mathfrak{B}_{\mathfrak{M}} \models \mathrm{Ord}(a).$$

Now let us return to consider the rationale behind the Δ_0 in Δ_0 Separation and Δ_0 Collection. We see from Corollary 8.6 that we have asserted separation and collection for absolute formulas, at least some of them. For example, if we form the set $b = \{x \in a \mid \varphi(x)\}$ in $\mathfrak{A}_{\mathfrak{M}}$, a model of KPU, (with φ a Δ_0 formula), then in any $\mathfrak{B}_{\mathfrak{N}} \supseteq_{\text{end}} \mathfrak{A}_{\mathfrak{M}}$, the equation for b will remain true.

Have we asserted separation and collection for all absolute formulas? Yes, but not explicitly. There are formulas $\varphi(x, y)$ which are absolute relative to KPU which are not Δ_0; separation for such φ is not an axiom of KPU. It is a *theorem* of KPU, though, as we see from the following result of Feferman-Kreisel [1966].

8.8 Theorem. *For any theory* T *of* L^*, *if $\varphi(x_1, \ldots, x_n)$ is persistent relative to* T *then there is a Σ formula $\psi(x_1, \ldots, x_n)$ such that*

$$\mathrm{T} \vdash \forall x_1, \ldots, x_n\, [\varphi(x_1, \ldots, x_n) \leftrightarrow \psi(x_1, \ldots, x_n)].$$

Hence, if φ is absolute relative to T, *there are Σ and Π formulas $\psi(x_1, \ldots, x_n)$, $\theta(x_1, \ldots, x_n)$ such that*

$$\mathrm{T} \vdash \forall x_1, \ldots, x_n\, [\varphi(x_1, \ldots, x_n) \leftrightarrow \psi(x_1, \ldots, x_n) \wedge \varphi(x_1, \ldots, x_n) \leftrightarrow \theta(x_1, \ldots, x_n)].$$

From the results in §4 it follows that we can prove separation in KPU for all formulas absolute relative to KPU and collection for all formulas persistent relative to KPU. Furthermore, if we later extend KPU to a stronger theory T (and we will from time to time) then we'll still have separation for all φ absolute relative to T and collection for all φ persistent relative to T. (If T is stronger then it has fewer models so, in general, it is easier for a formula to be persistent or absolute.) These results are not used in the actual study of KPU but they are reassuring.

We conclude this section with a lemma which will prove useful later on. We include it here so that the student can become familiar with the concept of absoluteness. First some remarks.

If $\mathfrak{B}_{\mathfrak{N}} \models \mathrm{KPU}$ and we use a phrase like "b is an ordinal of $\mathfrak{B}_{\mathfrak{N}}$", what we mean, of course, is that $b \in \mathfrak{B}_{\mathfrak{N}}$ and $\mathfrak{B}_{\mathfrak{N}} \models \mathrm{Ord}(b)$. The object b need not be a real ordinal at all. If $\mathfrak{A}_{\mathfrak{M}} \subseteq_{\mathrm{end}} \mathfrak{B}_{\mathfrak{N}}$ and $a \in \mathfrak{A}_{\mathfrak{M}}$ then, as we saw in Example 8.7, a is an ordinal of $\mathfrak{A}_{\mathfrak{M}}$ iff a is an ordinal of $\mathfrak{B}_{\mathfrak{N}}$. Furthermore, since $\mathfrak{A}_{\mathfrak{M}} \subseteq_{\mathrm{end}} \mathfrak{B}_{\mathfrak{N}}$ the ordinals of $\mathfrak{A}_{\mathfrak{M}}$ form an initial segment of the ordinals of $\mathfrak{B}_{\mathfrak{N}}$. (Why?) This initial segment may or may not exhaust the ordinals of $\mathfrak{B}_{\mathfrak{N}}$, even though $\mathfrak{A}_{\mathfrak{M}} \neq \mathfrak{B}_{\mathfrak{N}}$. In the case where it is a proper initial segment there need not be any ordinal b of $\mathfrak{B}_{\mathfrak{N}}$ which is the least upper bound of this segment.

8.9 Lemma. *Let $\mathfrak{A}_{\mathfrak{M}} \subseteq_{\mathrm{end}} \mathfrak{B}_{\mathfrak{M}}$ where $\mathfrak{B}_{\mathfrak{M}} \models \mathrm{KPU}$. Suppose that whenever $\mathfrak{B}_{\mathfrak{M}} \models \mathrm{rk}(a) = \alpha$ we have $a \in A$ iff $\alpha \in A$. Suppose further that there is no ordinal β of $\mathfrak{B}_{\mathfrak{M}}$ which is the least upper bound of the ordinals of $\mathfrak{A}_{\mathfrak{M}}$. Then, with the possible exception of foundation, all the axioms of* KPU *hold in $\mathfrak{A}_{\mathfrak{M}}$.*

Proof. We check three axioms and trust the student to verify the other two, *Extensionality* and *Union.*

Pair: Suppose, $x, y \in \mathfrak{A}_{\mathfrak{M}}$, let $\alpha, \beta \in A$ be such that

$$\mathfrak{B}_{\mathfrak{M}} \models \alpha = \mathrm{rk}(x) \wedge \beta = \mathrm{rk}(y).$$

Then, if $\mathfrak{B}_{\mathfrak{M}} \models \gamma = (\alpha+1) \cup (\beta+1)$, we have $\gamma \in A$ since otherwise γ would be the least upper bound of the ordinals of $\mathfrak{A}_{\mathfrak{M}}$. Thus if we choose $b \in \mathfrak{B}_{\mathfrak{M}}$ with $\mathfrak{B}_{\mathfrak{M}} \models b = \{x, y\}$ so that $\mathfrak{B}_{\mathfrak{M}} \models \mathrm{rk}(b) = \gamma$, then $b \in A$ and $\mathfrak{A}_{\mathfrak{M}} \models b = \{x, y\}$ by absoluteness of the formula $b = \{x, y\}$, from Table 1.

Δ_0 *Separation:* Suppose $a, y \in \mathfrak{A}_{\mathfrak{M}}$. Let $\varphi(x, y)$ be Δ_0. We want to find $a, b \in \mathfrak{A}_{\mathfrak{M}}$ such that

(1) $b = \{x \in a \mid \varphi(x, y)\}$

holds in $\mathfrak{A}_{\mathfrak{M}}$. Let $b \in \mathfrak{B}_{\mathfrak{M}}$ be such that (1) holds in $\mathfrak{B}_{\mathfrak{M}}$, using Δ_0 Separation in $\mathfrak{B}_{\mathfrak{M}}$. But since

$$\mathfrak{B}_{\mathfrak{M}} \models \mathrm{rk}(b) \leqslant \mathrm{rk}(a)$$

the set b is in $\mathfrak{A}_{\mathfrak{M}}$. It still satisfies (1) in $\mathfrak{A}_{\mathfrak{M}}$ by absoluteness.

Δ_0 *Collection:* Suppose that $a \in \mathfrak{A}_{\mathfrak{M}}$, the formula $\varphi(x, y)$ is Δ_0 with parameters from $\mathfrak{A}_{\mathfrak{M}}$ and that $\forall x \in a\, \exists y\, \varphi(x, y)$ holds in $\mathfrak{A}_{\mathfrak{M}}$. Then we have:

(2) for each $x \in a$ there is a $y \in A$ and $\alpha \in A$ such that $\mathfrak{A}_{\mathfrak{M}} \models \varphi(x, y)$ and $\mathfrak{B}_{\mathfrak{M}} \models \mathrm{rk}(y) = \alpha$, and hence, by absoluteness $\mathfrak{B}_{\mathfrak{M}} \models \varphi(x, y) \wedge \mathrm{rk}(y) = \alpha$.

Thus in $\mathfrak{B}_{\mathfrak{M}}$ we have $\forall x \in a\, \exists \alpha\, \exists y [\mathrm{rk}(y) = \alpha \wedge \varphi(x, y)]$. So, by Σ Reflection in $\mathfrak{B}_{\mathfrak{M}}$, there is an ordinal $\beta \in \mathfrak{B}_{\mathfrak{M}}$ such that

$$\forall x \in a\, \exists \alpha < \beta\, \exists y\, [\mathrm{rk}(y) = \alpha \wedge \varphi(x, y)],$$

and hence:

(3) $\forall x \in a\, \exists y\, [\mathrm{rk}(y) < \beta \wedge \varphi(x, y)]$

holds in $\mathfrak{B}_{\mathfrak{M}}$. In $\mathfrak{B}_{\mathfrak{M}}$ pick the least ordinal β satisfying (3): it exists by foundation. By (2), β is a sup of ordinals $\alpha \in \mathfrak{A}_{\mathfrak{M}}$, so $\beta \in \mathfrak{A}_{\mathfrak{M}}$. Apply Δ_0 Collection in $\mathfrak{B}_{\mathfrak{M}}$ to (3) to find a set $b \in \mathfrak{B}_{\mathfrak{M}}$ such that

$$\forall x \in a\, \exists y \in b\, [\varphi(x, y) \wedge \mathrm{rk}(y) < \beta]$$

holds in $\mathfrak{B}_{\mathfrak{M}}$. Since $\mathfrak{B}_{\mathfrak{M}} \models \mathrm{rk}(b) \leqslant \beta$, $b \in \mathfrak{A}_{\mathfrak{M}}$. But then the formula

$$\forall x \in a\, \exists y \in b\, \varphi(x, y)$$

is Δ_0, it holds in $\mathfrak{B}_{\mathfrak{M}}$, and it has all its parameters in $\mathfrak{A}_{\mathfrak{M}}$. Hence by absoluteness, it holds in $\mathfrak{A}_{\mathfrak{M}}$. □

8.10—8.12 Exercises

8.10. Given $\mathfrak{A}_{\mathfrak{M}} \subseteq \mathfrak{B}_{\mathfrak{N}}$, we write $\mathfrak{A}_{\mathfrak{M}} \prec_1 \mathfrak{B}_{\mathfrak{N}}$ if for all Σ_1 formulas $\varphi(x_1, \ldots, x_n)$ of L^* and all $x_1, \ldots, x_n$ in $\mathfrak{A}_{\mathfrak{M}}$:

$$\mathfrak{A}_{\mathfrak{M}} \models \varphi[x_1, \ldots, x_n] \quad \text{iff} \quad \mathfrak{B}_{\mathfrak{N}} \models \varphi[x_1, \ldots, x_n].$$

Show that if $\mathfrak{A}_{\mathfrak{M}} \subseteq_{\text{end}} \mathfrak{B}_{\mathfrak{N}}$, $\mathfrak{A}_{\mathfrak{M}} \prec_1 \mathfrak{B}_{\mathfrak{N}}$, and $\mathfrak{B}_{\mathfrak{N}} \models \mathrm{KPU}$ then, with the possible exception of foundation, all the axioms of KPU hold in $\mathfrak{A}_{\mathfrak{M}}$. (The end extension hypothesis is used to insure that Σ formulas persist from $\mathfrak{A}_{\mathfrak{M}}$ to $\mathfrak{B}_{\mathfrak{N}}$. Without this, the exercise is false.)

8.11. Given $\mathfrak{A}_{\mathfrak{M}} \subseteq \mathfrak{B}_{\mathfrak{N}}$, show that $\mathfrak{A}_{\mathfrak{M}} \prec_1 \mathfrak{B}_{\mathfrak{N}}$ iff for every Δ_0 formula $\varphi(v_1, \ldots, v_n, v_{n+1})$ and all $x_1, \ldots, x_n \in \mathfrak{A}_{\mathfrak{M}}$,

$$\mathfrak{B}_{\mathfrak{N}} \models \exists v_{n+1}\, \varphi[x_1, \ldots, x_n] \quad \text{implies} \quad \exists x_{n+1} \in \mathfrak{A}_{\mathfrak{M}} (\mathfrak{B}_{\mathfrak{N}} \models \varphi[x_1, \ldots, x_{n+1}]).$$

(We are not assuming $\mathfrak{A}_{\mathfrak{M}} \subseteq_{\text{end}} \mathfrak{B}_{\mathfrak{N}}$!)

8.12 (Schlipf). Find an example of two structures $\mathfrak{A}_{\mathfrak{M}}$ and $\mathfrak{B}_{\mathfrak{M}}$ satisfying the hypotheses of Lemma 8.9 but where $\mathfrak{A}_{\mathfrak{M}}$ fails to satisfy the axiom of foundation. [Let $\mathfrak{B}_{\mathfrak{M}}$ be a proper elementary extension of $\mathbb{HF}_{\mathfrak{M}}$. (Cf. § II.2.)]

8.13 Notes. The considerations involved in the choice of Δ_0 Separation and Δ_0 Collection are suggested by the informal notion of "predicative". Kripke, in fact, called his axioms for admissible sets PZF, for predicative ZF. As an explication of the intuitive idea of predicativity, however, KP has certain debatable features. See, for example, Feferman [1975] for a discussion and examples of set theories which are predicative in a more stringent sense.

9. *Additional Axioms*

There are certain extensions of KPU which surface from time to time. We have already defined KPU^+ and KP in § 2. We catalogue some of the others here.

9.1 Definition. *The axiom of infinity*, or *Infinity*, is the axiom:

$$\exists\alpha\, \mathrm{Lim}(\alpha)$$

where $\mathrm{Lim}(\alpha)$ is defined in Table 2. We use ω as a symbol for the first limit ordinal.

Note that $\neg$ (Infinity) asserts only that all ordinals are finite, not that all sets are finite.

The axiom of infinity is often used to form sets by taking $a = \bigcup_{n<\omega} b_n$ where b_n is defined by recursion on n. We saw one example where this would have been convenient in the proof of 6.1. For another example, define (in KPU)

$$F(a,0) = \{0\},$$
$$F(a,n+1) = \{b \cup \{x\} : b \in F(a,n),\ x \in a,\ x \notin b\},$$

by Σ recursion. We find that $F(a,n)$ is the set of n-element subsets of a. In KPU + (Infinity) we can introduce a new Σ operation symbol P_ω by

$$\mathsf{P}_\omega(a) = \bigcup_{n<\omega} F(a,n),$$

as the student should verify. We can use P_ω to convert quantifiers over finite subsets of a to bounded quantifiers:

$$\forall b\, [b \subseteq a \wedge b \text{ finite} \rightarrow (\ldots b \ldots)]$$

becomes

$$\forall b \in \mathsf{P}_\omega(a)\, (\ldots b \ldots)$$

in KPU + (Infinity). Since a is finite iff $a \in \mathsf{P}_\omega(a)$, we see that "$a$ is finite" is Δ_1 in KPU + (Infinity) (whereas it is only Σ_1 in KPU).

The remaining axioms will be of secondary importance for our study.

9.2 Definition. Σ_1 *Separation* is the set of axioms of the form

(i) $\exists b\, \forall x\, (x \in b \leftrightarrow x \in a \wedge \varphi(x))$,

where φ is a Σ_1 formula of L^*.

9.3 Definition. *Full separation* asserts 9.2 (i) for *all* formulas φ of L^*.

9.4 Definition. *Full collection* asserts the collection scheme

$$\forall x \in a\, \exists y\, \varphi(x,y) \rightarrow \exists b\, \forall x \in a\, \exists y \in b\, \varphi(x,y)$$

for all formulas φ of L^*.

9.5 Definition. *The axiom Beta.* A relation r is *well founded on a* if

$$\forall b\, [b \subseteq a \wedge b \neq 0 \rightarrow \exists x \in b\, \forall y \in b\, (\langle y, x\rangle \notin r)].$$

If $r \subseteq a \times a$ and r is well founded on a then we say that r is *well founded.* (If r is well founded on a, then $r \cap (a \times a)$ is well founded, but r itself may have some funny things going on outside the set a.) The axiom *Beta* asserts: for every well-founded relation $r \subseteq a \times a$ on a set a there is a function f, $\mathrm{dom}(f) = a$, satisfying:

(i) $f(x) = \{f(y) \colon y \in a \wedge \langle y, x\rangle \in r\}$,

for all $x \in a$. The function f is said to be *collapsing* for r.

The axiom Beta has the effect of making the Π_1 predicate "r is well founded on a" a Δ_1 predicate since it becomes equivalent to:

$$\exists f\, [\mathrm{dom}(f) = a \wedge f \text{ is collapsing for } r].$$

(See 9.8 (ii) (b).) Beta is not provable in KPU but it is provable if we add Σ_1 Separation.

9.6 Theorem. *Beta is provable in* KPU + (Σ_1 Separation).

Sketch of proof. Let us work in KPU+(Σ_1 Separation). Let r be well founded on a, and write $x \prec y$ for $\langle x, y\rangle \in r$. Define a operation F on the ordinals by Σ recursion:

$$\begin{aligned} F(\alpha) &= \{x \in a \mid \forall y \in a\, (y \prec x \rightarrow \exists \beta < \alpha\; y \in F(\beta))\} \\ &= \text{the set of all } x \in a \text{ such that } \{y \in a \mid y \prec x\} \subseteq \textstyle\bigcup_{\beta < \alpha} F(\beta). \end{aligned}$$

Note that $\alpha \leqslant \beta$ implies $F(\alpha) \subseteq F(\beta)$. Let us show that every $x \in a$ is in some $F(\alpha)$. If not, then the set $b = a - b_0$ is non empty, where

$$b_0 = \{x \in a \mid \exists \alpha\, (x \in F(\alpha))\},$$

this being the place where we need Σ_1 Separation. Let $x \in b$ be such that for all $y \in b$, we have $y \not\prec x$ (using the well-foundedness of r). Then

$$\forall y \in a\, [y \prec x \rightarrow \exists \beta\, (y \in F(\beta))]$$

so by Σ Reflection there is an α such that

$$\forall y \in a\,[y \prec x \to \exists \beta < \alpha\,(y \in F(\beta))], \text{ and hence}$$

$$\{y \in a: y \prec x\} \subseteq \bigcup_{\beta < \alpha} F(\beta).$$

So $x \in F(\alpha)$ by the definition of $F(\alpha)$, which contradicts $x \notin b_0$. Now since $a = \bigcup_\alpha F(\alpha)$ there is, by Σ Reflection, a γ such that $a = \bigcup_{\alpha < \gamma} F(\alpha)$.

The rest of the proof is easy. Define f_α, for $\alpha \leqslant \gamma$, by recursion on α: f_α is the function with domain $F(\alpha)$ such that

$$f_\alpha(x) = \{f_\beta(y) \mid \beta < \alpha \wedge y \in F(\beta) \wedge y \prec x\},$$

for all $x \in F(\alpha)$. These f_α are increasing ($\beta \leqslant \alpha$ implies $f_\beta \subseteq f_\alpha$, by induction on α), and $f = f_\gamma = \bigcup_{\alpha < \gamma} f_\alpha$ is the desired function satisfying

$$f(x) = \{f(y): y \in a \wedge y \prec x\}$$

for all $x \in a$. ▯

9.7 Definition. *The power set axiom.* We think of the power set operation as a primitive operation. When we use the power set axiom we will assume $\mathsf{L}^* = \mathsf{L}(\in, \mathsf{P}, \ldots)$ where P is a 1-place operation symbol. The power set axiom asserts

$$\forall x \forall y\,[x \in \mathsf{P}(y) \leftrightarrow (\mathsf{S}(x) \wedge x \subseteq y)]$$

where, as in Table 2, $\mathsf{S}(x)$ means "x is a set".

9.8—9.12 Exercises

9.8. Prove in KP (not KPU) that every set of finite rank is finite. Hence KP + ¬(Infinity) implies that every set is finite. This greatly limits KP as opposed to KPU, as we'll see in later chapters.

9.9. Let $r \subseteq a \times a$ and let f be a function with $\mathrm{dom}(f) = a$ which is collapsing for r. Prove the following in KPU:

(i) r is well founded;

(ii) $\mathrm{rng}(f)$ is transitive and has no urelements in it;

(iii) If g is a function with $\mathrm{dom}(g) = a$ and g is collapsing for r, then $f = g$. [Hint: Prove $\forall b\, \forall x \in a\,(f(x) = b \leftrightarrow g(x) = b)$ by ε-induction on b]

(iv) If for all $x, y \in a$, $x \neq y$, there is a $z \in a$ with $\neg(\langle z, x\rangle \in r \leftrightarrow \langle z, y\rangle \in r)$, then f is one-one and hence is an isomorphism of $\langle a, r\rangle$ with a transitive set $\langle b, \in \cap\, b^2\rangle$.

9.10. Show that in KPU + (Beta) we can introduce a Σ operation symbol B such that

$$\mathsf{B}(r, a) = 0 \quad \text{iff } r \text{ is not well founded on } a,$$

but if r is well founded on a, then $\mathsf{B}(r,a)$ is a (the) function f with $\mathrm{dom}(f)=a$ such that $f(x)=\{f(y): y\in a \wedge \langle y,x\rangle \in r\}$ for all $x\in a$. [Use 9.9.]

9.11. A relation $r\subseteq a\times a$ *well orders* a if it orders a linearly and is well founded. Show in KPU+(Beta), that if r well orders a, then there is a (unique) ordinal α such that $\langle a,r\rangle \cong \langle \alpha,\in\rangle$.

9.12. Show that adding Σ_1 Separation to KPU has the same effect (i.e. same theorems) as adding all the following axioms, where φ is Δ_0:

$$\exists b\,\forall x\in a\,[\exists y\,\varphi(x,y)\rightarrow \exists y\in b\,\varphi(x,y)].$$

9.13 Notes. In a theory like ZF containing Σ_1 Separation, Beta becomes a theorem and the collapsing lemma of § 7 is a consequence of it. In such theories Beta itself is often called *The Collapsing Lemma.* It is due to Mostowski [1949]. In KPU we must separate the two aspects since one is provable and the other is not. Beta is so named because Mostowski [1961] used the terminology "β-model" (with "β" for *bon ordre*) for models where well-orderings were absolute.

Chapter II

Some Admissible Sets

Having gained some feeling for the theory KPU we turn to its intended models, admissible sets. Admissible sets come in many sizes and shapes. In this chapter the student is introduced to some of the more attractive ones in a cursory fashion. We will delve into their structure and properties later.

1. The Definition of Admissible Set and Admissible Ordinal

It facilitates matters if we fix a largest possible universe of sets over an arbitrary collection M of urelements once and for all. We define by recursion:

$$\mathrm{V}_M(0)=0;$$

$$\mathrm{V}_M(\alpha+1)=\text{Power set of }(M\cup \mathrm{V}_M(\alpha));$$

$$\mathrm{V}_M(\lambda)=\bigcup_{\alpha<\lambda}\mathrm{V}_M(\alpha),\ \text{if } \lambda \text{ is a limit; and}$$

$$\mathbb{V}_M=\bigcup_{\alpha}\mathrm{V}_M(\alpha),$$

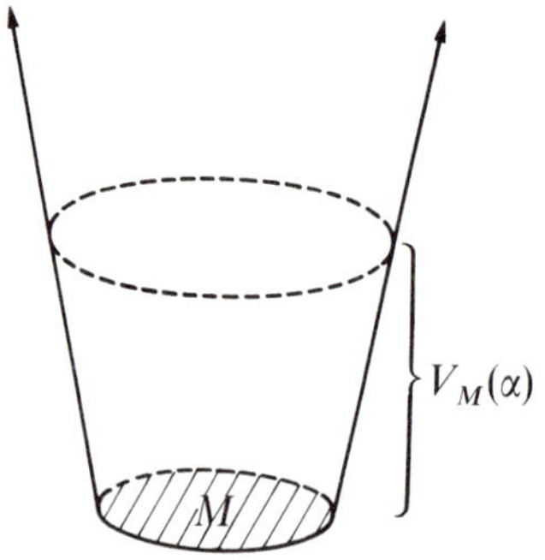

Fig. 1A. The universe $\mathbb{V}_M$ of sets on M

where the union in the last equation is taken over all ordinals α. (The reason for letting $V_M(0)=0$, rather than $V_M(0)=M$, is that $\mathbb{V}_M$ is to be a collection of *sets* on M.) We use $\in_M$ for the membership relation on $\mathbb{V}_M$, dropping the subscript if there is little room for confusion. If $\mathfrak{M}=\langle M,---\rangle$ we write $\mathbb{V}_{\mathfrak{M}}$ for $\mathbb{V}_M$. If M is the empty collection we write $V(\alpha)$ for $V_M(\alpha)$ and $\mathbb{V}$ for $\mathbb{V}_M$.

1.1 Definition. Let $\mathsf{L}^*=\mathsf{L}(\in,\ldots)$ and a structure $\mathfrak{M}$ for L be given. An *admissible set over* $\mathfrak{M}$ is a model $\mathfrak{A}_{\mathfrak{M}}$ of KPU of the form

$$\mathfrak{A}_{\mathfrak{M}}=(\mathfrak{M};A,\in,\ldots),$$

where $M\cup A$ is transitive in $\mathbb{V}_M$, and $\in$ is the restriction of $\in_M$ to $M\cup A$. The admissible set $\mathfrak{A}_{\mathfrak{M}}$ is *admissible above* $\mathfrak{M}$ if $M\in A$, i.e., if $\mathfrak{A}_{\mathfrak{N}}\models \mathrm{KPU}^+$. We use special Roman $\mathbb{A}$, $\mathbb{B}$, $\mathbb{C}$ to range over admissible sets. When we need to exhibit the underlying structure $\mathfrak{M}$ we write $\mathbb{A}_{\mathfrak{M}}$.

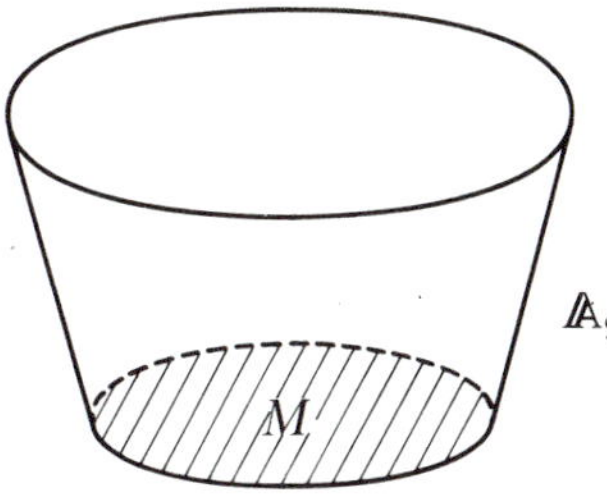

Fig. 1 B. A typical admissible set over $\mathfrak{M}$

In other words, admissible sets are models of KPU which are transitive hunks of $\mathbb{V}_M$ with the intended interpretation $\in_M$ of the membership symbol. *Warning:* While the interpretation of the membership symbol must be the natural one, Definition 1.1 makes no such demands on the interpretations of any other symbols in the list ... of $\mathsf{L}(\in,\ldots)$. They must fend for themselves. For example, if $\mathsf{L}^*=\mathsf{L}(\in,\mathsf{P})$ and the admissible set $\mathbb{A}_M=(\mathfrak{M};A,\in,P)$ is a model of Power, then there is nothing to guarantee that $P(a)$ is the real power set of a; it may very well be only a small subset of the real power set of a.

1.2 Lemma. *Suppose* $\mathfrak{A}_{\mathfrak{M}}=(\mathfrak{M};A,\in_M,\ldots)$ *and* $\mathfrak{B}_{\mathfrak{N}}=(\mathfrak{N};B,\in_{\mathfrak{N}},\ldots)$ *and* $\mathfrak{A}_{\mathfrak{M}}\subseteq\mathfrak{B}_{\mathfrak{N}}$. *If* $M\cup A$ *is transitive in* $\mathbb{V}_M$, *then* $\mathfrak{A}_{\mathfrak{M}}\subseteq_{\mathrm{end}}\mathfrak{B}_{\mathfrak{N}}$.

Proof. Recall the definition given in I.8.1. If $a\in A$ then $a_{\in_M}=a=a_{\in_N}$ since $M\cup A$ is transitive in $\mathbb{V}_M$. □

This lemma also holds for $\mathfrak{B}_{\mathfrak{N}}=\mathbb{V}_{\mathfrak{N}}$, except that $\mathbb{V}_{\mathfrak{N}}$ is not a proper structure. This trivial lemma is of real importance. With the results of I.8 it insures that Δ predicates and Σ operations of KPU have the same meaning in all admissible sets that they have in $\mathbb{V}_{\mathfrak{M}}$.

1.3 A Comparison. Consider the two operations TC and P (Power set) and an admissible set $\mathbb{A}=(\mathfrak{M}, A, \in, P)$ satisfying the Power set axiom. Given $a \in A$ the expressions

$$\mathrm{TC}(a), \quad \mathsf{P}(a)$$

each have *two* possible interpretations. For TC there are the sets b_0, b_1 such that

$$\mathbb{A} \models \mathrm{TC}(a)=b_0 \quad \text{and} \quad \mathbb{V}_{\mathfrak{M}} \models \mathrm{TC}(a)=b_1,$$

where $\mathbb{A}=(\mathfrak{M}; A, \in)$. For P there are the sets c_0, c_1 such that

$$\mathbb{A} \models \mathsf{P}(a)=c_0 \quad \text{and} \quad \mathbb{V}_{\mathfrak{M}} \models \mathsf{P}(a)=c_1.$$

Since $\mathbb{A}_{\mathfrak{M}} \subseteq_{\mathrm{end}} \mathbb{V}_{\mathfrak{M}}$ and TC is a Σ operation, we have $\mathbb{V}_{\mathfrak{M}} \models \mathrm{TC}(a)=b_0$; and so $b_0=b_1$. Thus b_0 is the real transitive closure of a, so that

$$b_0=\bigcap \{b \mid b \text{ transitive, } a \subseteq b\}.$$

For P, however, this fails. Since $x \subseteq y$ is Δ_0 we get $c_0 \subseteq c_1$ but that's all. Typically, c_0 will be a *proper* subset of the real power set c_1 of a.

1.4 Definitions. A *pure set* in $\mathbb{V}_M$ is a set a with empty support; i. e., one with $\mathrm{TC}(a) \cap M=0$. (For example, ordinals are pure sets.) A *pure admissible set* is an admissible set which is a model of KP; i. e., one without urelements. Pure admissibles can be written $\mathbb{A}=(A, \in, \ldots)$. If $\mathsf{L}^*=\{\in\}$ then we write A for $\mathbb{A}=\langle A, \in\rangle$.

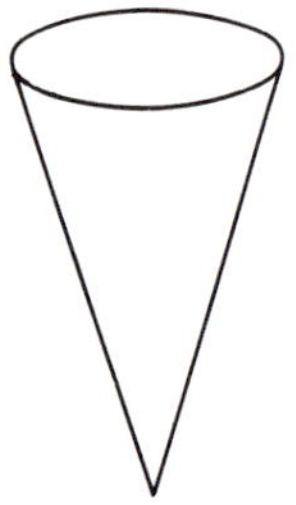

Fig. 1 C. A pure admissible set A

1.5 Theorem. *If* $\mathbb{A}_{\mathfrak{M}}=(\mathfrak{M}; A, \in)$ *is admissible and* $A_0=\{a \in A \mid a$ *is a pure set*$\}$, *then* A_0 *is a pure admissible set, called the* pure part *of* $\mathbb{A}_{\mathfrak{M}}$. (See Fig. 1 D.)

Proof. By 1.2 we have $A_0 \subseteq_{\mathrm{end}} \mathbb{A}_{\mathfrak{M}} \subseteq_{\mathrm{end}} \mathbb{V}_{\mathfrak{M}}$. By absoluteness, $\mathrm{sp}(a)$ has the same meaning in $\mathbb{A}_{\mathfrak{M}}$ and $\mathbb{V}_{\mathfrak{M}}$. Let us check Δ_0 Collection leaving the easier axioms as exercises to help the student master absoluteness arguments. Suppose $a, b \in A_0$ and suppose A_0 satisfies $\forall x \in a\, \exists y\, \varphi(x, y, b)$, where φ is Δ_0. If $\varphi(x, y, b)$ is true

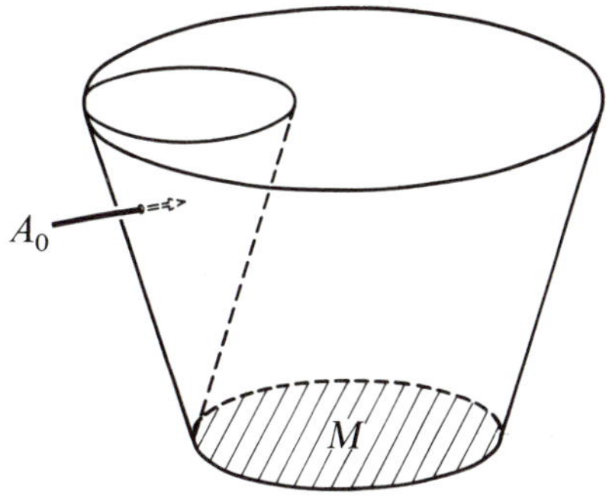

Fig. 1 D. The pure part A_0 of an admissible set $\mathbb{A}_{\mathfrak{M}}$

in A_0 it is also true in $\mathbb{A}_{\mathfrak{M}}$ by absoluteness so $\mathbb{A}_{\mathfrak{M}}$ satisfies:

$$\forall x \in a\, \exists y\, [\mathrm{sp}(y)=0 \wedge \varphi(x,y,b)].$$

Applying Σ collection in $\mathbb{A}_{\mathfrak{M}}$, we get a $c \in \mathbb{A}_{\mathfrak{M}}$ such that

(1) $\quad \forall x \in a\, \exists y \in c\, [\mathrm{sp}(y)=0 \wedge \varphi(x,y,b)]$, and

(2) $\quad \forall y \in c\, \exists x \in a\, [\mathrm{sp}(y)=0 \wedge \varphi(x,y,b)]$.

From (2) we get $\mathrm{sp}(c)=0$, since $\mathrm{sp}(c)=\bigcup\{\mathrm{sp}(y) \mid y \in c\}$; so $c \in A_0$. But then (1) is a Δ_0 formula with parameters from A_0, true in $\mathbb{A}_{\mathfrak{M}}$, hence true in A_0. □

1.6 Exercise. Verify Pair, Union and Δ_0 Separation for the proof of 1.5. Notice that Extensionality is trivial from the transitivity of A_0, and that Foundation is trivial by the well-foundedness of $\mathbb{A}_{\mathfrak{M}}$.

1.7 Definitions. The *ordinal of* an admissible set $\mathbb{A}_{\mathfrak{M}}$, denoted by $o(\mathbb{A}_{\mathfrak{M}})$, is the least ordinal not in $\mathbb{A}_{\mathfrak{M}}$; equivalently, it is the order type of the ordinals in $\mathbb{A}_{\mathfrak{M}}$. An ordinal α is *admissible* if $\alpha=o(\mathbb{A}_{\mathfrak{M}})$ for some $\mathfrak{M}$ and some admissible set $\mathbb{A}_{\mathfrak{M}}$. An ordinal α is *$\mathfrak{M}$-admissible* if $\alpha=o(\mathbb{A}_{\mathfrak{M}})$ for some $\mathbb{A}_{\mathfrak{M}}$ which is admissible above $\mathfrak{M}$ (in the sense of 1.1).

1.8 Corollary. *The ordinal α is admissible iff $\alpha=o(A)$ for some pure admissible set.*

Proof. If $\alpha=o(\mathbb{A}_{\mathfrak{M}})$ and $\mathbb{A}_{\mathfrak{M}}$ is admissible, then $\alpha=o(A_0)$, where A_0 is the pure part of $\mathbb{A}_{\mathfrak{M}}$. □

What kinds of ordinals are admissible? In the next section we will see that ω is admissible. From our development of ordinal arithmetic in Chapter I we see that if α is admissible then α is closed under ordinal successor, addition, multiplication, exponentiation and similar functions of ordinal arithmetic. Thus the least admissible $\alpha>\omega$ is bigger that

$$\omega+\omega,\ \omega\cdot\omega,\ \omega^{\omega},\ \omega^{\omega^{\omega}},\ldots,\varepsilon_0,\ldots,$$

where the operations are from ordinal (*not* cardinal) arithmetic. In § 3 we will prove that every infinite cardinal κ is admissible and that for any $\beta<\kappa$, there are κ admissible ordinals α between β and κ. (Thus, κ is a limit of admissibles.)

1.9 Definitions. Let $\mathbb{A}=\mathbb{A}_{\mathfrak{M}}=(\mathfrak{M}; A,\in,\ldots)$. We often use the following notation and terminology. An object x is *in* $\mathbb{A}$ if $x\in M\cup A$, and we write $x\in\mathbb{A}$. A *relation on* $\mathbb{A}$ is a relation on $M\cup A$. An n-ary relation S on $\mathbb{A}$ is Σ_1 *on* $\mathbb{A}$ if there is a Σ_1 formula φ, possibly having constants $y_1,\ldots,y_k$ from $\mathbb{A}$, such that

$$(1)\qquad S(x_1,\ldots,x_n)\quad\text{iff}\quad \mathbb{A}\models\varphi[x_1,\ldots,x_n]$$

for all $x_1,\ldots,x_n\in\mathbb{A}$. The relation S is Π_1 *on* $\mathbb{A}$ if (1) holds for some Π_1 formula φ, and S is Δ_1 *on* $\mathbb{A}$ if S is both Σ_1 and Π_1 on $\mathbb{A}$. A function F *on* $\mathbb{A}$ is a function with domain a subset of $(M\cup A)^n$ for some n and range a subset of $M\cup A$. We say F is Σ_1 *on* $\mathbb{A}$ if its graph is Σ_1 on $\mathbb{A}$.

1.10 Proposition. *Let* $\mathbb{A}$ *be admissible.*
(i) *If* $a\in\mathbb{A}$ *then* a *is* Δ_1 *on* $\mathbb{A}$.
(ii) *If* $x\in\mathbb{A}$ *then* $\{x\}$ *is* Δ_1 *on* $\mathbb{A}$.
(iii) *The* Σ_1 *relations of* $\mathbb{A}$ *are closed under* $\wedge$, $\vee$, $\exists x\in a$, $\forall x\in a$, $\exists x$.

Proof. (i) $x\in a$ iff $\mathbb{A}\models x\in a$; so a is Δ_1 as a subset of $\mathbb{A}$. Part (ii) follows from (i). Part (iii) is immediate from the fact that every Σ formula is equivalent, over $\mathbb{A}$, to a Σ_1 formula and the Σ formulas are closed under the operations mentioned. ▯

1.11 Exercise. Let $\mathbb{A}=\mathbb{A}_{\mathfrak{M}}$ be admissible and let G be an operation defined on all triples in $\mathbb{A}_{\mathfrak{M}}$ whose restriction to $\mathbb{A}_{\mathfrak{M}}$ is Σ_1 definable on $\mathbb{A}_{\mathfrak{M}}$. Define, in $\mathbb{V}_{\mathfrak{M}}$,

$$F(x,y)=G(x,y,\{F(x,z)\mid z\in\mathrm{TC}(y)\})$$

by Σ recursion. Show that $x\in\mathbb{A}_{\mathfrak{M}}$ implies $F(x)\in\mathbb{A}_{\mathfrak{M}}$, and that $F\restriction\mathbb{A}_{\mathfrak{M}}\times\mathbb{A}_{\mathfrak{M}}$ is Σ_1 on $\mathbb{A}_{\mathfrak{M}}$. (This should be easy if the student has understood what has come before.)

2. *Hereditarily Finite Sets*

A set $a\in\mathbb{V}_{\mathfrak{M}}$ is *hereditarily finite* if $\mathrm{TC}(a)$ is finite. $\mathbb{HF}_M$ is the set of hereditarily finite sets of $\mathbb{V}_{\mathfrak{M}}$. It can also be defined by:

$$\begin{aligned}\mathrm{HF}_M(0)&=0;\\ \mathrm{HF}_M(n+1)&=\text{set of all \textit{finite} subsets of } (M\cup\mathrm{HF}_M(n));\\ \mathbb{HF}_M&=\textstyle\bigcup_{n<\omega}\mathrm{HF}_M(n).\end{aligned}$$

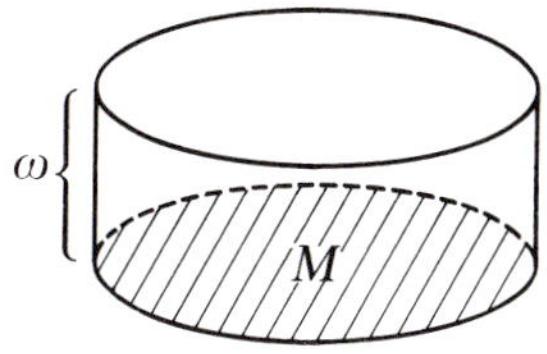

Fig. 2A. $\mathbb{HF}_{\mathfrak{M}}$

2.1 Theorem. *$\mathbb{HF}_{\mathfrak{M}}$ is the smallest admissible set over $\mathfrak{M}$. More precisely, let $\mathsf{L}^* = \mathsf{L}(\in, \ldots)$ and let $\mathbb{HF}_{\mathfrak{M}} = (\mathfrak{M}; \mathrm{HF}_M, \in, \ldots)$ be an L^*-structure.*

(i) *$\mathbb{HF}_{\mathfrak{M}}$ is admissible.*

(ii) *If $\mathbb{A}_{\mathfrak{M}} = (\mathfrak{M}; A, \in, \ldots)$ is admissible, then $\mathbb{HF}_{\mathfrak{M}} \subseteq \mathbb{A}_{\mathfrak{M}}$.*

There is a difference between $\mathbb{HF}_{\mathfrak{M}}$ as a set and as an L^*-structure, but it is usually clear which we have in mind.

Proof of 2.1. (ii) is trivial since A must be closed under pair and union so that $\mathrm{HF}_M(n) \subseteq A$ for all n, by induction on n. Let us prove that $\mathbb{HF}_{\mathfrak{M}}$ is admissible. Since $\mathbb{HF}_{\mathfrak{M}}$ is transitive in $\mathbb{V}_{\mathfrak{M}}$ we get extensionality and foundation for free. Note that each $\mathrm{HF}_{\mathfrak{M}}(n)$ is also transitive. If $x, y \in \mathrm{HF}_{\mathfrak{M}}(n)$ then $\{x, y\} \in \mathrm{HF}_{\mathfrak{M}}(n+1)$ so we have Pair. If $a \in \mathrm{HF}_{\mathfrak{M}}(n)$ then $\bigcup a$ is a finite subset of $\mathrm{HF}_{\mathfrak{M}}(n)$ so is an element of $\mathrm{HF}_{\mathfrak{M}}(n+1)$, and we have Union. If $a \subseteq b \in \mathrm{HF}_{\mathfrak{M}}(n)$ then $a \in \mathrm{HF}_{\mathfrak{M}}(n)$ since a subset of a finite set is finite, so we have full separation, hence Δ_0 Separation. Similarly, we have full collection for if $a \in \mathbb{HF}_{\mathfrak{M}}$ has say k elements $y_1, \ldots, y_k$ and for each of these y_i there is an x_i such that $\varphi(x_i, y_i)$ holds, then all $x_1, \ldots, x_k$ occur in some $\mathrm{HF}_{\mathfrak{M}}(n)$, hence $\{x_1, \ldots, x_k\} \in \mathrm{HF}_{\mathfrak{M}}(n+1)$. □

2.2 Corollary. *The smallest admissible set is*

$$\mathbb{HF} = \{a \in \mathbb{V} \mid a \text{ is a pure hereditarily finite set}).$$

The smallst admissible ordinal is ω.

Proof. $\mathbb{HF}$ is the pure part of any $\mathbb{HF}_{\mathfrak{M}}$, *and* $o(\mathbb{HF}) = \omega$. □

$\mathbb{HF}$ is really where the study of admissible sets began. It was in attempting to generalize recursion theory on the integers that admissible sets developed (by a rather tortuous route) and, as we now show, recursion theory on the integers amounts to the study of Σ_1 and Δ_1 on $\mathbb{HF}$.

2.3 Theorem. *Let S be a relation on natural numbers.*

(i) *S is* r.e. *iff S is Σ_1 on* $\mathbb{HF}$.

(ii) *S is recursive iff S is Δ_1 on* $\mathbb{HF}$.

There are relativized versions of 2.3 that are just as easy to prove. For example, S is recursive *in f* iff S is Δ_1 on $\langle \mathbb{HF}, \in, f \rangle$, which by 2.1 is admissible.

For the proof of 2.3 we assume familiarity with the elements of ordinary recursion theory.

Proof of 2.3 ($\Rightarrow$). Note that (i) implies (ii) since S is recursive iff S and $\neg S$ are r.e. Nevertheless, first we prove the ($\Rightarrow$) part of (ii) to help us prove the corresponding half of (i). It clearly suffices to show that every recursive total function on the integers f can be extended to a Σ_1 function $\hat{f}$ on $\mathbb{HF}$ by the definition:

$$\hat{f}(x) = f(x), \quad \text{for} \quad x \in \omega$$

$$= 0, \quad \text{for} \quad x \notin \omega.$$

To prove this we take a definition of recursive function where one starts with basic total functions and closes under some operations which take one from total functions to total functions. We choose the one given in Shoenfield [1967], though any other will go through just as easily. Thus, the (total) recursive functions are the smallest class containing $+$, $\cdot$, $K_<$ (the characteristic function of $<$), $F(x_1, \ldots, x_n) = x_i$ (the projection functions), closed under composition and closed under the μ-operation (if G is a recursive function such that $\forall \vec{n}\, \exists m\, [G(\vec{n}, m) = 0]$ and for all $\vec{n}$,

$$F(\vec{n}) = \mu m[G(\vec{n}, m) = 0], \quad \textit{the least } m \textit{ such that} \quad G(\vec{n}, m) = 0,$$

then F is recursive).

We have already defined Σ_1 operations $+$ and $\cdot$ in § I.6 and the Δ_0 relation $\alpha < \beta$ in Table 2. The composition of total Σ_1 functions is total and Σ_1 so we need only verify that the class of f with Σ_1 $\hat{f}$ are closed unter the μ-operator. Suppose $\forall \vec{n}\, \exists m\, (G(\vec{n}, m) = 0)$, that G is recursive, that $\hat{G}$ is Σ_1 on $\mathbb{HF}$ by the inductive hypothesis and that $F(\vec{n}) = \mu m[G(\vec{n}, m) = 0]$.

Then $\hat{F}(\vec{x}) = y$ iff

> Some x_i is not a natural number $\wedge\, y = 0$; or all x_i and y are natural numbers and $G(\vec{x}, y) = 0$ and $\forall z < y\, \exists n\, [n \neq 0 \wedge G(\vec{x}, z) = n]$.

This is Σ (since G is Σ_1), and hence it is Σ_1 by I.4.3. Thus every recursive function and predicate on ω is Δ_1 on $\mathbb{HF}$. But every r.e. predicate $S(\vec{x})$ can be written in the form $\exists n\, R(\vec{x}, n)$, where R is recursive by a standard result of recursion theory; so every r.e. predicate is Σ_1 on $\mathbb{HF}$. ☐

To prove the other half of 2.3 we need the following lemma.

2.4 Lemma. *There is a function* $e\colon \omega \to \mathbb{HF}$ *with the following properties:*

(i) *e is a bijection (e is one-one and onto);*

(ii) *e is Σ_1 on $\langle \mathbb{HF}, \in \rangle$;*

(iii) *$n = e(m)$ is a recursive relation of m, n; and*

(iv) *for any Δ_0 formula $\varphi(x_1, \ldots, x_k)$ the relation $\langle \mathbb{HF}, \in \rangle \models \varphi(e(n_1), \ldots, e(n_k))$ of $n_1, \ldots, n_k$ is recursive.*

Proof. Let us define:

$$
\begin{array}{ll}
e(0)=0 & \\
e(1)=\{e(0)\}=\{0\} & (1=2^0) \\
e(2)=\{e(1)\} & (2=2^1) \\
\vdots & \vdots \\
e(5)=\{e(2), e(0)\} & (5=2^2+2^0) \\
\vdots & \vdots \\
e(2^{n_1}+2^{n_2}+\cdots+2^{n_k})=\{e(n_1),\ldots,e(n_k)\} & (n_1>n_2>\cdots>n_k).
\end{array}
$$

We are using the binary expansion of integers, so $e(n)$ is defined for all n by Σ recursion. Hence e is Σ_1 by I.6.4. and 2.16. An easy induction shows that e is one-one and onto. To prove (iii), note that if $e(k)$ is an integer n, then $e(k+2^k)=n+1$. To prove (iv), note that $e(n)\in e(m)$ iff n is an exponent in the binary expansion $2^{k_1}+\cdots+2^{k_l}$ of m. Other Δ_0 formulas follow by induction on Δ_0 formulas using familiar closure properties of the recursive predicates. □

Proof of 2.3 ($\Leftarrow$). Now suppose S in Σ_1 on $\mathbb{HF}$, say $S(n)$ iff $\mathbb{HF}\models\exists y\,\varphi(n,y)$, where φ is Δ_0, the case where S has more than one argument being similar. Then $S(n)$ iff $\exists k\exists m\,[e(k)=n\wedge\varphi(e(k),e(m))]$. The part within brackets is recursive by 2.4 (iii) and 2.4 (iv), so S is r.e. □

There is another way one might want to consider ordinary recursion theory. Suppose we think of the natural numbers not as finite ordinals but as primitive objects (urelements) given to us with some structure, say

$$\mathfrak{N}=\langle N,\otimes,\oplus\rangle$$

where we use $\mathbf{0},\mathbf{1},\mathbf{2},\ldots$ for these natural numbers, $N=\{\mathbf{0},\mathbf{1},\mathbf{2},\ldots\}$, and $\otimes$, $\oplus$ for addition and multiplication in $\mathfrak{N}$.

2.5 Theorem. *Let S be a relation on $\mathfrak{N}=\langle N,\otimes,\oplus\rangle$. Then*
(i) *S is* r.e. *iff S is Σ_1 on $\mathbb{HF}_{\mathfrak{N}}$;*
(ii) *S is recursive iff S is Δ_1 on $\mathbb{HF}_{\mathfrak{N}}$.*

The proof is similar to 2.3. For a different proof one can use Theorem VI.4.12. We include 2.5 because it suggests that one might consider Δ_1 and Σ_1 on $\mathbb{HF}_{\mathfrak{M}}$ as definitions of recursive and r.e. on $\mathfrak{M}$, for an arbitrary structure $\mathfrak{M}$. This is, in effect, what Montague suggested in Montague [1968] for the case of what he calls $\aleph_0$-recursion theory.

Another definition of a recursion theory over an arbitrary structure $\mathfrak{M}$ was presented in Moschovakis [1969a], the generalizations of recursive and r.e. being called *search computable* and *semi-search computable*. What Moschovakis did was this. He started with $\mathfrak{M}=\langle M,R_1,\ldots,R_k\rangle$, chose a new object $0\notin M$ and closed $M\cup\{0\}$ under an ordered pair function, calling the result M^*. Then in M^* he introduced, via an inductive definition similar to Kleene's for higher type recursion theory, the class of search computable functions. Theorem 2.6

below, due to Gordon [1970] shows that these two approaches coincide. This result will not be used in this book. The reader unfamiliar with search computability should consider 2.6 as a *definition.* A proof is sketched in the notes for those familiar with the notions involved.

2.6 Theorem. *Let* $\mathfrak{M}=\langle M,R_1,\ldots,R_k\rangle$, *and let S be a relation on* $\mathfrak{M}$.
(i) *S is semi-search computable on* $\mathfrak{M}$ *iff S is* Σ_1 *on* $\mathbb{HF}_{\mathfrak{M}}$.
(ii) *S is search computable on* $\mathfrak{M}$ *iff S is* Δ_1 *on* $\mathbb{HF}_{\mathfrak{M}}$.

In the context of recursion theory one often works with $\mathbb{HF}_{\mathfrak{M}}$ as opposed to $\mathfrak{M}$ itself since the relations on $\mathfrak{M}$ which are semi-search computable are not always definable at all over $\mathfrak{M}$ itself. The trouble with your average structure $\mathfrak{M}$ is that it lacks coding ability. This lack is what rests behind the need for the following class of formulas. We will not use them until Chapters IV and VI.

2.7 Definition. The *extended first order formulas* of $\mathsf{L}^*=\mathsf{L}(\in,\ldots)$ form the smallest collection containing:
(i) all formulas of L,
(ii) all Δ_0 formulas of L^*,
and closed under:
(iii) $\wedge$, $\vee$, $\forall u\in v$, $\exists u\in v$ (u, v any kind of variables), $\forall p$, $\exists p$,
(iv) $\exists a$.
The *coextended first order formulas* of L^* form the smallest collection containing (i) and (ii) and closed under (iii) and under:
(v) $\forall a$.

The extended first order formulas do not allow unbounded universal quantifiers over sets. The coextended formulas form the dual collection. That these collections are more natural than they seem at first is shown by the next result and the fact that its converse also holds. The converse is a theorem of Feferman [1968] and will not be needed here.

2.8 Proposition. *Let* $\varphi(v_1,\ldots,v_n)$ *be extended first order. For any structures* $\mathfrak{A}_{\mathfrak{M}}\subseteq_{\text{end}}\mathfrak{B}_{\mathfrak{M}}$, *and any* $x_1,\ldots,x_n\in\mathfrak{A}_{\mathfrak{M}}$:
(i) $\mathfrak{A}_{\mathfrak{M}}\models\varphi[x_1,\ldots,x_n]$ *implies* $\mathfrak{B}_{\mathfrak{M}}\models\varphi[x_1,\ldots,x_n]$.

Proof. The difference between this and Lemma I.8.4 rests in the fact that these structures have the same urelement base $\mathfrak{M}$. The proof is a trivial proof by induction. □

2.9 Example. Let L be the language of number theory with $\mathbf{0}, \mathbf{1}, \otimes, \oplus$. In a model $\mathfrak{N}$ of arithmetic the set of standard finite integers is defined in $\mathbb{HF}_{\mathfrak{N}}$ by the extended first-order formula $\psi(x)$ shown here:

$$\exists a\,[x\in a\wedge\forall z\in a\,[z\neq\mathbf{0}\rightarrow\exists y\in a\,((y\oplus\mathbf{1})=z)]].$$

This formula is Σ_1, in fact, so that *the set of finite integers is semi-search computable over* $\mathfrak{N}$. The sentence $\forall p\, \psi(p)$ is extended first order, and $\mathbb{HF}_{\mathfrak{N}} \models \forall p\, \psi(p)$ iff $\mathfrak{N}$ is the standard model of arithmetic.

The extended and coextended first order formulas of $\mathsf{L}(\in)$, when interpreted over $\mathbb{HF}_{\mathfrak{M}}$, form a very small fragment of so called weak second-order logic. *Weak second-order logic over* $\mathfrak{M}$ just consists of the language $\mathsf{L}(\in)$ interpreted in $\mathbb{HF}_{\mathfrak{M}}$. At least that is one way of describing it.

2.10—2.16 Exercises

2.10. Prove that $\mathbb{HF}_{\mathfrak{M}} \subseteq \mathrm{V}_{\mathfrak{M}}(\omega)$, and that $\mathbb{HF}_{\mathfrak{M}} = \mathrm{V}_{\mathfrak{M}}(\omega)$ iff $\mathfrak{M}$ is finite.

2.11. If $\mathbb{A}$ is a pure admissible set, $\mathbb{A} \neq \mathbb{HF}$, then $\omega \in \mathbb{A}$.

2.12. If $\mathbb{A}_{\mathfrak{M}}$ is admissible and $o(\mathbb{A}_{\mathfrak{M}}) = \omega$ then the pure part of $\mathbb{A}_{\mathfrak{M}}$ is $\mathbb{HF}$.

2.13. Prove that $\mathbb{HF}$ is a Δ_1 subset of any admissible set.

2.14. Let X be Σ_1 on $\mathbb{HF}$. Prove that X is Σ_1 on every admissible set.

2.15. Prove that $\mathrm{V}_M(\omega)$ is admissible iff M is finite.

2.16. Prove that $H(l) = \{n_1, \ldots, n_k\}$, where $l = 2^{n_1} + \cdots + 2^{n_k}$, $n_1 > \cdots > n_k$, is a Σ_1 operation of l.

2.17 Notes. Theorem 2.3 is a standard result of recursion theory, as is 2.5. Theorem 2.6 is due to Gordon [1970]. The class of extended first order formulas, introduced in 2.7, will be quite important in Chapters IV and VI when dealing with structures without much coding machinery built into them.

We conclude the notes to this section with a sketch of a proof of Theorem 2.6. The proof uses results from later chapters. We first show that every semi-search computable relation on $\mathfrak{M}$ is Σ_1 on $\mathbb{HF}_{\mathfrak{M}}$. The basic relation of the theory is

$$\{e\}\,(\vec{x}) \to y$$

and it is defined by means of a first order positive Σ inductive definition and so, by the main result of § VI.2, is Σ_1 on $\mathbb{HF}_{\mathfrak{M}}$.

To prove the other half, it suffices to show that some complete Σ_1 relation on $\mathbb{HF}_{\mathfrak{M}}$ is semi-search computable. Let T be the diagram of $\mathfrak{M}$ plus the axioms KPU coded up on M^* by means of the pairing function and let $S(x)$ iff "*x codes a sentence provable from T*".

It is implicit in Chapter V (and explicit in Chapter VIII) that S is a complete Σ_1 prediciate. But the relation "*p is a proof of x from axioms in T*" must be search computable (if the notion is to make any sense).

Hence the relation $\exists p$ ("*p is a proof of x from axioms in T*") is semi-search computable, since the semi-search computable relations are closed under $\exists$. Note that this gives another proof of 2.3 and 2.5.

3. Sets of Hereditary Cardinality Less Than a Cardinal κ

The next admissible set we come across is a simple generalization of $\mathbb{HF}_{\mathfrak{M}}$. Let κ be any infinite cardinal and define

$$H(\kappa)_M = \{a \in \mathbb{V}_M \mid \mathrm{TC}(a) \text{ has cardinality less than } \kappa\}.$$

In particular $H(\omega)_M = \mathbb{HF}_M$. If M is empty then we write $H(\kappa)$ for $H(\kappa)_M$. If κ is regular then we can also characterize $H(\kappa)_M$ as follows:

$$G(0) = 0;$$
$$G(\alpha+1) = \{a \subseteq M \cup G(\alpha) \mid \operatorname{card}(a) < \kappa\};$$
$$G(\lambda) = \bigcup_{\alpha<\lambda} G(\alpha), \text{ if } \lambda \text{ is a limit ordinal;}$$
$$H(\kappa)_M = \bigcup_{\alpha} G(\alpha) = \bigcup_{\alpha<\kappa} G(\alpha).$$

For singular κ this characterization fails: a bad set sneaks into $G(\kappa+1)$, if not before (see Exercise 3.7). *We use the axiom of choice in this section.*

3.1 Theorem. *For all infinite cardinals κ, the set $H(\kappa)_{\mathfrak{M}} = (\mathfrak{M}; H(\kappa)_M, \in)$ is admissible. It is admissible above $\mathfrak{M}$ iff $\kappa > \operatorname{card}(M)$.*

The proof of this is not as simple as one might expect in the case when κ is a singular cardinal. For κ regular, though, it is a trivial result. We will return to the proof of 3.1 after Theorem 3.3.

3.2 Theorem. *Let κ be regular. If $(\mathfrak{M}; H(\kappa)_M, \in, \ldots)$ is a structure for $\mathsf{L}(\in, \ldots)$, then it is admissible.*

Proof. Just like for the case $\kappa = \omega$. In fact, we get full separation and full collection. □

The next result, besides giving us a lot of new examples of admissible sets, also allows us to prove Theorem 3.1 for singular κ. By $\operatorname{card}(\mathsf{L}^*)$ we mean the cardinality of the set of symbols of L^*.

3.3 Theorem (A Löwenheim-Skolem Lemma). *Let $\mathsf{L}^* = \mathsf{L}(\in, \ldots)$ and let $\mathbb{A}_{\mathfrak{M}} = (\mathfrak{M}; A, \in, \ldots)$ be admissible. Let $A_0 \subseteq M \cup A$ be transitive and let κ be a cardinal with $\kappa \geq \operatorname{card}(\mathsf{L}^*) \cup \operatorname{card}(A_0)$. There is an admissible set $\mathbb{B}_{\mathfrak{N}} = (\mathfrak{N}; B, \in, \ldots)$ with the following properties:*

(i) $\mathfrak{N} \prec \mathfrak{M}$ *($\mathfrak{N}$ is an elementary submodel of $\mathfrak{M}$);*
(ii) $\operatorname{card}(N \cup B) \leq \kappa$;
(iii) $A_0 \subseteq N \cup B$;
(iv) *For any φ of L^* and any $x_1, \ldots, x_n \in A_0$, $\mathbb{B}_{\mathfrak{N}} \models \varphi[x_1, \ldots, x_n]$ iff $\mathbb{A}_{\mathfrak{M}} \models \varphi[x_1, \ldots, x_n]$; and*
(v) *In particular, $\mathbb{A}_{\mathfrak{M}} \equiv \mathbb{B}_{\mathfrak{N}}$ ($\equiv$ indicates elementary equivalence).*

Proof. (Note that it is not asserted that $\mathbb{B}_{\mathfrak{N}} \subseteq \mathbb{A}_{\mathfrak{M}}$!) Think of $\mathbb{A}_{\mathfrak{M}}$ as a single sorted structure

$$\mathfrak{A} = \langle M \cup A, M, A, \in, ---, \cdots \rangle,$$

where $\mathfrak{M} = \langle M, --- \rangle$. Find $\mathfrak{A}_1 \prec \mathfrak{A}$ with $A_0 \subseteq \mathfrak{A}_1$ and $\operatorname{card}(\mathfrak{A}_1) \leqslant \kappa$ by the usual Löwenheim-Skolem-Tarski Theorem. $\mathfrak{A}_1$ has the form:

$$\mathfrak{A}_1 = \langle N \cup A_1, N, A_1, \in \cap (N \cup A_1)^2, ---, \cdots \rangle.$$

Since there are no urelements in A_1, $\operatorname{clpse}(N \cup A_1) \subseteq V_N$. Let $B = \operatorname{clpse}(N \cup A_1) \cap V_N$ (i. e. B is the set of sets in $\operatorname{clpse}(N \cup A_1)$) and note that the set of urelements in $\operatorname{clpse}(N \cup A_1)$ is just N. Let $f = c_{N \cup A_1}$ in the notation of I.7. Since $N \cup A_1$ is extensional, f establishes an isomorphism between $\mathfrak{A}_1$ and a structure $\mathfrak{B} = \langle N \cup B; N, B, \in, ---, \cdots \rangle$, by the collapsing lemma. The isomorphism f is the identity for $x \in A_0$ by Lemma I.7.1. Let $\mathfrak{N} = \langle N, --- \rangle$ and $\mathbb{B}_{\mathfrak{N}} = (\mathfrak{N}; B, \in, \ldots)$, and all the properties of the theorem are clear. ▯

Proof of 3.1. It remains to show that if κ is singular then $H(\kappa)_{\mathfrak{M}} = (\mathfrak{M}; H(\kappa)_M, \in)$ is admissible over $\mathfrak{M}$. Let κ^+ be the next cardinal $> \kappa$. The only axiom which is not immediate is Δ_0 Collection. ($H(\kappa)_M$ still satisfies full separation since $a \subseteq b \in H(\kappa)_M \Rightarrow a \in H(\kappa)_M$.) Suppose

(1) $\qquad \forall x \in a\, \exists y\, \varphi(x, y, z)$

is true in $H(\kappa)_{\mathfrak{M}}$, where $z \in H(\kappa)_{\mathfrak{M}}$. Now φ has only a finite number of symbols of L^* in it, so we may ignore the rest of L^* in what follows. Thus we assume $\operatorname{card}(\mathsf{L}^*) \leqslant \aleph_0 < \kappa$. Let $a, z \in X$, X transitive, $\operatorname{card}(X) < \kappa$; say $X = \operatorname{TC}(\{a, z\})$. Since (1) is true in $H(\kappa)_{\mathfrak{M}}$, it persists to $H(\kappa^+)_{\mathfrak{M}}$, which is admissible by 3.2. Using 3.3 we can get an admissible $\mathbb{A}_{\mathfrak{N}}$, with $\mathfrak{N} \subseteq \mathfrak{M}$, so that $X \subseteq \mathbb{A}_{\mathfrak{N}}$, $\operatorname{card}(\mathbb{A}_{\mathfrak{N}}) < \kappa$ and (1) holds in $\mathbb{A}_{\mathfrak{N}}$. By Δ_0 Collection in $\mathbb{A}_{\mathfrak{N}}$ there is a $b \in \mathbb{A}_{\mathfrak{N}}$ so that

(2) $\qquad \forall x \in a\, \exists y \in b\, \varphi(x, y)$

holds in $\mathbb{A}_{\mathfrak{N}}$. But $\mathbb{A}_{\mathfrak{N}} \subseteq_{\text{end}} H(\kappa)_{\mathfrak{M}}$, so (2) holds in $H(\kappa)_{\mathfrak{M}}$ by persistence. ▯

3.4 Corollary. *Every infinite cardinal is an admissible ordinal. For every uncountable cardinal κ and $\beta < \kappa$, there is an admissible α where $\beta < \alpha < \kappa$.*

Proof. $\kappa = o(H(\kappa))$ proves the first assertion in view of 3.1. The second assertion follows from 3.3 by setting $\mathbb{A}_{\mathfrak{M}} = H(\kappa)$, $A_0 = \beta + 1$ and $\kappa = \operatorname{card}(\beta)$. ▯

We could have also proved 3.1 by using the following result of Lévy [1965] (proved there for $M = 0$).

3.5 Theorem. *For all uncountable cardinals $\kappa < \lambda$ we have $H(\kappa)_{\mathfrak{M}} \prec_1 H(\lambda)_{\mathfrak{M}}$. That is, any Σ_1 sentence with constants from $H(\kappa)_{\mathfrak{M}}$ true in $H(\lambda)_{\mathfrak{M}}$ is already true in $H(\kappa)_{\mathfrak{M}}$.*

Proof. This is really just like the proof of 3.1. Suppose the formula $\exists y\, \varphi(x,y)$ holds in $H(\lambda)_{\mathfrak{M}}$, where $x \in H(\kappa)_{\mathfrak{M}}$. As in 3.1 we find an admissible set $\mathbb{A}_{\mathfrak{N}}$, with $\mathfrak{N} \subseteq \mathfrak{M}$, such that the formula holds in $\mathbb{A}_{\mathfrak{N}}$ and $\mathrm{card}(\mathbb{A}_{\mathfrak{N}}) < \kappa$. But then $\mathbb{A}_{\mathfrak{N}} \subseteq_{\mathrm{end}} H(\kappa)_{\mathfrak{M}}$; and so the formula holds in $H(\kappa)_{\mathfrak{M}}$ by persistence. □

One of the earliest generalizations of ordinary recursion theory on the integers goes back to papers of Takeuti where he defines recursive functions on ordinals less than some cardinal κ. When one looks for the analogue of $\mathbb{HF}$ for ordinal recursion theory on κ, the proper structure turns out to be $L(\kappa)$, the set of sets constructible before κ, rather than $H(\kappa)$. The reason is that one needs to be able to code up the sets by ordinals in some way analogous to Lemma 2.4, if one is to prove a result like Theorem 2.3. We will study the constructible sets in § 5 and again in Chapter V.

3.6—3.7 Exercises

3.6. Let $\kappa < \lambda$ be infinite cardinals and let X be a transitive subset of $H(\lambda)$ with $\mathrm{card}(X) = \kappa$. Prove that there is an admissible set A of cardinality κ with $X \subseteq A$ such that $A \prec_1 H(\lambda)$, where $\prec_1$ is explained in I.8.10 and I.8.11. [Iterate 3.2.]

3.7. Let κ be a singular cardinal, let M be a set of urelements of cardinality κ and define G as above. Show that already in $G(2)$ there is a set not in $H(\kappa)_M$. [G is defined just before 3.1.]

3.8 Notes. The technique of following an application of the Downward Lowenheim-Skolem Theorem with an application of the Collapsing Lemma (as in 3.3) is extremely important. In some sense, it goes back to Gödel's original proof that the GCH holds in L, the constructible universe. It was later used implicitly by Takeuti when proving, in our terminology, that uncountable cardinals are stable. Theorem 3.5 is due to Lévy [1965]. Theorem 3.1 is due to Kripke and Platek.

4. *Inner Models: The Method of Interpretations*

We assume that the reader understands the notion of an *interpretation*, say I, of one theory T_1 (formulated in a language L_1) in another theory T_2 (formulated in a possibly different language L_2). Readable accounts of this can be found in Enderton [1972] and Shoenfeld [1967]. We use φ^I for the interpretation of φ given by I. Thus φ is in L_1, φ^I is in L_2; and if φ is an axiom of T_1, then φ^I is a theorem of T_2. If $\mathfrak{M}$ is a model of T_2, then we use $\mathfrak{M}^{-I}$ for the L_1-structure given by $\mathfrak{M}$ and I; $\mathfrak{M}^{-I}$ is a model of T_1. Note that Enderton uses ${}^{\pi}\mathfrak{M}$ for our $\mathfrak{M}^{-I}$; while Shoenfield doesn't make explicit the model theoretic counterpart of the syntactic transformation I.

We give a simple example. We can interpret Peano arithmetic in KPU by having I define

"natural number"	by	"finite ordinal",
"addition"	by	"ordinal addition",
"multiplication"	by	"ordinal multiplication",
"$x<y$"	by	"$x\in y$".

Then every axiom φ of Peano arithmetic (in $+, \cdot, <$) goes over to a theorem φ^I of KPU (formulated in $\mathsf{L}(\in,\dots)$). If $\mathfrak{A}_{\mathfrak{M}}\models$KPU then $\mathfrak{A}_{\mathfrak{M}}^{-I}=\langle N', +, \cdot, <\rangle$ is the model of Peano arithmetic whose domain N' is the set of finite ordinals of $\mathfrak{A}_{\mathfrak{M}}$, and where $+, \cdot, <$ are the restrictions of the corresponding functions and relations of $\mathfrak{A}_{\mathfrak{M}}$ to N'. Rather than launch into a discussion of just how we use interpretations to construct admissible sets, we give a straight-forward illustration. The following result is a generalization of Theorem 1.5.

4.1 Theorem. *Let $\mathbb{A}_{\mathfrak{M}}=(\mathfrak{M}; A,\in)$ be admissible and let $\mathfrak{M}_0\subseteq\mathfrak{M}$ be a substructure of $\mathfrak{M}$ whose universe M_0 is Σ_1 definable on $\mathbb{A}_{\mathfrak{M}}$. If $\mathbb{B}_{\mathfrak{M}_0}=(\mathfrak{M}_0; B,\in)$ is defined by $B=\{a\in A \mid \mathrm{sp}(a)\subseteq M_0\}$, then $\mathbb{B}_{\mathfrak{M}_0}$ is admissible over $\mathfrak{M}_0$.*

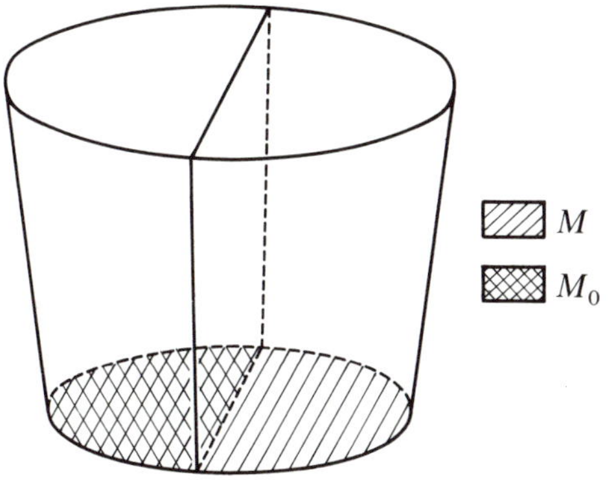

Fig. 4A. $\mathbb{B}_{\mathfrak{M}_0}$, the left half of $\mathbb{A}_{\mathfrak{M}}$

Proof. B is transitive so extensionality and foundation come for free. Pair, Union and Δ_0 Separation are routine. We prove Δ_0 Collection. Suppose $\mathbb{B}_{\mathfrak{M}_0}$ satisfies $\forall x\in a\, \exists y\, \varphi(x,y)$, where a and any other parameters in φ are in $\mathbb{B}_{\mathfrak{M}_0}$. For fixed $y\in\mathbb{B}_{\mathfrak{M}_0}$, we find in $\mathbb{A}_{\mathfrak{M}}$ that $\forall p\in\mathrm{sp}(y)\, \theta(p)$, where $\theta(p)$ is the Σ_1 formula defining $\mathfrak{M}_0$ in $\mathbb{A}_{\mathfrak{M}}$. Hence $\mathbb{A}_{\mathfrak{M}}$ satisfies the formula:

$$\forall x\in a\, \exists y\, [\varphi(x,y) \wedge \forall p\in\mathrm{sp}(y)\, \theta(p)].$$

By Σ Collection in $\mathbb{A}_{\mathfrak{M}}$, there is a b in $\mathbb{A}_{\mathfrak{M}}$ so that

(1) $\qquad \forall x\in a\, \exists y\in b\, \varphi(x,y) \quad \text{and} \quad \forall y\in b\, \forall p\in\mathrm{sp}(y)\, \theta(p).$

But then $\mathrm{sp}(b) \subseteq M_0$. So $b \in B$, and (1) holds in $\mathbb{B}_{\mathfrak{M}_0}$ by absoluteness. ☐

Properly viewed, Theorem 4.1 is a trivial application of an interpretation I. If $\theta(p)$ defines $\mathfrak{M}_0$ then I, in effect, simply redefines:

"x is an urelement" by "x is an urelement $\wedge\, \theta(x)$",
"x is a set" by "x is a set $\wedge\, \forall p \in \mathrm{sp}(x)\, \theta(p)$",

and leaves $\in$ and the symbols of L unchanged. The proof that every axiom φ of KPU becomes a theorem φ^I of KPU′ is just like the proof of 4.1 (where KPU′ is KPU with axioms asserting θ is closed under any function symbols of L). Hence, for every model $\mathfrak{A}_{\mathfrak{M}}$ of KPU′, the structure $\mathfrak{A}_{\mathfrak{M}}^{-I}$ is also a model of KPU. In Theorem 4.1 we have $\mathbb{B}_{\mathfrak{M}_0} = \mathfrak{A}_{\mathfrak{M}}^{-I}$. In this example we don't gain much by looking at it from the point of view of interpretation, but we will in more complicated situations.

The interpretation we just used has some important features in common with most of the interpretations we use. They are what Shoenfield [1967, § 9.5] calls transitive $\in$-interpretations.

4.2 Definition. Let $\mathsf{L}^* = \mathsf{L}(\in)$ and let I be an interpretation of L^* into KPU (as formulated in L^*). I is a *transitive $\in$-interpretation* if I leaves the symbols of L and $\in$ unchanged and merely "cuts down on the urelements and sets" so that the following are provable in KPU:

(i) if (x is an urelement)^I then x is an urelement;

(ii) if (x is a set)^I, then x is a set and for all $y \in x$, (y is an urelement)^I or (y is a set)^I.

If I is a transitive $\in$-interpretation and $\mathfrak{A}_{\mathfrak{M}} \models \mathrm{KPU}$ then $\mathfrak{A}_{\mathfrak{M}}^{-I}$ is called the *inner submodel* of $\mathbb{A}_{\mathfrak{M}}$ given by I.

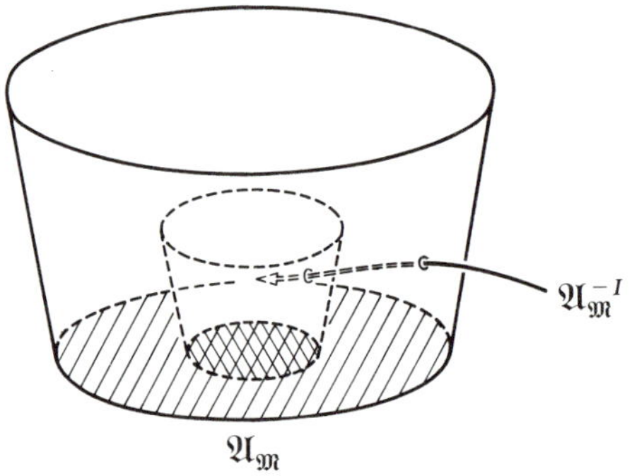

Fig. 4 B. A model $\mathfrak{A}_{\mathfrak{M}}$ and an inner submodel

The conditions in 4.2 guarantee that $\mathfrak{A}_{\mathfrak{M}}^{-I} \subseteq_{\mathrm{end}} \mathfrak{A}_{\mathfrak{M}}$. Fig. 4 B indicates the idea behind transitive $\in$-interpretations and inner submodels.

The following lemma is useful to keep in mind.

4.3 Lemma. *Let I be a transitive $\in$-interpretation.*

(i) $\mathrm{KPU} \vdash (\text{Extensionality})^I$;

(ii) *For each instance of foundation φ, we have* $\mathrm{KPU} \vdash \varphi^I$;

(iii) *For each Σ formula $\varphi(x)$:*

$$\mathrm{KPU} \vdash \mathrm{Urelement}(x)^I \vee \mathrm{Set}(x)^I \rightarrow [\varphi(x)^I \rightarrow \varphi(x)],$$

(iv) *For each Δ_0 formula φ:*

$$\mathrm{KPU} \vdash \mathrm{Urelement}(x)^I \vee \mathrm{Set}(x)^I \rightarrow [\varphi(x)^I \leftrightarrow \varphi(x)].$$

Proof. (i) $(\text{Extensionality})^I$ can be written as:

$$\mathrm{Set}^I(a) \wedge \mathrm{Set}^I(b) \wedge a \neq b \rightarrow \exists x\, [(\mathrm{Set}^I(x) \vee \mathrm{Urelement}^I(x)) \wedge \neg(x \in a \leftrightarrow x \in b)].$$

This follows immediately from property 4.2 (ii). To prove (ii) let φ be $\exists a\, \psi(a) \rightarrow \exists a\, [\psi(a) \wedge \neg \exists b \in a\, \psi(b)]$. Then φ^I states: If $\exists a\, [\mathrm{Set}^I(a) \wedge \psi^I(a)]$, then there is an a such that $\mathrm{Set}^I(a)$ and $\psi^I(a)$; but there is no b with $\mathrm{Set}^I(b)$ such that $b \in a$ and $\psi^I(b)$. This follows immediately by applying foundation to the formula: $\mathrm{Set}^I(a) \wedge \psi^I(a)$. Part (iii) follows model theoretically by the comment above about $\mathfrak{A}_{\mathfrak{M}}^{-I} \subseteq_{\mathrm{end}} \mathfrak{A}_{\mathfrak{M}}$, for all $A_M \models \mathrm{KPU}$. It can also be proved directly by induction on φ. Part (iv) follows from (iii). $\square$

4.4 Exercise. Verify that the specific I defined on p. 56 is a transitive $\in$-interpretation.

5. *Constructible Sets with Urelements; $\mathbb{HYP}_{\mathfrak{M}}$ Defined*

In this section we construct most of the more important admissible sets in one fell swoop by means of Gödel's hierarchy of constructible sets. For reasons which will become apparent, we restrict ourselves to the case where the language L has only a finite number of nonlogical symbols and where $\mathsf{L}^* = \mathsf{L}(\in)$. For simplicity we assume the symbols of L are relation symbols: a simple modification will extend the results to languages with function and constant symbols.

5.1 Apologia. There are two well known ways of defining the constructible sets in a theory without urelements, both developed by Gödel. The most intuitive is by iterating definability through the ordinals; the other uses some form of Gödel's $\mathscr{F}_1, \ldots, \mathscr{F}_8$. We have always preferred the former method but find ourselves forced to use the latter here. The reason is simple enough, but is one that doesn't arise in ZF. Many admissible sets $\mathbb{A}_{\mathfrak{M}}$ have ordinal $o(\mathbb{A}_{\mathfrak{M}}) = \omega$, i. e., are models of $\neg$Infinity, whereas natural ways of iterating first order definability need ω.

Even though we give up the iteration of full first order definability, we modify the usual approach (along lines used by Gandy [1975] and Jensen [1972]) via the $\mathscr{F}_i$'s to make it as similar to the definability approach as possible.

5.2 Assumption. *For the rest of § 5 we assume that* $\mathscr{F}_1,\ldots,\mathscr{F}_N$ *are* Σ_1 *operations (of two arguments each) introduced into* KPU *so that the following hold, where we define* $\mathscr{D}(b)=b\cup\{\mathscr{F}_i(x,y)\mid x,y\in b, 1\leqslant i\leqslant N\}$.

(i) $\mathscr{F}_1(x,y)=\{x,y\}$;

(ii) $\mathscr{F}_2(x,y)=\bigcup x$;

(iii) $\mathrm{KPU}\vdash \mathrm{sp}(\mathscr{F}_i(x,y))\subseteq \mathrm{sp}(x)\cup\mathrm{sp}(y)$, *for all* $i\leqslant N$;

(iv) $\mathrm{KPU}\vdash[\mathrm{Tran}(b)\rightarrow\mathrm{Tran}(\mathscr{D}(b))]$;

(v) *For each* Δ_0 *formula* $\varphi(x_1,\ldots,x_n)$ *with free variables among* $x_1,\ldots,x_n$ *and each variable* x_i, $i\leqslant n$, *there is a term* $\mathscr{F}$ *of* n *arguments built from* $\mathscr{F}_1,\ldots,\mathscr{F}_N$ *so that:*

$$\mathrm{KPU}\vdash \mathscr{F}(a,x_1,\ldots,x_{i-1},x_{i+1},\ldots,x_n)=\{x_i\in a\mid \varphi(x_1,\ldots,x_n)\}.$$

There are many ways of fulfilling the assumptions. We will return to give a specific solution in § 6. Next, with 5.2 firmly in mind, we return to the development of set theory in KPU begun in Chapter I. First note that $\mathscr{D}$ is a Σ operation since $\mathscr{F}_1,\ldots,\mathscr{F}_N$ are. Define, in KPU, a Σ operation $\mathrm{L}(\cdot,\cdot)$ by recursion over the second argument:

$$\mathrm{L}(a,0)=\mathrm{TC}(a),$$

$$\mathrm{L}(a,\alpha+1)=\mathscr{D}(\mathscr{S}(\mathrm{L}(a,\alpha)))=\mathscr{D}(\mathrm{L}(a,\alpha)\cup\{\mathrm{L}(a,\alpha)\}),$$

$$\mathrm{L}(a,\lambda)=\textstyle\bigcup_{\alpha<\lambda}\mathrm{L}(a,\alpha)\quad\text{if}\quad \mathrm{Lim}(\lambda).$$

5.3 Definition. An object x is *constructible from* a, written $x\in\mathrm{L}(a)$, if $\exists\alpha[x\in\mathrm{L}(a,\alpha)]$. If x is constructible from 0, we say x is *constructible* and write $x\in\mathrm{L}$.

5.4 Lemma (of KPU). *For all sets* a *and ordinals* α:

(i) $a\in\mathrm{L}(a,1)$ *if* a *is transitive;*

(ii) $\mathrm{L}(a,\alpha)$ *is transitive;*

(iii) $\alpha<\beta$ *implies* $\mathrm{L}(a,\alpha)\subseteq\mathrm{L}(a,\beta)$;

(iv) $\mathrm{L}(a,\alpha)\in\mathrm{L}(a,\alpha+1)$;

(v) $x,y\in\mathrm{L}(a,\alpha)$ *implies* $\mathscr{F}_i(x,y)\in\mathrm{L}(a,\alpha+1)$, $1\leqslant i\leqslant N$;

(vi) $\alpha\in\mathrm{L}(a,\beta)$ *for some* β;

(vii) *An urelement* p *is in* $\mathrm{L}(a)$ *iff* $p\in\mathrm{sp}(a)$.

Proof. (i), (iii), (iv), (v) follow from the definition of $\mathrm{L}(a,\alpha)$ directly. Part (ii) is by induction on α using Assumption 5.2 (iv). Part (vii) is proved by showing that $p\in\mathrm{L}(a,\alpha)$ iff $p\in\mathrm{sp}(a)$ by induction on α (using 5.2 (iii)). This leaves (vi) which is also proved by induction on α. By the induction hypothesis we have $\forall\gamma<\alpha\,\exists\delta\,[\gamma\in\mathrm{L}(a,\delta)]$. So, by Σ Reflection, there is an ordinal λ such that

$\forall\gamma<\alpha\,\exists\delta<\lambda\,[\gamma\in L(a,\delta)]$. But then by (iii), every $\gamma<\alpha$ is in $L(a,\lambda)$; that is, $\alpha\subseteq L(a,\lambda)$. Now, applying Assumption 5.2(v) for the first time, we see that the set $b=\{x\in L(a,\lambda)\mid \mathrm{Ord}(x)\}$ is in $L(a,\lambda')$ for some $\lambda'\geqslant\lambda$. Since $L(a,\lambda)$ is transitive, b is an ordinal β and $\alpha\leqslant\beta$. Again, since $L(a,\lambda')$ is transitive, $\alpha\in L(a,\lambda')$, because either $\alpha=\beta$ or $\alpha\in\beta$. □

We now define a transitive $\in$-interpretation $\varphi^{L(a)}$ by the following:

$$(x \textit{ is a urelement})^{L(a)} \quad \text{is} \quad (x\in \mathrm{sp}(a)),$$

$$(x \textit{ is a set})^{L(a)} \quad \text{is} \quad (x \textit{ is a set } \wedge x\in L(a)),$$

leaving $\in$ and all symbols of the original language L unchanged. (We apologize for the two L's, but note that one is sanserif.) Note that this is indeed a transitive $\in$-interpretation in the sense of § 4.

5.5 Theorem. *For every axiom φ of* KPU$^+$, *we have* KPU$\vdash\varphi^{L(a)}$.

Proof. We run through the axioms of KPU$^+$. Extensionality and Foundation follows from 4.3. Pair and Union follow from 5.2 (i), (ii), and 4.3 (iv). Δ_0 separation follows from 5.2 (v) and 4.3 (iv).

Δ_0 *Collection:* Suppose that $\varphi(x,y,z)$ is Δ_0. Working in KPU assume $a_0\in L(a)$, $z\in L(a)$ and $\forall x\in a_0\,\exists y\in L(a)\,[\varphi(x,y,z)^{L(a)}]$.

We suppress mention of z. Writing out $y\in L(a)$ and using 4.3 (iv) on $\varphi(x,y)$ we get $\forall x\in a_0\,\exists\alpha\,[\exists y\in L(a,\alpha)\,\varphi(x,y)]$. By Σ collection there is a β such that

$$\forall x\in a_0\,\exists\alpha<\beta\,[\exists y\in L(a,\alpha)\,\varphi(x,y)].$$

So, by 5.4 (iii), $\forall x\in a_0\,\exists y\in L(a,\beta)\,\varphi(x,y)$. Using 4.3 (iv) again, setting $b=L(a,\beta)$, we find:

$$[\forall x\in a_0\,\exists y\in b\,\varphi(x,y)]^{L(a)}.$$

Thus, the interpretation of Δ_0 Collection is provable.

Finally, we need to prove $[\exists b\,\forall x\,(x\in b\leftrightarrow\exists p\,(x=p))]^{L(a)}$. By Δ Separation it suffices to prove $[\exists b\,\forall p\,(p\in b)]^{L(a)}$. Let $b=\mathrm{TC}(a)=L(a,0)$. By definition, $b\in L(a,1)$ and $(x \textit{ is an urelement})^{L(a)}$ is just $x\in\mathrm{sp}(a)$; but $\mathrm{sp}(a)\subseteq b$. □

5.6 Definition. $L(\alpha)_{\mathfrak{M}}=(\mathfrak{M};\, L(M,\alpha)\cap\mathbb{V}_M,\in)$.

$L(\alpha)_{\mathfrak{M}}$ is a structure (for the language $\mathsf{L}^*=\mathsf{L}(\in)$) which may or may not be admissible. We use the intersection with $\mathbb{V}_M$ in 5.6 is just to take out the urelements in strict accord with our definition of structure for L^*.

5.7 Theorem. *If there is an admissible set* $\mathbb{A}=\mathbb{A}_{\mathfrak{M}}$ *above* $\mathfrak{M}$ *with* $o(\mathbb{A}_{\mathfrak{M}})=\alpha$, *then* $L(\alpha)_{\mathfrak{M}}$ *is the smallest such. In other words* $L(\alpha)_{\mathfrak{M}}$ *is admissible,* $L(\alpha)_{\mathfrak{M}}\subseteq\mathbb{A}_{\mathfrak{M}}$, $M\in L(\alpha)_{\mathfrak{M}}$ *and* $o(L(\alpha)_{\mathfrak{M}})=\alpha$.

Proof. For $\beta<\alpha$, $L(M,\beta)$ has the same meaning in $\mathbb{A}_{\mathfrak{M}}$ and $\mathbb{V}_{\mathfrak{M}}$ by absoluteness. Thus $L(\alpha)_{\mathfrak{M}}$ is the inner model of $\mathbb{A}_{\mathfrak{M}}$ given by the interpretation defined above. Thus, in particular, $L(\alpha)_{\mathfrak{M}}\subseteq\mathbb{A}_{\mathfrak{M}}$. By Theorem 5.5, $L(\alpha)_{\mathfrak{M}}$ is admissible, and $M\in L(\alpha)_{\mathfrak{M}}$. We see that $o(L(\alpha)_{\mathfrak{M}})=\alpha$ from 5.4 (vi). □

If we had the option, the following definition would be printed in red. It introduces one of the principal objects of our study. Recall that $\mathbb{A}_{\mathfrak{M}}$ is admissible *above* $\mathfrak{M}$ if $\mathbb{A}_{\mathfrak{M}}\models \mathrm{KPU}^+$, the "+" being the part that gives "above".

5.8 Definition (The Next Admissible).

(i) $\mathbb{HYP}_{\mathfrak{M}}=(\mathfrak{M}; A,\in)$, where $A=\bigcap\{B\mid(\mathfrak{M}; B,\in)$ is admissible above $\mathfrak{M}\}$.

(ii) $O(\mathfrak{M})=o(\mathbb{HYP}_{\mathfrak{M}})$.

5.9 Theorem. (i) $\mathbb{HYP}_{\mathfrak{M}}$ *is the smallest admissible set above* $\mathfrak{M}$.

(ii) $\mathbb{HYP}_{\mathfrak{M}}=L(\alpha)_{\mathfrak{M}}$ *for* $\alpha=O(\mathfrak{M})$.

Proof. We need only see that $\mathbb{HYP}_{\mathfrak{M}}$ is *admissible* over $\mathfrak{M}$, since it is certainly contained in all other admissibles over $\mathfrak{M}$ with M an element. There is an admissible $\mathbb{A}_{\mathfrak{M}}$ with $M\in\mathbb{A}_{\mathfrak{M}}$ by 3.1. Let α be the least ordinal of the form $o(\mathbb{A}_{\mathfrak{M}})$, where $\mathbb{A}_{\mathfrak{M}}$ is admissible above $\mathfrak{M}$. Apply 5.7 to α and $\mathbb{A}_{\mathfrak{M}}$. □

We will study the structure of $\mathbb{HYP}_{\mathfrak{M}}$ off and on in Chapters IV, VI, VII, VIII. For now we will simply state without proof, for the reader who understands the notions involved, that *if* $\mathcal{N}=\langle N,+,\cdot\rangle$ *is the usual structure of the natural numbers, then for any relation* R *on* $\mathcal{N}$, R *is hyperarithmetic iff* $R\in\mathbb{HYP}_{\mathcal{N}}$, *and* R *is* Π_1^1 *on* $\mathcal{N}$ *iff* R *is* Σ_1 *on* $\mathbb{HYP}_{\mathcal{N}}$. *Furthermore,* $O(\mathcal{N})=\omega_1^c=$ *the least non-recursive ordinal.* Proofs will appear later.

For the next result recall that $L(\alpha)=L(0,\alpha)$; so $L(\alpha)$ is a pure set. The proof is immediate from 1.8 and 5.7 with $M=0$.

5.10 Corollary. *An ordinal* α *is admissible iff* $L(\alpha)$ *is a pure admissible set.*

An urelement free version of 5.9 is given below; the proof is similar.

5.11 Theorem. *Let* a *be a pure transitive set,* $A=\bigcap\{B\colon B$ *admissible,* $a\in B\}$. *Then* A *is admissible; it is of the form* $L(a,\alpha)$ *for some admissible* α*; and it is the smallest admissible set with* a *as an element.*

5.12 Corollary. *If* α *is admissible and* $a\in L(\alpha)$, *then* $L(a,\alpha)=L(\alpha)$.

Proof. Both $L(a,\alpha)$ and $L(\alpha)$ are the smallest admissible sets with a an element and ordinal α. □

The final results of this section will appear rather technical at present, but they are extremely important for much that is to follow.

5.13 Definition. Let $\mathbb{A}_{\mathfrak{M}}$ be admissible. Let $\varphi(v)$ be a Σ_1 formula with one free variable but with parameters from some set $X \subseteq \mathbb{A}_{\mathfrak{M}}$.

(i) If $\mathbb{A}_{\mathfrak{M}} \models \exists! v\, \varphi(v)$ and $\mathbb{A}_{\mathfrak{M}} \models \varphi[a]$, then $\varphi(v)$ is a Σ_1 *definition of a with parameters from X*.

(ii) If, in addition to (i), for every $\mathfrak{B}_{\mathfrak{M}} \supseteq_{\text{end}} \mathbb{A}_{\mathfrak{M}}$ which is a model of KPU we have $\mathfrak{B}_{\mathfrak{M}} \models \exists! v\, \varphi(v)$, then $\varphi(v)$ is a *good* Σ_1 definition of a with parameters from X.

5.14 Theorem. *Let $M = \mathrm{sp}(a)$ where a is transitive and let α be the least ordinal such that $\mathbb{A} = (\mathfrak{M}; \mathrm{L}(a,\alpha) \cap \mathbb{V}_M, \in)$ is admissible. Every $x \in \mathbb{A}$ has a good Σ_1 definition on $\mathbb{A}$ with parameters from $a \cup \{a\}$.*

Proof. Let B be the set of $x \in \mathbb{A}$ which have good Σ_1 definitions on $\mathbb{A}$ with parameters from $\mathscr{S}(a)$. Note the following:

(1) $\mathscr{S}(a) \subseteq B$; and

(2) $x, y \in B$ implies $\mathscr{F}_i(x,y) \in B$ for $1 \leqslant i \leqslant N$.

For (2) we need the fact that $\mathscr{F}_i$ is Σ_1 definable in KPU without parameters, which was implicit in 5.2.

(3) *If $b \subseteq B$, then $\mathscr{D}(b) \subseteq B$.*

This follows from the fact that $\mathscr{D}(b) = b \cup \{\mathscr{F}_i(x,y) \mid x, y \in b, 1 \leqslant i \leqslant N\}$ and from (2). Next since $\mathrm{L}(\cdot,\cdot)$ is a Σ_1 operation of KPU we find:

(4) *If $\beta \in B$, then* $\mathrm{L}(a,\beta) \in B$.

We now prove, by induction on $\beta < \alpha$, that $\beta \in B$ and $\mathrm{L}(a,\beta) \subseteq B$.

Case 1. $\beta = 0$. 0 has a good Σ_1 definition and $\mathrm{L}(a,0) = a \subseteq B$ by (1).

Case 2. $\beta = \gamma + 1$. By induction hypothesis $\gamma \in B$ and $\mathrm{L}(a,\gamma) \subseteq B$. But if $\gamma \in B$ so is $\gamma + 1$. $\mathrm{L}(a,\gamma) \in B$ by (4). Thus

$$\mathscr{S}(\mathrm{L}(a,\gamma)) = \mathrm{L}(a,\gamma) \cup \{\mathrm{L}(a,\gamma)\} \subseteq B;$$

so $\mathrm{L}(a,\gamma+1) = \mathscr{D}(\mathscr{S}(\mathrm{L}(a,\gamma))) \subseteq B$, by (3).

Case 3. β is a limit ordinal. By the induction hypothesis we have $\beta \subseteq B$ and $\mathrm{L}(a,\beta) \subseteq B$, since $\beta = \{\gamma \mid \gamma < \beta\}$ and $\mathrm{L}(a,\beta) = \bigcup_{\gamma<\beta} \mathrm{L}(a,\gamma)$. Thus we need only prove $\beta \in B$. This, however, is the main point of the proof. By our choice of α and $\beta < \alpha$ we have $\mathrm{L}(a,\beta)_{\mathfrak{M}} = (\mathfrak{M}; \mathrm{L}(a,\beta) \cap \mathbb{V}_M, \in)$ is not admissible so there is a Δ_0 formula $\varphi(x,y,z)$ and there are objects $z, b \in \mathrm{L}(a,\beta)_{\mathfrak{M}}$ so that

(5) $\mathrm{L}(a,\beta)_{\mathfrak{M}} \models \forall x \in b\, \exists y\, \varphi(x,y,z)$, and

(6) $\mathrm{L}(a,\beta)_{\mathfrak{M}} \models \neg \exists c\, \forall x \in b\, \exists y \in c\, \varphi(x,y,z)$.

(Since $\mathrm{Lim}(\beta)$ holds, Δ_0 Collection is the only way for $\mathrm{L}(a,\beta)_{\mathfrak{M}}$ to fail to be admissible by Exercise 5.16). Now $b, z \in B$, so they have good Σ_1 definitions $\sigma(u)$, $\psi(w)$ with parameters from $\mathscr{S}(a)$. Consider the following Σ formula $\theta(\beta)$:

$$\mathrm{Ord}(\beta) \wedge \exists b\, \exists z\, [\sigma(b) \wedge \psi(z) \wedge \forall x \in b\, \exists y \in \mathrm{L}(a,\beta)\, \varphi(x,y,z) \\ \wedge \forall \gamma < \beta\, \exists x \in b\, \forall y \in \mathrm{L}(a,\gamma) \neg\varphi(x,y,z)].$$

Now clearly $\mathbb{A} \models \theta(\beta)$ so every end extension $\mathfrak{B}_{\mathfrak{M}} \models \theta(\beta)$. If $\mathfrak{B}_{\mathfrak{M}} \models \mathrm{KPU}$ then no "ordinal" of $\mathfrak{B}_{\mathfrak{M}}$ greater than β can satisfy θ by (5), (6). Similarly, no ordinal smaller can satisfy θ. Thus $\theta(\beta)$ defines β in every end extension of $\mathbb{A}_{\mathfrak{M}}$ satisfying KPU, so $\beta \in B$. □

5.15 Corollary. *Every* $a \in \mathbb{HYP}_{\mathfrak{M}}$ *has a good* Σ_1 *definition on* $\mathbb{HYP}_{\mathfrak{M}}$ *with no parameters other than* M *and some* $p_1, \ldots, p_k \in M$.

5.16—5.20 Exercises

5.16. Let $M \subseteq \mathrm{sp}(a)$ and let λ be a limit ordinal. Show that $(\mathfrak{M}; \mathrm{L}(a,\lambda) \cap \mathbb{V}_M, \in)$ satisfies all the axioms of KPU except, possibly, Δ_0 Collection.

5.17. If κ is a cardinal, $\kappa > \mathrm{card}(M)$, then $\mathrm{L}(\kappa)_{\mathfrak{M}}$ is admissible above $\mathfrak{M}$.

5.18. If κ, λ are uncountable cardinals, $\kappa < \lambda$ then $\mathrm{L}(\kappa)_{\mathfrak{M}} \prec_1 \mathrm{L}(\lambda)_{\mathfrak{M}}$.

5.19. $\mathrm{L}(\kappa)$ is admissible for all cardinals $\kappa \geqslant \omega$.

5.20. Improve 5.4 (vi) by proving that $\alpha \in \mathrm{L}(M, \alpha + \omega)$, assuming ω exists.

5.21 Notes. The constructible sets were first used by Gödel [1939] in his famous proof of the consistency of the generalized continuum hypothesis. In this paper, Gödel used iterated first-order definability. In the proof of Gödel [1940] the fundamental operations were introduced and used to generate the constructible sets. The approach to the constructible sets taken here borrows some ideas from Jensen [1972], but it is a little more complicated due to the presence of urelements and relations on them. We shall see that the complications only come up in fulfilling Assumption 5.2 in the next section.

6. Operations for Generating the Constructible Sets

We now turn to the task of finding $\mathscr{F}_1, \ldots, \mathscr{F}_N$ satisfying Assumption 5.2. We will see that we can get by with especially simple functions (substitutable functions). This will prove useful in understanding the sets constructible in ω steps.

The real strength of 5.2 resides in the requirement that for each Δ_0 formula $\varphi(x_1,\ldots,x_n)$, there is a term $\mathscr{F}$ built from the symbols $\mathscr{F}_1,\ldots,\mathscr{F}_N$ so that

$$\mathrm{KPU} \vdash \mathscr{F}(a, x_1,\ldots,x_{i-1}, x_{i+1},\ldots,x_n) = \{x_i \in a \mid \varphi(x_1,\ldots,x_n)\}.$$

We take care of this condition first.

We already have by 5.2:

($\mathscr{F}$ 1) $\qquad \mathscr{F}_1(x,y) = \{x,y\}$, and

($\mathscr{F}$ 2) $\qquad \mathscr{F}_2(x,y) = \bigcup x$.

From these we obtain by various simple compositions the following:

$$\{x\} = \mathscr{F}_1(x,x),$$
$$x \cup y = \bigcup\{x,y\} = \mathscr{F}_2(\mathscr{F}_1(x,y),y),$$
$$\mathscr{S}(x) = x \cup \{x\},$$
$$\langle x,y\rangle = \{\{x\},\{x,y\}\},$$
$$\langle x_1,\ldots,x_n\rangle = \langle x_1, \langle x_2,\ldots,x_n\rangle\rangle.$$

The function $\mathscr{F}_2$ corresponds (in Lemma 6.1) to $\vee$ in Δ_0 formulas. To handle negations we need to define:

($\mathscr{F}$ 3) $\qquad \mathscr{F}_3(x,y) = x - y$.

From this we get, by composition, $x \cap y = x - (x - y) = \mathscr{F}_3(x, \mathscr{F}_3(x,y))$.

The need to treat quantifiers leads us to the following more complicated functions:

($\mathscr{F}$ 4) $\qquad \mathscr{F}_4(x,y) = x \times y$,

($\mathscr{F}$ 5) $\qquad \mathscr{F}_5(x,y) = \mathrm{dom}(x) = \{1^{\mathrm{st}}(z) \mid z \in x,\ z \text{ an ordered pair}\}$,

($\mathscr{F}$ 6) $\qquad \mathscr{F}_6(x,y) = \mathrm{rng}(x) = \{2^{\mathrm{nd}}(z) \mid z \in x,\ z \text{ an ordered pair}\}$,

($\mathscr{F}$ 7) $\qquad \mathscr{F}_7(x,y) = \{\langle u,v,w\rangle \mid \langle u,v\rangle \in x,\ w \in y\}$,

($\mathscr{F}$ 8) $\qquad \mathscr{F}_8(x,y) = \{\langle u,w,v\rangle \mid \langle u,v\rangle \in x,\ w \in y\}$.

The functions $\mathscr{F}_7$, $\mathscr{F}_8$ are annoying. They arise from the peculiar nature of the ordered n-tuple. We tend to think of $\langle x_1,x_2,x_3,x_4\rangle$ as a rather symmetric object but it is, in fact, far from it. We can form it from x_1 and $\langle x_2,x_3,x_4\rangle$ (since it is just $\langle x_1,\langle x_2,x_3,x_4\rangle\rangle$) but we cannot form it from, say x_4 and $\langle x_1,x_2,x_3\rangle$ or from x_3 and $\langle x_1,x_2,x_4\rangle$ using $\mathscr{F}_1,\ldots,\mathscr{F}_6$. This accounts for the appearance of $\mathscr{F}_7$ and $\mathscr{F}_8$.

It now remains only to add the functions which correspond to atomic formulas:

($\mathcal{F}$ 9) $\mathcal{F}_{\mathsf{U}}(x,y) = \{z \in x \mid z \text{ is an urelement}\}$,

($\mathcal{F}$ 10) $\mathcal{F}_{=}(x,y) = \{\langle v,u\rangle \in y \times x \mid u = v\}$,

($\mathcal{F}$ 11) $\mathcal{F}_{\in}(x,y) = \{\langle v,u\rangle \in y \times x \mid u \in v\}$,

and for each relation symbol $\mathsf{R}(x_1,\dots,x_n)$ of L an operation:

($\mathcal{F}$ 12)—($\mathcal{F}$ K) $\mathcal{F}_{\mathsf{R}}(x,y) = \{\langle p_n,\dots,p_1,v\rangle \mid \langle p_n,\dots,p_1\rangle \in x, \mathsf{R}(p_1,\dots,p_n), \text{ and } v \in y\}$

In order to prove the desired result we prove something a little more general. It gives us a better inductive hypothesis in our proof which uses induction on Δ_0 formulas. For technical reasons, we have inverted the order of the variables in 6.1. For the same reason, there is an inversion taking place in lines ($\mathcal{F}$ 10), ($\mathcal{F}$ 11), ..., ($\mathcal{F}$ K).

6.1 Lemma. *For every Δ_0 formula $\varphi(x_1,\dots,x_n)$ with free variables among $x_1,\dots,x_n$, there is a term $\mathcal{F}_\varphi$ built up from the symbols $\mathcal{F}_1,\dots,\mathcal{F}_K$ so that*

$$\text{KPU} \vdash \mathcal{F}_\varphi(a_1,\dots,a_n) = \{\langle x_n,\dots,x_1\rangle \in a_n \times \cdots \times a_1 \mid \varphi(x_1,\dots,x_n)\}.$$

Proof. We treat $\mathsf{L}^* = \mathsf{L}(\in)$ as a single sorted language with symbols U (for urelement) and S (for set), $\in$, $=$, $\mathsf{R}_1,\dots,\mathsf{R}_l$, and variables $x_1,x_2,x_3,\dots$. Whenever we write a formula φ as $\varphi(x_1,\dots,x_n)$ we mean that all the free variables of are among $x_1,\dots,x_n$, but not all of these variables need actually appear as free variables in φ. For the purpose of this proof we need two special definitions. We call a formula of L^* an *orderly* formula if it satisfies the following condition: whenever a quantifier $\exists x_j$ or $\forall x_j$ occurs in φ, the index j is the largest index of all the free variables in the scope of the quantifier. By simply renaming bound variables systematically, we have:

(a) *Every Δ_0 formula of L^* is logically equivalent to an orderly Δ_0 formula with the same free variables.*

We call a formula $\varphi(x_1,\dots,x_n)$ a termed-formula, or *t-formula*, if there is a term $\mathcal{F}_\varphi$ such that the conclusion of 6.1 holds. Note that there is a possible ambiguity here since a formula with free variables among x_1,x_2 is also a formula with free variables among x_1,x_2,x_3 and so could be written as $\varphi(x_1,x_2)$ or as $\varphi(x_1,x_2,x_3)$. To be completely precise, we should say that φ with free variables among $x_1,\dots,x_n$ is a t-formula. Line (e) below will show us that we don't have to be this careful.

Our goal is to prove that every Δ_0 formula is a t-formula. We want to prove this by induction on Δ_0 formulas, but we must dispose of certain logical trivialities before we can treat even the atomic formulas. These trivialities are handled in $(b)-(j)$ below.

(b) *If* KPU $\vdash \varphi(x_1, \ldots, x_n) \leftrightarrow \psi(x_1, \ldots, x_n)$ *and* ψ *is a t-formula then so is* φ.

This last is clear. Combining (a) and (b) allows us to restrict attention to orderly Δ_0 formulas, so Lemma 6.1 follows finally from (z) below.

(c) *If* $\varphi(x_1, \ldots, x_n)$ *is* $\psi(x_1, \ldots, x_{n-1})$ *and* ψ *is a t-formula then so is* φ.

Define $\mathscr{F}_\varphi(a_1, \ldots, a_n) = a_n \times \mathscr{F}_\psi(a_1, \ldots, a_{n-1})$. This proves (c).

(d) *If* $\varphi(x_1, \ldots, x_n)$ *is* $\psi(x_1, \ldots, x_{n+1})$ *and* ψ *is a t-formula then so is* φ.

Note that $\{0\} = \{\mathscr{F}_3(a_1, a_1)\} = \mathscr{F}_1(\mathscr{F}_3(a_1, a_1), \mathscr{F}_3(a_1, a_1))$, so we may use $\{0\}$ inside terms. Define next:

$$\begin{aligned}\mathscr{F}_\varphi(a_1, \ldots, a_n) &= \operatorname{rng}(\mathscr{F}_\psi(a_1, \ldots, a_n, \{0\}))\\ &= \operatorname{rng}(\{\langle 0, x_n, \ldots, x_1\rangle \mid x_i \in a_i \text{ and } \psi(x_1, \ldots, x_n, 0)\})\\ &= \{\langle x_n, \ldots, x_1\rangle \in a_n \times \cdots \times a_1 \mid \varphi(x_1, \ldots, x_n)\}.\end{aligned}$$

This proves (d).

(e) *If* $\varphi(x_1, \ldots, x_n)$ *is* $\psi(x_1, \ldots, x_m)$ *and* ψ *is a t-formula, then so is* φ.

For $n > m$ this follows by induction on n using (c). For $m > n$ this follows by induction on $m - n$ using (d). For $m = n$ there is nothing to prove.

(f) *If* $\varphi(x_1, \ldots, x_n)$ *is a t-formula, so is* $\neg\varphi$.

Define $\mathscr{F}_{\neg\varphi}(a_1, \ldots, a_n) = a_n \times \cdots \times a_1 - F_\varphi(a_1, \ldots, a_n)$. This proves (f).

(g) *If* $\varphi(x_1, \ldots, x_n)$ *and* $\psi(x_1, \ldots, x_n)$ *are t-formulas so is* $\varphi \wedge \psi$.

Define $\mathscr{F}_{\varphi \wedge \psi}(a_1, \ldots, a_n) = \mathscr{F}_\varphi(a_1, \ldots, a_n) \cap \mathscr{F}_\psi(a_1, \ldots, a_n)$. This proves (g).

(h) *The t-formulas are closed under propositional connectives.*

This follows by (b), (e), (f) and (g). In the following we use $\varphi(x/y)$ to denote the result of replacing all free occurrences of y by x.

(i) *If* $\psi(x_1, \ldots, x_n)$ *is a t-formula and* $\varphi(x_1, \ldots, x_{n+1})$ *is* $\psi(x_1, \ldots, x_{n-1}, x_{n+1}/x_n)$, *then* φ *is a t-formula.*

If $n = 1$, define $\mathscr{F}_\varphi(a_1, a_2) = \mathscr{F}_\psi(a_2) \times a_1$. If $n > 1$, define:

$$\begin{aligned}\mathscr{F}_\varphi(a_1, \ldots, a_{n+1}) &= \mathscr{F}_8(\mathscr{F}_\psi(a_1, \ldots, a_{n-1}, a_{n+1}), a_n)\\ &= \{\langle x_{n+1}, \ldots, x_1\rangle \mid x_n \in a_n \text{ and}\\ &\qquad \langle x_{n+1}, x_{n-1}, \ldots, x_1\rangle \in \mathscr{F}_\psi(a_1, \ldots, a_{n-1}, a_{n+1})\}.\end{aligned}$$

(j) *If* $\psi(x_1,x_2)$ *is a t-formula and* $\varphi(x_1,\ldots,x_n)$ *is* $\psi(x_{n-1}/x_1,x_n/x_2)$, *then* φ *is a t-formula.*

This makes sense only if $n\geqslant 2$ and is non-trivial only if $n>2$. To prove (j) define:

$$\begin{aligned}\mathscr{F}_\varphi(a_1,\ldots,a_n)&=\mathscr{F}_7(\mathscr{F}_\psi(a_{n-1},a_n),a_{n-2}\times\ldots\times a_1)\\&=\{\langle x_n,\ldots,x_1\rangle\in a_n\times\ldots\times a_1\,|\,\langle x_n,x_{n-1}\rangle\in\mathscr{F}_\psi(a_{n-1},a_n)\}\,.\end{aligned}$$

In (k)—(v) we prove that atomic formulas are t-formulas.

(k) *For all n, if* $\varphi(x_1,\ldots,x_n)$ *is* $\mathsf{U}(x_n)$ *then* φ *is a t-formula.*

For (k) define $\mathscr{F}_\varphi(a_1,\ldots,a_n)=\mathscr{F}_{\mathsf{U}}(a_n,a_n)\times a_{n-1}\times\cdots\times a_1$.

(l) $(x_1=x_2)$ *is a t-formula by* ($\mathscr{F}$ 10).

(m) $(x_n=x_{n+1})$ *is a t-formula by* (l) *and* (j).

(n) $(x_n=x_m)$ *is a t-formula for all* $m>n$.

This follows by induction on m using (m) for the base and (i) for the induction step.

(p) $(x_n=x_m)$ *is a t-formula for all* n,m.

For $n<m$, this is (n). For $n=m$, take $\mathscr{F}_\varphi(a_1,\ldots,a_n)=a_n\times\cdots\times a_1$. For $n>m$, note that $(x_n=x_m)$ iff $(x_m=x_n)$, so the result follows from (b) and (n).

(q) $(x_1\in x_2)$ *is a t-formula by* ($\mathscr{F}$ 11).

(r) $(x_{n+1}\in x_{n+2})$ *is a t-formula by* (q) *and* (j).

(s) *If* $\varphi(x_1,\ldots,x_n)$ *is* $(x_i\in x_j)$, *then* φ *is a t-formula.*

Let $\psi(x_1,\ldots,x_{n+2})$ be $(x_i=x_{n+1})\wedge(x_j=x_{n+2})\wedge(x_{n+1}\in x_{n+2})$, so that ψ is a t-formula by (p), (r), (e), (q). Hence we define:

$$\begin{aligned}&\mathscr{F}_\psi(a_1,\ldots,a_n,a_i,a_j)\\&=\{\langle x_{n+2},\ldots,x_1\rangle\in a_j\times a_i\times a_n\times\cdots\times a_1\,|\,x_i=x_{n+1},x_j=x_{n+2},x_i\in x_j\}\end{aligned}$$

We now use $\mathscr{F}_6$ to obtain the proof of (s):

$$\mathscr{F}_\varphi(a_1,\ldots,a_n)=\operatorname{rng}\operatorname{rng}(F_\psi(a_1,\ldots,a_n,a_i,a_j)))\,.$$

(t) *If* $\varphi(x_1,\ldots,x_{k+m})$ *is* $\mathsf{R}(x_{k+1},\ldots,x_{k+m})$, *where* R *is an m-ary relation symbol of* L *and* $k>1$, *then* φ *is a t-formula.*

Define $\mathscr{F}_\varphi(a_1, \ldots, a_{k+m}) = \mathscr{F}_{\mathsf{R}}(a_{k+m} \times \cdots \times a_{k+1}, a_k \times \cdots \times a_1)$. This proves (t).

(u) *If* R *is an m-ary relation symbol of* L *and* $\varphi(x_1, \ldots, x_n)$ *is* $\mathsf{R}(x_{i_1}, \ldots, x_{i_m})$, *then* φ *is a t-formula.*

Let $\psi(x_1, \ldots, x_n, x_{n+1}, \ldots, x_{n+m})$ be $\mathsf{R}(x_{n+1}, \ldots, x_{n+m}) \wedge (x_{i_1} = x_{n+1}) \wedge \cdots \wedge (x_{i_m} = x_{n+m})$. Thus ψ is a t-formula by (t), (p), (e) and (g). Define

$$\mathscr{F}_\varphi(a_1, \ldots, a_n) = \mathrm{rng}^m(F_\psi(a_1, \ldots, a_m, a_{i_1}, \ldots, a_{i_m})),$$

where we apply rng m-times. This proves (u).

(v) *All atomic formulas are t-formulas.*

The only ones not covered by earlier cases are those of the form $\mathsf{S}(x_i)$, but $\mathsf{S}(x_i) \leftrightarrow \neg \mathsf{U}(x_i)$ so this follows from (b), (f) and (k). We have not only shown that every atomic formula is a t-formula, but also that the t-formulas are closed under propositional connectives. We now turn to bounded quantifiers.

(w) *If* $\psi(x_1, \ldots, x_{n+1})$ *is a t-formula and* $\varphi(x_1, \ldots, x_n)$ *is* $\exists x_{n+1} \in x_j \, \psi(x_1, \ldots, x_{n+1})$, *then* φ *is a t-formula.*

Let $\theta(x_1, \ldots, x_{n+1})$ be $(x_{n+1} \in x_j)$ so $\psi \wedge \theta$ is a t-formula $\sigma(x_1, \ldots, x_{n+1})$. Note that

$$\mathscr{F}_\sigma(a_1, \ldots, a_n, \bigcup a_j) = \{\langle x_{n+1}, \ldots, x_1 \rangle \mid x_{n+1} \in x_j, x_i \in a_i \text{ for } 1 \leqslant i \leqslant n, \text{ and } \psi(x_1, \ldots, x_{n+1})\}.$$

So we may define $\mathscr{F}_\varphi$ by $\mathscr{F}_\varphi(a_1, \ldots, a_n) = \mathrm{rng}(\mathscr{F}_\psi(a_1, \ldots, a_n, \bigcup a_j))$. This proves (w).

(x) *If* $\psi(x_1, \ldots, x_k)$ *is a t-formula and* $\varphi(x_1, \ldots, x_n)$ *is* $\exists x_k \in x_j \psi(x_1, \ldots, x_k)$, *where* $k > n$, *then* φ *is a t-formula.*

The proof of (x) is just like that for (w) except we must apply rng $k - n$ times.

(y) *If* $\psi(x_1, \ldots, x_n)$ *is a t-formula and* $\varphi(x_1, \ldots, x_n)$ *is* $\forall x_k \in x_j \psi$, *where* $k > n$, *then* $\varphi(x_1, \ldots, x_n)$ *is a t-formula.*

This follows from (b), (f) and (x) since

$$\forall x_k \in x_j \psi \leftrightarrow \neg \exists x_k \in x_j \neg \psi .$$

(z) *All orderly* Δ_0 *formulas are t-formulas by* (v), (h), (x) and (y). □

6.2 Corollary. $\mathscr{F}_1, \ldots, \mathscr{F}_K$ *satisfy Assumption* 5.2(v).

Proof. Let $\varphi(x_1, \ldots, x_n)$ be a Δ_0 formula. We need a term $\mathscr{F}$ so that

$$\mathscr{F}(a, x_1, \ldots, x_{i-1}, x_{i+1}, \ldots, x_n) = \{x_i \in a \mid \varphi(x_1, \ldots, x_n)\}.$$

But we can form this set from $\mathscr{F}_\varphi(\{x_1\}, \ldots, \{x_{i-1}\}, a, \{x_{i+1}\}, \ldots, \{x_n\})$ by using $\mathscr{F}_6$ (rng) $n-i$ times and then $\mathscr{F}_5$ (dom). □

It may seem discouraging, but we are not through yet because $\mathscr{F}_1, \ldots, \mathscr{F}_K$ do not give us the transitivity condition demanded by 5.2(iv). Recall that we want to show that Tran(b) implies Tran$(\mathscr{D}(b))$, where:

$$\mathscr{D}(b) = b \cup \{\mathscr{F}_i(x, y) \mid x, y \in b, 1 \leqslant i \leqslant N\}.$$

This reduces to showing that for $1 \leqslant i \leqslant N$ we have:

(*) b transitive and $x, y \in b$ implies $\mathrm{TC}(\mathscr{F}_i(x, y)) \subseteq \mathscr{D}(b)$.

The only functions among $\mathscr{F}_1, \ldots, \mathscr{F}_K$ for which condition (*) could fail are those involving n-tuples. To satisfy (*) for these functions define, for each $n \geqslant 2$, functions $\mathscr{G}_n^1, \mathscr{G}_n^2$, and $\mathscr{H}_1, \mathscr{H}_2, \mathscr{H}_3$ by:

$$\begin{aligned} \mathscr{G}_n^1(x, y) &= \langle x_n, \ldots, x_1, y\rangle &&, \text{ if } x = \langle x_n, \ldots, x_1\rangle \\ &= 0 &&, \text{ otherwise;} \end{aligned}$$

$$\begin{aligned} \mathscr{G}_n^2(x, y) &= \{x_n, \langle x_{n-1} \ldots x_1, y\rangle\} &&, \text{ if } x = \langle x_n \ldots x_1\rangle \\ &= 0 &&, \text{ otherwise;} \end{aligned}$$

$$\mathscr{H}_1(x, y) = \langle x, y\rangle;$$

$$\begin{aligned} \mathscr{H}_2(x, y) &= \langle u, y, v\rangle &&, \text{ if } x = \langle u, v\rangle \\ &= 0 &&, \text{ otherwise;} \end{aligned}$$

$$\begin{aligned} \mathscr{H}_3(x, y) &= \{u, \langle y, v\rangle\} &&, \text{ if } x = \langle u, v\rangle \\ &= 0 &&, \text{ otherwise.} \end{aligned}$$

6.3 Definition. Let J be the largest number of places of a symbol of L. The functions $\mathscr{F}_1, \ldots, \mathscr{F}_N$ use to generate L consist of $\mathscr{F}_1, \ldots, \mathscr{F}_K$ together with $\mathscr{G}_n^1, \mathscr{G}_n^2$, for all $n \leqslant J$, plus $\mathscr{H}_1, \mathscr{H}_2, \mathscr{H}_3$.

6.4 Theorem. *The functions* $\mathscr{F}_1, \ldots, \mathscr{F}_N$ *satisfy Assumption* 5.2.

Proof. We need to see that condition (*) holds for those functions involving n-tuples. Let us check $\mathscr{F}_7$ in some detail.

Suppose x, y are in the transitive set b. Let us list the members of $\mathrm{TC}(\mathscr{F}_7(x, y))$ which are not in b, together with the reason they are in $\mathscr{D}(b)$. Recall that $\mathscr{F}_7(x, y) = \{\langle u, v, w\rangle \mid \langle u, v\rangle \in x, w \in y\}$

Members of $\mathrm{TC}(\mathscr{F}_7(x,y))$	*Excuse for appearing in* $\mathscr{D}(b)$
$\langle u,v,w\rangle$ with $\langle u,v\rangle \in x,\ w\in y$	$\mathscr{G}_2^1(\langle u,v\rangle,w)$
$\{u\}$	$\mathscr{F}_1(u,u)$
$\{u,\langle v,w\rangle\}$	$\mathscr{G}_2^2(\langle u,v\rangle,w)$
$\langle v,w\rangle$	$\mathscr{H}_1(v,w)$
$\{v\}$	$\mathscr{F}_1(v,v)$
$\{v,w\}$	$\mathscr{F}_1(v,w)$.

Anything else in $\mathrm{TC}(\mathscr{F}_7(x,y))$ is in b, since b is transitive. $\mathscr{F}_8$ and the $\mathscr{F}_R$ are similar. The others are simpler. □

6.5—6.7 Exercises

6.5. Show that each of $\mathscr{F}_{K+1},\ldots,\mathscr{F}_N$ can be written as a term in $\mathscr{F}_1,\ldots,\mathscr{F}_K$. [Hint: This is fairly easy using 6.1.]

6.6. Define $\mathrm{L}'(a,\lambda)$ using only $\mathscr{F}_1\ldots\mathscr{F}_K$. Show that for limit ordinals λ, $\mathrm{L}'(a,\lambda)=\mathrm{L}(a,\lambda)$. The only point of using $\mathscr{F}_{K+1},\ldots,\mathscr{F}_N$ was to make each $\mathrm{L}(a,\alpha)$ transitive.

6.7. Verify condition (*) in the proof of 6.2 for $\mathscr{F}_8$.

6.8 Notes. The proof of 6.1 is one of the few places where the addition of urelements and relations on them causes extra work. Neither space nor memory permit us to list all the people who have found gaps in earlier proofs of this lemma.

When used in a class or seminar, section 6 should be supplemented with coffee (*not* decaffeinated) and a light refreshment. We suggest *Heatherton Rock Cakes.* (*Recipe:* Combine 2 cups of self-rising flour with 1 *t.* allspice and a pinch of salt. Use a pastry blender or two cold knives to cut in 6 *T* butter. Add $\frac{1}{3}$ cup each of sugar and raisins (or other urelements). Combine this with 1 egg and enough milk to make a stiff batter (3 or 4 *T* milk). Divide this into 12 heaps, sprinkle with sugar, and bake at 400 °F. for 10—15 minutes. They taste better than they sound.)

7. *First Order Definability and Substitutable Functions*

The functions $\mathscr{F}_1,\ldots,\mathscr{F}_N$ defined in 6.3 are actually quite simple compared with some Σ operations we might have used to satisfy Assumption 5.2. We will exploit this to prove the following theorem; the first corollary is of special importance.

7.1 Theorem. *Let* $\mathfrak{M}=\langle M,R_1,\ldots,R_l\rangle$, *let* a *be transitive in* V_M *with* $M\subseteq a$. *Let* $A=a\cap V_M$ *and let* $\mathbb{A}_{\mathfrak{M}}=(\mathfrak{M};A,\in)$. *Then a relation* S *on* $\mathbb{A}_{\mathfrak{M}}$ *is first-order definable using parameters from a iff* $S\in \mathrm{L}(a,\omega)$.

7.2 Corollary. *If* $O(\mathfrak{M})=\omega$, *then the relations on* $\mathfrak{M}$ *in* $\mathbb{HYP}_{\mathfrak{M}}$ *are just the first-order relations.*

Proof. If $o(\mathbb{HYP}_{\mathfrak{M}})=\alpha$ *then* $\mathbb{HYP}_{\mathfrak{M}}=\mathrm{L}(\alpha)_{\mathfrak{M}}$. □

7.3 Corollary. *The relations on* $\mathrm{L}(a,\alpha)$ *in* $\mathrm{L}(a,\alpha+\omega)$ *are the relations first order definable over* $(\mathfrak{M}_0;\mathrm{L}(a,\alpha)\cap V_{M_0},\in)$, *where* $\mathfrak{M}_0$ *is the substructure of* $\mathfrak{M}$ *with domain* $\mathrm{Sp}(a)$.

Proof. Apply 7.1, reading $\mathrm{L}(a,\alpha)$ for a and $\mathfrak{M}_0$ for $\mathfrak{M}$. □

We begin the proof of 7.1 by studying substitutable functions.

7.4 Definition. A Σ operation symbol F of n-arguments is *substitutable* if the Δ_0 formulas are closed under substitution by F; that is, if for each Δ_0 formula $\varphi(w,v_1,\ldots,v_k)$, there is a Δ_0 formula $\psi(u_1,\ldots,u_n,v_1,\ldots,v_k)$ not involving F so that $\mathrm{KPU}\vdash\varphi(\mathsf{F}(\vec{u}),\vec{v})\leftrightarrow\psi(\vec{u},\vec{v})$.

7.5 Lemma. (i) *The substitutable operations are closed under composition.*

(ii) *If* $\mathrm{KPU}\vdash\forall\vec{u}(\mathsf{F}(\vec{u})$ *is a set), then* F *is substitutable iff for each* Δ_0 *formula* φ, *the formula* $\exists x\in\mathsf{F}(\vec{u})\,\varphi(x,\vec{v})$ *is equivalent (in* KPU) *to a* Δ_0 *formula* $\psi(\vec{u},\vec{v})$.

(iii) *If* F *is substitutable, so is* G *defined by* $\mathsf{G}(x,\vec{y})=\{\mathsf{F}(z,\vec{y})\mid z\in x\}$.

Proof: (i) is more or less obvious. For example, if $\varphi(\mathsf{F}(u))\leftrightarrow\psi(u)$ and $\psi(\mathsf{G}(x))\leftrightarrow\theta(x)$, then $\varphi(\mathsf{F}(\mathsf{G}(x)))\leftrightarrow\psi(\mathsf{G}(x))\leftrightarrow\theta(x)$.

The necessity in (ii) is a special case of Definition 7.4. To prove the other half note that

$$
\begin{array}{rll}
y\in\mathsf{F}(\vec{x}) & \text{iff} & \exists z\in\mathsf{F}(\vec{x})(y=z)\,,\\
a=\mathsf{F}(\vec{x}) & \text{iff} & \forall z\in a(z\in\mathsf{F}(\vec{x}))\wedge\forall z\in\mathsf{F}(\vec{x})(z\in a)\,,\\
\mathsf{F}(\vec{x})\in a & \text{iff} & \exists b\in a[\mathsf{F}(\vec{x})=b]\,,\\
p=\mathsf{F}(\vec{x}) & \text{iff} & p\neq p\,,\\
\mathsf{R}(\ldots\mathsf{F}(\vec{x})\ldots) & \text{iff} & x_1\neq x_1\,.
\end{array}
$$

So all atomic formulas involving F are Δ_0. A simple induction on Δ_0 formulas, using the hypothesis of (ii), shows that F is substitutable in each of them.

To prove (iii) note that $\exists u\in\mathsf{G}(x,\vec{y})\,\varphi(u)\leftrightarrow\exists z\in x\,\varphi(\mathsf{F}(z,\vec{y}))$; so G is substitutable by (ii). □

7.6 Lemma. *Each of the operations* $\mathscr{F}_1,\ldots,\mathscr{F}_N,\mathscr{D}$ *is substitutable.*

Proof. We run through a few cases, using 7.5(ii) quite heavily.

$\mathscr{F}_1$: $\exists u \in \{x, y\}\, \varphi(u) \leftrightarrow \varphi(x) \vee \varphi(y)$;

$\mathscr{F}_2$: $\exists u \in \bigcup x\, \varphi(u) \leftrightarrow \exists z \in x\, \exists u \in z\, \varphi(u)$;

$\mathscr{H}_1$: $\exists z \in \langle x, y\rangle\, \varphi(z) \leftrightarrow \varphi(\{x\}) \vee \varphi(\{x, y\})$, which is Δ_0 since $\mathscr{F}_1$ is substitutable;

$\mathscr{F}_4$: $\mathscr{F}_4(x, y) = x \times y = \{\langle u, v\rangle \mid u \in x, v \in y\} = \bigcup\{\{\langle u, v\rangle \mid u \in x\} \mid v \in y\}$.

Thus we see that $\mathscr{F}_4$ is substitutable, since $\mathscr{F}_2$ and $\mathscr{H}_1$ are, by composition and 7.5 (iii).

$\mathscr{F}_5$: $\exists u \in \mathrm{dom}(x)\, \varphi(u) \leftrightarrow \exists u, v \in \bigcup\bigcup x\, [\langle u, v\rangle \in x \wedge \varphi(u)]$, so $\mathscr{F}_5$ follows from $\mathscr{H}_1$.

$\mathscr{D}$: $\exists x \in \mathscr{D}(b)\, \varphi(x) \leftrightarrow \exists x \in b\, \varphi(x) \vee \exists y, z \in b\, [\bigvee_{i \leqslant n} \varphi(\mathscr{F}_i(x, y))]$.

The other $\mathscr{F}_i$ are just as routine. □

For the remainder of the section fix $\mathfrak{M}, a$ and $\mathbb{A}_{\mathfrak{M}}$ as in the statement of the theorem to be proved, Theorem 7.1.

7.7 Lemma. *For every element $x \in \mathrm{L}(a, \omega)$ there is a term $\mathscr{F}$ in the symbols $\mathscr{F}_1, \ldots, \mathscr{F}_N, \mathscr{D}$ and $y_1, \ldots, y_m \in a \cup \{a\}$ such that $x = \mathscr{F}(y_1, \ldots, y_m)$.*

Proof. Note that $\mathrm{L}(\alpha, n) = \mathscr{D}\mathscr{S}(\mathscr{D}\mathscr{S}(\ldots(a)\ldots))$ for n repetitions of $\mathscr{D} \circ \mathscr{S}$ ($\mathscr{S}$ is a term in $\mathscr{F}_1, \mathscr{F}_2$ as we saw in § 6) so each $\mathrm{L}(a, n)$ is of the appropriate form. We now show that each $x \in \mathrm{L}(a, n)$ is of the appropriate form by induction on n. Since $\mathrm{L}(a, \omega) = \bigcup_{n<\omega} \mathrm{L}(a, n)$ the result follows.

For $n=0$ we have $\mathrm{L}(a, 0) = a$, since a is transitive, so the result is trivial. If $x \in \mathrm{L}(a, n+1) - \mathrm{L}(a, n)$, then $x = \mathrm{L}(a, n)$ or $x = \mathscr{F}_i(z, y)$ for some $y, z \in \mathrm{L}(a, n) \cup \{\mathrm{L}(a, n)\}$. The first case is taken care of by the first part of the proof. If $x = \mathscr{F}_i(y, z)$ with $y, z \in \mathrm{L}(a, n) \cup \{\mathrm{L}(a, n)\}$, then y, z are of the appropriate form. Hence, x is also of the correct form. □

7.8 Lemma. *If $\varphi(x_1, \ldots, x_n, y)$ is Δ_0 without parameters, then the relation*

$$\{(x_1, \ldots, x_n) \in \mathbb{A}_{\mathfrak{M}} \mid \mathrm{L}(a, \omega) \models \varphi(x_1, \ldots, x_n, a)\}$$

is first-order definable over $\mathbb{A}_{\mathfrak{M}}$.

Proof. A trivial induction on Δ_0 formulas; just replace $\forall x \in a$ by $\forall x$, etc. □

Proof of Theorem 7.1. Suppose $S \subseteq a^n$, $S \in \mathrm{L}(a, \omega)$. Then, by 7.7, there is a term $\mathscr{F}$ in $\mathscr{F}_1, \ldots, \mathscr{F}_N, \mathscr{D}$ such that $S = \mathscr{F}(x_1, \ldots, x_k, a)$ for some $x_1, \ldots, x_k \in a$. But then $S(y_1, \ldots, y_n)$ iff $\langle y_1, \ldots, y_n\rangle \in \mathscr{F}(x_1, \ldots, x_k, a)$.

The right hand side is equivalent to a Δ_0 formula $\varphi(y_1, \ldots, y_n, x_1, \ldots, x_k, a)$ by the substitutability of $\mathscr{F}$ (using 7.6 and 7.5(i)) and $\langle\ \rangle$. The relation $\varphi(y_1, \ldots, y_n, x_1, \ldots, x_k, a)$ is definable on $\mathbb{A}_{\mathfrak{M}}$, and hence S is definable using the parameters $x_1, \ldots, x_k$. The converse is trivial since every definable relation S on $\mathbb{A}_{\mathfrak{M}}$ is Δ_0 on $\mathrm{L}(a, 1)$ and so is in $\mathrm{L}(a, \omega)$ by, say, Exercise 5.16. □

7.9—7.10 Exercises

7.9. F is effectively substitutable if the ψ of 7.4 can be found effectively from φ. Show that each $\mathscr{F}_1, \ldots, \mathscr{F}_N, \mathscr{D}$ is effectively substitutable. [Use Church's Thesis.]

7.10. Verify that the effective version of 7.8 holds.

7.11 Notes. It seems to be an open problem whether the converse of 7.2 is true in general. The study of substitutable functions goes back to Levy [1965]. He called them "admissible terms", terminology clearly inadmissible in our context. They were used by Gandy [1975] and Jensen [1972] (written later than Gandy [1975]) to prove the urelementless version of Corollary 7.3. Gandy called them "substitutable", Jensen called them "simple".

8. *The Truncation Lemma*

Recall (from I.9.5) that a binary relation E on a set X is well founded iff for all nonempty $Y \subseteq X$ there is an $x \in Y$ such that for all $y \in Y$ we have $\neg(yEx)$. The notion is what we have tried to capture in the axiom of foundation, but of course we fail since it is just not expressible in the first-order language of set theory. A *nonstandard* model of KPU is one of the form $\mathfrak{A}_{\mathfrak{M}} = (\mathfrak{M}; A, E, \ldots)$, where E is not well founded; the other models are the standard, or intended models since, by the next result, they are isomorphic to admissible sets. The proof is essentially the same as that of I.9.6.

8.1 Proposition. *If $\mathfrak{A}_{\mathfrak{M}} = (\mathfrak{M}; A, E, \ldots)$ is a well-founded model of extensionality then, it is isomorphic to a structure of the form $\mathbb{B}_M = (\mathfrak{M}; B, \in, \ldots)$ with $M \cup B$ transtitive. Both $\mathbb{B}_{\mathfrak{M}}$ and the isomorphism f are unique, and f satisfies*

$$f(p) = p, \qquad \text{for} \quad p \in M;$$
$$f(a) = \{f(b) \mid bEa\}, \quad \text{for} \quad a \in A.$$

Now let $\mathfrak{A}_{\mathfrak{M}} = (\mathfrak{M}; A, E)$ be any structure and let $\mathscr{W} = \{\mathfrak{B}_{\mathfrak{M}} \subseteq_{\text{end}} \mathfrak{A}_{\mathfrak{M}} \mid \mathfrak{B}_{\mathfrak{M}}$ is well founded$\}$. Assume $\mathscr{W} \neq 0$, which is the case iff $\mathfrak{A}_{\mathfrak{M}} \models \exists x \, \forall x \, (y \notin x)$.

8.2 Lemma. *There is a largest $\mathfrak{B}_{\mathfrak{M}} \in \mathscr{W}$ (one which is an end extension of all other members of $\mathscr{W}$).*

Proof. Let $\mathfrak{B}_{\mathfrak{M}}$ be the union of all structures in $\mathscr{W}$. It is easy to check that $\mathfrak{B}_{\mathfrak{M}} \subseteq_{\text{end}} \mathfrak{A}_{\mathfrak{M}}$ and $\mathfrak{C}_{\mathfrak{M}} \subseteq_{\text{end}} \mathfrak{B}_{\mathfrak{M}}$ for all $\mathfrak{C}_{\mathfrak{M}} \in \mathscr{W}$. To see that $\mathfrak{B}_{\mathfrak{M}}$ is well founded, let X be a non empty subset of $M \cup B$. We must find an $x \in X$ such that $y \in X$ implies $\neg yEx$. Since $\mathfrak{B}_{\mathfrak{M}}$ is the union of $\mathscr{W}$, there is a $\mathfrak{C}_{\mathfrak{M}} \in \mathscr{W}$ such that $X' = X \cap (M \cup C)$ is nonempty. Since $\mathfrak{C}_{\mathfrak{M}}$ is well founded there is an $x \in X'$ such that $y \in X'$ implies $\neg yEx$. But yEx implies $y \in M \cup C$ for all $y \in M \cup A$ (by $\mathfrak{C}_{\mathfrak{M}} \subseteq_{\text{end}} \mathfrak{A}_{\mathfrak{M}}$), so we have $\neg yEx$ for all $y \in X$. $\square$

8.3 Definition. The largest well-founded $\mathfrak{B}_{\mathfrak{M}}$ such that $\mathfrak{B}_{\mathfrak{M}} \subseteq_{\text{end}} \mathfrak{A}_{\mathfrak{M}}$ is called the *well-founded part* of $\mathfrak{A}_{\mathfrak{M}}$ and is denoted by $\mathscr{W}\!f(\mathfrak{A}_{\mathfrak{M}})$.

Note that this makes sense whether or not $\mathfrak{A}_{\mathfrak{M}}$ is not well founded. If $\mathfrak{A}_{\mathfrak{M}}$ is well founded, then $\mathscr{W}\!f(\mathfrak{A}_{\mathfrak{M}}) = \mathfrak{A}_{\mathfrak{M}}$. If $\mathfrak{A}_{\mathfrak{M}}$ is a model of extensionality, so is $\mathscr{W}\!f(\mathfrak{A}_{\mathfrak{M}})$, since $\mathscr{W}\!f(\mathfrak{A}_{\mathfrak{M}}) \subseteq_{\text{end}} \mathfrak{A}_{\mathfrak{M}}$. In this case *we often identity $\mathscr{W}\!f(\mathfrak{A}_{\mathfrak{M}})$ with the unique transitive structure isomorphic to it*, as given by 8.1. We make this identification in the next result, for example, which is an example of one way in which KPU is better behaved than stronger theories like ZF. It gives us a new method of constructing admissible sets, which accounts for its occurrence in this chapter.

8.4 Truncation Lemma. *Let* $\mathfrak{A}_{\mathfrak{M}} = (\mathfrak{M}; A, E, \ldots)$ *and* $\mathbb{B}_{\mathfrak{M}} = (\mathfrak{M}; B, \in, \ldots)$ *be* L^*-*structures with* $\mathfrak{A}_{\mathfrak{M}} \models \text{KPU}$ and $\mathbb{B}_{\mathfrak{M}} \subseteq_{\text{end}} \mathfrak{A}_{\mathfrak{M}}$, *where* $(\mathfrak{M}; B, \in) = \mathscr{W}\!f(\mathfrak{M}; A, E)$. *Then* $\mathbb{B}_{\mathfrak{M}}$ *is admissible over* $\mathfrak{M}$.

Proof. We need to show that the hypotheses of Lemma I.8.9 are satisfied, for then we get all the axioms of KPU except Foundation true in $\mathbb{B}_{\mathfrak{M}}$. But $\mathbb{B}_{\mathfrak{M}}$ is well founded, so it certainly satisfies Foundation. First note:

(1) If $a \in A$ and $a_E \subseteq B$, then $a \in B$.

This follows from the maximality of $\mathbb{B}_{\mathfrak{M}} \in \mathscr{W}$.

(2) If $a \in B$ and $\mathfrak{A}_{\mathfrak{M}} \models \text{rk}(a) = \alpha$, then $\alpha \in B$.

This follows by $\in$ induction on a, using (1), since $\mathfrak{A}_{\mathfrak{M}} \models \alpha = \sup\{\text{rk}(x) + 1 \mid x \in a\}$.

(3) If $\alpha \in B$ and $\mathfrak{A}_{\mathfrak{M}} \models \text{rk}(a) = \alpha$, then $a \in B$.

This follows by induction on α using (1). Thus we see that if $\mathfrak{A}_{\mathfrak{M}} \models \text{rk}(a) = \alpha$, then $a \in B$ iff $\alpha \in B$.

(4) There is no sup in $\mathfrak{A}_{\mathfrak{M}}$ for the ordinals of $\mathbb{B}_{\mathfrak{M}}$.

This follows from (1). Thus, we have what we need to apply I.8.9. □

We have worded 8.4 in a roundabout way because of the functions which might appear in the list.... The universe of $\mathscr{W}\!f(\mathfrak{M}; A, E)$ might not be closed under them. Perhaps it is worth stating a special case of 8.4, the one we usually apply. It follows at once from 8.4.

8.5 Corollary. *If* $\mathfrak{A}_{\mathfrak{M}} = (\mathfrak{M}; A, E)$ *is a model of* KPU *then its wellfounded part is an admissible set over* $\mathfrak{M}$.

8.6 Theorem. *Let* $\mathfrak{M} = \langle M, R_1, \ldots, R_l \rangle$. *The admissible set* $\mathbb{HYP}_{\mathfrak{M}}$ *is the intersection of all models* $\mathfrak{A}_{\mathfrak{M}}$, *well-founded or not, of* KPU^+. *More accurately, given any model* $\mathfrak{A}_{\mathfrak{M}}$ *of* KPU^+, *there is a unique embedding of* $\mathbb{HYP}_{\mathfrak{M}}$ *onto an initial substructure of* $\mathfrak{A}_{\mathfrak{M}}$.

Proof. By 8.5, $\mathscr{W}\!f(\mathfrak{A}_{\mathfrak{M}})$ is admissible above $\mathfrak{M}$ and hence $\mathbb{HYP}_{\mathfrak{M}} \subseteq \mathscr{W}\!f(\mathfrak{A}_{\mathfrak{M}}) \subseteq_{\text{end}} \mathfrak{A}_{\mathfrak{M}}$, the first inclusion being correct up to the unique embedding discussed above. ☐

Recall that $O(\mathfrak{M})$ is, by definition, $o(\mathbb{HYP}_{\mathfrak{M}})$. Structures $\mathfrak{M}$ such that $O(\mathfrak{M})=\omega$ are going to play an interesting role in our study of admissible sets and structures. We call such structures *recursively saturated.* This terminology will be justified in Chapter IV (cf. Definition IV.5.1 and Theorem IV.5.3). In the next theorem we use the truncation lemma to prove that there are lots of recursively saturated structures; that is, structures $\mathfrak{M}$ with $O(\mathfrak{M})=\omega$.

8.7 Theorem. *For every structure $\mathfrak{M}=\langle M, R_1, \ldots, R_k\rangle$ there is a recursively saturated elementary extension $\mathfrak{N}$ of $\mathfrak{M}$ of the same cardinality.*

Proof. Consider $\mathbb{HYP}_{\mathfrak{M}}$ as a single-sorted structure of the form:

$$\mathfrak{A}=\langle M\cup A, M, A, R_1, \ldots, R_l, \in\rangle,$$

and let $\mathfrak{B}=\langle N\cup B, N, B, R'_1, \ldots, R'_l, E\rangle$ be an elementary extension with non-standard natural numbers. This exists by the ordinary Compactness Theorem. Let $\mathfrak{N}=\langle N, R'_1, \ldots, R'_l\rangle$, and let $\mathfrak{B}_{\mathfrak{N}}=(\mathfrak{N}; B, E)$, which is a model of KPU$^+$. The well-founded part of $\mathfrak{B}_{\mathfrak{N}}$ is an admissible set $\mathbb{B}'_{\mathfrak{N}}$ with $N\in\mathfrak{B}'$, since $\mathrm{rk}(N)=1$. Also $o(\mathbb{B}'_{\mathfrak{N}})=\omega$, since $\mathfrak{B}_{\mathfrak{N}}$ has non-standard integers. Thus $o(\mathbb{HYP}_{\mathfrak{N}})=\omega$ by Theorem 5.9. The cardinality considerations are routine. ☐

This shows that we cannot expect $\mathfrak{M}\prec\mathfrak{N}$ and $O(\mathfrak{N})=\omega$ together to imply $O(\mathfrak{M})=\omega$.

Finally, we use 8.5 to get a rather technical looking results. The real content of 8.8 will emerge gradually throughout the book.

8.8. Proposition. *Let S be an n-ary relation on a structure $\mathfrak{M}=\langle M, R_1, \ldots, R_l\rangle$. If S is Σ_1 on $\mathbb{HYP}_{\mathfrak{M}}$ then there is a Σ_1 formula $\varphi(x_1, \ldots, x_n, p_1, \ldots, p_k, \mathfrak{M})$, with only constants $p_1, \ldots, p_k \in M$ such that for all $q_1, \ldots, q_n \in M$ the following are equivalent:*

(i) $S(q_1, \ldots, q_n)$;

(ii) $\mathbb{HYP}_{\mathfrak{M}} \models \varphi(\vec{q}, \vec{p}, M)$;

(iii) *For all models of* KPU$^+$ *of the form* $\mathfrak{A}_{\mathfrak{M}}=(\mathfrak{M}; A, E)$ *we have* $\mathfrak{A}_{\mathfrak{M}} \models \varphi(\vec{q}, \vec{p}, M)$.

Proof. By 6.4 every $a\in\mathbb{HYP}_{\mathfrak{M}}$ can be defined by a Σ_1 formula with constants from $M\cup\{M\}$. Thus we may replace any of these a's by its definition to get a φ of the appropriate kind such that (i)$\Longleftrightarrow$(ii). Since $\mathbb{HYP}_{\mathfrak{M}}\models$ KPU$^+$, we see that (iii)$\Rightarrow$(ii). To see that (ii)$\Rightarrow$(iii) note that any such $\mathfrak{A}_{\mathfrak{M}}$ is (isomorphic to) an end extension of $\mathbb{HYP}_{\mathfrak{M}}$, by 8.6. Hence if $\varphi(q, p, M)$ holds in $\mathbb{HYP}_{\mathfrak{M}}$, it holds in $\mathfrak{A}_{\mathfrak{M}}$, since it is Σ_1. Of course, we need to know that the isomorphism is the identity on $M\cup\{M\}$, but this follows from 8.1. ☐

8.9—8.15 Exercises

8.9. Let $\mathfrak{M}=\langle M,R_1,\ldots,R_l\rangle$ be such that $\mathfrak{M}\prec\mathfrak{N}$, and $\mathrm{card}(\mathfrak{M})=\mathrm{card}(\mathfrak{N})$ implies $\mathfrak{M}\cong\mathfrak{N}$ (equivalently, $\mathrm{Th}(\mathfrak{M})$ is $\mathrm{card}(\mathfrak{M})$-categorical). Show that $o(\mathbb{HYP}_{\mathfrak{M}})=\omega$, and hence the relations S on $\mathfrak{M}$ in $\mathbb{HYP}_{\mathfrak{M}}$ are just the ones first-order definable over $\mathfrak{M}$.

8.10. Let $\mathfrak{M}=\langle M,=\rangle$ be infinite. Show that a subset $X\subseteq M$ is in $\mathbb{HYP}_{\mathfrak{M}}$ iff X or $M-X$ is finite.

8.11. (F. Ville) Suppose α is not admissible and $\langle \mathrm{L}(\alpha),\in\rangle\subseteq_{\mathrm{end}}\langle A,E\rangle$, where $\langle A,E\rangle\models \mathrm{KP}$. Show that, up to a unique isomorphism, $\langle \mathrm{L}(\beta),\in\rangle\subseteq_{\mathrm{end}}\langle A,E\rangle$, where β is the least admissible ordinal greater than α.

8.12. Use the notation of 8.11. Let S be a relation on $\mathrm{L}(\alpha)$, S Σ_1 on $\mathrm{L}(\beta)$. Find a Σ_1 formula $\varphi(x_1,\ldots,x_n,a)$ with $\vec{a}\in\mathrm{L}(\alpha)$ and no other constants such that the following are equivalent:

(i) $S(\vec{x})$;

(ii) $\mathrm{L}(\beta)\models\varphi(\vec{x},\vec{a})$;

(iii) For all models $\mathfrak{A}=\langle A,E\rangle$ of KP if $\langle\mathrm{L}(\alpha),\in\rangle\subseteq_{\mathrm{end}}\mathfrak{A}$, then $\mathfrak{A}\models\varphi(\vec{x},\vec{a})$.

[Hint: Find a good Σ_1 definition of α to get rid of $\mathrm{L}(\alpha)$ in φ.]

8.13. If $\mathfrak{A}_{\mathfrak{M}}=(M;A,E,P)$ is a model for KPU + Power and $(\mathfrak{M};B,\in)=\mathscr{W}\!\!\not{f}(\mathfrak{M};A,E)$, then $\mathbb{B}_{\mathfrak{M}}=(\mathfrak{M},B,\in,P\restriction B)$ is admissible and a model of Power.

8.14. Show that the well-founded part of a model $\mathfrak{A}_{\mathfrak{M}}$ of KPU + Beta need not satisfy Beta. (Not for the beginner.) The well-founded part of a model $\langle A,E\rangle$ of all of ZF need not satisfy Beta.

8.15. (For those familiar with Π_1^1.) Let $\mathfrak{N}=\langle N,+,\cdot\rangle$ and let S be a relation on $\mathfrak{N}$. Show that if S is Σ_1 on $\mathbb{HYP}_{\mathfrak{N}}$, then S is Π_1^1.

8.16 Notes. The history of the Truncation Lemma is more complicated than the lemma itself. Starting from the fact that every ω-model of second-order arithmetic contains all hyperarithmetic sets of natural numbers, Mlle. F. Ville generalized this by proving Exercise 8.11. This was in 1966 and her proof remains unpublished. Barwise [1969] generalized this to obtain a $V=L$ or $V=L(x)$ version of the Truncation Lemma. It is not clear to the present author who first thought of the trick (used back in Lemma I.8.9) that allows the full result to go through.

9. The Lévy Absoluteness Principle

We have been rather free wheeling with our metatheory, for example in § 1 and § 3 of this chapter. We used the power set axiom, results on cardinal numbers and even the axiom of choice (in the guise of the Downward Löwenheim-Skolem theorem) in § 3. It should be clear, though, that everything we have done could be formalized within ZFC, Zermelo-Fraenkel set theory with choice. (Given a structure $\mathfrak{M}=\langle M,\ldots\rangle$ for example, with $M\in\mathbb{V}$, we can define $\mathbb{V}_{\mathfrak{M}}$ as a class in $\mathbb{V}$ without difficulty as long as we remember that $\in_M$ is distinct from $\in$.) Weaker theories would suffice; but, because it is familiar to almost everyone, *we fix* ZFC *as our metatheory for this book*, unless some other theory like KPU is specified the way it was in Chapter I.

The following version of the Löwenheim-Skolem Theorem, implicit in 3.4, will be of considerable use to us in what follows, though we usually use the simple parameter-free version given in 9.2.

9.1 Theorem. *Let* $\varphi(x_1,\ldots,x_n,y_1,\ldots,y_m)$ *be a* Π *formula in the language of* ZFC *(with only* $\in$ *and* $=$*) with the free variables only as shown. The following sentence is a theorem of* ZFC:

$$\forall y_1,\ldots,y_m\in H(\aleph_1)[\forall x_1,\ldots,x_n\in H(\aleph_1)\ \varphi(\vec{x},\vec{y})\rightarrow\forall x_1,\ldots,x_n\ \varphi(\vec{x},\vec{y})].$$

9.2 Corollary. *Let* $\varphi(x)$ *be a* Π *formula in the language of* ZFC *with only the one free variable x. Then* $\mathrm{ZFC}\vdash\forall x\in H(\aleph_1)\ \varphi(x)\rightarrow\forall x\ \varphi(x)$.

Proof of 9.1. *Since* $\mathrm{KP}\subseteq\mathrm{ZFC}$ *we may assume* $\varphi(x,y)$ *is* Π_1, that is, of the form $\forall z\,\psi(\vec{x},\vec{y},z)$, where ψ is Δ_0 by I.4.3. We work within ZFC and prove the sentence in question by contraposition. Let $y_1,\ldots,y_m\in H(\aleph_1)$, and suppose there are $x_1,\ldots,x_n$, such that $\neg\varphi(\vec{x},\vec{y})$, i.e. there is a z such that $\neg\psi(\vec{x},\vec{y},z)$. Pick $\kappa\geqslant\aleph_1$ so large that $x_1,\ldots,x_n,z\in H(\kappa)$. Then, since $\neg\psi$ is Δ_0 we have $\langle H(\kappa),\in\rangle\models\neg\psi(\vec{x},\vec{y},z)$ by absoluteness. By 3.4, $\langle H(\aleph_1),\in\rangle\prec_1\langle H(\kappa),\in\rangle$, so we find $\langle H(\aleph_1),\in\rangle\models\exists x_1,\ldots,x_n z\,\neg\psi(\vec{x},\vec{y},z)$. Pick $x_1,\ldots,x_n\in H(\aleph_1)$ so that $\langle H(\aleph_1),\in\rangle\models\exists z\neg\psi(\vec{x},\vec{y},z)$. Then by Lemma I.4.2, $\exists z\neg\psi(\vec{x},\vec{y},z)$ is true, which means that $\neg\varphi(\vec{x},\vec{y})$. Since $x_1,\ldots,x_n\in \mathrm{H}(\aleph_1)$, this proves our result. □

We conclude this section with a simple example of the use of the Absoluteness Principle.

9.3 Proposition. *Let* $\mathfrak{M}=\langle M\rangle$ *be a structure with no relations. If* $X\subseteq M$ *is constructible from* $\mathfrak{M}$, $X\in\mathrm{L}(\mathfrak{M})$, *then* X *or* $M-X$ *is finite.*

Proof. The statement to be proved has the form:

$$\forall M\forall X\forall\alpha\,[X\subseteq M\wedge X\in\mathrm{L}(M,\alpha)\rightarrow X\ \textit{is finite}\ \vee\ M-X\ \textit{is finite}].$$

In ZF, or even in KPU + Infinity, this is a Π statement (by use of P_ω from I.9) so it suffices to prove it for countable M and α. We may assume M is infinite since otherwise the result is trivial.

Let σ be any one-one map of M onto M. We can extend σ to an automorphism $\bar{\sigma}$ of $\mathbb{V}_M$ onto $\mathbb{V}_M$ by recursion on $\in$:

$$\bar{\sigma}(p)=\sigma(p),$$

$$\bar{\sigma}(a)=\{\bar{\sigma}(x)\mid x\in a\}.$$

Note that $\bar{\sigma}(\mathscr{F}_i(x,y))=\mathscr{F}_i(\bar{\sigma}(x),\bar{\sigma}(y))$, whenever $1\leqslant i\leqslant \mathscr{N}$, by inspection. A simple proof by induction shows that $\bar{\sigma}(\mathrm{L}(M,\alpha))=\mathrm{L}(M,\alpha)$ for all α.

Now suppose that M and α are countable but that there is an $X\in \mathrm{L}(M,\alpha)$ with $X\subseteq M$ such that X and $M-X$ are both infinite. Then, for any $Y\subseteq M$ with Y and $M-Y$ infinite, there is a one-one map σ mapping X onto Y so that $\bar{\sigma}(X)=Y$. But then $X\in \mathrm{L}(M,\alpha)$ implies $\bar{\sigma}(\mathrm{L}(M,\alpha))$; so $Y\in \mathrm{L}(M,\alpha)$. But there are $2^{\aleph_0}$ such X, whereas $\mathrm{L}(M,\alpha)$ is countable. $\square$

9.4—9.7 Exercises

9.4. Let ZFU$^+$ be KPU$^+$ plus full separation, full collection, Power and Infinity. Prove that for each $\varphi\in$ ZFU$^+$, we have ZFU $\vdash \varphi^{\mathrm{L}(M)}$.

9.5. Show that if M is as in 9.3 then $\mathrm{L}(M)$ is a model of ZFU$^+$ plus "all subsets of M are finite or cofinite". This shows that choice fails very badly in this particular $\mathrm{L}(M)$.

9.6. Let $\mathfrak{M}=\langle M,R_1,\ldots,R_l\rangle$ and let σ be an automorphism of $\mathfrak{M}$. Extend σ to a $\bar{\sigma}:\mathbb{V}_{\mathfrak{M}}\to\mathbb{V}_{\mathfrak{M}}$ as in 9.3. Show that $\bar{\sigma}:\mathrm{L}(\alpha)_{\mathfrak{M}}\to\mathrm{L}(\alpha)_{\mathfrak{M}}$, one-one and onto, for all α.

9.7 Notes. The Lévy Absoluteness Principle was first proved by Lévy [1965]. See the notes from § 3 for more details on the general argument. One of the main features of this book (at least from our point of view) is the systematic use of the Lévy Absoluteness Principle to simplify results by reducing them to the countable case. This is particularly true of Part *B* of the book.

We will see, as a by product of § V.8, that the axiom of choice is not needed in the proof of 9.1. See the proof of V.8.10, in particular.

Chapter III
Countable Fragments of $\mathsf{L}_{\infty\omega}$

In this chapter the student is introduced to the infinitary logic $L_{\infty\omega}$ and its countable fragments. The reason for treating infinitary logic so early in the book is two-fold. In the first place it offers a nice application of the very notion of admissible set, since the fragments of $L_{\infty\omega}$ most like ordinary logic are those given by countable admissible sets. More important, however, is the powerful tool that infinitary logic gives us in our study of admissible sets. The results from model theory presented in this chapter are all chosen because of their applicability to the theory of admissible sets and generalized recursion theory.

1. Formalizing Syntax and Semantics in KPU

In § I.3 we formalized informal notions of mathematics in KPU, notions like "function", "natural number", and "ordinal". In this section we do the same thing for informal notions of logic, notions like "language", "structure", "formula".

In this section we work in KPU but we suppose that among the atomic predicates of our metalanguage L^* are the following:

$$\mathsf{Relation\text{-}symbol}\ (x),$$
$$\mathsf{Function\text{-}symbol}\ (x),$$
$$\mathsf{Constant\text{-}symbol}\ (x),$$
$$\mathsf{Variable}\ (x)$$

and among the operation symbols of our metalanguage are two unary ones:

$$\mathsf{v} \quad \text{and} \quad \# .$$

We use $\mathsf{r}, \mathsf{r}_1, \ldots$ to vary over objects x satisfying $\mathsf{Relation\text{-}symbol}\ (x)$. Similarly $\mathsf{h}, \mathsf{h}_1, \ldots$ for function symbols and $\mathsf{c}, \mathsf{c}_1, \mathsf{d}, \ldots$ for constant symbols. We also assume that among the constant symbols of our *metalanguage* L^* are

$$\neg,\ \bigwedge,\ \bigvee,\ \forall,\ \exists,\ \equiv .$$

These twelve symbols may be part of our original metalanguage L* or they may be defined symbols introduced into KPU as in § I.5. In applications, the latter is almost always the case.

We assume the following *axioms on syntax:*

(1) An axiom asserting that the classes of variables, function symbols, relation symbols, contant symbols are all disjoint, and that none of the six constants displayed above are in any of these classes.

(2) An axiom on variables which asserts, writing v_α for $\mathsf{v}(\alpha)$,

$$\alpha \neq \beta \Longrightarrow v_\alpha \neq v_\beta,$$
$$\mathsf{Variable}\,(x) \Longleftrightarrow \exists\alpha\,(x = v_\alpha)$$

(3) An axiom on #, which tells us the "arity" of relation and function symbols:

> if x is a relation or function symbol then $\#(x)$ is a positive natural number.

A set L is a *language* if L is a *set* of relation, function, and constant symbols.

The predicates "t is a term" and "t is a term of L" are defined by recursion on TC(t):

1.1 Definition. t is a *term* (*of* L)$\leftrightarrow t$ is a variable, or t is a constant symbol (in L), or $t = \langle \mathsf{h}, y\rangle$ where h is a function symbol (in L), $y = \langle y_1, \ldots, y_{\#(h)}\rangle$ and each y_i is a term (of L).

These two definitions ("t is a term", and "t is a term of L") are of the type permitted by I.6.6 so they define Δ predicates. (The only sticky point comes in checking that the predicates $P(y,n)$ iff "y is a sequence of length n" and $Q(y,n,x,i)$ iff "$P(y,n)$ and $1 \leq i \leq n$ and x is the ith term in the sequence y" are Δ predicates. This also follows from I.6.6 by recursion on n. For example, $P(y,n)$ iff n is a natural number ≥ 1 and, if $n > 1$ then there exist $z_1, z_2 \in \mathrm{TC}(y)$ such that $y = \langle z_1, z_2\rangle$ and $P(z_2, n-1)$.)

1.2 Definition. An *atomic formula* (of L) is a set of one of the following forms:

(i) $\langle \equiv, t_1, t_2\rangle$ where t_1, t_2 are terms (of L); we write $(t_1 \equiv t_2)$ or even $(t_1 = t_2)$.

(ii) $\langle \mathsf{r}, t_1, \ldots, t_n\rangle$ where r is a relation symbol (in L), $n = \#(\mathsf{r})$ and $t_1, \ldots, t_n$ are terms (of L); we write $\mathsf{r}(t_1, \ldots, t_n)$ for $\langle \mathsf{r}, t_1, \ldots, t_n\rangle$.

1.3 Definition. A set φ is a *finite formula* (of L) iff

φ is an atomic formula (of L), or
φ is $\langle \neg, \psi\rangle$ and ψ is a finite formula (of L), or
φ is $\langle \bigwedge, \{\psi, \theta\}\rangle$ or $\langle \bigvee, \{\psi, \theta\}\rangle$ where ψ, θ are finite formulas (of L), or
φ is $\langle \exists, v, \psi\rangle$ or $\langle \forall, v, \varphi\rangle$ where v is a variable and ψ is a finite formula (of L).

We write $\neg\psi$ for $\langle\neg,\psi\rangle$, $\psi\wedge\theta$ for $\bigwedge\{\psi,\theta\}$ and $\exists v\psi$ for $\langle\exists,v,\psi\rangle$; similarly for $\vee,\forall$. We use the usual abbreviations like $\varphi\rightarrow\psi$ for $((\neg\varphi)\vee\psi)$. All of the above predicates are Δ predicates, the last again by I.6.6.

1.4 Proposition. *If* Infinity *is true then for any language* L *there is a set*

$$\mathsf{L}_{\omega\omega}=\{\varphi\,|\,\varphi \textit{ a finite formula of } \mathsf{L} \textit{ with only variables of the form } v_n \textit{ occurring in } \varphi, n<\omega\}.$$

Proof. We first show that

$$\mathit{Terms}=\{t\,|\,t \text{ a term of } \mathsf{L} \text{ with only variables of the form } v_n \text{ in } t\}$$

is a set. Define

$$\mathit{Terms}(0)=\{\mathsf{c}\in\mathsf{L}\,|\,\mathsf{c} \text{ a constant symbol}\}\cup\{\mathsf{v}^{(n)}\,|\,n<\omega\},$$

$$\mathit{Terms}(n+1)=\{\langle\mathsf{h},t_1,\ldots,t_k\rangle\,|\,h\in\mathsf{L},\ \mathsf{h} \text{ a constant symbol}, k=\#(\mathsf{h}), t_1,\ldots,t_k\in\mathit{Terms}(n)\}\cup\mathit{Terms}(n).$$

by induction on n. This makes sense if ω exists, by replacement for $\mathit{Terms}(0)$, as does

$$\mathit{Terms}=\bigcup\nolimits_{n<\omega}\mathit{Terms}(n)\,.$$

A similar proof shows that $\mathsf{L}_{\omega\omega}$ is a set. $\square$

For the past twenty years, and more, logicians have been working to find manageable strengthenings of $\mathsf{L}_{\omega\omega}$. It has turned out that languages with expressions of infinite length are one of the best lines of attack. These languages allow us to form expressions like the following

$$\forall x \bigvee\nolimits_{n<\omega}[x\equiv\underbrace{\mathsf{h}(\mathsf{h}(\ldots(\mathsf{h}}_{n}(\mathsf{c})\ldots)]$$

which says that every element is of the form $c, h(c), h(h(c))$, etc.; or

$$\forall x \bigvee\nolimits_{\varphi\in\mathsf{L}_{\omega\omega}}[\varphi(x)\wedge\exists!y\,\varphi(y/x)]$$

which says that every x is definable by some finite formula; or $\varphi_\alpha(x)$ defined by recursion on α by:

$$\varphi_0(v_0) \quad \text{is} \quad \forall y\neg(y<v_0),$$

$$\varphi_\alpha(v_\alpha) \quad \text{is} \quad \forall y\,(y<v_\alpha\leftrightarrow\bigvee\nolimits_{\beta<\alpha}\varphi_\beta(y/v_\beta))\,.$$

Then $\varphi_\alpha(x)$ is going to be true in a linearly ordered structure iff the predecessors of x have order type exactly α.

1.5 Definition. A set φ is an *infinitary formula* if one of the following hold:

φ is a finite formula,
φ is $\neg\psi$ where ψ is an infinitary formula,
φ is $\exists v\psi$ or $\forall v\psi$ where v is a variable and ψ is an infinitary formula,
φ is $\langle\bigwedge,\Phi\rangle$ or φ is $\langle\bigvee,\Phi\rangle$ where Φ is a nonempty set of infinitary formulas.

Again this definition is justified by I.6.6. We write

$$\bigwedge\Phi \quad \text{for} \quad \langle\bigwedge,\Phi\rangle,$$
$$\bigvee\Phi \quad \text{for} \quad \langle\bigvee,\Phi\rangle;$$

$\bigwedge\Phi$ is called the *conjunction* of the formulas in Φ, $\bigvee\Phi$ the *disjunction*. The notion of infinitary formula of a language L is defined in a parallel way.

We assume that the reader can carry out all the syntactic definitions (free and bound variable, substitution of a term t for a free variable in φ, for example) only noting that substitution must be defined by recursion over TC(φ). We denote the result of substituting t for v in φ by $\varphi(t/v)$. A *sentence* is a formula with no free variables.

We define the set sub(φ) of *subformulas* of φ by recursion over TC as follows:

$$\begin{aligned} \mathrm{sub}(\varphi) &= \{\varphi\} && \text{if } \varphi \text{ is atomic} \\ &= \{\varphi\}\cup \mathrm{sub}(\psi) && \text{if } \varphi \text{ is } \neg\psi,\ \exists v\psi \text{ or } \forall v\psi \\ &= \{\varphi\}\cup \textstyle\bigcup_{\psi\in\Phi}\mathrm{sub}(\psi) && \text{if } \varphi \text{ is } \bigwedge\Phi \text{ or } \bigvee\Phi. \end{aligned}$$

1.6 Lemma. *If φ has a finite number of free variables so does any $\psi\in\mathrm{sub}(\varphi)$. In particular, if ψ is a subformula of some sentence then ψ has a finite number of free variables.*

Lemma 1.6 is proved by a routine induction on formulas, and motivates the following definition.

A *proper infinitary formula* is one with only a finite number of free variables. The notion of "proper infinitary formula" is a Δ notion, since

$$\begin{aligned} \varphi \text{ is proper iff } &\{v \mid v \text{ a free variable in } \varphi\} \text{ is finite} \\ \text{iff } &\{\alpha \mid v_\alpha \text{ is free in } \varphi\} \text{ is finite,} \end{aligned}$$

and the notion "a is a finite set of ordinals" is Δ by Exercise I.7.6. Since we will only be discussing proper infinitary formulas, we might just as well drop the adjective "proper" once and for all. We use the symbol $\mathsf{L}_{\infty\omega}$ to denote the class of all (proper) infinitary formulas of L.

1.7 Definition. A *structure* $\mathfrak{M}$ for a language L is a pair $\mathfrak{M}=\langle M,f\rangle$ such that, writing $x^{\mathfrak{M}}$ for $f(x)$ we have:

(i) M is a nonempty set,
(ii) f is a function with $\mathrm{dom}(f) = \mathsf{L}$,
(iii) $\mathsf{r} \in \mathsf{L}$ implies $\mathsf{r}^{\mathfrak{M}}$ is a subset of $M^{\#(\mathsf{r})}$,
(iv) $\mathsf{h} \in \mathsf{L}$ implies $\mathsf{h}^{\mathfrak{M}}$ is a function with domain $M^{\#(\mathsf{h})}$ and range $\subseteq M$,
(v) if $\mathsf{c} \in \mathsf{L}$ then $\mathsf{c}^{\mathfrak{M}} \in M$.

This too is a Δ predicate of $\mathfrak{M}$ and L.

An *assignment* in $\mathfrak{M}$ is a function s with $\mathrm{dom}(s)$ a finite set of variables and $\mathrm{rng}(s) \subseteq M$. Given a structure $\mathfrak{M}$ for L, a term t of L and an assignment s in $\mathfrak{M}$ with the variables of t contained in $\mathrm{dom}(s)$, we let $t^{\mathfrak{M}}(s)$ be the value of t in $\mathfrak{M}$ at s. This is defined by recursion:

$$\begin{aligned} t^{\mathfrak{M}}(s) &= \mathsf{c}^{\mathfrak{M}} && \text{if } t \text{ is the constant symbol } \mathsf{c} \\ &= s(v) && \text{if } t \text{ is the variable } v \\ &= \mathsf{h}^{\mathfrak{M}}(t_1^{\mathfrak{M}}(s), \ldots, t_k^{\mathfrak{M}}(s)) && \text{if } t \text{ is } \mathsf{h}(t_1, \ldots, t_k). \end{aligned}$$

Our next goal is to formalize the notion of satisfaction:

$$\mathfrak{M} \models \varphi[s]$$

where $\mathfrak{M}$ is a structure for L, φ is a formula of L and s is an assignment to the free variables of φ. In order to make the definition fit into the form of definition by recursion available to us, we have to be a little awkward. Since there is no set of all variables, there can't be a set of all assignments to $\mathfrak{M}$. There is, however, a Σ operation G such that for all L, all L structures $\mathfrak{M}$ and all $\varphi \in \mathsf{L}_{\infty\omega}$

$$G(\mathfrak{M}, \varphi) = \{s \mid s \text{ an assignment in } \mathfrak{M} \text{ with } \mathrm{dom}(s) = \text{free variables of } \varphi\}.$$

We outline the definition of G in Exercise 1.11. It is then a routine matter to define

$$\mathrm{Sat}_{\mathsf{L}}(\mathfrak{M}, \varphi) = \{s \in G(\mathfrak{M}, \varphi) \mid \mathfrak{M} \models \varphi[s]\}$$

for languages L, structures $\mathfrak{M}$ for L and formulas φ of $\mathsf{L}_{\infty\omega}$, by recursion over $\mathrm{TC}(\varphi)$. We give some of the clauses of this recursive definition:

$$\mathrm{Sat}_{\mathsf{L}}(\mathfrak{M}, \mathsf{r}(t_1, t_2)) = \{s \in G(\mathfrak{M}, \mathsf{r}(t_1, t_2)) \mid \langle t_1^{\mathfrak{M}}(s), t_2^{\mathfrak{M}}(s) \rangle \in \mathsf{r}^{\mathfrak{M}}\},$$

$$\mathrm{Sat}_{\mathsf{L}}(\mathfrak{M}, \neg\varphi) = \{s \in G(\mathfrak{M}, \neg\varphi) \mid s \notin \mathrm{Sat}_{\mathsf{L}}(\mathfrak{M}, \varphi)\},$$

$$\mathrm{Sat}_{\mathsf{L}}(\mathfrak{M}, \bigwedge\Phi) = \{s \in G(\mathfrak{M}, \bigwedge\Phi) \mid \text{for all } \varphi \in \Phi,\ s \restriction \text{Free-Var}(\varphi) \in \mathrm{Sat}_{\mathsf{L}}(\mathfrak{M}, \varphi)\},$$

$$\begin{aligned} \mathrm{Sat}_{\mathsf{L}}(\mathfrak{M}, \exists v\, \varphi) &= \{s \in G(\mathfrak{M}, \exists v\, \varphi) \mid \text{for some } x \in M,\ s \cup \{\langle v, x \rangle\} \in \mathrm{Sat}_{\mathsf{L}}(\mathfrak{M}, \varphi)\} \\ &\qquad \text{if } v \text{ is free in } \varphi \\ &= \mathrm{Sat}_{\mathsf{L}}(\mathfrak{M}, \varphi) \quad \text{if } v \text{ is not free in } \varphi. \end{aligned}$$

We can now define, for L-structures $\mathfrak{M}$, formulas φ of L, and assignments s, the predicate $\mathfrak{M}\models\varphi[s]$ by

$$\mathfrak{M}\models\varphi[s] \quad \text{iff} \quad s\restriction\text{Free-Var}(\varphi)\in\text{Sat}_{\mathsf{L}}(\mathfrak{M},\varphi)\,.$$

For sentences φ we have $\mathfrak{M}\models\varphi$ if the empty function 0 is in $\text{Sat}_{\mathsf{L}}(\mathfrak{M},\varphi)$. Since $\text{Sat}_{\mathsf{L}}(\mathfrak{M},\varphi)$ is a Σ operation, by I.6.4, we see that $\mathfrak{M}\models\varphi[s]$ *is a* Δ *predicate of* $\mathfrak{M}, \varphi, s$ and the suppressed L. (Note that L can be recovered from $\mathfrak{M}$; $\mathsf{L}=\text{dom}(2^{\text{nd}}\mathfrak{M})$.) If the free variables of φ are among $v_1,\ldots,v_n$ we write

$$\mathfrak{M}\models\varphi[a_1,\ldots,a_n]$$

for $\mathfrak{M}\models\varphi[s]$ where $s=\{\langle v_1,a_1\rangle,\ldots,\langle v_n,a_n\rangle\}$.

Given structures $\mathfrak{M},\mathfrak{N}$ for a language L we write

$$\mathfrak{M}\equiv\mathfrak{N}\ (\mathsf{L}_{\omega\omega}) \quad \text{if, for all finite sentences } \varphi, \quad \mathfrak{M}\models\varphi \text{ iff } \mathfrak{N}\models\varphi; \quad \text{and}$$

$$\mathfrak{M}\equiv\mathfrak{N}\ (\mathsf{L}_{\infty\omega}) \quad \text{if, for all sentences } \varphi \text{ of } \mathsf{L}_{\infty\omega}, \quad \mathfrak{M}\models\varphi \text{ iff } \mathfrak{N}\models\varphi\,.$$

As written both of these are Π_1 predicates of $\mathfrak{M},\mathfrak{N}$ (and L). By Proposition 1.1, $\mathsf{L}_{\omega\omega}$ is a set if Infinity holds; in fact the operation which takes L to the set $\mathsf{L}_{\omega\omega}$ is a Σ operation on L. Thus, in the presence of the axiom of infinity we can rewrite $\mathfrak{M}\equiv\mathfrak{N}\ (\mathsf{L}_{\omega\omega})$ to see that it is an absolute predicate of $\mathfrak{M},\mathfrak{N}$ and L. In the presence of Σ_1 Separation $\mathfrak{M}\equiv\mathfrak{N}\ (\mathsf{L}_{\infty\omega})$ also becomes absolute, but for entirely different reasons. More on that in Chapter VII.

1.8—1.12 Exercises. Work in KPU

1.8. Show that a^n ($=a\times a\times\cdots\times a$, n-times) is a Σ operation of a and n.

1.9. Prove that the predicate Q, defined in the parenthetical remark following Definition 1.1, is indeed a Δ predicate.

1.10. Show that the operation $a\mapsto\text{card}(a)$, defined on finite sets a of variables, is a Σ operation. [Use the collapsing lemma.]

1.11. Using 1.10 and the Σ operation

$$S(a,n)=\{b\subseteq a\,|\,\text{card}(b)=n\}\,,$$

show that the operation G used above is a Σ operation.

1.12. Write out the few remaining details needed for the definition of $\text{Sat}_{\mathsf{L}}(\mathfrak{M},\varphi)$.

2. Consistency Properties

There is a very general method for constructing models which has evolved into what Keisler [1971] calls the "Model Existence Theorem". We will prove this theorem here in KPU + Infinity. To prove it in this weak metatheory we must be a little more careful than usual. Among notions which are equivalent in ZF we must choose those which avoid unnecessary uses of Power and Choice. This explains why our presentation must diverge in minor ways from Keisler's.

The collection of infinitary formulas of a language L never forms a set but we must usually deal with a set of formulas. Hence the next definition. We repeat, for emphasis, that *we work in* KPU + Infinity *in this section.*

2.1 Definition. Let L be a language. A *fragment* of $\mathsf{L}_{\infty\omega}$ is a set L_A of infinitary formulas and variables such that

(i) every finite formula of $\mathsf{L}_{\omega\omega}$ is in L_A,
(ii) if $\varphi \in \mathsf{L}_A$ then every subformula and variable of φ is in L_A,
(iii) if $\varphi(v) \in \mathsf{L}_A$ and t is a term of L all of whose variables lie in L_A then $\varphi(t/v)$ is in L_A, and
(iv) if φ, ψ, v are in L_A so are

$$\neg\varphi,\ \sim\varphi,\ \exists v\,\varphi,\ \forall v\,\varphi,\ \varphi \vee \psi,\ \varphi \wedge \psi\,.$$

At this stage, the subscript A serves merely as an index. It will serve a more useful purpose later.

We have used an undefined notion in 2.1, a silly technical device $\sim\varphi$. It is defined by:

$$\begin{array}{lll} \sim\varphi & \text{is} & \neg\varphi \quad \text{if } \varphi \text{ is atomic,} \\ \sim(\neg\varphi) & \text{is} & \varphi, \\ \sim(\bigwedge\Phi) & \text{is} & \bigvee_{\varphi\in\Phi}\neg\varphi, \\ \sim(\bigvee\Phi) & \text{is} & \bigwedge_{\varphi\in\Phi}\neg\varphi, \\ \sim(\exists v\,\varphi) & \text{is} & \forall v\,\neg\varphi, \\ \sim(\forall v\,\varphi) & \text{is} & \exists v\,\neg\varphi\,. \end{array}$$

We see that $\sim$ has an explicit Σ definition, not a recursive one, since $\sim$ does not occur on the right hand side of the above. Note that $\sim\varphi$ is logically equivalent to $\neg\varphi$. (Keisler uses $\varphi\neg$ for our $\sim\varphi$.)

Let K be a language and $C = \{\mathsf{c}_n : n < \omega\}$ a countable set of constant symbols not in K. We keep K, C and $\mathsf{L} = \mathsf{K} \cup C$ fixed for the rest of this section.

If K_A is a fragment of $\mathsf{K}_{\infty\omega}$ then there is a natural fragment $\mathsf{L}_A = \mathsf{K}_A(C)$ of $\mathsf{L}_{\infty\omega}$ associated with it; namely the set of all formulas of the form $\varphi(\mathsf{c}_{i_1}, \ldots, \mathsf{c}_{i_n})$ which result by replacing a finite number of free variables by constants from C. Fix these fragments K_A and $\mathsf{L}_A = \mathsf{K}_A(C)$ for the rest of this section. A term t of L_A is *basic* if it is in C or if it's of the form $\mathsf{h}(\mathsf{c}_{i_1}, \ldots, \mathsf{c}_{i_n})$ for $\mathsf{h} \in \mathsf{K}$ and the c_i's in C.

Next comes the cumbersome but crucial definition.

2.2 Definition. A *consistency property* for L_A is a set S of sets s such that each $s \in S$ is a set of sentences of L_A and such that all the following hold for every $s \in S$:

(C0) (*Triviality rule*) $0 \in S$; if $s \subseteq s' \in S$ then $s \cup \{\varphi\} \in S$ for each $\varphi \in s'$.
(C1) (*Consistency rule*) If $\varphi \in s$ then $\neg\varphi \notin s$.
(C2) ($\neg$-*rule*) If $\neg\varphi \in s$ then $s \cup \{\sim\varphi\} \in S$.
(C3) ($\bigwedge$-*rule*) If $\bigwedge \Phi \in s$ then for all $\varphi \in \Phi$, $s \cup \{\varphi\} \in S$.
(C4) ($\forall$-*rule*) If $(\forall v\, \varphi(v)) \in s$ then for each $\mathsf{c} \in C$, $s \cup \{\varphi(\mathsf{c}/v)\} \in S$.
(C5) ($\bigvee$-*rule*) If $\bigvee \Phi \in s$ then for some $\varphi \in \Phi$, $s \cup \{\varphi\} \in S$.
(C6) ($\exists$-*rule*) If $(\exists v\, \varphi(v)) \in s$ then for some $\mathsf{c} \in C$, $s \cup \{\varphi(\mathsf{c}/v)\} \in S$.
(C7) (*Equality rules*). Let t be a basic term of L_A and $\mathsf{c}, \mathsf{d} \in C$.

i) If $(\mathsf{c} \equiv \mathsf{d}) \in s$ then $s \cup \{(\mathsf{d} \equiv \mathsf{c})\} \in S$.
ii) If $\varphi(t), (\mathsf{c} \equiv t) \in s$ then $s \cup \{\varphi(\mathsf{c})\} \in S$.
iii) For some $\mathsf{e} \in C$, $s \cup \{\mathsf{e} \equiv t\} \in S$.

The rule (C0) was not included in Keisler's definition. It really is a triviality though.

2.3 Lemma. *If S satisfies all of 2.2 except* (C0) *then there is a smallest consistency property* $S' \supseteq S$.

Proof. Define

$$f(0) = S \cup \{0\},$$
$$f(n+1) = f(n) \cup \{s \cup \{\varphi\} \mid s \in f(n) \wedge \exists s' \in f(n)[s \subseteq s' \wedge \varphi \in s']\},$$
$$S' = \bigcup_{n<\omega} f(n).$$

This is easily seen to be a consistency property. If $S \subseteq S''$ and S'' is a consistency property then $f(n) \subseteq S''$ by induction on n. ☐

2.4 Lemma. *Let S be a consistency property, $s \in S$.*

i) $\varphi, (\varphi \to \psi) \in s$ *implies* $s \cup \{\psi\} \in S$.
ii) $\mathsf{c} \in C$ *implies* $s \cup \{(\mathsf{c} \equiv \mathsf{c})\} \in S$.
iii) $\mathsf{c}, \mathsf{d}, \mathsf{e} \in C$, $(\mathsf{c} \equiv \mathsf{d}) \in s$, $(\mathsf{d} \equiv \mathsf{e}) \in s$ *implies* $s \cup \{\mathsf{c} \equiv \mathsf{e}\} \in S$.

Proof. These are all similar. Assume $\varphi, (\varphi \to \psi) \in s$. Since $(\varphi \to \psi)$ is really $(\neg\varphi \vee \psi)$ we see by (C5) that either $s \cup \{\neg\varphi\} \in S$ or $s \cup \{\psi\} \in S$. The first possibility is ruled out by (C1). Next assume the hypothesis of (iii). By (C7i) we have $s' = s \cup \{\mathsf{e} \equiv \mathsf{d}\} \in S$. By (C7ii) we have $s' \cup \{\mathsf{c} \equiv \mathsf{e}\} \in S$ so, by (C0), $s \cup \{\mathsf{c} \equiv \mathsf{e}\} \in S$. To prove (ii) let $\mathsf{c} \in C$. By (C7iii) there is an $\mathsf{e} \in C$ with $s \cup \{\mathsf{c} \equiv \mathsf{e}\} \in S$. By (C7i), $s \cup \{\mathsf{c} \equiv \mathsf{e}, \mathsf{e} \equiv \mathsf{c}\} \in S$. Applying (iii) we have $s \cup \{\mathsf{c} \equiv \mathsf{e}, \mathsf{e} \equiv \mathsf{c}, \mathsf{c} \equiv \mathsf{c}\} \in S$, so, by (C0), $s \cup \{\mathsf{c} \equiv \mathsf{c}\} \in S$. ☐

The point of Definition 2.2 is that it exactly isolates the principles needed to carry out the "Henkin argument". To be more specific, it allows us to prove the Model Existence Theorem. A structure $\mathfrak{M}$ for L is a *canonical structure* if every element of $\mathfrak{M}$ is of the form $\mathsf{c}^{\mathfrak{M}}$ for some $\mathsf{c} \in C$.

2.5 Model Existence Theorem. *Let L_A be a countable fragment and let S be a consistency property for L_A. For every $s \in S$ there is a canonical structure $\mathfrak{M}$ for L such that $\mathfrak{M}$ is a model of s, i.e., for every $\varphi \in s$, $\mathfrak{M} \models \varphi$.*

Proof. We can't be quite as free wheeling as Keisler [1971, p. 13] since we have a rather limited metatheory. We have already taken care of most of the difficulties, though, by careful choice of definitions and by the wording of the theorem. Since L_A is countable we can enumerate its sentences:

$$\varphi_0, \varphi_1, \ldots, \varphi_n, \ldots \qquad (n < \omega) ;$$

and the terms occurring in L_A:

$$t_0, t_1, \ldots, t_n, \ldots \qquad (n < \omega) .$$

We shall construct a sequence

$$s_0 \subseteq s_1 \subseteq \cdots \subseteq s_n \subseteq \cdots$$

of elements of S as follows. We take s_0 to be the s of the theorem. Given s_n we define s_{n+1} by adding on one, two, or three sentences of L_A.

Step 1. Find the first constant symbol c of C, in the list of terms, such that $s_n \cup \{\mathsf{c} \equiv t_n\} \in S$ and let $s_n' = s_n \cup \{\mathsf{c} \equiv t_n\}$.

Step 2. If $s_n' \cup \{\varphi_n\} \notin S$ let $s_{n+1} = s_n'$. If $s_n' \cup \{\varphi_n\} \in S$ then let $s_n'' = s_n' \cup \{\varphi_n\}$.

There now are three distinct cases to consider, depending on the principal connective in φ_n.

Step 3. If φ_n does not begin with $\exists$ or $\bigvee$ let $s_{n+1} = s_n''$.

Step 4. If φ_n is $\exists v \psi$ then find the first $\mathsf{c} \in C$ in the list (2), by (C6), such that $s_n'' \cup \{\psi(\mathsf{c}/v)\} \in S$ and let s_{n+1} be this element of S.

Step 5. If φ_n is $\bigvee \Phi$ then use (C5) to find the least $\psi \in \Phi$, least in the list (1), such that $s_n'' \cup \{\psi\} \in S$ and let s_{n+1} be this element of S.

Now let $s_\omega = \bigcup_{n<\omega} s_n$. The rest of the proof is exactly as in Keisler [1971]. We define an equivalence relation on C by

$$\mathsf{c} \approx \mathsf{d} \quad \text{iff} \quad (\mathsf{c} \equiv \mathsf{d}) \in s_\omega$$

and let $M = \{c/\approx : c \in C\}$. By (C7), if $\varphi(c_1, \ldots, c_n) \in s_\omega$ and $c_i \approx d_i$ then $\varphi(d_1, \ldots, d_n) \in s_\omega$. This tells us how to interpret the relation and function symbols of L:

$$\langle c_1/\approx, \ldots, c_n/\approx \rangle \in r^{\mathfrak{M}} \quad \text{iff} \quad r(c_1, \ldots, c_n) \in s_\omega,$$

$$h^{\mathfrak{M}}(c_1/\approx, \ldots, c_n/\approx) \text{ is } d/\approx \text{ for that } d \text{ such that } (h(c_1, \ldots, c_n) \equiv d) \in s_\omega.$$

A simple proof by induction on formulas of L_A shows that $\mathfrak{M} \models \varphi$ for every $\varphi \in s_\omega$. One uses the properties (C0)—(C7) of course. □

2.6 Extended Model Existence Theorem. *Let L_A and S be as in the model existence theorem. If T is a set of sentences of L_A such that*

$$s \in S, \ \varphi \in T \quad \textit{implies} \quad s \cup \{\varphi\} \in S,$$

then for any $s \in S$, $T \cup s$ has a canonical model.

Proof. Let $S' = \{T \cup s : s \in S\}$. While S' is not a consistency property, it almost is one. It satisfies (C1)—(C7) so we can apply Lemma 2.3 to get a consistency property $S'' \supset S'$. Apply the Model Existence Theorem to $(T \cup s) \in S''$. □

Note, in passing, that canonical structures for countable fragments are countable structures.

2.7—2.8 Exercises

2.7. Prove, in KPU + Infinity, that if φ is an infinitary sentence with sub(φ) countable then there is a countable fragment L_A with $\varphi \in L_A$.

2.8. Use the Model Existence Theorem to show that if L_A is a countable fragment and if $\varphi \in L_A$ has a model then it has a countable model.

2.9 Notes. A history of the Model Existence Theorem can be found in the Preface and Lecture 3 of Keisler [1971].

The completeness and compactness theorems of § 5 cannot be proved in KPU + Infinity. The main reason for working in KPU + Infinity in this section is that it allows us to pinpoint the exact place where stronger principles are needed in these other theorems.

3. $\mathfrak{M}$-Logic and the Omitting Types Theorem

As a first, and important, application of the Model Existence Theorem, we prove a general version of the Henkin-Orey ω-Completeness Theorem. We then use the same proof to obtain the Omitting Types Theorem for countable fragments.

Let $\mathfrak{M}=\langle M,f\rangle$ be a structure for a countable language L and let L^+ be some language containing L, a unary relation symbol $\bar{\mathsf{M}}$, and for each $m\in M$ a constant symbol $\bar{\mathsf{m}}$, and possibly other symbols.

3.1 Definition. An $\mathfrak{M}$*-structure for* L^+ is a structure $\mathfrak{N}=\langle N,g\rangle$ satisfying

i) The interpretation of $\bar{\mathsf{M}}$ in $\mathfrak{N}$ is M;
ii) The interpretation of $\bar{\mathsf{m}}$ in $\mathfrak{N}$ is m, for all $m\in M$; and
iii) $\mathfrak{M}$ is a substructure of $\langle N,g\restriction\mathsf{L}\rangle$.

3.2 Examples. i) *ω-logic*: Let $M=\omega=\{0,1,\dots\}$ and $\mathfrak{M}=\langle M\rangle$. In this case $\mathfrak{M}$-logic is usually called *ω-logic*. Thus, in ω-logic one adds a symbol $\bar{\omega}(x)$, and constant symbols $\bar{0},\bar{1},\bar{2},\dots$ An ω-model is a structure $\mathfrak{N}$ with $\bar{\omega}^{\mathfrak{N}}=\omega$ and $\bar{n}^{\mathfrak{N}}=n$. To study ω-logic one adds the *ω-rule* to the usual rules of proof:

If you can prove $\varphi(\bar{n}/v_0)$ for each $n<\omega$ then conclude

$$\forall v_0[\bar{\omega}(v_0)\rightarrow\varphi(v_0)].$$

The $\mathfrak{M}$-rule below is the natural generalization of this.

ii) Let $\mathfrak{M}=\langle M,\text{---}\rangle$ be a fixed structure for L and let $\mathsf{L}^*=\mathsf{L}(\in,\dots)$ be as usual. Treat L^* as a single sorted language with a symbol U for the collection of urelements (as we have done from time to time). Among structures $\mathfrak{A}_{\mathfrak{N}}$ for L^* we want to single out those with $\mathfrak{N}=\mathfrak{M}$. Let $\bar{\mathsf{M}}$ be U and add a constant symbol $\bar{\mathsf{p}}$ for each $p\in M$. Let $\mathsf{L}^+=\mathsf{L}^*\cup\{\bar{\mathsf{p}}\mid p\in M\}$. An $\mathfrak{M}$-structure for L^+ has, by 3.1, the form

$$\langle M\cup A, M,\text{---},A,E,\dots\rangle$$

(with $\bar{\mathsf{p}}$ interpreted by p for $p\in M$), for some $A,E,\dots$. If it is a model of that part of KPU contained in the definition of L^*-structure (cf. I.2.6) then we can write it as $(\mathfrak{M};A,E,\dots)$. In particular, *any $\mathfrak{M}$-structure for L^+ which is a model of* KPU *has the form* $\mathfrak{A}_{\mathfrak{M}}=(\mathfrak{M};A,E,\dots)$, with $\bar{\mathsf{p}}$ interpreted by p for each $p\in M$.

These two examples are the most important ones but what we do in this section is entirely general.

Thus let $\mathfrak{M}$ be a structure for L and let L^+ be as described prior to Definition 3.1. We wish to find a set of axioms and rules which generate the finite sentences of L^+ which hold in all $\mathfrak{M}$-structures or, more generally, in all models of some theory T which are $\mathfrak{M}$-structures. We can do this only if $\mathfrak{M}$ is countable. Given a set of sentences T we write $T\models_{\mathfrak{M}}\varphi$ if φ holds in all $\mathfrak{M}$-structures which are models of T and $\models_{\mathfrak{M}}\varphi$ if φ is true in all $\mathfrak{M}$-structures.

The results of this section must be carried out in a set theory a little more powerful than KPU. Our metatheory is discussed in the notes for this section.

3.3 Axioms for $\mathfrak{M}$-logic. Let L and L^+ be as above.

i) $\bar{\mathsf{M}}(\bar{\mathsf{m}})$ is an axiom of $\mathfrak{M}$-logic, for all m in $\mathfrak{M}$.

ii) Every atomic or negated atomic sentence of $\mathsf{L}\cup\{\bar{\mathsf{m}}: m\in\mathfrak{M}\}$ true in $\mathfrak{M}$ is an axiom of $\mathfrak{M}$-logic.

iii) The usual axioms for $\mathsf{L}^+_{\omega\omega}$ are all axioms of $\mathfrak{M}$-logic.

3.4 Definition. Let T be a set of finitary sentences of L^+. A finite formula φ is a *consequence* of T by the $\mathfrak{M}$-rule, written

$$T \vdash_{\mathfrak{M}} \varphi ,$$

if φ is in the smallest set of formulas containing T and the axioms of $\mathfrak{M}$-logic and closed under the following rules:

i) (*Modus ponens*) If $T \vdash_{\mathfrak{M}} \varphi$ and $T \vdash_{\mathfrak{M}} (\varphi \to \psi)$ then $T \vdash_{\mathfrak{M}} \psi$.

ii) (*Generalization*) If $T \vdash_{\mathfrak{M}} (\varphi \to \psi(v_n))$ and v_n not free in φ then $T \vdash_{\mathfrak{M}} (\varphi \to \forall v_n \psi(v_n))$.

iii) ($\mathfrak{M}$-*rule*) If $T \vdash_{\mathfrak{M}} \varphi(\bar{\mathsf{m}}/v_0)$ for every $m \in \mathfrak{M}$ then $T \vdash_{\mathfrak{M}} \forall v_0(\bar{\mathsf{M}}(v_0) \to \varphi(v_0))$.

A sentence is provable by the $\mathfrak{M}$-rule, written $\vdash_{\mathfrak{M}} \varphi$, if $T \vdash_{\mathfrak{M}} \varphi$ for $T = 0$.

Notice that we have made no mention of the phrase "proof by the $\mathfrak{M}$-rule". Instead we gave an inductive definition of $T \vdash_{\mathfrak{M}} \varphi$ directly. A straightforward proof by induction shows that

$$T \vdash_{\mathfrak{M}} \varphi(v_1, \ldots, v_n) \quad \text{implies} \quad T \models_{\mathfrak{M}} \forall v_1, \ldots, \forall v_n \varphi(v_1, \ldots, v_n) .$$

If $\mathfrak{M}$ and L^+ are countable, then the converse holds. It is known as the ω-Completeness, or $\mathfrak{M}$-Completeness, Theorem.

A set of sentences T is *consistent in* $\mathfrak{M}$-*logic* if

$$T \vdash_{\mathfrak{M}} (\varphi \wedge \neg \varphi)$$

is false for all formulas φ of L^+. (Note the $\vdash_{\mathfrak{M}}$ as opposed to $\models_{\mathfrak{M}}$.)

3.5 $\mathfrak{M}$-Completeness Theorem. *Let L^+ and $\mathfrak{M}$ be countable and let T be a set of finitary sentences of L^+. If φ is a finite sentence of L^+ then*

$$T \vdash_{\mathfrak{M}} \varphi \quad \textit{iff} \quad T \models_{\mathfrak{M}} \varphi .$$

Proof. We can assume L^+ has a countable set $\{\mathsf{c}_n : n < \omega\}$ of constant symbols not in $\mathsf{L} \cup \{\bar{\mathsf{m}} : m \in \mathfrak{M}\}$ and not mentioned in T for otherwise we simply enlarge L^+ a little more. This enlargement would not enlarge the set of theorems of $\mathfrak{M}$-logic of the original L^+. Suppose φ is not a theorem of T in $\mathfrak{M}$-logic. Our goal is to construct, via the Model Existence Theorem, an $\mathfrak{M}$-structure $\mathfrak{N}$ which is a model of T and $\neg \varphi$.

Let L_A^+ be a countable fragment of $\mathsf{L}_{\infty\omega}^+$ with the sentence

$$\forall v_0 \bigvee_{m \in M} [\neg \bar{\mathsf{M}}(v_0) \vee v_0 \equiv \bar{\mathsf{m}}]$$

in L_A^+. Let S consist of all sets s of the form

$$s_0 \cup s_1$$

where s_0 is a set of finitary sentences of L^+ such that $T\cup s_0$ is consistent in $\mathfrak{M}$-logic, and s_1 is a finite set of infinitary sentences of the form

$$(1) \qquad \bigvee_{m\in M}[\neg\bar{\mathsf{M}}(\mathsf{c}_n)\vee \mathsf{c}_n\equiv\bar{\mathsf{m}}].$$

Note that $s=\{\neg\varphi\}\in S$ by hypothesis. The only nontrivial step in showing that S is a consistency property is to show that S satisfies the $\bigvee$-rule, (C5). Suppose $\bigvee\Phi\in s\in S$ where $s=s_0\cup s_1$ is partitioned as above. If $\bigvee\Phi$ is in s_0 then it is just a binary disjunction $\psi\vee\theta$. If neither ψ nor θ were consistent in $\mathfrak{M}$-logic with $T\cup s_0$ then $\neg\psi\wedge\neg\theta$ would be a consequence of $T\cup s_0$ in $\mathfrak{M}$-logic, hence $T_0\cup s_0$ would not be consistent, since $\psi\vee\theta\in s_0$. If $\bigvee\Phi\in s_1$ then it is of the form (1) for some $n<\omega$. We need to show that either

$$(2) \qquad T\cup s_0\cup\{\neg\bar{\mathsf{M}}(\mathsf{c}_n)\}$$

or one of

$$(3)_m \qquad T\cup s_0\cup\{(\mathsf{c}_n\equiv\bar{\mathsf{m}})$$

for some $m\in M$, is consistent in $\mathfrak{M}$-logic so that one of

$$s\cup\{\neg\bar{\mathsf{M}}(\mathsf{c}_n)\},\qquad s\cup\{\mathsf{c}_n\equiv\bar{\mathsf{m}}\}$$

is in S. Suppose that none of the $(3)_m$ are consistent in $\mathfrak{M}$-logic. Write $\psi(\mathsf{c}_n)$ for the conjunction of s_0. It follows that

$$T\vdash_{\mathfrak{M}}\neg\psi(\bar{\mathsf{m}})$$

for each $m\in M$, so, by the $\mathfrak{M}$-rule

$$T\vdash_{\mathfrak{M}}\forall v_0[\bar{\mathsf{M}}(v_0)\rightarrow\neg\psi(v_0)]$$

and hence

$$T\vdash_{\mathfrak{M}}\bar{\mathsf{M}}(\mathsf{c}_n)\rightarrow\neg\psi(\mathsf{c}_n).$$

Taking the contrapositive we get

$$T\vdash_{\mathfrak{M}}\bigwedge s_0\rightarrow\neg\bar{\mathsf{M}}(\mathsf{c}_n)$$

so $\neg\bar{\mathsf{M}}(\mathsf{c}_n)$ is a consequence of $T\cup s_0$ in $\mathfrak{M}$-logic, so (2) is consistent in $\mathfrak{M}$-logic. This completes the proof of (C5). Let $T'=T+$ all sentences of the form (1). Then if $s\in S$ and $\psi\in T'$ then $s\cup\{\psi\}\in S$. Thus by the Extended Model Existence Theorem, $T'\cup\{s\}$ has a canonical model whenever $s\in S$. But a canonical model of all of (1) is isomorphic to an $\mathfrak{M}$-structure so every $s\in S$ is true in some $\mathfrak{M}$-structure, in particular $s=\{\neg\varphi\}$. ☐

3.6 Corollary. *If $\mathfrak{M}$ and L^+ are countable then a theory T of L^+ has an $\mathfrak{M}$-structure for a model iff T is consistent in $\mathfrak{M}$-logic.* ☐

3.7 Corollary. *If $\mathfrak{M}$ and L^+ are countable and φ is a sentence of L^+ then $\vdash_{\mathfrak{M}} \varphi$ iff $\models_{\mathfrak{M}} \varphi$.* □

In applications in this book, L^+ will usually be as given in Example 3.2(ii) and T will usually be KPU or KPU$^+$.

The fraternal twin of the ω-Completeness Theorem is the so-called Omitting Types Theorem, a result which helps us construct models which "omit" elements not satisfying certain infinite disjunctions.

3.8 Omitting Types Theorem. *Let L_A be a countable fragment of $\mathsf{L}_{\infty\omega}$, and let T be a set of sentences of L_A which has a model. For each n let Φ_n be a set of formulas of L_A with free variables among $v_1, \ldots, v_{k_n}$. Assume that for each n and each formula $\psi(v_1, \ldots, v_{k_n})$ of L_A: if*

$$T + \exists v_1, \ldots, v_{k_n} \psi$$

has a model, so does

$$T + \exists v_1, \ldots, v_{k_n} (\psi \wedge \varphi)$$

for some $\varphi(v_1, \ldots, v_{k_n}) \in \Phi_n$. Given this hypothesis, there is a model $\mathfrak{M}$ of T such that for each $n < \omega$,

$$\mathfrak{M} \models \forall v_1, \ldots, v_{k_n} \bigvee_{\varphi \in \Phi_n} \varphi(v_1, \ldots, v_{k_n}).$$

Proof. A simple modification of the proof of the $\mathfrak{M}$-Completeness Theorem suffices. We first expand L to a language $\mathsf{L}' = \mathsf{L} \cup \{\mathsf{c}_n : n < \omega\}$. Let L_B be a countable fragment containing L_A and each of the sentences

$$\forall v_1, \ldots, v_{k_n} \bigvee_{\varphi \in \Phi_n} \varphi(v_1, \ldots, v_{k_n})$$

and let $\mathsf{L}'_A, \mathsf{L}'_B$ be the natural fragment of $\mathsf{L}'_{\infty\omega}$ associated with L_A and L_B as in § 2. Let S consist of all finite sets s of the form

$$s_0 \cup s_1$$

where s_0 is a finite set of sentences of L'_A with $T \cup s_0$ having a model and where s_1 is a finite set of sentences of the form

(1) $$\bigvee_{\varphi \in \Phi_n} \varphi(\mathsf{c}_{i_1}/v_1, \ldots, \mathsf{c}_{i_{k_n}}/v_{k_n}).$$

The proof that S is a consistency property is just like the proof in 3.5 except we use "has a model" for "consistent in $\mathfrak{M}$-logic". If $\varphi \in T$ or φ is of the form (1) then for each $s \in S$, $s \cup \{\varphi\} \in S$ so there is a canonical model $\mathfrak{M}$ of T and each of (1). The student who has trouble filling in the details is referred to Lecture 11 of Keisler [1971]. □

3.9 Exercise. Show how the $\mathfrak{M}$-completeness theorem can be derived from the Omitting Types Theorem.

3.10 Notes. The ω-Completeness Theorem goes back to Henkin [1954], [1957] and Orey [1956]. The extension of the Omitting Types Theorem to arbitrary countable fragments is due to Keisler [1971].

We cannot carry out the $\mathfrak{M}$-Completeness Theorem in KPU or KPU + Infinity. The problems arises from the definition of "the set of consequences f of T by the $\mathfrak{M}$-rule". This inductive definition would need something like Σ_1 Separation to justify it by proving that there *is* a smallest set of the kind described in 3.4. In Chapter VI we will step back and look at such inductive definitions.

4. A Weak Completeness Theorem for Countable Fragments

Let L_A be a fragment of $\mathsf{L}_{\infty\omega}$. A sentence φ of L_A is *valid*, written $\models\varphi$, if

$$\mathfrak{M}\models\varphi$$

for every structure $\mathfrak{M}$ for L. We would like to prove a generalization of the ordinary completeness theorem for L by showing that

$$\models\varphi \quad \text{iff} \quad \exists P\,[P \text{ is a proof of } \varphi]$$

for some notion of "proof". For such a result to be of any use there must be something "effective" about the notion of proof (otherwise we could take as proofs all valid sentences) and there should be a relation between L_A and the "size" of proofs of sentences in L_A.

We approach the notion of "proof" in a tentative fashion so that we can see exactly what it is that forces us to consider admissible fragments for the eventual result, Theorem 5.5.

After the brief respite of § 3 we return to use KPU + Infinity as our metatheory in this section.

4.1 Definition. Let L_A be a fragment of L. A set Γ of formulas of L_A is a *validity property* for L_A if Γ contains (A1)—(A7) below, is closed under (R1)—(R3), and does not contain $\varphi\wedge\neg\varphi$, for any $\varphi\in\mathsf{L}_A$.

(A1) Any instance of a tautology of finitary propositional logic.
(A2) $(\neg\varphi)\leftrightarrow(\sim\varphi)$.
(A3) $\bigwedge\Phi\rightarrow\varphi$, if $\varphi\in\Phi$.
(A4) $v_\alpha=v_\alpha$.
(A5) $v_\alpha=v_\beta\rightarrow v_\beta=v_\alpha$.
(A6) $\forall v\,\varphi(v)\rightarrow\varphi(t/v)$, t any term free for v in $\varphi(v)$.
(A7) $\varphi(v)\wedge v=t\rightarrow\varphi(t/v)$, t any term free for v in $\varphi(v)$.

(R1) (*Modus Ponens*). If φ and $(\varphi \to \psi)$ are in Γ so is ψ.
(R2) (*Generalization*). If $(\varphi \to \psi(v))$ is in Γ and v is not free in φ then $(\varphi \to \forall v\, \psi(v))$ is in Γ.
(R3) (*Conjunction*). If $\bigwedge \Phi \in \mathsf{L}_A$ and $(\psi \to \varphi)$ is in Γ for each $\varphi \in \Phi$, then $(\psi \to \bigwedge \Phi)$ is in Γ.

All formulas in the above are assumed to be elements of L_A.

4.2 Example. (i) *Let $\mathfrak{M}$ be a structure for L and let $\Gamma_{\mathfrak{M}}$ be the set of all $\varphi(v_1,\ldots,v_n) \in \mathsf{L}_A$ such that*

$$\mathfrak{M} \models \forall v_1,\ldots,v_n\, \varphi(v_1,\ldots,v_n).$$

$\Gamma_{\mathfrak{M}}$ is a set by Δ_1 Separation. It is clearly a validity property.
(ii) *If $\mathscr{X}$ is a set of validity properties then*

$$\bigcap \mathscr{X} = \bigcap \{\Gamma : \Gamma \in \mathscr{X}\}$$

is a validity property.

As one might guess from the way the definition and examples were given, we cannot prove (in our current metatheory KPU + Infinity) that there is a smallest validity property, though it is very instructive to try. It is also useful to think of the members of Γ as the "provable" formulas in the next lemma.

Let us fix for the rest of this section, a fragment L_A, a set $C = \{\mathsf{c}_n : n < \omega\}$ of new constant symbols, and let $\mathsf{K} = \mathsf{L} \cup C$ and $\mathsf{K}_A = \mathsf{L}_A(C)$ be the natural fragment of $\mathsf{K}_{\infty\omega}$ associated with L_A: $\varphi \in K_A$ iff there is a $\psi \in \mathsf{L}_A$ such that φ results from ψ by replacing some free variables by constant symbols throughout,

$$\varphi = \psi(\mathsf{c}_{i_1}/v_{\alpha_1},\ldots,\mathsf{c}_{i_k}/v_{\alpha_k}).$$

We say that φ is a *free substitution instance* of ψ. We have purposely interchanged the roles of K and L from § 2.

The following proposition allows us to apply the Extended Model Existence Theorem.

4.3 Proposition. *Let Γ_0 be a validity property for L_A and let Γ be the set of all free substitution instances of formulas in Γ_0.*
Define

$$S = \{s \mid s \text{ a finite set of } \mathsf{K}_A\text{-sentences, } (\neg \bigwedge s) \notin \Gamma\}$$

Then:
(i) *S is a consistency property for K_A.*
(ii) *if $\varphi \in \Gamma$, $s \in S$, then $s \cup \{\varphi\} \in S$.*

Proof. We first observe that Γ is a validity property for K_A. Two examples should suffice.

(A 7) Consider a sentence of K_A of the form

(1) $$\varphi(\mathsf{c}_1,\ldots,\mathsf{c}_n,v)\wedge v=t\rightarrow\varphi(\mathsf{c}_1,\ldots,\mathsf{c}_n,t/v),$$

where t also may contain $\mathsf{c}_1,\ldots,\mathsf{c}_n$. Now this is a free substitution of some formula of the form

(2) $$\varphi(w_1,\ldots,w_n,v)\wedge v=t'\rightarrow\varphi(w_1,\ldots,w_n,t'/v)$$

where $t=t'(\mathsf{c}_1/w_1,\ldots,\mathsf{c}_n/w_n)$. By (A 7) for Γ_0, (2) is in Γ_0 so (1) is in Γ.

(R 3) Suppose $(\psi\rightarrow\bigwedge\Phi)\in\mathsf{K}_A$ and that $(\psi\rightarrow\varphi)\in\Gamma$ for each $\varphi\in\Phi$. We need to see that $(\psi\rightarrow\bigwedge\Phi)\in\Gamma$. Now $(\psi\rightarrow\bigwedge\Phi)$ is a free substitution instance of some formula in L_A, say it is of the form

$$\psi(\mathsf{c}_1/v_1)\rightarrow\bigwedge\{\varphi(\mathsf{c}_1/v_1)\,|\,\varphi(v_1)\in\Phi_0\}.$$

Since $(\psi(\mathsf{c}_1)\rightarrow\varphi(\mathsf{c}_1))\in\Gamma$ it is a free substitution instance of, say,

$$\psi(\mathsf{c}_1/v_\beta)\rightarrow\varphi(\mathsf{c}_1/v_\beta)$$

where $\psi(v_\beta)\rightarrow\varphi(v_\beta)\in\Gamma_0$. But then using (R 1), (R 2) and (A 6) for Γ_0 we get $(\psi(v_1)\rightarrow\varphi(v_1))\in\Gamma_0$, for each $\varphi(v_1)\in\Phi_0$. By (R 3) for Γ_0, $\psi(v_1)\rightarrow\bigwedge\{\varphi(v_1)\,|\,\varphi(v_1)\in\Phi_0\}$ is in Γ_0, and hence $\psi\rightarrow\bigwedge\Phi$ is in Γ. We leave the other clauses to the student, and assume Γ is a validity property for K_A. The verification of (ii) is entirely routine, for suppose $\varphi\in\Gamma$, $s\in S$ but $s\cup\{\varphi\}\notin S$. Then $\neg(\bigwedge s\wedge\varphi)\in\Gamma$, so by (A 1) and (R 1), $(\varphi\rightarrow\neg\bigwedge s)$. But then, by (R 1), $\neg\bigwedge s\in\Gamma$, so $s\notin S$, which is a contradiction. The various cases in the verification that S is a consistency property are similar, with one slight twist for (C 6). Suppose $\exists v\,\varphi(v)\in s\in S$ but that for each $\mathsf{c}\in C$, $s\cup\{\varphi(\mathsf{c}/v)\}\notin S$. Hence $\bigwedge s\rightarrow\neg\varphi(\mathsf{c}/v)$ is in Γ for each $\mathsf{c}\in C$ and, in particular, for some c not appearing in S. Since c does not appear in S,

$$\bigwedge s\rightarrow\neg\varphi(v)$$

is also a free substitution instance of something in Γ_0, so it is in Γ, and hence, so are all the following:

$$\bigwedge s\rightarrow\forall v\neg\varphi(v)\quad\text{(by (R 2))},$$
$$\bigwedge s\rightarrow\neg\exists v\,\varphi(v)\quad\text{(by (R 1), (A 1), (A 2))},$$
$$\neg\bigwedge s\quad\text{(by (A 1), (R 1))}$$

which contradicts $s\in S$. The other clauses are left to the student. □

4.4 Definition. A sentence φ of L_A is a *theorem of* L_A if φ is in every validity property Γ for L_A.

A word of warning: The predicate

φ is a theorem of L_A

is Π_1 in KPU but not in general Δ_1 in KPU. Thus we cannot assert (in KPU) that there is a set of all theorems of L_A.

4.5 Weak Completeness Theorem for Countable Fragments. *Let L_A be a countable fragment. A sentence φ of L_A is valid iff it is a theorem of L_A.*

Proof. Assume φ is a theorem of L_A. Let $\mathfrak{M}$ be any model and let $\Gamma_{\mathfrak{M}}$ be as in Example 4.2 (i). $\Gamma_{\mathfrak{M}}$ is a validity property for L_A so $\varphi \in \Gamma_{\mathfrak{M}}$, i. e. $\mathfrak{M} \models \varphi$.

Now assume φ is not a theorem of L_A. Hence there is a validity property Γ_0 with $\varphi \notin \Gamma_0$. Let $\Gamma_0 \subseteq \Gamma$ and S be as in 4.3. Then $\{\neg\varphi\} \in S$ by 4.3 (ii) and S is a consistency property so $\neg\varphi$ has a model, by the Model Existence Theorem. □

The word "weak" in Theorem 4.5 is there because we still have no notions of proof compatable with L_A such that

(3) $\qquad \models \varphi \quad \text{iff} \quad \exists P\,[P \textit{ is a proof of } \varphi].$

All we have managed to do so far is replace one Π_1 notion ($\models \varphi$) with another Π_1 notion (φ is a theorem of L_A). We want a Δ_1 notion of proof so that line (3) gives a Σ_1 form for $\models \varphi$, and we want a proof of φ to be essentially the same "size" as the members of L_A.

4.6 Exercise. Define: *φ is a theorem of T* iff φ is in every validity property containing T as a subset. Show that if $T \subseteq \mathsf{L}_A$ where L_A is countable, then φ is a theorem of T iff every model of T is a model of φ.

4.7 Note. The weak completeness theorem is one form of the Karp Completeness Theorem for $\mathsf{L}_{\omega_1\omega}$.

5. *Completeness and Compactness for Countable Admissible Fragments*

In this section we prove the completeness theorem alluded to in the previous section. We have reduced the task to finding a suitable notion of L_A-*proof* to go along with the notion of theorem of L_A introduced in Definition 4.4.

The first notion of proof one thinks of in this setting is: an L_A-proof is a well-ordered sequence

$$\varphi_0, \varphi_1, \ldots, \varphi_\beta$$

such that each φ_α, for $\alpha \leqslant \beta$, is either an axiom (A 1)—(A 7) of L_A or is a consequence of earlier φ_γ's $(\gamma < \alpha)$ by one of the rules (R 1), (R 2) or (R 3). We can of course prove (given a strong enough metatheory) that φ is a theorem of L_A iff there is such a sequence with $\varphi_\beta = \varphi$. This notion of proof is too restrictive, however, and does not have the nice properties we need for applications. We need a notion which does not have the axiom of choice built into its very definition. There are many ways of doing this. We simply choose one.

5.1 Definition. An ordered pair P is an *infinitary proof* iff one of the following holds:

(A 1)—(A 7) $P = \langle n, \varphi \rangle$ where $1 \leqslant n \leqslant 7$ and φ is an axiom of $\mathsf{L}_{\infty\omega}$ by An of Definition 4.1.

(R 1) $P = \langle f, \psi \rangle$ where f is a function, $\mathrm{dom}(f) = \{0, 1\}$, $f(0)$ is an infinitary proof P_0 with $2^{\mathrm{nd}} P_0$ of the form $(\varphi \to \psi)$, and $f(1)$ is an infinitary proof P_1 with $2^{\mathrm{nd}} P_1 = \varphi$.

(R 2) $P = \langle P_0, (\varphi \to \forall v_\alpha \psi(v_\alpha)) \rangle$ where P_0 is an infinitary proof with $2^{\mathrm{nd}} P_0$ of the form $(\varphi \to \psi(v_\alpha))$ where v_α is not free in φ.

(R 3) $P = \langle f, (\psi \to \bigwedge \Phi) \rangle$ where f is a function with domain Φ such that for each $\varphi \in \Phi$, $f(\varphi)$ is a nonempty set of infinitary proofs, and for each $P_0 \in f(\varphi)$, $2^{\mathrm{nd}} P_0 = (\psi \to \varphi)$.

If P is an infinitary proof and $\varphi = 2^{\mathrm{nd}} P$ then P is said to be a *proof of* φ. See Fig. 5A, B, C.

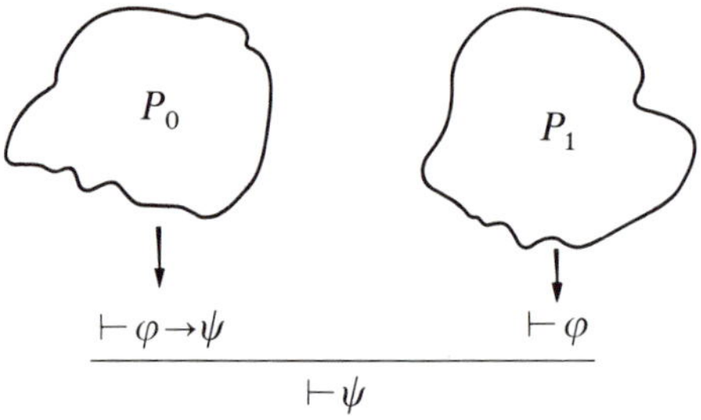

Fig. 5A. A proof P ending with an application of R1

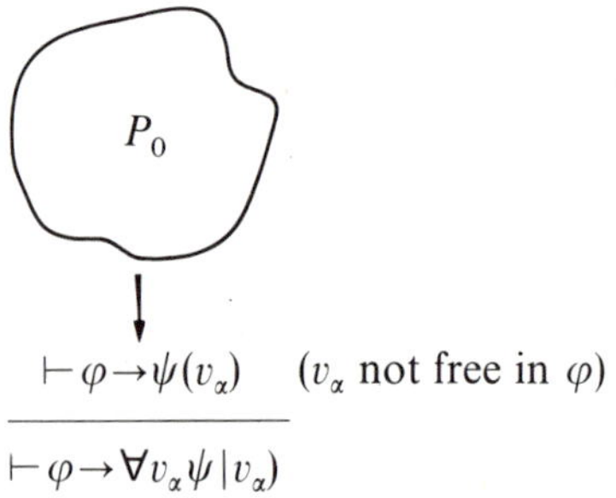

Fig. 5B. A proof P ending with an application of R2

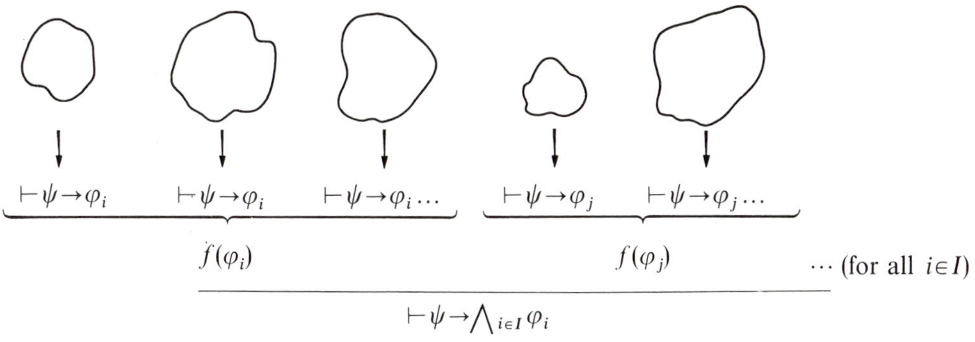

Fig. 5C. A proof P ending with an application of R3

Definition 5.1 can be given in KPU by recursion over TC(P) and consequently results in a Δ_1 predicate in KPU. Consequently,

$$\exists P\,[P \textit{ is an infinitary proof of } \varphi]$$

is a Σ_1 predicate of φ.

We now leave our weak metatheory and step out into the universe $\mathbb{V}_M$ of all sets on M (which is of course a "model" of KPU). It is the interplay between this universe and admissible sets which is of interest.

Let $\mathbb{A} = \mathbb{A}_{\mathfrak{M}}$ be an admissible set with constants $\bigwedge, \bigvee, \neg, \exists, \forall, \equiv$, and the predicates and functions ($\mathsf{v}, \#$) mentioned in § 1 satisfying the axioms on syntax found there. We can interpret the results of § 1 and Definition 5.1 in either $\mathbb{A}_{\mathfrak{M}}$ or $\mathbb{V}_{\mathfrak{M}}$. As long as we are dealing with Δ_1 notions, we know that the results are the same, by absoluteness. For example, we have, for $\varphi \in \mathbb{A}_{\mathfrak{M}}$

"φ is an infinitary formula"

true in $\mathbb{A}_{\mathfrak{M}}$ iff it is true (in $\mathbb{V}_{\mathfrak{M}}$). If, moreover, $\mathfrak{N} \in \mathbb{A}_{\mathfrak{M}}$ then $\mathfrak{N} \models \varphi$ holds in $\mathbb{A}_{\mathfrak{M}}$ iff it is true (in $\mathbb{V}_{\mathfrak{M}}$) and, if $P \in \mathbb{A}_{\mathfrak{M}}$ then

"*P is an infinitary proof of φ*"

is true interpreted in $\mathbb{A}_{\mathfrak{M}}$ iff it is true (in $\mathbb{V}_{\mathfrak{M}}$).

5.2 Definition. (i) If $\mathbb{A} = \mathbb{A}_{\mathfrak{M}} = (\mathfrak{M}; A, \in, \ldots)$ is admissible and L is a language which is Δ_1 on $\mathbb{A}$ then

$$\begin{aligned} \mathsf{L}_{\mathbb{A}} &= \{\varphi \in \mathbb{A} \mid \varphi \text{ is an infinitary formula of } \mathsf{L}_{\infty\omega}\} \\ &= \{\varphi \in \mathbb{A} \mid \mathbb{A} \models \text{"}\varphi \text{ is an infinitary formula of } \mathsf{L}_{\infty\omega}\text{"}\} \end{aligned}$$

is called the *admissible fragment* of $\mathsf{L}_{\infty\omega}$ given by $\mathbb{A}$.

(ii) If $P \in \mathbb{A}$ and P is an infinitary proof then P is said to be an $\mathsf{L}_{\mathbb{A}}$-*proof*.

It is a trivial matter to check that an admissible fragment $\mathsf{L}_{\mathbb{A}}$ really is a fragment in the sense of Definition 2.1.

5.3 Theorem. *Let $\mathsf{L}_{\mathbb{A}}$ be an admissible fragment of $\mathsf{L}_{\infty\omega}$ and let φ be a sentence of $\mathsf{L}_{\mathbb{A}}$. The following are equivalent:*

(i) $\exists P$ [*P is an $\mathsf{L}_{\mathbb{A}}$-proof of φ*],

(ii) $\exists P$ [*P is an infinitary proof of φ*],

(iii) *φ is a theorem of $\mathsf{L}_{\mathbb{A}}$, i.e., φ is in every validity property for $\mathsf{L}_{\mathbb{A}}$.*

(Warning: One can*not* in general add a (iv) asserting that φ is in every validity property *which is an element of* $\mathbb{A}$. This (iv) is usually much weaker than (iii).)

Proof. (i)$\Rightarrow$(ii) is trivial.

(ii)$\Rightarrow$(iii). Let Γ be any validity property for $\mathsf{L}_{\mathbb{A}}$. A routine proof by induction on TC(P) shows that if

$$P \text{ is a proof of } \varphi$$

then $\varphi \in \Gamma$ since Γ contains (A 1)—(A 7) and is closed under (R 1)—(R 3).

(iii)$\Rightarrow$(i). We need to show that the set

$$\Gamma = \{\psi \in \mathsf{L}_{\mathbb{A}} \mid \exists P \in \mathbb{A}\ (P \textit{ a proof of } \varphi)\}$$

is a validity property for $\mathsf{L}_{\mathbb{A}}$, for then $\varphi \in \Gamma$ since φ is in all validity properties. We need to see, then, that Γ contains (A 1)—(A 7) and is closed under (R 1)—(R 3). The first part is obvious so let us check (R 1) and (R 3), (R 2) being similar to (R 1).

(R 1) Suppose φ, $(\varphi \to \psi) \in \Gamma$. There are P_0, $P_1 \in \mathbb{A}$ with

$$2^{\text{nd}} P_0 = (\varphi \to \psi), \qquad 2^{\text{nd}} P_1 = \varphi.$$

Let $f(0) = P_0$, $f(1) = P_1$. Then $P = \langle f, \psi \rangle \in \mathbb{A}$ and P is a proof of ψ, hence $\psi \in \Gamma$.

(R 3) This is where admissibility and our careful choice of the notion of proof come into play. Suppose $(\psi \to \bigwedge \Phi) \in \mathsf{L}_{\mathbb{A}}$ and that $(\psi \to \varphi) \in \Gamma$ for all $\varphi \in \Phi$. Thus, for each $\varphi \in \Phi$ there is a $P \in \mathbb{A}$ such that P is a proof of $(\psi \to \varphi)$. Apply strong Σ Replacement in $\mathbb{A}$ to get a function $f \in \mathbb{A}$, $\text{dom}(f) = \Phi$ so that for each $\varphi \in \Phi$:

$$f(\varphi) \neq 0, \quad \text{and}$$

$$\text{if} \quad P \in f(\varphi) \quad \text{then } P \text{ is a proof of} \quad (\psi \to \varphi).$$

Then $\langle f, (\psi \to \bigwedge \Phi) \rangle \in \mathbb{A}$ and it is a proof of $(\psi \to \bigwedge \Phi)$. Thus $(\psi \to \bigwedge \Phi)$ is in Γ. $\square$

5.4 Corollary. *If $\mathsf{L}_{\mathbb{A}}$ is an admissible fragment then the set of theorems of $\mathsf{L}_{\mathbb{A}}$ is a Σ_1 subset of $\mathbb{A}$. Moreover, the Σ_1 definition has no parameters in it and is independent of $\mathbb{A}$.*

Proof. The Σ_1 formula is

$$\exists P\, [P \text{ is a proof of } \varphi]. \quad \square$$

Let us write

$$\vdash \varphi$$

for the Σ_1 formula

$$\exists P\, [P \text{ is a proof of } \varphi].$$

By combining 5.3 with the Weak Completeness Theorem of 4.5, we obtain the desired result.

5.5 Barwise Completeness Theorem. *Let $\mathsf{L}_{\mathbb{A}}$ be a countable admissible fragment. Then for all $\varphi \in \mathsf{L}_{\mathbb{A}}$, the following are equivalent:*
(i) $\models \varphi$,
(ii) $\vdash \varphi$,
(iii) $\mathbb{A}$ *satisfies* $\vdash \varphi$.
Thus the set of valid sentences of $\mathsf{L}_{\mathbb{A}}$ is Σ_1 on $\mathbb{A}$. $\square$

It is this completeness theorem which accounts for the tractable nature of countable, admissible fragments. It is used to prove many of the results in this book. Before going on though, we pause to point out one thing that the theorem most emphatically does *not* say, but which is sometimes mistaken for the conclusion of the theorem. It does not say that the sentence

$$\models \varphi \quad \text{iff} \quad \vdash \varphi$$

is true *in* the countable admissible set $\mathbb{A}$. This, together with 5.5, would imply that if $\varphi \in \mathsf{L}_{\mathbb{A}}$ and φ has a model then φ has a model $\mathfrak{N}$, $\mathfrak{N} \in \mathbb{A}$. This is false for most $\mathbb{A}$. See Exercises 5.11—5.14.

Now we turn to our first application of the completeness theorem.

5.6 Barwise Compactness Theorem. *Let $\mathsf{L}_{\mathbb{A}}$ be a countable admissible fragment of $\mathsf{L}_{\infty\omega}$. Let T be a set of sentences of $\mathsf{L}_{\mathbb{A}}$ which is Σ_1 on $\mathbb{A}$. If every $T_0 \subseteq T$ which is an element of $\mathbb{A}$ has a model, then T has a model.*

Proof. Expand L to $\mathsf{K} = \mathsf{L} \cup \{\mathsf{c}_n \mid n < \omega\}$ as usual but do it so that K remains Δ_1 on $\mathbb{A}$. Let $\mathsf{K}_{\mathbb{A}} = \mathsf{L}_{\mathbb{A}}(C)$ be the usual fragment of $\mathsf{L}_{\mathbb{A}}$ associated with K. Thus $\mathsf{K}_{\mathbb{A}}$ is the set of all sentences of $\mathsf{K}_{\infty\omega}$ which are elements of $\mathbb{A}$ and have only a finite number of c's in them. We use the Model Existence Theorem for $\mathsf{K}_{\mathbb{A}}$.

Let S be the set of all finite sets s of sentences of $\mathsf{K}_{\mathbb{A}}$ such that for all $T_0 \subseteq T$ with $T_0 \in \mathbb{A}$,

$$T_0 \cup s \quad \text{has a model.}$$

Note that if $s \in S$ and $\varphi \in T$ then $s \cup \{\varphi\} \in S$ so we are all set to apply the Extended Model Existence Theorem to get a model of T once we show that S is a consistence property. As usual it is (C 5) that causes the problems. So suppose $\bigvee \Phi \in s \in S$ but that for each $\varphi \in \Phi$, $s \cup \{\varphi\} \notin S$. Thus, for each $\varphi \in \Phi$ there is a $T_0 \subseteq T$, $T_0 \in \mathbb{A}$ such that

$$T_0 \cup s \cup \{\varphi\} \quad \text{has no model.}$$

Let $\theta(x)$ be the Σ_1 definition of T on $\mathbb{A}$. The following Σ sentence is true in $\mathbb{A}$ by the Completeness Theorem for $\mathsf{K}_{\mathbb{A}}$:

$$\text{(1)} \qquad \forall \varphi \in \Phi\, \exists T_0 \left[\forall \psi \in T_0\, \theta(\psi) \wedge \vdash (\bigwedge (T_0 \cup s) \rightarrow \neg \varphi)\right].$$

By Σ Reflection there is a set $a \in \mathbb{A}$ such that (1) holds relativized to a. We can assume a is transitive by I.4.2. Let

$$T_1 = \{\psi \in a \mid \theta^{(a)}(\psi)\}.$$

by Δ_0 Separation. Then $T_1 \in \mathbb{A}$, $T_1 \subseteq T$ by I.4.2 and, for each $\varphi \in \Phi$,

$$\models (\bigwedge (T_1 \cup s) \rightarrow \neg \varphi)$$

since there is some $T_0 \subseteq T_1$ with

$$\models \bigwedge (T_0 \cup s) \rightarrow \neg \varphi.$$

But then $s \cup T_1$ can have no model since $\bigvee \Phi \in s$. This contradicts the assumption that $s \in S$. $\square$

Combining Completeness and Compactness we obtain the following extension of the Completeness Theorem. We use $T \models \varphi$ to indicate that every model of T is a model of φ.

5.7 Extended Completeness Theorem. *Let $\mathsf{L}_{\mathbb{A}}$ be a countable admissible fragment. Let T be a set of sentences of $L_{\mathbb{A}}$ which is Σ_1 on $\mathbb{A}$. The set*

$$\{\varphi \in \mathsf{L}_{\mathbb{A}} : T \models \varphi\}$$

is Σ_1 on $\mathbb{A}$.

Proof. If φ is a sentence of L_A then

$$T \models \varphi \quad \text{iff} \quad \exists T_0 \in \mathbb{A}\left[T_0 \subseteq T \wedge T_0 \models \varphi\right]$$

by the Barwise Compactness Theorem (applied to $T \cup \{\neg \varphi\}$) so $T \models \varphi$ iff the following is true in $\mathbb{A}$, where $\theta(x)$ defines T,

$$\exists T_0 \left[\forall \psi \in T_0\, \theta(\psi) \wedge \vdash \bigwedge T_0 \rightarrow \varphi\right]$$

which gives a Σ definition of $T \models \varphi$. Note that it depends only on the definition θ of T, not on $\mathbb{A}$, and that is has only the same parameters occuring in it that occur in θ. $\square$

One peculiar instance of the Compactness Theorem deserves special mention because it comes up frequently. It applies, for example, to $\mathbb{A} = \mathbb{HYP}_{\mathfrak{M}}$ when $\mathfrak{M}$ is recursively saturated.

5.8 Theorem. *Let $\mathbb{A} = \mathbb{A}_{\mathfrak{M}}$ be a countable admissible set with $o(\mathbb{A}_{\mathfrak{M}}) = \omega$. Let T, T' be theories of $\mathsf{L}_{\mathbb{A}}$ which are Σ_1 on $\mathbb{A}_{\mathfrak{M}}$ such that every $\varphi \in T$ is a pure set. (Hence T is a set of finitary sentences.) If for each finite $T_0 \subseteq T$,*

$$T_0 \cup T' \quad \text{has a model,}$$

then $T \cup T'$ has a model.

Proof. If $T \cup T'$ has no model then, by the Compactness Theorem, $\models \neg \bigwedge \Phi$, for some $\Phi \in \mathbb{A}$, $\Phi \subseteq T \cup T'$. Now, if we write θ_1, θ_2 for the Σ_1 definitions of T and T', respectively, then we have $\forall x \in \Phi [\theta_1(x) \vee \theta_2(x)]$, so by Σ Reflection there is an $a \in \mathbb{A}$ such that $\forall x \in \Phi [\theta_1^{(a)}(x) \vee \theta_2^{(a)}(x)]$.

Thus, if we use Δ_0 Separation to form

$$\Phi_1 = \{x \in \Phi : \theta_1^{(a)}(x)\},$$

$$\Phi_2 = \{x \in \Phi : \theta_2^{(a)}(x)\}$$

then $\Phi = \Phi_1 \cup \Phi_2$, $\Phi_1 \subseteq T$, $\Phi_2 \subseteq T'$ and $\Phi_1 \cup \Phi_2$ has no model. But Φ_1 is a set of pure sets, hence a pure set, hence finite since $o(\mathbb{A}) = \omega$. $\square$

There is a question that often comes up. Let $\mathbb{A}_{\mathfrak{M}} = \mathbb{HF}_{\mathfrak{M}}$. To what extent does the Compactness Theorem 5.6 give us the full compactness theorem for $\mathsf{L}_{\mathbb{A}} = \mathsf{L}_{\omega\omega}$? In other words, how does the requirement that T be Σ_1 on $\mathbb{A}_{\mathfrak{M}}$ affect us. If $\mathbb{HF}_{\mathfrak{M}}$ is countable (i. e. if $\mathfrak{M}$ is countable) then 5.6 gives us the full compactness for $\mathsf{L}_{\omega\omega}$. For let T be any theory of $\mathsf{L}_{\omega\omega}$. T is Σ_1 on $(\mathfrak{M}; \mathbb{HF}_M, \in, T)$ which is admissible by Theorem II.2.1 so we can apply 5.6 to this admissible set.

5.9–5.14 Exercises

5.9. Define "P is a proof from axioms in T" parallel to Definition 5.1. You must build in a Σ_1 definition of T. Use this to prove the Extended Completeness Theorem and Compactness Theorem in one fell swoop, without using the Completeness Theorem. [You will need to use Exercise 4.6.]

5.10. Let $\mathsf{L}_{\mathbb{A}}$ be an admissible fragment. Show that if P is an $\mathsf{L}_{\mathbb{A}}$-proof then all formulas in the proof P are $\mathsf{L}_{\mathbb{A}}$-formulas.

5.11. Let $\mathbb{A}$, $\mathbb{B}$ be admissible sets with $\mathbb{A} \in \mathbb{B}$ and

$$\text{“}\mathbb{A} \text{ is countable”}$$

true in $\mathbb{B}$. Let $\mathsf{L}_{\mathbb{A}}$ be an admissible fragment given by $\mathbb{A}$. Show that

$$\mathbb{B} \models \text{“}\varphi \text{ is valid”}$$

iff

$$\mathbb{A} \models \exists P\, [P \text{ is a proof of } \varphi].$$

Conclude that if $\varphi \in \mathsf{L}_{\mathbb{A}}$ has a model then it has a model in $\mathbb{B}$.

5.12. Let $\mathbb{A}$ be admissible and satisfy the following:

(i) (locally countable). $\mathbb{A} \models \forall a$ (a is countable)

(ii) (recursively inaccessible). $\forall a \in \mathbb{A}$ there is an admissible $\mathbb{B} \in \mathbb{A}$ with $a \in \mathbb{B}$.

Show that for any sentence φ of $\mathsf{L}_{\mathbb{A}}$

$$(\models \varphi) \leftrightarrow (\vdash \varphi)$$

holds in $\mathbb{A}$. Show that $\mathbb{HC}_{\mathfrak{M}}$ is locally countable and recursively inaccessible. (Of course, for $\mathbb{HC}_{\mathfrak{M}}$ the conclusion is trivial since every countable structure for a countable language is isomorphic to a structure in $\mathbb{HC}_{\mathfrak{M}}$.)

5.13. Let $\mathbb{A}$ be a countable transitive model of ZFC (or enough of it to insure that $\mathbb{A}$ is admissible and prove that $\aleph_1$ exists). Let α be the ordinal of $\mathbb{A}$ which satisfies

$$\mathbb{A} \models \text{“}\alpha \text{ is the first uncountable ordinal”}.$$

Write a sentence φ of $\mathsf{L}_{\mathbb{A}}$ which asserts that α is countable. Thus φ has a model but does not have a model $\mathfrak{N} \in \mathbb{A}$. In other words, $\neg\varphi$ is valid in the sense of $\mathbb{A}$ but it is not provable since φ does indeed have a model.

5.14. Let α be the first admissible ordinal $>\omega$. Let $\mathbb{A} = \mathrm{L}(\alpha)$. Unlike the $\mathbb{A}$ in 5.13, this $\mathbb{A}$ is a model of

$$\forall a\, [a \text{ is countable}].$$

Find a sentence φ of $\mathsf{L}_{\mathbb{A}}$ which has a model but none in $\mathbb{A}$.

5.15 Notes. The completeness and compactness theorems of this section are due to Barwise [1967] and appeared in Barwise [1969]. The terminology “Barwise Completeness Theorem” and “Barwise Compactness Theorem” have become so standard that it would be false modesty (and confusing) to give them some other name here.

The observation that these theorems go through unchanged in the presence of urelements was first made in Barwise [1973], though it was really clear all along. The odd 5.8 first appears in Barwise [1973].

6. The Interpolation Theorem

The Interpolation Theorem is one of the results which holds for countable admissible fragments but not for arbitrary countable fragments; the proof requires the Completeness Theorem of § 5.

6.1 Theorem. *Let $\mathsf{L}_{\mathbb{A}}$ be a countable admissible fragment, φ, ψ sentences of $\mathsf{L}_{\mathbb{A}}$ such that*

$$\models \varphi \to \psi.$$

There is a sentence θ of $\mathsf{L}_{\mathbb{A}}$ whose relation, function and constant symbols are common to those of both φ and ψ such that

$$\models \varphi \to \theta \quad \textit{and} \quad \models \theta \to \psi.$$

Note. Equality is not treated as a relation symbol. It may appear in θ while appearing in only one of φ, ψ.

Proof. Let L^0 be the set of those symbols occurring in φ and let L^1 be the set of those symbols occurring in ψ. Let C be a countable set of new constant symbols coded as a Δ_1 subset of $\mathbb{A}$ and let $\mathsf{L}_{\mathbb{A}}(C)$ be the set of free substitution instances of formulas of $\mathsf{L}_{\mathbb{A}}$ by a finite number of symbols from C. Define $\mathsf{L}^0_{\mathbb{A}}(C)$ and $\mathsf{L}^1_{\mathbb{A}}(C)$ similarly. We define the consistency property S to be the set of all finite sets s of $\mathsf{L}_{\mathbb{A}}(C)$ which can be written as a union $s = s_0 \cup s_1$ satisfying the following conditions:

(1) s_0 is a set of sentences of $\mathsf{L}^0_{\mathbb{A}}(C)$, and similarly for s_1;

(2) If $\theta_0, \theta_1 \in \mathsf{L}^0_{\mathbb{A}}(C) \cap \mathsf{L}^1_{\mathbb{A}}(C)$ are such that $s_0 \models \theta_0$ and $s_1 \models \theta_1$, then the sentence $\theta_0 \wedge \theta_1$ has a model.

The verification that S is indeed a consistency property is routine (indeed, it is just like the $\mathsf{L}_{\omega_1\omega}$ case in Keisler [1971]) except for the $\bigvee$-rule, (C5). It is in the verification of this rule that we need the Completeness Theorem for $\mathsf{L}_{\mathbb{A}}(C)$. So suppose $s = s_0 \cup s_1$ is as above and that $\bigvee \Phi \in s$. Since the two cases are symmetric, we may assume that $\bigvee \Phi \in s_0$. We want to prove that for some $\sigma \in \Phi$,

$$s_0 \cup \{\sigma\} \cup s_1 \in S.$$

Suppose this is not the case. Then for every $\sigma \in \Phi$, there is a pair θ_0, θ_1 such that

(3) $\models \bigwedge s_0 \wedge \sigma \to \theta_0$, $\models \bigwedge s_1 \to \theta_1$, and $\models \neg(\theta_0 \wedge \theta_1)$, and the constants from C in θ_0 and θ_1 are in $s_0 \cup s_1$.

Let us indulge in a little wishful thinking and suppose that there are functions f, g which are *elements of our admissible set* $\mathbb{A}$ with $\mathrm{dom}(f) = \mathrm{dom}(g) = \Phi$ such that, for each $\sigma \in \Phi$, $\langle f(\sigma), g(\sigma) \rangle$ is a pair $\langle \theta_0, \theta_1 \rangle$ satisfying (3). Then we can let

$$\theta_0' = \bigvee \{f(\sigma) \mid \sigma \in \Phi\}, \quad \theta_1' = \bigwedge \{g(\sigma) \mid \sigma \in \Phi\}.$$

By Σ Replacement, θ'_0 and θ'_1 are elements of $\mathbb{A}$ so they are both sentences of the languages $\mathsf{L}^0_{\mathbb{A}}(C)$ and $\mathsf{L}^1_{\mathbb{A}}(C)$. Furthermore, $s_0 \models \theta'_0$, $s_1 \models \theta'_1$ and $\models \neg(\theta'_0 \wedge \theta'_1)$, which contradicts $s_0 \cup s_1 \in S$. But what about our bit of wishful thinking? At first it seems exactly that, since we have no choice principle holding in $\mathbb{A}$. Once again is it Strong Σ Replacement which which comes to the rescue. By the Completeness Theorem for $\mathsf{L}_{\mathbb{A}}(C)$, line (3) can be expressed by a Σ_1 formula. By Strong Σ Replacement there is a function $h \in \mathbb{A}$ with $\mathrm{dom}(h) = \Phi$ such that, for each $\sigma \in \Phi$, $h(\sigma)$ is a nonempty set of pairs $\langle \theta_0, \theta_1 \rangle$ satisfying (3). Define f and g by

$$f(\sigma) = \bigwedge \{1^{\mathrm{st}} h(\sigma) \mid \sigma \in \Phi\},$$

$$g(\sigma) = \bigwedge \{2^{\mathrm{nd}} h(\sigma) \mid \sigma \in \Phi\}.$$

Then $f, g \in \mathbb{A}$ and our wish has come true since $\theta_0 = f(\sigma)$ and $\theta_1 = g(\sigma)$ also satisfy (3). Thus S is a consistency property.

The conclusion of the theorem now follows easily from the observation that $\{\varphi, \neg\psi\} \notin S$. Just quantify out the finite number of new constant symbols in the sentence $\theta(\mathsf{c}_1, \ldots, \mathsf{c}_n)$ $(= \theta_0$ in the notation used above):

$$\models \varphi \to \forall v_1, \ldots, v_n \, \theta(v_1, \ldots, v_n),$$

$$\models \forall v_1, \ldots, v_n \, \theta(v_1, \ldots, v_n) \to \psi. \quad \square$$

We could use the interpolation theorem for $\mathsf{L}_{\mathbb{A}}$ to prove Beth's Theorem for $\mathsf{L}_{\mathbb{A}}$, but we will not be needing this result.

6.2 – 6.6 Exercises

6.2 (Hard). Let I_1, I_2 be interpretations of a language L^0 in consistent infinitary theories, T_1, T_2 formulated in languages L^1, L^2 respectively, and suppose that there are no two models $\mathfrak{M}_1$, $\mathfrak{M}_2$ of T_1, T_2 respectively such that $\mathfrak{M}_1^{-I_1} = \mathfrak{M}_2^{-I_2}$.

If L^0, L^1, L^2 are Δ_1 on the countable admissible $\mathbb{A}$, T_1, T_2 are Σ_1 theories of $\mathsf{L}^1_{\mathbb{A}}$ and $\mathsf{L}^2_{\mathbb{A}}$ and the interpretation I_1, I_2 are Σ_1 functions on $\mathbb{A}$ then there is a sentence φ of $\mathsf{L}^0_{\mathbb{A}}$ such that

$$\mathfrak{M}_1 \models T_1 \text{ implies } \mathfrak{M}_1^{-I_1} \models \varphi, \quad \text{for all } \mathsf{L}^1\text{-structures } \mathfrak{M}_1;$$

$$\mathfrak{M}_2 \models T_2 \text{ implies } \mathfrak{M}_2^{-I_2} \models \neg\varphi, \quad \text{for all } \mathsf{L}^2\text{-structures } \mathfrak{M}_2.$$

6.3. Let $\mathsf{L}_{\mathbb{A}}$, $\mathsf{K}_{\mathbb{A}}$ be countable admissible fragments and let I be an interpretation of L into a theory T of $\mathsf{K}_{\mathbb{A}}$, T and I being Σ_1 on $\mathbb{A}$. Let φ be a sentence of $\mathsf{K}_{\mathbb{A}}$ such that for all models $\mathfrak{N}_1$, $\mathfrak{N}_2$ of T with $\mathfrak{N}_1^{-I} = \mathfrak{N}_2^{-I}$, $\mathfrak{N}_1 \models \varphi$ iff $\mathfrak{N}_2 \models \varphi$, Then there is a $\psi \in \mathsf{L}_{\mathbb{A}}$ such that $T \vdash (\varphi \leftrightarrow \psi^I)$.

6.4. Let I be an interpretation of a complete theory T of $\mathsf{L}_{\omega\omega}$ in an incomplete theory T_1 of $\mathsf{K}_{\omega\omega}$. Show that if $\varphi \in \mathsf{K}_{\omega\omega}$ is not decided by T_1 then there are models $\mathfrak{M}$, $\mathfrak{N}$ of T with $\mathfrak{M} \models \varphi$, $\mathfrak{N} \models \neg\varphi$ and $\mathfrak{M}^{-I} = \mathfrak{N}^{-I}$.

6.5. Show that there are models $\mathfrak{N}_1=\langle N,+,x_1\rangle$, $\mathfrak{N}_2=\langle N,+,x_2\rangle$ of Peano arithmetic, with the same integers and addition, but $\mathfrak{N}_1 \not\equiv \mathfrak{N}_2$.

6.6. Show that there are models $\mathfrak{M}=\langle M,E\rangle$, $\mathfrak{N}=\langle N,F\rangle$ of ZF+V=L with the same ordinals, $\langle \mathrm{Ord}^{\mathfrak{M}}, E\restriction \mathrm{Ord}^{\mathfrak{M}}\rangle = \langle \mathrm{Ord}^{\mathfrak{N}}, F\restriction \mathrm{Ord}^{\mathfrak{N}}\rangle$, but with different sets of hereditarily finite sets. [Use Gödel's Incompleteness Theorem, 6.4 and the fact that every true sentence about the ordinals (with $<$) is provable in ZF.]

6.7 Notes. The interpolation theorem for $\mathsf{L}_{\omega\omega}$ is due to Craig [1957]. For the full $\mathsf{L}_{\omega_1\omega}$ it is due to Lopez-Escobar [1965]. Theorem 6.1 is due to Barwise [1969]. Exercise 6.2, a generalization of 6.1, is useful in abstract logic and is due to Barwise [1973]. References for the other exercises can be found there.

7. *Definable Well-Orderings*

In this section we prove a model theoretic result which will have applications to $\mathbb{HYP}_{\mathfrak{M}}$ in Chapter IV. The basic question is: What ordinals can be define in an admissible fragment? We solve this problem here for countable fragments. The uncountable case is taken up in Chapters VII and VIII.

7.1 Example. Define $\theta_\alpha(x)$ by recursion over α as follows:

$$\theta_0(x) \quad \text{is} \quad \forall y \neg (y<x),$$

$$\theta_\alpha(x) \quad \text{is} \quad \forall y\,(y<x \leftrightarrow \bigvee_{\beta<\alpha}\theta_\beta(y/x)).$$

Let $\mathfrak{M}=\langle M,<\rangle$ be a linear ordering. A simple proof by induction shows that if $\mathfrak{M}\models\theta_\alpha[x]$ then $\{y\in\mathfrak{M} \mid y<x\}$ is well ordered and has order type α. Hence $\mathfrak{M}\models\forall x \bigvee_{\beta<\alpha}\theta_\beta(x)$ iff $\mathfrak{M}$ is well ordered and has order type $\leqslant\alpha$. Thus, if we set σ_α equal to

$$\forall x \bigvee_{\beta<\alpha}\theta_\beta(x) \wedge \bigwedge_{\beta<\alpha} \exists x\, \theta_\beta(x)$$

then $\mathfrak{M}\models\sigma_\alpha$ iff $\mathfrak{M}$ has order type exactly α. □

The formulas from Example 7.1 were defined by recursion on α so the definition can be phrased as a Σ recursion in KPU. Hence, if α is in an admissible set $\mathbb{A}$ then $\varphi_\alpha(x)$, $\sigma_\alpha\in\mathsf{L}_{\mathbb{A}}$ (as long as the symbol $<$ is in L and $\mathbb{A}$).

7.2 Definition. Let L have a binary symbol $<$ and possible other symbols. A sentence $\varphi(<)$ of an admissible fragment $\mathsf{L}_{\mathbb{A}}$ *pins down the ordinal* α if

(i) $\mathfrak{N}\models\varphi$ implies $<^{\mathfrak{N}}$ is a well-ordering of its field.

(ii) φ has a model $\mathfrak{N}$ with $<^{\mathfrak{N}}$ of order type α.

A theory T of $\mathsf{L}_{\mathbb{A}}$ pins down α if $\bigwedge T$ pins down α.

The example above shows that every ordinal in the admissible set A can be pinned down by a sentence of $\mathsf{L}_{\mathbb{A}}$; in fact,

$$(\text{“} < \textit{ is a linear ordering”}) \wedge \sigma_\alpha$$

has only models of order type α. For countable admissible fragments, no other ordinal can be pinned down, as we show below.

For uncountable admissible fragments one can often pin down an ordinal $\alpha > o(\mathbb{A})$ (though one cannot give an explicit definition like σ_α above). The least ordinal which cannot be pinned down plays a key role in the model theory of uncountable fragments. We will go into this further in Chapter VII.

Note that if $\varphi(<_1)$ pins down α then $\psi(<_1, <, \mathsf{f})$ defined by

$$\varphi(<_1) \wedge \text{“}\mathsf{f} \textit{ maps } < \textit{ into } <_1 \textit{ in an order preserving fashion”},$$

as a sentence about $<$, pins down all ordinals $\leqslant \alpha$. Thus we can always work with sentences which pin down an initial segment of the ordinals. We will use this implicitly in the proof of the next theorem and several times in Chapter VII.

7.3 Theorem. *Let $\mathsf{L}_{\mathbb{A}}$ be a countable admissible fragment, $\varphi(\prec)$ a sentence which pins down ordinals. There is an ordinal α in the admissible set $\mathbb{A}$ such that every ordinal pinned down by φ is less than α.*

Proof. Suppose, to prove the contrapositive, that for every $\alpha \in \mathbb{A}$, $\varphi(\prec)$ has a model $\mathfrak{N}$ with $\prec^{\mathfrak{N}}$ of order type α. We prove that $\varphi(\prec)$ has a model $\mathfrak{N}$ where $\prec^{\mathfrak{N}}$ is not well ordered. It is instructive to split into cases, though not really necessary.

Case 1. $o(\mathbb{A}) = \omega$. If $\mathbb{A} = \mathbb{HF}_{\mathfrak{M}}$ then $\varphi(\prec)$ is just a finitary sentence and the result is well known to follow from the compactness theorem. But even if $\mathbb{A} \neq \mathbb{HF}_{\mathfrak{M}}$, the proof goes through. Let $<$, $\mathsf{c}_0, \mathsf{c}_1, \ldots$ be new symbols in the pure part of $\mathbb{A}$ and let T be the theory

$$\mathsf{c}_{n+1} \prec \mathsf{c}_n \; (n = 0, 1, 2, \ldots).$$

Let ψ be

$$\varphi(\prec) \wedge \text{“}\mathsf{f} \textit{ maps } < \textit{ order preserving into } \prec\text{”}.$$

We need only show $T + \psi$ has a model. Since T is a set of pure sets and $o(\mathbb{A}) = \omega$ we need (by 5.8) only see that ψ is consistent with every finite subset of T, which is obvious.

Case 2. $o(\mathbb{A}) > \omega$. The basic idea is the same, but we must work a little harder. Let $a_0 = \mathrm{TC}(\varphi)$. We may assume that $\mathbb{A}$ is the smallest admissible set with a_0 as an element. Introduce the following new symbols into L: $\in$, unary U, S, N for urelement, set and member of N respectively, a function symbol f, a constant c and, for each $x \in \mathbb{A}$ a constant $\overline{\mathsf{x}}$. Let T be the following set of sentences of $\mathsf{L}_{\mathbb{A}}$:

(0) "U, S, N *are disjoint and their union is everything*",
(1) φ^{N}, the relativization of $\varphi(\prec)$ to N,
(2) KPU formulated in terms of U, S, $\in$,
(3) diagram ($\mathbb{A}$),
(4) $\forall x[x \in \bar{\mathsf{a}} \rightarrow \bigvee_{b \in a} x \equiv \bar{\mathsf{b}}]$, for each $a \in \mathbb{A}$,
(5) "$\bar{\beta} \in \mathsf{c} \wedge \mathsf{c}$ *is an ordinal*", for each $\beta < o(\mathbb{A})$,
(6) $\forall x \leqslant \mathsf{c} \bigvee_{\psi \in \mathbf{KPU}} \neg \psi^{\mathsf{L}(a_0, x)}$,
(7) "f *maps the* $\in$ *predecessors of* c *into* $\prec$ *so that*

$$x < y \Rightarrow \mathsf{f}(x) \prec \mathsf{f}(y)\text{''}.$$

This theory is Σ_1 (in fact Δ_0) on $\mathbb{A}$ so we can apply the Compactness Theorem. If $T_0 \subseteq T$, $T_0 \in \mathbb{A}$ then T_0 will have a model of the form

$$(\mathbb{A}, \beta, \mathfrak{N}, f)$$

where $\beta \in \mathbb{A}$, f maps the ordinals $<\beta$ into $\prec^{\mathfrak{N}}$, and $\mathfrak{N} \models \varphi$. Let

$$(\mathfrak{B}, c, \mathfrak{N}, f)$$

be a model of the whole theory T. By (1), $\mathfrak{N} \models \varphi$. By (2)—(5), $\mathbb{A} \subseteq_{\text{end}} \mathfrak{B}$, but, by (6), c is not in the well-founded part of $\mathfrak{B}$. Hence, by (7), $\prec^{\mathfrak{N}}$ cannot be a well-ordering. ☐

7.4 Corollary. *Let* $\mathsf{L}_{\mathbb{A}}$ *be a countable admissible fragment,* T *a* Σ_1 *theory of* $\mathsf{L}_{\mathbb{A}}$ *which pins down ordinals. There is an* $\alpha < o(\mathbb{A})$ *such that every ordinal pinned down by* T *is less than* α.

Proof. If not, then every $T_0 \subseteq T$, $T_0 \in \mathbb{A}$ would be consistent with

$$\{\mathsf{c}_{n+1} \prec \mathsf{c}_n : n = 0, 1, 2, \ldots\}$$

by 7.3 and hence T would also be consistent with this set by the Barwise Compactness Theorem. ☐

We can use Theorem 7.4 to prove a general version of a theorem of Friedman on models of set theories $T \supseteq \text{KP}$. Note that if $<$ is a linear ordering then, by the definition in II.8.3, $\mathcal{W}\!f(<)$ is the largest well-ordered initial segment of $<$. We can identify $\mathcal{W}\!f(<)$ with an ordinal without confusion. In Theorem 7.5, L is a language containing a binary symbol $<$ among its symbols.

7.5 Theorem. *Let* $\mathsf{L}_{\mathbb{A}}$ *be a countable admissible fragment of* $\mathsf{L}_{\infty\omega}$ *and let* $\alpha = o(\mathbb{A})$. *Let* T *be a* Σ_1 *theory of* $\mathsf{L}_{\mathbb{A}}$ *such that:*
i) $T \models$ "$<$ *is a linear ordering*";
ii) *for each* $\beta < \alpha$, T *has a model* $\mathfrak{M}$ *with*

$$\mathcal{W}\!f(<^{\mathfrak{M}}) \geqslant \beta.$$

Then T *has a model* M *with* $\mathcal{W}\!f(<^{\mathfrak{M}}) = \alpha$.

Proof. Notice that T is consistent with the set

$$\{\exists x\, \theta_\beta(x) \mid \beta < \alpha\},$$

where θ_β is as in 7.1, by the Barwise Compactness Theorem, so we may as well assume that the sentences are actually in T. This insures that any model $\mathfrak{M}$ of T has $\mathscr{W}\!f(<^{\mathfrak{M}}) \geqslant \alpha$. By Theorem 7.4, T also has a model $\mathfrak{M}$ where $<^{\mathfrak{M}}$ is not well ordered. Let T' be the theory

$$T \cup \{(\mathsf{d}_{n+1} < \mathsf{d}_n) \mid n < \omega\}$$

where $\{\mathsf{d}_n \mid n < \omega\}$ is a new set of constant symbols. Thus T' is consistent. Let $\mathsf{K} = \mathsf{L} \cup \{\mathsf{d}_n \mid n < \omega\}$ and let $\mathsf{K}_{\mathbb{A}}$ be the corresponding admissible fragment. Let $\mathsf{K}^0_{\mathbb{A}}$ be the set of formulas in which at most a finite member of d_n's occur. Thus T' is a theory of $\mathsf{K}^0_{\mathbb{A}}$. We are going to use the Omitting Types Theorem for $\mathsf{K}^0_{\mathbb{A}}$ to find a model $(\mathfrak{M}, d_1, \ldots, d_n, \ldots)$ of T' and the sentence

$$\forall v\, [\text{“}v \notin \text{Field}(<)\text{”} \vee \bigvee_{\beta<\alpha} \theta_\beta(v) \vee \bigvee_{m<\omega} (\mathsf{d}_m < v)]$$

where “$v \notin \text{Field}(<)$” stands for $\forall x\,(v \nless x \wedge x \nless v)$. Such a model $\mathfrak{M}$ must have $\mathscr{W}\!f(<^{\mathfrak{M}}) = \alpha$. Suppose there were no such model. Then by the Omitting Types Theorem there is a $\sigma(v, \mathsf{d}_1, \ldots, \mathsf{d}_n) \in \mathsf{K}^0_{\mathbb{A}}$ such that $T' + \exists v\, \sigma(v, \mathsf{d}_1, \ldots, \mathsf{d}_n)$ is consistent, but such that all the following are theorems of T':

(8) $\forall v\, [\sigma(v, \vec{\mathsf{d}}) \rightarrow v \in \text{Field}(<)]$,
(9) $\forall v\, [\sigma(v, \vec{\mathsf{d}}) \rightarrow \neg\theta_\beta(v)]$,
(10) $\forall v\, [\sigma(v, \vec{\mathsf{d}}) \rightarrow v \leqslant \mathsf{d}_m]$, for all $m < \omega$.

Note that, by (10),

$$T + \exists v\, [\sigma(v, \mathsf{d}_1, \ldots, \mathsf{d}_n) \wedge v < \mathsf{d}_n < \cdots < \mathsf{d}_1]$$

is consistent. Let c be a new constant symbol and let T'' be

$$T \cup \{\sigma(\mathsf{c}, \mathsf{d}_1, \ldots, \mathsf{d}_n) \wedge \mathsf{c} < \mathsf{d}_n < \cdots < \mathsf{d}_1\}$$

which is consistent since $T \subseteq T'$. We claim that

(11) *For every model $\mathfrak{M}$ of T'', the $<^{\mathfrak{M}}$ predecessors of $c = \mathsf{c}^{\mathfrak{M}}$ are well ordered.*

If not, then there is a descending sequence below c, so we can name its members $d_{n+1}, \ldots$, giving us

$$d_1 > \cdots > d_n > c > d_{n+1} > \cdots > \cdots$$

which makes $(\mathfrak{M}, d_1, \ldots, d_n, \ldots) \models T'$, where the element c violates (10).

From (11) and Theorem 7.4 we obtain

(12) *there is a* $\gamma<\alpha$ *such that*

$$T''\models\bigvee_{\beta<\gamma}\theta_\beta(\mathsf{c}).$$

(To see this consider $T'''=T''+$"$\prec=<\upharpoonright$ the predecessors of c" and apply Theorem 7.4 to T''' as a theory about $\prec$.)

Thus we have

$$T\models\sigma(\mathsf{c},\mathsf{d}_1,\ldots,\mathsf{d}_n)\wedge\mathsf{c}<\mathsf{d}_n<\cdots<\mathsf{d}_1\rightarrow\bigvee_{\beta<\gamma}\theta_\beta(\mathsf{c})$$

and hence

$$T'\models\sigma(\mathsf{c},\mathsf{d}_1,\ldots,\mathsf{d}_n)\rightarrow\bigvee_{\beta<\gamma}\theta_\beta(\mathsf{c})$$

contradicting (9). ☐

7.6—7.7 Exercises

7.6 (Friedman). Assume ZF has an uncountable transitive model. Show that the order types of the ordinals in countable nonstandard ω-models of ZF are exactly the order types $\alpha(1+\eta)$ for α a countable admissible ordinal, $\alpha>\omega$, and η the order type of the rationals. [Use Theorem 7.5.]

7.7. Let α be nonadmissible. Let $\mathrm{T}=\mathrm{KP}+$"$<$ is $\in\upharpoonright$ordinals". Show that there can be no model $\mathfrak{M}$ of T with $\alpha=\mathscr{W}\!\mathscr{f}(<)$.

7.8 Notes. Theorem 7.3 is from Barwise [1969]. It refines older results of Lopez-Escobar [1966] and Morley [1965]. Theorem 7.5 is new here. It generalizes a result in Friedman [1973].

8. *Another Look at Consistency Properties*

There is room for a lot of creativity inside the proof of the Model Existence Theorem. Let S be a consistency property, $s_0\in S$, and recall the way we constructed a model of s_0. As we built our sequence $s_0\subseteq s_1\subset\cdots\subseteq s_n\subseteq\cdots\subseteq s_\omega$ (and in so doing built a canonical model) there was freedom in defining s_{n+1} that we didn't use. At the nth stage, after defining s_n we could first enlarge s_n to some other $s_n^*\supseteq s_n$ before going on to get $s_{n+1}\supseteq s_n^*$, as long as $s_n^*\in S$. The resulting

$$s_0\subseteq s_0^*\subseteq s_1\subseteq s_1^*\subseteq\cdots\subseteq s_n\subseteq s_n^*\subseteq\cdots$$

with

$$s_\omega=\bigcup_{n<\omega}s_n=\bigcup_{n<\omega}s_n^*$$

would give rise to a canonical model $\mathfrak{M}$ of each s_n and each s_n^*. We give a modest illustration of this technique here, just to illustrate the general method. We will return to it in § IV.4.

Let $\mathsf{L}_A \subseteq \mathsf{K}_B$ be fragments. A theory T of K_B is *complete for* L_A if for each sentence φ of L_A, $T \models \varphi$ or $T \models \neg\varphi$, but not both. Given a structure $\mathfrak{M}$ for K we define

$$\mathrm{Th}_{\mathsf{L}_A}(\mathfrak{M}) = \{\varphi \in \mathsf{L}_A \mid \varphi \text{ is a sentence true in } \mathfrak{M}\}.$$

Note that this is a complete L_A-theory.

8.1 Theorem. *Let* $\mathsf{L}_A \subseteq \mathsf{K}_B$ *be countable fragments, T a consistent set of sentences of* K_B *such that for each sentence ψ of K_B, $T \cup \{\psi\}$ is not complete for* L_A. *There are* $2^{\aleph_0}$ *distinct* L_A *theories of the form* $\mathrm{Th}_{\mathsf{L}_A}(\mathfrak{M})$ *for models $\mathfrak{M}$ of T.*

Proof. Let $\mathsf{K}' = \mathsf{K} \cup C = \mathsf{K} \cup \{\mathsf{c}_n : n < \omega\}$ and let K'_B be the set of free substitution instances of φ's in K_B. Note that there is no sentence $\psi(\mathsf{c}_1, \ldots, \mathsf{c}_n) \in \mathsf{K}'_A$ such that $T + \varphi(\mathsf{c}_1, \ldots, \mathsf{c}_n)$ is complete for L_A, for if it were then $T + \exists v_1, \ldots, \exists v_n\, \psi(v_1, \ldots, v_n)$ would be complete for L_A. Consequently there is no finite set s of K'_A such that $T \cup s$ is complete for L_A (for otherwise we would form $\psi = \bigwedge s$). Define

$$S_0 = \{T \cup s \mid s \text{ a finite set of sentences of } \mathsf{K}_B \text{ such that } T \cup s \text{ is consistent}\}.$$

The set S_0 obviously satisfies (C1)—(C7) so let S be the smallest consistency property containing S_0, by 2.3. To simplify notation let us suppress T altogether and write s for $T \cup s$ in what follows. We wish to construct a "tree" of members of S such that

(1) any "branch" through the tree gives us a theory T_0 of L_A consistent with T,
(2) distinct branches lead to incompatible theories, and
(3) there are $2^{\aleph_0}$ distinct branches.

Since this is our first tree argument, and since this is an important kind of argument, we give the proof in more detail than is usual. Our tree consists of all finite sequences d of 0's and 1's arranged as illustrated below.

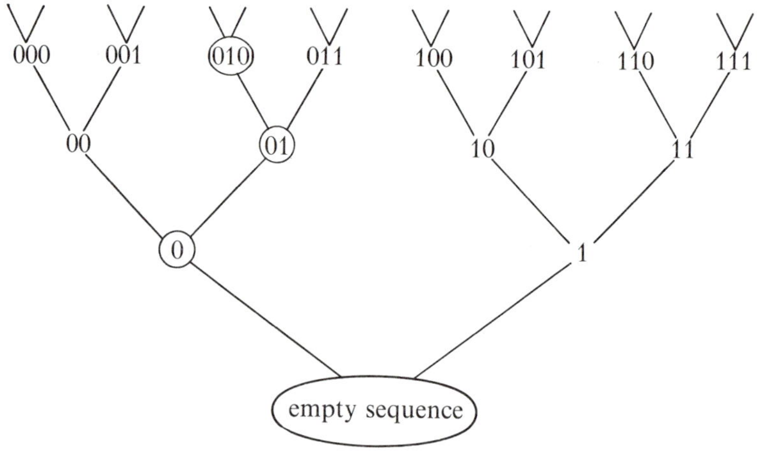

A branch through this tree is just an infinite sequence b of 0's and 1's. (The branch $\langle 01000\ldots\rangle$ has its nodes circled in the tree drawn above.) We wish to place at each node d of length n an element $s^d \in S$ which is one of the s_n^*'s referred to in the introduction to this section. At the empty node place T and let $s_0 = T$ in the notation from the proof of the Model Existence Theorem. Before defining s_1, pick some $\varphi \in \mathsf{L}_A$ not decided by s_0, i. e.

$$s_0^* = s_0 \cup \{\varphi\} \quad \text{and} \quad s_0^{**} = s_0 \cup \{\neg\varphi\}$$

are consistent and hence in S. Let $s^d = s_0^*$ for $d = \langle 0\rangle$, $s^d = s_0^{**}$ for $d = \langle 1\rangle$.

Given d of length n we go on to find $s_{n+1} \supseteq s^d$ just as in the proof of the Model Existence Theorem. Then, given s_{n+1}, choose a $\varphi \in \mathsf{L}_A$ such that

$$s_{n+1}^* = s_{n+1} \cup \{\varphi\}, \qquad s_{n+1}^{**} = s_{n+1} \cup \{\neg\varphi\}$$

are in S, and let

$$\begin{aligned} s^{d'} &= s_{n+1}^*, \quad \text{if} \quad d' = d0 \\ &= s_{n+1}^{**}, \quad \text{if} \quad d' = d1, \end{aligned}$$

where $d0$ is the sequence d followed by 0. Now let b be any branch through the tree. The set $s^b = \bigcup\{s^d \mid d \text{ a node on } b\}$ is one of our s_ω's so it has a model $\mathfrak{M}_b$. Since $s_0 = T$, $\mathfrak{M}_b \models T$. If b_1, b_2 are distinct branches then there is a $\varphi \in \mathsf{L}_A$ such that we have put $\varphi \in s^{b_1}$, $\neg\varphi \in s^{b_2}$ at the point where b_1 and b_2 split. Thus $\mathfrak{M}_{b_1} \models \varphi$ and $\mathfrak{M}_{b_2} \models \neg\varphi$, where $\varphi \in \mathsf{L}_A$. There are $2^{\aleph_0}$ branches so the sets

$$\mathrm{Th}_{\mathsf{L}_A}(\mathfrak{M}_b)$$

form $2^{\aleph_0}$ distinct complete theories of L_A. $\square$

We will apply the following corollary of Theorem 8.1 in Chapter IV. Let $\mathsf{L}_{\mathbb{A}}$ be an admissible fragment of $\mathsf{L}_{\infty\omega}$. A structure $\mathfrak{M}$ is *decidable* for $\mathsf{L}_{\mathbb{A}}$ if $\mathrm{Th}_{\mathsf{L}_{\mathbb{A}}}(\mathfrak{M})$ is Δ_1 on $\mathbb{A}$. The structure $\mathfrak{M}$ could be a structure for some language K properly containing L.

8.2 Corollary. *Let $\mathsf{L}_{\mathbb{A}} \subseteq \mathsf{K}_{\mathbb{A}}$ be countable admissible fragments, let T be a consistent theory of $\mathsf{K}_{\mathbb{A}}$ which is Σ_1 on the admissible set $\mathbb{A}$ such that T has no model which is decidable for $\mathsf{L}_{\mathbb{A}}$. There are $2^{\aleph_0}$ distinct theories of the form $\mathrm{Th}_{\mathsf{L}_{\mathbb{A}}}(\mathfrak{M})$ with $\mathfrak{M} \models T$.*

Proof. If there are fewer than $2^{\aleph_0}$ such sets then there is a $\psi \in \mathsf{K}_{\mathbb{A}}$ such that $T + \psi$ is complete for $\mathsf{L}_{\mathbb{A}}$. But then any model $\mathfrak{M}$ of $T + \psi$ is decidable for $\mathsf{L}_{\mathbb{A}}$ since

$$\begin{aligned} \mathfrak{M} \models \varphi \quad &\text{iff} \quad T \models \psi \to \varphi, \\ \mathfrak{M} \not\models \varphi \quad &\text{iff} \quad T \models \psi \to \neg\varphi \end{aligned}$$

which makes $\mathrm{Th}_{\mathsf{L}_{\mathbb{A}}}(\mathfrak{M})$ a Δ_1 set by the Extended Completeness Theorem. $\square$

For $\mathsf{L}_{\omega\omega}$ and $\mathsf{K}_{\omega\omega}$ the theorem and its corollary are old indeed. Here the proof of the theorem is even easier since one no longer has to go back to the Model Existence Theorem but can use the Compactness Theorem for $\mathsf{K}_{\omega\omega}$.

8.3—8.4 Exercises

8.3. Show that if $\mathsf{K}_{\mathbb{B}}=\mathsf{K}_{\omega\omega}$, $\mathsf{L}_{\mathbb{A}}=\mathsf{L}_{\omega\omega}$ then the hypothesis of Theorem 8.1 can be weakened to:

$$\forall\psi\in\mathsf{L}_{\mathbb{A}}\ (T+\psi \text{ not complete for } \mathsf{L}_{\mathbb{A}}).$$

Prove this directly from the Compactness Theorem for $\mathsf{K}_{\omega\omega}$.

8.4. Let L have constant symbols $\overline{\mathbf{0}},\overline{\mathbf{1}},\ldots,\overline{\mathsf{n}},\ldots$ and a unary predicate P. Find a consistent theory $T=\{\varphi\}$ of a countable fragment L_A such that ψ has only $\aleph_0$ non-isomorphic models, but for each $\psi\in\mathsf{L}_{\omega\omega}$, $T+\psi$ is not complete for $\mathsf{L}_{\omega\omega}$. This shows that the strengthening of 8.1 carried out in 8.3 is not possible in general.

8.5 Notes. The results of this section are new here. They are suggested by, and imply, the theorem of recursion theory that any Σ_1^1 set of subsets of ω with less than $2^{\aleph_0}$ members is actually a subset of $\mathbb{H}\mathrm{YP}$. See § IV.4 for proofs of this and related results.

Chapter IV
Elementary Results on $\mathbb{HYP}_{\mathfrak{M}}$

We have seen, in Chapter III, how admissible sets provide a tool for the study of infinitary logic by giving rise to those countable fragments which are especially well-behaved. In this chapter we begin the study of $\mathbb{HYP}_{\mathfrak{M}}$ by means of the logical tools developed in Chapter III.

1. On Set Existence

Given $\mathfrak{M}$ we form the universe of sets $\mathbb{V}_{\mathfrak{M}}$ on $\mathfrak{M}$ and speak glibly about arbitrary sets $a \in \mathbb{V}_{\mathfrak{M}}$. In practice, however, one seldom considers the impalpable sets of extremely high rank. There is even a feeling that these sets have a weaker claim to existence than the sets one normally encounters. Without becoming too philosophical, we want to touch here on the question: If we assume $\mathfrak{M}$ as given, to the existence of what sets are we more or less firmly committed?

$\mathbb{HYP}_{\mathfrak{M}}$ is the intersection of all models $\mathfrak{A}_{\mathfrak{M}}$ of KPU$^+$ and is an admissible set above $\mathfrak{M}$. There appears to be a certain *ad hoc* feature to $\mathbb{HYP}_{\mathfrak{M}}$, however, since it might depend on the exact axioms of KPU$^+$ in a sensitive way. You would expect that if you took a stronger theory than KPU$^+$ (say throw in Power, or Infinity or Full Separation) that more sets from $\mathbb{V}_{\mathfrak{M}}$ would occur in all models of this stronger theory. That, for $\mathfrak{M}$ countable, this cannot happen, lends considerable weight to the contension that $\mathbb{HYP}_{\mathfrak{M}}$ is here to stay.

Of the two results which follow, the second implies the first. We present them in the opposite order for expository and historical reasons.

A set $S \subseteq \mathfrak{M}$ is *internal* for $\mathfrak{A}_{\mathfrak{M}} = (\mathfrak{M}; A, E, \ldots)$ if there is an $a \in A$ such that $S = a_E = \{x \in \mathfrak{A}_{\mathfrak{M}} \mid x E a\}$.

1.1 Theorem. *Let* $\mathfrak{M} = \langle M, R_1, \ldots, R_l \rangle$ *be a countable structure for* L. *Let* T *be a consistent theory (finitary or infinitary) which is* Σ_1 *on* $\mathbb{HYP}_{\mathfrak{M}}$ *and which has a model of the form* $\mathfrak{A}_{\mathfrak{M}} = (\mathfrak{M}; A, E, \ldots)$. *Let* $S \subseteq M$ *be such that* S *is internal for every such model of* T. *Then* $S \in \mathbb{HYP}_{\mathfrak{M}}$.

Proof. The proof is a routine application of Completeness and Omitting Types. Given the above assumptions we see that there can be no model $\mathfrak{A}_{\mathfrak{M}}$ of

$$T' + \forall v \bigvee \Phi(v)$$

where T' is T plus

(1) $\forall v\,[\mathsf{U}(v) \to \bigvee_{p \in M} v = \overline{\mathsf{p}}]$
Diagram $(\mathfrak{M})$

and Φ is the set of formulas

$$\{\overline{\mathsf{p}} \notin v \mid p \in S\} \cup \{\overline{\mathsf{p}} \in v \mid p \notin S\},$$

for then S would not be internal for $\mathfrak{A}_{\mathfrak{M}}$. The formulas in T' and in $\Phi(v)$ are members of the admissible fragment $\mathsf{L}^*_{\mathbb{A}}$ of $\mathsf{L}^*_{\infty\omega}$ where $\mathbb{A} = \mathbb{HYP}_{\mathfrak{M}} = (\mathfrak{M}; A, \in)$, and where we have introduced $\overline{\mathsf{p}}$ by some convention like $\overline{\mathsf{p}} = \langle 0, p \rangle$. By the Omitting Types Theorem there is a formula $\sigma(v)$ of $\mathsf{L}^*_{\mathbb{A}}$ such that $T' + \exists v\, \sigma(v)$ is consistent but such that:

$$T' \models \forall v\,[\sigma(v) \to \overline{\mathsf{p}} \in v], \quad \text{for all} \quad p \in S;$$

$$T' \models \forall v\,[\sigma(v) \to \overline{\mathsf{p}} \notin v], \quad \text{for all} \quad p \notin S.$$

But then

$$S = \{p \in \mathfrak{M} \mid T' \models \forall v\,(\sigma(v) \to \overline{\mathsf{p}} \in v)\}$$

so S is Σ_1 on $\mathbb{HYP}_{\mathfrak{M}}$ by the Extended Completeness Theorem for $\mathsf{L}^*_{\mathbb{A}}$. Similarly $\neg S$ is Σ_1 on $\mathbb{HYP}_{\mathfrak{M}}$ so S is Δ_1 on $\mathbb{HYP}_{\mathfrak{M}}$. Thus $S \in \mathbb{HYP}_{\mathfrak{M}}$ by Δ_1 Separation. □

Before stating our next result we need a more sophisticated notion of what it means for a set $a \in \mathbb{V}_{\mathfrak{M}}$ to be *internal* for $\mathfrak{A}_{\mathfrak{M}} = (\mathfrak{M}; A, E, \ldots)$.

1.2 Definition. A set $a \in \mathbb{V}_{\mathfrak{M}}$ is *internal for* $\mathfrak{A}_{\mathfrak{M}} = (\mathfrak{M}; A, E, \ldots)$ if $a \in \mathcal{W}\!f(\mathfrak{M}; A, E)$, where we again identify $\mathcal{W}\!f(\mathfrak{M}; A, E)$ with its transitive collapse.

Note that for $a \subseteq M$ this is equivalent to the existence of an $x \in A$ with $a = x_E$. Also notice that if a is internal and $b \in a$ then b is internal.

1.3 Theorem. *Let $\mathfrak{M}$ be countable and let $a \in \mathbb{V}_{\mathfrak{M}}$ be a set which is internal for every model*

$$\mathfrak{A}_{\mathfrak{M}} = (\mathfrak{M}; A, E, \ldots)$$

of some consistent theory T, finitary or not, formulated in $\mathsf{L}^* = \mathsf{L}(\in, \ldots)$, $\mathrm{KPU}^+ \subseteq T$. If T is Σ_1 on $\mathbb{HYP}_{\mathfrak{M}}$, then $a \in \mathbb{HYP}_{\mathfrak{M}}$.

Proof. We prove the theorem by $\in$-induction. By the comment above, if a is internal for every model $\mathfrak{A}_{\mathfrak{M}}$ of T, so is every $b \in a$. By $\in$-induction, each of these b is in $\mathbb{HYP}_{\mathfrak{M}}$. That is, $a \subseteq \mathbb{HYP}_{\mathfrak{M}}$. A routine modification of the proof of 1.1 shows that a is Δ_1 on $\mathbb{HYP}_{\mathfrak{M}}$. If we can prove that $a \subseteq L(\mathfrak{M}, \beta)$ for some $\beta < o(\mathbb{HYP}_{\mathfrak{M}})$ then, by Δ_1 separation, $a \in \mathbb{HYP}_{\mathfrak{M}}$. Assume, on the contrary, that

$$O(\mathfrak{M}) = \text{the least ordinal } \beta \text{ such that } a \subseteq L(\mathfrak{M}, \beta).$$

In any model $\mathfrak{A}_{\mathfrak{M}}$ of T there would be a unique ordinal x such that

$$\mathfrak{A}_{\mathfrak{M}} \models \text{“}x = \text{least ordinal } \beta \text{ such that } a \subseteq L(\mathfrak{M}, \beta)\text{”}.$$

By Σ Reflection in $\mathfrak{A}_{\mathfrak{M}}$ and, by the absoluteness of $L(\cdot,\cdot)$, this x must be $O(\mathfrak{M})$. Hence $T+$ the following theory pins down $O(\mathfrak{M})$, contrary to Corollary III.7.4.

Diagram $(\mathfrak{M})$,

$\forall x\, [\mathsf{U}(x) \to \bigvee_{p \in M} x = \mathsf{p}]$,

“$<$ *is the order type of the* $\in$*-precedessors of* c”,

(2) “c *is the first ordinal such that* $L(\mathfrak{M}, \mathsf{c})$ *is admissible*“ (if $\alpha > \omega$)

or

(3) “c *is the first limit ordinal*” (if $\alpha = \omega$).

This theory is formulated in $\mathsf{L}(\in, \ldots, <, \mathsf{c}, \mathsf{p})_{p \in M}$. (The reason for the two cases is that we do not yet know how to write “x is admissible” by a finite formula.) We can write (2) as

$$\forall x\, [x < \mathsf{c} \to \bigvee_{\varphi \in \mathrm{KPU}} \neg \varphi^{L(\mathfrak{M}, x)}]. \quad \square$$

Thus we see that no matter how we strengthen KPU$^+$ to an axiomitizable theory T, we cannot assure that any set in $\mathbb{V}_{\mathfrak{M}} - \mathbb{HYP}_{\mathfrak{M}}$ should be internal to every model $\mathfrak{A}_{\mathfrak{M}}$ of T.

One could consider $\mathbb{HYP}_{\mathfrak{M}}$ as a new structure $\mathfrak{N}$ and form $\mathbb{HYP}_{\mathfrak{N}}$ but it is more natural, and essentially equivalent, to procced differently.

1.4 Definition. Let $\mathbb{A}_{\mathfrak{M}} = (\mathfrak{M}; A, \in)$ be transitive in $\mathbb{V}_{\mathfrak{M}}$. Then $\mathbb{HYP}(\mathbb{A}_{\mathfrak{M}})$ is the structure $(\mathfrak{M}; B, \in)$ where

$$B = \bigcap \{B' \mid (M \cup A) \in B', (\mathfrak{M}; B', \in) \text{ admissible}\}.$$

We consider $\mathbb{HYP}_{\mathfrak{M}}$ as a special case of $\mathbb{HYP}(\mathbb{A}_{\mathfrak{M}})$.

1.5—1.9 Exercises

1.5. Show that $\mathbb{HYP}(\mathbb{A}_{\mathfrak{M}})$ is admissible.

1.6. Show that every element $a \in \mathbb{HYP}(\mathbb{A}_{\mathfrak{M}})$ has a good Σ_1 definition with parameters from $M \cup A \cup \{M, A\}$.

1.7. Show that the obvious generalizations of 1.1 and 1.3 are true.

1.8. Let $\mathcal{N}=\langle\omega,+,\cdot,0\rangle$ and let $X\subseteq\omega$. Show that there is a $T\subseteq L_{\omega\omega}$, $\mathrm{KPU}^+\subseteq T$ such that X is in every model $\mathfrak{A}_{\mathcal{N}}$ of T. This shows that the condition that T be Σ_1 on $\mathbb{HYP}_{\mathfrak{M}}$ is necessary in 1.1 and 1.3.

1.9. Show that the hypothesis $\mathrm{KPU}^+\subseteq T$ can be dropped from Theorem 1.3. [Hint: add a new $\in$-symbol and a function symbol used to denote an $\in$-isomorphism.]

1.10 Notes. Theorem 1.1 is a modern version of the Gandy-Kreisel-Tait Theorem: For any consistent Π_1^1 T set of axioms for second order number theory, if $a\subseteq\omega$ is internal to every model of T, then a is hyperarithmetic.

Theorem 1.3 was announced by Barwise in Barwise-Gandy-Moschovakis [1971]. The part of it contained in Theorem 1.1 is due independently to Grilliot [1972]. The improvement in 1.9 is due to Ville [1974].

2. *Defining* Π_1^1 *and* Σ_1^1 *Predicates*

Let $\mathfrak{M}=\langle M,R_1,\ldots,R_l\rangle$ be a fixed infinite structure for a language L. An n-ary relation S on $\mathfrak{M}$ is Π_1^1 *on* $\mathfrak{M}$ if it can be defined by a second order formula of the form

$$S(p_1,\ldots,p_n) \quad \text{iff} \quad \forall\mathsf{T}_1,\ldots,\forall\mathsf{T}_k\ \varphi(p_1,\ldots,p_n,\mathsf{T}_1,\ldots,\mathsf{T}_k),$$

where φ is a first order formula of $\mathsf{L}(\mathsf{T}_1,\ldots,\mathsf{T}_k)$, possibly containing parameters from $\mathfrak{M}$. More formally we should write this as: for all $p_1,\ldots,p_n\in M$, $S(p_1,\ldots,p_n)$ holds iff for all relations $T_1,\ldots,T_k$ on $\mathfrak{M}$,

$$(\mathfrak{M},T_1,\ldots,T_k)\models\varphi[p_1,\ldots,p_n].$$

The negation of a Π_1^1 relation is called Σ_1^1 *on* $\mathfrak{M}$. Thus S is Σ_1^1 iff it can be defined by

$$S(\vec{p}) \quad \text{iff} \quad \exists\mathsf{T}_1,\ldots,\exists\mathsf{T}_k\ \psi(\vec{p},\mathsf{T}_1,\ldots,\mathsf{T}_k)$$

for some first order ψ. If S is both Π_1^1 and Σ_1^1 on $\mathfrak{M}$ then S is said to be Δ_1^1 *on* $\mathfrak{M}$.

This section is primarily concerned with techniques that can be used to show that predicates are Π_1^1 or Σ_1^1 on $\mathfrak{M}$. The reason for discussing this material can be seen by glancing at the next section.

2.1 Examples. (i) *If* $\mathcal{N}=\langle\omega,0,+,\cdot\rangle$, *then a set is* Δ_1^1 *over* $\mathcal{N}$ *iff it is hyperarithmetic.* (This is the classical Souslin-Kleene theorem. See, e.g., Shoenfield [1967].)

(ii) *If* $\mathfrak{N}=\langle N,0,+,\cdot\rangle$ *is a nonstandard model of arithmetic then the standard integers form a* Π_1^1 *set but not, in general, a* Δ_1^1 *set:*

$$x \quad \text{is standard iff} \quad \forall S\,[S(0)\wedge\forall y\,(S(y)\to S(y+1))\to S(x)].$$

(iii) *If* $\mathfrak{M}=\langle G,0,+\rangle$ *is an abelian group then the torsion part* T *of* G, the set of elements of G of finite order, *is* Π_1^1 *on* G*:*

$$x\in T \quad \text{iff} \quad \forall S\,[S(x)\wedge\forall y\,(S(y)\to S(x+y))\to S(0)].$$

(iv) *If* $\mathfrak{M}=\langle G,0,+\rangle$ *is an abelian group then the largest divisible subgroup* D *of* G *is* Σ_1^1, but this time it is not so obvious.

$$x\in D \quad \text{iff} \quad \exists H\,[H \text{ a subgroup } \wedge H \text{ divisible } \wedge H(x)]$$

but the clause "H is divisible", meaning

$$\textit{for all integers } n,\ \forall y\in H\ \exists z\in H,\ nz=y$$

cannot be expressed by a single first order sentence. It is still possible, though, to write D out as a Σ_1^1 predicate. The student should try this before going on in order to appreciate the machinery developed below. ▯

The last example is just the tip of an iceberg. In writing out Π_1^1 predicates we frequently discover that we would like to use an extended first order formula as defined in § II.2. (In writing out the Σ_1^1 predicate in 2.1 (iv) we need the co-extended predicate "H is divisible".) It turns out we can allow ourselves this freedom without changing the class of Π_1^1 predicates.

2.2 Definition. (i) An *extended* Π_1^1 *predicate over* $\mathfrak{M}$ is a predicate $\mathbf{S}(p_1,\ldots,p_i, S_1,\ldots,S_m, a_1,\ldots,a_j, P_1,\ldots,P_n)$ defined by

$$(\mathfrak{M},S_1,\ldots,S_m;\ \mathbb{HF}_{\mathfrak{M}},\in,P_1,\ldots,P_n)\models\forall \mathsf{T}_1,\ldots,\forall \mathsf{T}_m\ \forall \mathsf{Q}_1,\ldots,\forall \mathsf{Q}\ \varphi(\vec{p},\vec{a},\vec{\mathsf{S}},\vec{\mathsf{T}},\vec{\mathsf{P}},\vec{\mathsf{Q}}),$$

for some extended first order formula φ which may have parameters in it from $M\cup\mathbb{HF}_{\mathfrak{M}}$. (We use S, T for relations over M; P, Q for relations over $M\cup\mathbb{HF}_{\mathfrak{M}}$.)

(ii) $\mathbf{S}$ is *co-extended* Σ_1^1 if it is in the dual class; that is, if it can be defined by

$$(\mathfrak{M},\vec{S};\ \mathbb{HF}_{\mathfrak{M}},\in,\vec{P})\models\exists\vec{\mathsf{T}}\ \exists\vec{\mathsf{Q}}\ \varphi(\vec{p},\vec{a},\vec{\mathsf{S}},\vec{\mathsf{T}},\vec{\mathsf{P}},\vec{\mathsf{Q}})$$

where φ is co-extended.

Thus extended Π_1^1 predicates over $\mathfrak{M}$ are not really predicates over $\mathfrak{M}$; they are predicates of points in $\mathfrak{M}$, relations on $\mathfrak{M}$, sets in $\mathbb{HF}_{\mathfrak{M}}$ and relations on $\mathbb{HF}_{\mathfrak{M}}$. They are important as a tool for showing predicates over $\mathfrak{M}$ are Π_1^1. For example, in 2.1 (iv), it is clear that D is co-extended Σ_1^1, so that D is Σ_1^1 over G by 2.8 below.

2.3 Lemma. *If* $\mathbf{S}_1, \mathbf{S}_2$ *are extended* Π_1^1 *(respectively co-extended* Σ_1^1*) so are* $(\mathbf{S}_1 \vee \mathbf{S}_2)$ *and* $(\mathbf{S}_1 \wedge \mathbf{S}_2)$.

Proof. For example,

$$\forall \mathsf{T}\, \psi(\_\_, \mathsf{T}) \wedge \forall \mathsf{T}'\, \forall \mathsf{Q}\, \theta(\_\_, \mathsf{T}', \mathsf{Q})$$

is equivalent to

$$\forall \mathsf{T}\, \forall \mathsf{T}'\, \forall \mathsf{Q}\, [\psi(\_\_, \mathsf{T}) \wedge \theta(\_\_, \mathsf{T}'\mathsf{Q})]$$

as long as we first make sure T and T' are distinct symbols. The part inside the brackets is still extended first order. ☐

2.4 Lemma. *If* $\mathbf{S}$ *is extended* Π_1^1 *(respectively, co-extended* Σ_1^1*) then* $\neg\mathbf{S}$ *is co-extended* Σ_1^1 *(respectively, extended* Π_1^1*).* ☐

2.5 Lemma. *If* $\mathbf{S} = \mathbf{S}(p_1, \ldots, p_i, \_\_)$ *is extended* Π_1^1 *(co-extended* Σ_1^1*) then so are*

$$\mathbf{S}_1(p_1, \ldots, p_{i-1}, \_\_) \quad \textit{iff} \quad \forall p_i\, \mathbf{S}(p_1, \ldots, p_{i-1}, p_i, \_\_),$$
$$\mathbf{S}_2(p_1, \ldots, p_{i-1}, \_\_) \quad \textit{iff} \quad \exists p_i\, \mathbf{S}(p_1, \ldots, p_{i-1}, p_i, \_\_).$$

Proof. It is hard to see the extended Π_1^1 case directly, but we can prove the co-extended Σ_1^1 case and then apply 2.4. If

$$\mathbf{S}(\vec{p}, \_\_) \quad \text{iff} \quad \exists \mathsf{Q}\, \psi(\vec{p}, \_\_, \mathsf{Q})$$

then

$$\mathbf{S}_1(p_1, \ldots, p_{i-1}, \_\_) \quad \text{iff} \quad \exists \mathsf{Q}\, \exists p_i\, \psi(\vec{p}, \_\_, \mathsf{Q})$$

and

$$\mathbf{S}_2(p_1, \ldots, p_{i-1}, \_\_) \quad \text{iff} \quad \forall p_i\, \exists \mathsf{Q}\, \psi(p_1, \ldots, p_i, \_\_, \mathsf{Q})$$
$$\text{iff} \quad \exists \mathsf{Q}'\, \forall p_i\, \psi(p_1, \ldots, p_i, \_\_, \mathsf{Q}'(\ldots, p_i))$$

where the notation indicates that we have replaced the n-ary relation $\mathsf{Q}(t_1, \ldots, t_n)$ by the new $n+1$-ary $\mathsf{Q}'(t_1, \ldots, t_n, p_i)$ throughout ψ. ☐

2.6 Lemma. *If* $\mathbf{S} = \mathbf{S}(a_1, \ldots, a_j, \_\_)$ *is extended* Π_1^1 *then*

$$\mathbf{S}_1(a_1, \ldots, a_{j-1}, \_\_) \quad \textit{iff} \quad \exists a_j\, \mathbf{S}(a_1, \ldots, a_{j-1}, a_j, \_\_)$$

is extended Π_1^1. *If* $\mathbf{S}$ *is co-extended* Σ_1^1 *then*

$$\mathbf{S}_2(a_1, \ldots, a_{j-1}, \_\_) \quad \textit{iff} \quad \forall a_j\, \mathbf{S}(a_1, \ldots, a_{j-1}, a_j, \_\_)$$

is co-extended Σ_1^1.

Proof. Again we do the extended Σ_1^1 case and then apply 2.4. The proof is just like the "hard" half of 2.5. Note that the easy half does not go through! ☐

2.7 Lemma. *If* $\mathbf{S}=\mathbf{S}(\vec{p},S_1,\ldots,S_m,\vec{a},P_1,\ldots,P_n)$ *is extended* Π_1^1 *then so are*

$$\forall S_m\,\mathbf{S}(\_\,S_m\_) \quad \textit{and} \quad \forall P_n\,\mathbf{S}(\_,P_n).$$

If $\mathbf{S}$ *is co-extended* Σ_1^1 *then so are*

$$\exists S_m\,\mathbf{S}(\_,S_m,\_) \quad \textit{and} \quad \exists P_n\,\mathbf{S}(\_,P_n). \quad \square$$

2.8 Proposition. *If* $S=S(p_1,\ldots,p_i)$ *is extended* Π_1^1 *(co-extended* Σ_1^1*) and is really a predicate over* $\mathfrak{M}$*; i.e.* $S\subseteq M^i$*, then* S *is* Π_1^1 *over* $\mathfrak{M}$ *(*Σ_1^1 *over* $\mathfrak{M}$*).*

Proof. It suffices to prove one of these and take negations, so we prove the Σ_1^1 case. Typically S has a definition of the form

$$S(\vec{p}) \quad \text{iff} \quad \exists\vec{T}\,\exists\vec{Q}\,\varphi(\vec{p},\vec{q},\vec{a},\vec{T},\vec{Q})$$

where $\vec{a}$ are some parameters from $\mathbb{HF}_{\mathfrak{M}}$, $\vec{q}\in\mathfrak{M}$, and φ is co-extended. The quantifiers $\exists T_i$ can alway be treated as quantifiers over relations on $\mathbb{HF}_{\mathfrak{M}}$, since we can always say in φ that T_i is a relation of urelements, so we restrict ourselves to

$$S(\vec{p}) \quad \text{iff} \quad \exists Q\,\varphi(\vec{p},q,a,Q)$$

where φ is co-extended. First we need to get rid of the parameter a. But every $a\in\mathbb{HF}_{\mathfrak{M}}$ can be defined over $\mathbb{HF}_{\mathfrak{M}}$ by some extended formula $\psi(x,q_1,\ldots,q_r)$ so

$$S(\vec{p}) \quad \text{iff} \quad \forall x\,[\psi(x,q_1,\ldots,q_r)\to\exists Q\,\varphi(\vec{p},q,x,Q)]$$

and the right hand side, by the above rules, is extended Σ_1^1. We are therefore down to the case

$$S(p) \quad \text{iff} \quad \exists Q\,\varphi(p,q,Q)$$

where Q is, say, 3-ary and φ is co-extended. Now the following are equivalent, where ψ is the conjunction of the axioms of extensionality, pair and union and the empty set axiom:

$$S(p),$$

$$\mathbb{HF}_{\mathfrak{M}}\models\exists \mathrm{Q}\,\varphi(p,q,\mathrm{Q}),$$

$$(\mathbb{HF}_{\mathfrak{M}},Q)\models\varphi(p,q,\mathrm{Q}), \quad \text{for some } Q,$$

$$(\mathfrak{A}_{\mathfrak{M}},Q)\models\varphi(p,q,\mathrm{Q}), \quad \text{for some } (\mathfrak{A}_{\mathfrak{M}},Q) \text{ with } \quad \mathbb{HF}_{\mathfrak{M}}\subseteq_{\text{end}}\mathfrak{A}_{\mathfrak{M}},$$

$$(\mathfrak{A}_{\mathfrak{M}},Q)\models\varphi(p,q,\mathrm{Q}), \quad \text{for some } (\mathfrak{A}_{\mathfrak{M}},Q) \text{ with } \quad \mathfrak{A}_{\mathfrak{M}}\models\psi.$$

The structure $\mathfrak{A}_{\mathfrak{M}}$ can be have the same cardinality as $\mathfrak{M}$ in the last two lines since $\mathfrak{M}$ is infinite. The equivalence of the third and fourth lines follows from the fact that φ is co-extended so it drops down from $\mathfrak{A}_{\mathfrak{M}}$ to $\mathbb{HF}_{\mathfrak{M}}$ by II.2.8. The equivalence of the fourth and last lines in a consequence of the fact that $\mathfrak{A}_{\mathfrak{M}}$ must be isomorphic to an end extension of $\mathbb{HF}_{\mathfrak{M}}$ if $\mathfrak{A}_{\mathfrak{M}}$ is a model of the axioms mentioned. The last line can be rewritten as a Σ_1^1 relation on $\mathfrak{M}$ without much trouble. Let's assume that $\mathfrak{M}=\langle M,R\rangle$ with R binary, to simplify things. We introduce a lot of new relation symbols and define $\mathbf{S}_1(M',R',A,E,F,Q)$ by

$$M'\subseteq M,$$

$$R'\subseteq M'\times M',$$

$$A\subseteq M,\ A\cap M'=0,$$

$$E\subseteq(M'\cup A)\times A,$$

$$F\subseteq M\times M',$$

"*F establishes an isomorphism between* $\langle M,R\rangle$ *and* $\langle M',R'\rangle$",

$$Q\subseteq(M'\cup A)^3.$$

Thus $\mathbf{S}_1$ insures that $(\langle M',R'\rangle; A,E,Q)$ is isomorphic to an $(\mathfrak{A}_{\mathfrak{M}},Q)$. Let $\mathbf{S}_2(M',A,E)$ assert that this structure satisfies *Extensionality, Pair, Union* and *Empty set;* e. g. *Pair* can be expressed by

$$\forall x\,\forall y\,[\mathsf{A}(x)\vee \mathsf{M}'(x))\wedge(\mathsf{A}(y)\vee \mathsf{M}'(y)) \\ \rightarrow \exists z(\mathsf{A}(z)\wedge \forall w\,[w\,\mathsf{E}z\leftrightarrow w=x\vee w=y])].$$

Both $\mathbf{S}_1$, $\mathbf{S}_2$ can be defined by first order sentences over $\mathfrak{M}$ in the additional symbols. Finally, we let $\varphi'(x,y)$ result from $\varphi(x,y)$ by rewritting it in terms of the structure $(\langle M',R'\rangle, A,E,Q)$. For example $\in$ is replaced by E throughout. Then we have

$S(p)$ iff there are M',R',A,E,F,Q such that

$$\mathbf{S}_1(M',R',A,E,F,Q),$$

$$\mathbf{S}_2(M',A,E)\quad\text{and}$$

$$\exists p'\,\exists q'\,(F(p,p')\wedge F(q,q')\wedge\varphi'(p',q'))$$

which makes S Σ_1^1 on $\mathfrak{M}$. □

2.9 Examples. (i) It is worthwhile going back to look at some of the examples in 2.1. In 2.1 (ii) and 2.1 (iii) the Π_1^1 predicates are actually extended first order. For example, in 2.1 (iii),

$$x \text{ is torsion iff}\quad \mathbb{HF}_{\mathfrak{M}}\models\exists n\,(nx=0)$$

where nx is defined by recursion in $\mathbb{HF}_{\mathfrak{M}}$ just as usual:

$$0x=0,$$

$$(n+1)x=nx+x$$

where the 0 and + on the right hand side are the group 0 and group addition. In 2.1 (iv), D is not co-extended but it is co-extended Σ_1^1, hence Σ_1^1 by 2.8.

(ii) Another example that will come up later is where $\mathfrak{M} = \langle M, \sim \rangle$ with $\sim$ an equivalence relation. Define

$$x < y \quad \text{iff} \quad \operatorname{card}(x/\sim) < \operatorname{card}(y/\sim).$$

This relation is Π_1^1. (This is so simple that the above machinery is of little use.) *If each equivalence class is finite then* $<$ *is also* Σ_1^1:

$$\neg(x<y) \quad \text{iff} \quad \mathbb{HF}_{\mathfrak{M}} \models \exists a\, \exists b\, (a = x/\sim \wedge b = y/\sim \wedge \operatorname{card}(b) \leqslant \operatorname{card}(a)),$$

which is extended first order so $\neg(x<y)$ is Π_1^1 so $x<y$ is Σ_1^1. ☐

Let $\mathsf{S}(\vec{p}, \vec{S})$ be a predicate of i-tuples $\vec{p}$ from $\mathfrak{M}$ and m-tuples $\vec{S}$ of relations over $\mathfrak{M}$. S is Π_1^1 on $\mathfrak{M}$ if there is a $\varphi(\vec{p}, \vec{S}, \vec{T})$ such that

$$\mathsf{S}(\vec{p}, \vec{S}) \quad \text{iff} \quad (\mathfrak{M}, \vec{S}) \models \forall \mathsf{T}_1, \ldots, \forall \mathsf{T}_m\, \varphi(\vec{p}, \vec{\mathsf{S}}, \vec{\mathsf{T}}).$$

Some authors refer to such predicates as second order Π_1^1 predicates. The proof of 2.8 may be modified in an obvious way to yield a little more.

2.10 Proposition. *If* $\mathsf{S}(\vec{p}, \vec{S})$ *is extended* Π_1^1 *then* S *is* Π_1^1 *on* $\mathfrak{M}$.

Proof. The extra relations $\vec{S}$ ride along for free. ☐

Probably the most familiar example of a Δ_1^1 non-elementary set over $\mathcal{N}$ is the set of (Gödel numbers of) true sentences of arithmetic. This kind of example is very important. It is contained in the following proposition. Here K is some finite language which is coded up in $\mathbb{HF}$. To keep the notation (barely) manageable, we restrict the statement of the propositions to the case where K has one binary symbol r.

2.11 Proposition. *Define a predicate* $\mathsf{S}(N, R, \varphi, s)$ *by the conjunction:*

(i) $N \subseteq M$; $R \subseteq N \times N$; $\varphi, s \in \mathbb{HF}_{\mathfrak{M}}$;

(ii) φ *is a formula of* $\mathsf{K}_{\omega\omega}$, s *is a function with* $\operatorname{dom}(s) \supseteq$ free variables (φ);

(iii) $\forall x \in \operatorname{rng}(s)\, N(x)$;

(iv) $\langle N, R \rangle \models \varphi[s]$.

Then S *is both extended* Π_1^1 *and co-extended* Σ_1^1.

Proof. There is no trouble with (i)—(iii) since (i), (ii) are Δ_1 on $\mathbb{HF}_{\mathfrak{M}}$ and (iii) is both extended and co-extended first order. The work comes in with (iv). Note, however, that if this particular S is co-extended Σ_1^1 then it is also extended Π_1^1 since

$$\mathsf{S}(N, R, \varphi, s) \quad \text{iff} \quad \text{(i)} \wedge \text{(ii)} \wedge \text{(iii)} \wedge \exists x\, [x = \langle \neg, \varphi \rangle \wedge \neg \mathsf{S}(N, R, x, s)]$$

and the right hand side is extended Π_1^1 by the various lemmas above. We prove that **S** is co-extended Σ_1^1 by introducing another binary relation symbol Sat and finding a co-extended first order $\mathbf{S}^*(N,R,Sat)$ such that for N,R,φ,s satisfying (i)—(iii),

$$\langle N,R\rangle \models \varphi[s] \quad \text{iff} \quad \exists\, Sat\,[\mathbf{S}^*(N,R,Sat) \wedge Sat(\varphi,s)]\,.$$

To write out $\mathbf{S}^*$ we use $s(p/v_i)$ for

$$s\upharpoonright(\mathrm{dom}(s)-\{v_i\}) \cup \{\langle v_i,p\rangle\},$$

this being a Δ_1 operation of s, p and v_i. Now define $\mathbf{S}^*(N,R,Sat)$ by

$$\forall\varphi\,\forall s\,[(\mathrm{i}) \wedge (\mathrm{ii}) \wedge (\mathrm{iii}) \rightarrow$$

$$\text{if } \varphi \text{ is atomic, say } \mathsf{r}(v_i,v_j), \text{ then } R(s(v_i),s(v_j)) \leftrightarrow Sat(\varphi,s),$$
$$\text{if } \varphi \text{ is } \langle \wedge, \{\psi,\theta\}\rangle \text{ then } Sat(\varphi,s) \leftrightarrow Sat(\psi,s) \wedge Sat(\theta,s),$$
$$\text{if } \varphi \text{ is } \langle \neg,\psi\rangle \text{ then } Sat(\varphi,s) \leftrightarrow \neg Sat(\psi,s),$$
$$\text{if } \varphi \text{ is } \langle \exists, v_i, \psi\rangle \text{ then } Sat(\varphi,s) \leftrightarrow \exists p\,[N(p) \wedge Sat(\psi,s(p/v_i))]]$$

with similar clauses for equality, $\bigvee$, $\forall$. Note that the only unbounded existential quantifier comes from the last clause and that quantifier is over urelements so $\mathbf{S}^*$ is co-extended first order. It clearly has the properties needed to finish our proof. □

3. Π_1^1 and Δ_1^1 on Countable Structures

We continue to consider a fixed infinite structure $\mathfrak{M} = \langle M, R_1, \ldots, R_l\rangle$. Our goal here is to show that if $\mathfrak{M}$ is countable then the Δ_1^1 relations over $\mathfrak{M}$ are exactly those relations in $\mathbb{HYP}_{\mathfrak{M}}$. In view of II.5, this shows that the Δ_1^1 relations over $\mathfrak{M}$ are exactly those which are constructible from $\mathfrak{M}$ by the time you come to the first $\mathfrak{M}$-admissible ordinal.

We split the result in half to isolate the role of countability.

3.1 Theorem. *Let $\mathfrak{M}$ be countable. If S is a Π_1^1 relation on $\mathfrak{M}$ then S is Σ_1 on* $\mathbb{HYP}_{\mathfrak{M}}$.

Proof. Consider the language $\mathsf{L} \cup \{\mathsf{P}\}$ as coded in $\mathbb{HYP}_{\mathfrak{M}}$ with $\overline{\mathsf{p}}$ a distinct constant symbol for each $p \in M$. Suppose $S(p)$ iff $\mathfrak{M} \models \forall \mathsf{P}\ \varphi(p,q,\mathsf{P})$. Then $S(p)$ holds iff $(\sigma \rightarrow \varphi(\overline{\mathsf{p}},\overline{\mathsf{q}},\mathsf{P}))$ is valid, where σ is the conjunction of the diagram of $\mathfrak{M}$ and $\forall x \bigvee_{p\in M}(x=\overline{\mathsf{p}})$.

Thus $S(p)$ holds iff the following is true in $\mathbb{HYP}_{\mathfrak{M}}$:

$$\vdash (\sigma \to \varphi(\overline{\mathsf{p}}, \overline{\mathsf{q}}, \mathsf{P}))$$

by the Completeness Theorem for countable, admissible fragments. Thus S is Σ_1 on $\mathbb{HYP}_{\mathfrak{M}}$. ▯

3.2 Corollary. *Let $\mathfrak{M}$ be countable. If S is Δ_1^1 on $\mathfrak{M}$ then $S \in \mathbb{HYP}_{\mathfrak{M}}$.*

Proof. Immediate from 3.1 and Δ_1 separation in $\mathbb{HYP}_{\mathfrak{M}}$. ▯

The converse does not need the countability assumption.

3.3 Theorem. *Let S be a relation on $\mathfrak{M}$. If S is Σ_1 on $\mathbb{HYP}_{\mathfrak{M}}$ then S is Π_1^1 on $\mathfrak{M}$.*

3.4 Corollary. *If a relation S on $\mathfrak{M}$ is in $\mathbb{HYP}_{\mathfrak{M}}$ then S is Δ_1^1 on $\mathfrak{M}$.*

Proof. If $S \in \mathbb{HYP}_{\mathfrak{M}}$ then S and $\neg S$ are Σ_1 on $\mathbb{HYP}_{\mathfrak{M}}$. (Remember that parameters from $\mathbb{HYP}_{\mathfrak{M}}$ are allowed in Σ_1 definitions.) ▯

Proof of 3.3. Let $S(p)$ be Σ_1 on $\mathbb{HYP}_{\mathfrak{M}}$. By Proposition II.8.8 we can find a Σ_1 formula $\varphi(x, q, M)$ such that the following are equivalent:

$S(p)$,

$\mathbb{HYP}_{\mathfrak{M}} \models \varphi[p, q, M]$,

(1) $\mathfrak{A}_{\mathfrak{M}} \models \varphi[p, q, M]$ for every model $\mathfrak{A}_{\mathfrak{M}}$ of KPU^+ (of cardinality $\mathrm{card}(M)$).

The last line is true with or without the parenthetical phrase since $\mathrm{card}(M) = \mathrm{card}(\mathbb{HYP}_{\mathfrak{M}})$. Now code up the language $\mathsf{L}(\in)$ in $\mathbb{HF}$. Call the resultung set K, $\mathsf{K} \in \mathbb{HF}$. Let kpu^+ be the set of codes of KPU^+ and let $\varphi = \varphi(v_1, v_2, v_3)$ denote the code of itself. Thus $\varphi \in \mathbb{HF}$ and kpu^+ is a Δ_1 subset ot $\mathbb{HF}$ by Theorem II.2.3. Our plan is to rewrite (1) as a Π_1^1 relation over $\mathfrak{M}$ with the aid of 2.10 and 2.8. Again we simplify notation by assuming $\mathfrak{M} = \langle M, R \rangle$ with R binary. Now (1) is equivalent to:

For all M, R, F and all A, E,

(2) *if* $\langle M', R' \rangle \overset{F}{\cong} \langle M, R \rangle$
and $(\langle M', R' \rangle; A, E)$ *is a structure of the appropriate kind, and*

(3) *if* $\langle M', R', A, E \rangle \models \mathrm{kpu}^+$,

(4) *then for some* p', q', m, $\langle M', R', A, E \rangle \models \varphi(p', q', m)$ *where* $F(p) = p'$, $F(q) = q'$ *and* $m \in A$ *is such that* $x\, E m \leftrightarrow M'(x)$ *for all* x.

Let $\mathbf{S}_1(M', R', A, E, F)$ be just as in the proof of 2.8 so that $\mathbf{S}_1$ is first order in the symbols and $\mathbf{S}_1$ expresses line (2). Let $\mathbf{S}_2(M', R', A, E)$ hold if

$$\forall \psi\, [\psi \in \mathrm{kpu}^+ \to (M', R', A, E) \models \psi].$$

$\mathbf{S}_2$ expresses (3) and is co-extended Σ_1^1 by 2.11, 2.6 and other lemmas. (It is not necessarily extended Π_1^1, though, due to the $\forall\psi$ in front.) Line (4) can be written in extended Π_1^1 form by 2.11. This makes $S(p)$ of the form

$$\forall M', R', A, E, F\,[\mathbf{S}_1 \wedge \mathbf{S}_2 \rightarrow \mathbf{S}_3]$$

where $\mathbf{S}_1$ is first order, $\mathbf{S}_2$ is co-extended Σ_1^1 and $\mathbf{S}_3$ extended Π_1^1 so S is extended Π_1^1 and hence Π_1^1 by 2.8. □

3.5 Corollary. *For any structure $\mathfrak{M} = \langle M, R_1, \ldots, R_l \rangle$, countable or not, the relations S on $\mathfrak{M}$ in $\mathbb{HYP}_M$ are exactly the relations definable over $\mathfrak{M}$ by some formula $\varphi(v_1, \ldots, v_n, q_1, \ldots, q_m)$ of the admissible fragment $\mathsf{L}_{\mathbb{A}}$ where $\mathbb{A} = \mathbb{HYP}_{\mathfrak{M}}$.*

Proof. If S is defined by

$$S(p_1, \ldots, p_n) \quad \text{iff} \quad \mathfrak{M} \models \varphi(p_1, \ldots, p_n, q_1, \ldots, q_m)$$

where $\varphi \in \mathbb{HYP}_{\mathfrak{M}}$ then S is Δ_1 since $\models$ is Δ_1. Thus $S \in \mathbb{HYP}_{\mathfrak{M}}$ by Δ_1 separation.

To prove the converse, first assume $\mathfrak{M}$ is countable. Since $S \in \mathbb{HYP}_{\mathfrak{M}}$ we can write

$$\begin{aligned} S(\vec{p}) \quad &\text{iff} \quad \mathfrak{M} \models \forall \mathsf{T}\; \varphi(\mathsf{T}, \vec{p}) \\ &\text{iff} \quad \mathfrak{M} \models \exists \mathsf{T}'\; \psi(\mathsf{T}', \vec{p}) \end{aligned}$$

for some first order formulas φ, ψ possibly with constants $\overline{\mathsf{q}}_1, \ldots, \overline{\mathsf{q}}_m$. We may assume T, T' are distinct symbols. Let σ be the sentence

$$\bigwedge \text{Diagram}(\mathfrak{M}) \wedge \forall x \bigvee_{p \in M} x = \overline{\mathsf{p}}.$$

The sentence

$$\forall v_1, \ldots, v_n\,[(\sigma \wedge \psi(\mathsf{T}', v_1, \ldots, v_n)) \rightarrow \varphi(\mathsf{T}, v_1, \ldots, v_n)]$$

is logically valid since for any T' on $\mathfrak{M}$, $(\mathfrak{M}, T') \models \psi(p_1, \ldots, p_n)$ implies $S(p_1, \ldots, p_n)$, which in turn implies $(\mathfrak{M}, T) \models \varphi(p_1, \ldots, p_n)$ for any T on $\mathfrak{M}$. By the Interpolation Theorem of III.6.1 there is a formula $\theta(v_1, \ldots, v_n) \in \mathbb{HYP}_{\mathfrak{M}}$, θ involving only the symbols of L and any constants $\overline{\mathsf{q}}$ in φ such that both

$$\sigma \wedge \psi(\mathsf{T}', v_1, \ldots, v_n) \rightarrow \theta(v_1, \ldots, v_n) \quad \text{and}$$

$$\theta(v_1, \ldots, v_n) \rightarrow \varphi(\mathsf{T}, v_1, \ldots, v_n)$$

are valid. But then

$$S(p_1, \ldots, p_n) \quad \text{iff} \quad \mathfrak{M} \models \theta[p_1, \ldots, p_n].$$

Thus the result holds if $\mathfrak{M}$ is countable.

To prove the result for uncountable $\mathfrak{M}$ we apply the Lévy Absoluteness Principle of II.9. The theorem to be proved can be written out as

$$\forall \mathfrak{M}\, \forall S\, [S \subseteq M^n \wedge S \in \mathbb{H}\mathrm{YP}_{\mathfrak{M}} \rightarrow \exists \theta \in \mathbb{H}\mathrm{YP}_{\mathfrak{M}} (\forall \vec{p} \in M(S(\vec{p}) \leftrightarrow \mathfrak{M} \models \theta(\vec{p})))]$$

so we need to see that the part within brackets can be written as a Π predicate in ZFC. Recalling that $\mathbb{H}\mathrm{YP}_{\mathfrak{M}} = \mathrm{L}(\alpha)_{\mathfrak{M}}$ for the first α to make $\mathrm{L}(\alpha)_{\mathfrak{M}}$ admissible, we can rewrite it as

$$S \subseteq M^n \wedge \forall \alpha [\mathrm{L}(\alpha)_M \models \mathrm{KPU}^+ \wedge \forall \beta < \alpha (\mathrm{L}(\beta)_{\mathfrak{M}} \not\models \mathrm{KPU}^+)$$
$$\wedge\, S \in \mathrm{L}(\alpha)_{\mathfrak{M}} \rightarrow \exists \theta (v_1, \ldots, v_n) \in \mathrm{L}(\alpha)_{\mathfrak{M}} (\forall \vec{p} \in M^n, \vec{p} \in S \leftrightarrow \mathfrak{M} \models \theta[\vec{p}])]$$

The part within brackets here is clearly Δ_1 since $\models$ is Δ_1. Thus the theorem is a Π sentence and so it suffices to prove it for countable structures $\mathfrak{M}$. □

There are useful second order generalizations of the above theorems. For example, generalizing 3.1 we get the following result.

3.6 Theorem. *Let $\mathbf{S}(\vec{p}, \vec{S})$ be a Π_1^1 predicate on a countable structure $\mathfrak{M}$. For every admissible set $\mathbb{A}$ with $M \in \mathbb{A}$, $S \cap \mathbb{A}$ is Σ_1 on $\mathbb{A}$. The Σ_1 definition is independent of $\mathbb{A}$.*

Proof. If $\mathbf{S}(\vec{p}, S)$ holds iff $(\mathfrak{M}, S) \models \forall \mathsf{T}\, \varphi(\vec{p}, \mathsf{S}, \mathsf{T})$, then $\mathbf{S}(\vec{p}, S)$ holds iff $(\sigma(\mathsf{S}) \rightarrow \varphi(\bar{\mathsf{p}}, \mathsf{S}, T))$ is valid, where $\sigma(\mathsf{S})$ is

$$\bigwedge \mathrm{diagram}(\mathfrak{M}, S) \wedge \forall x \bigvee_{p \in M} (x = \bar{\mathsf{p}}).$$

This is a countable sentence of $\mathsf{L}_{\infty\omega}$ so the proof given in 3.1 carries over. □

The second order generalization of 3.3 is not quite the converse of 3.6.

3.7 Theorem. *Let $\mathbf{S} = \mathbf{S}(\vec{p}, \vec{S})$ be a second order predicate on $\mathfrak{M}$ which is a Σ_1 subset of $\mathbb{H}\mathrm{YP}_{\mathfrak{M}}$. Then $\mathbf{S}$ is Π_1^1 on $\mathfrak{M}$.*

Proof. A simple modification of the proof of 3.3 suffices. Line (1) becomes

(1′) $(\mathfrak{A}_{\mathfrak{M}}, S) \models \varphi[p, q, \mathsf{S}, M]$, for every model $\mathfrak{A}_{\mathfrak{M}}$ of KPU^+ and every S

which results in a modification of (4) to

(4′) *then for some* p', q', m, s, $\langle M', R', A, E \rangle \models \varphi(p', q', s, m,)$ *where*

$$F(p) = p',\ F(q) = q',\ A(m) \wedge \forall x\, [xEm \leftrightarrow M'(x)] \wedge A(s),$$
$$\forall x\, [S(x) \leftrightarrow \exists y (F(x) = y \wedge yEs)].\ \square$$

3.8 Corollary. *The set* $\mathbf{S}$ *defined by*

$$\mathbf{S} = \{S \subseteq M^n : S \in \mathbb{H}\mathrm{YP}_{\mathfrak{M}}\}$$

is Π_1^1 on $\mathfrak{M}$ (as a second order predicate).

Proof. $\mathbf{S}$ is Δ_0 on $\mathbb{HYP}_{\mathfrak{M}}$ since

$$x \in \mathbf{S} \quad \text{iff} \quad \mathbb{HYP}_{\mathfrak{M}} \models \text{“}x \text{ is a subset of } M\text{”}$$

so $\mathbf{S}$ is Π_1^1 on $\mathfrak{M}$ by 3.7. Note, however, that 3.7 will not allow us to conclude that $\neg\mathbf{S}$ is Π_1^1 on $\mathfrak{M}$ since $\neg\mathbf{S}$ is not a subset of $\mathbb{HYP}_{\mathfrak{M}}$; far from it. ▯

3.9 Example. Let us return to consider nonstandard models of arithmetic. We showed in § 3 that the set of standard integers in a nonstandard model $\mathfrak{M} = \langle M, 0, +, \cdot \rangle$ is Π_1^1 on $\mathfrak{M}$. Sometimes it is Σ_1^1 hence Δ_1^1, sometimes not. Recall that $\mathcal{N} = \langle \omega, 0, +, x \rangle$.

i) For an $\mathfrak{M}$ where the set of standard integers *is* Σ_1^1 let $\mathfrak{M}$ be a minimal elementary extension of $\mathcal{N}$: i.e., $\mathcal{N} \prec \mathfrak{M}$ but $\mathcal{N} \prec \mathfrak{N} \prec \mathfrak{M}$ implies $\mathcal{N} = \mathfrak{N}$ or $\mathfrak{N} = \mathfrak{M}$. Such $\mathfrak{M}$ exist by results of Gaifman [1970]. In such an $\mathfrak{M}$ we can define, for $x \in \mathfrak{M}$,

x is standard iff $\exists M_0$ [M_0 is the universe of a proper elementary submodel of $\mathfrak{M}$ and $M_0(x)$].

This is extended Σ_1^1 by 3.10, hence Σ_1^1 by 3.8.

ii) For an $\mathfrak{M} \succ \mathcal{N}$ where the set of standard integers is not Δ_1^1 hence not Σ_1^1, choose a countable $\mathfrak{M}$ with $O(\mathfrak{M}) = \omega$ (by II.8.7). The subsets of $\mathfrak{M}$ in $\mathbb{HYP}_{\mathfrak{M}}$ are exactly the first order definable sets (by II.6.7) so the set of standard integers are not in $\mathbb{HYP}_{\mathfrak{M}}$ and hence, by the results of this section, they are not Δ_1^1 on $\mathfrak{M}$. In fact, we see that for countable M, the set of standard integers is Δ_1^1 on $\mathfrak{M}$ iff $O(\mathfrak{M}) > \omega$. We will return to this example later. ▯

3.10—3.12 Exercises

3.10. Let $\mathfrak{M}$ be countable and let $\mathbf{S}_1(p,P)$, $\mathbf{S}_2(p,P)$ be predicates of $p \in M$, $P \subseteq M^2$, each Σ_1^1 on $\mathfrak{M}$. Assume that no pair (p,P) satisfies both $\mathbf{S}_1$ and $\mathbf{S}_2$. Show that there is a Δ_1^1 predicate $\mathbf{S}(p,P)$ containing $\mathbf{S}_1$ but disjoint from $\mathbf{S}_2$. [Copy the proof of 3.5 to find a $\theta(p,P)$ in $\mathsf{L}_{\mathbb{A}}$ such that $\mathbf{S}(p,P)$ iff $(\mathfrak{M},P) \models \theta(p,P)$ and then show that $\mathbf{S}$ is Δ_1^1.]

3.11. Recall Example 2.1 (iv). Let $\alpha > \omega$ be any countable admissible ordinal. Let p be any prime. Show that there is a countable p-group G with length $(G) = \alpha$ such that G has a proper divisible subgroup but none in $\mathbb{HYP}_G$. For such a G the largest divisible subgroup of G is thus Σ_1^1 but not Π_1^1. [Use the $\overline{YY}$-Compactness Theorem.]

3.12. Generalize the results of this section to show, for $\mathbb{A}_{\mathfrak{M}}$ transitive, $\mathbb{HF}_{\mathfrak{M}} \subseteq \mathbb{A}_{\mathfrak{M}}$:

i) If S is a relation on $\mathbb{A}_{\mathfrak{M}}$ and S is Σ_1 on $\mathbb{HYP}(\mathbb{A}_{\mathfrak{M}})$ then S is Π_1^1 on $\mathbb{A}_{\mathfrak{M}}$.

ii) If $\mathbb{A}_{\mathfrak{M}}$ is countable then the converse of i) holds.

3.13 Notes. Kripke and Platek proved that a subset X of $\mathbb{HF}$ is Π_1^1 over $\mathbb{HF}$ iff X is Σ_1 over $\mathbb{HYP}(\mathbb{HF})$ and hence that X is Δ_1^1 over $\mathbb{HF}$ iff $X \in \mathbb{HYP}(\mathbb{HF})$. This was generalized in Barwise-Gandy-Moschovakis [1971] by replacing $\mathbb{HF}$ by any countable transitive set A closed under pairs. It is clear from the proof given there that Theorem 3.1 holds. It came as somewhat of a surprise that its converse, Theorem 3.3, holds without any coding assumptions about the structure $\mathfrak{M}$, since the inductive definability approach (discussed in Chapter VI) does not work in this complete generality.

4. Perfect Set Results

In this section we give a more sophisticated example of the interplay of model theory and recursion theory showing how each subject can shed light on the other and how logic on admissible sets sheds light on both. The results themselves will not be used in the remainder of the book.

The following, a classical result on hyperarithmetic sets, is the effective version (due to Harrison) of an even older result in descriptive set theory.

4.1 Theorem. *If* $\mathbf{S} \subseteq \text{Power}(\omega)$ *is* Σ_1^1 *on* $\mathcal{N} = \langle \omega, 0, +, \cdot \rangle$ *and* $\text{card}(\mathbf{S}) < 2^{\aleph_0}$ *then* $\mathbf{S}$ *is a set of hyperarithmetic sets.*

Compare this with two results from model theory. The first is due to Kueker [1968].

4.2 Theorem. *Let* $\mathfrak{M} = \langle M, R_1, \ldots, R_l \rangle$ *be a countable structure for a language* L *and let* P *be an n-ary relation on* M. *If the set*

$$\mathbf{S} = \{Q \mid (\mathfrak{M}, P) \cong (\mathfrak{M}, Q)\}$$

has $\text{card}(\mathbf{S}) < 2^{\aleph_0}$ *then*

$$P = \{(x_1, \ldots, x_n) \mid \mathfrak{M} \models \varphi[x_1, \ldots, x_n, q_1, \ldots, q_m]\}$$

for some formula φ *of* $\mathsf{L}_{\omega_1\omega}$ *and some* $q_1, \ldots, q_m \in \mathfrak{M}$.

(A formula φ is in $\mathsf{L}_{\omega_1\omega}$ if it is in $\mathsf{L}_{\mathbb{A}}$ for some countable fragment $\mathsf{L}_{\mathbb{A}}$ of $\mathsf{L}_{\infty\omega}$.)

The next result is a theorem of Chang [1964], Makkai [1964], and Reyes [1968]. Chang and Makkai had a stronger hypothesis.

4.3 Theorem. *Let* $\varphi(\mathsf{P})$ *be a finitary sentence of* $\mathsf{L} \cup \{\mathsf{P}\}$. *Assume that for each countable model* $\mathfrak{M}$ *there are fewer than* $2^{\aleph_0}$ *relations* P *such that*

$$(\mathfrak{M}, P) \models \varphi(\mathsf{P}).$$

Then there are finitary formulas $\psi_1(\vec{x}, y_1, \ldots, y_k), \ldots, \psi_m(\vec{x}, y_1, \ldots, y_{k_m})$ *of* $\mathsf{L}_{\omega\omega}$ *such that for every model* $(\mathfrak{M}, P)$ *of* $\varphi(\mathsf{P})$, *there is an* i, $1 \leqslant i \leqslant m$, *and* $q_1, \ldots, q_{k_i} \in \mathfrak{M}$ *such that*

$$P = \{(x_1, \ldots, x_n) \mid \mathfrak{M} \models \psi_i[\vec{x}, q_1, \ldots, q_{k_i}]\}.$$

The conclusion of 4.3 can be restated as: *the sentence*

$$\varphi(\mathsf{P}) \to \bigvee_{1 \leqslant i \leqslant m} \exists y_1, \ldots, y_{k_i} \forall \vec{x}[\mathsf{P}(\vec{x}) \leftrightarrow \psi_i(\vec{x}, y_1, \ldots, y_{k_i})]$$

is logically valid.

These three results, while incomparable, are obviously quite similar. They all begin with the assumption that a certain definable or Σ_1^1 class **S** has fewer than $2^{\aleph_0}$ elements and conclude that each element of **S** is definable in some way. We want to show these results are more than merely analogous, that they are in fact shadows of a single definability result about logic on admissible sets. First, though, we prove a generalization of 4.1, because the proof is relevant to our general result.

4.4 Theorem. *Let* $\mathfrak{M} = \langle M, R_1, \ldots, R_l \rangle$ *be a countable structure and let* **S** *be a second order* Σ_1^1 *predicate on* $\mathfrak{M}$. *If* $\operatorname{card}(\mathbf{S}) < 2^{\aleph_0}$ *then* $\mathbf{S} \subseteq \mathbb{HYP}_{\mathfrak{M}}$ *(and hence* **S** *is countable).*

Proof. After a trick the result falls right out of III.8.2. Assume $\mathbf{S} \nsubseteq \mathbb{HYP}_{\mathfrak{M}}$. Then by 3.8 (and this is the trick), $\mathbf{S}_0 = \mathbf{S} - \mathbb{HYP}_{\mathfrak{M}}$ is Σ_1^1 and non-empty. We prove that $\mathbf{S}_0$ (and hence **S**) has cardinality $2^{\aleph_0}$. Let us handle the case where $\mathbf{S}_0$ is a predicate of one relation:

$$\mathbf{S}_0(S) \text{ iff } (\mathfrak{M}, S) \models \exists \mathsf{T}\, \varphi(\mathsf{S}, \mathsf{T}).$$

Let $\mathsf{L}' = \mathsf{L} \cup \{\bar{\mathsf{p}} : p \in M\} \cup \{\mathsf{S}\}$, $\mathsf{K} = \mathsf{L}' \cup \{\mathsf{T}\}$ and let $\mathsf{L}'_{\mathbb{A}}, \mathsf{K}_{\mathbb{A}}$ be the countable admissible fragments given by $\mathbb{HYP}_{\mathfrak{M}}$. If σ is

$$\operatorname{Diagram}(\mathfrak{M}) \wedge \forall x \bigvee_{p \in M} (x = \bar{\mathsf{p}})$$

then $\sigma \wedge \varphi(\mathsf{S}, \mathsf{T})$ is in $\mathsf{K}_{\mathbb{A}}$. We claim that σ can have no model which is decidable for $\mathsf{L}'_{\mathbb{A}}$. Such a model would be isomorphic to some structure of the form $(\mathfrak{M}, S, T)$, where S is Δ_1 on $\mathbb{HYP}_{\mathfrak{M}}$ and hence $S \in \mathbb{HYP}_{\mathfrak{M}}$, whereas $(\mathfrak{M}, S, T) \models \varphi(\mathsf{S}, \mathsf{T})$, implies $S \in \mathbf{S}_0$. Thus the result follows from III.8.2. □

We now turn to consider the relationship between 4.2 and 4.4. If we apply 4.4 to the situation described in Theorem 4.2 we learn that if there are $< 2^{\aleph_0}$ Q's with $(\mathfrak{M}, P) \cong (\mathfrak{M}, Q)$, then each of these is Δ_1^1 on $(\mathfrak{M}, P)$ which (while interesting and not obvious from 4.2) says nothing about the original P. There are examples $(\mathfrak{M}, P)$ satisfying 4.2 but where $P \notin \mathbb{HYP}_{\mathfrak{M}}$, i.e., is not Δ_1^1 on $\mathfrak{M}$, which rules out one possible strengthing of 4.4 that would yield 4.2. To find the correct generalization of 4.2, 4.3 and 4.4 we need a new definition.

4.5 Definition. A Σ_1^1 sentence of an admissible fragment $\mathsf{L}_{\mathbb{A}}$ is a second order infinitary sentence of the form

$$\exists \mathcal{Q}\, \varphi$$

where $\mathcal{Q}$ is a set of symbols of L, $\mathcal{Q} \in \mathbb{A}$, and $\varphi \in \mathsf{L}_{\mathbb{A}}$.

If $\mathcal{Q}$ is finite, the requirement $\mathcal{Q} \in \mathbb{A}$ is automatically true, and we could write

$$\exists \vec{\mathrm{Q}} \varphi(\vec{\mathrm{Q}})$$

or

$$\exists \mathrm{Q}_1, \ldots, \exists \mathrm{Q}_n \varphi(\mathrm{Q}_1, \ldots, \mathrm{Q}_n).$$

In the infinite case, however, we should not think of $\mathcal{Q}$ as being a well ordered sequence of symbols. Note that even though we have written $\mathcal{Q}$, the definition actually permits function and constant symbols to occur in $\mathcal{Q}$ as well as relations symbols.

The following result has 4.2—4.4 as consequences. For ordinary (as opposed to Σ_1^1) sentences of $\mathsf{L}_{\mathbb{A}}$ it is due to Makkai [1973]. For 4.4, though, it is the Σ_1^1 version which matters. The proof is a minor variation on Makkai's theme, the Interpolation Theorem taking the part formerly played by Beth's theorem.

4.6 Theorem. *Let $\exists \mathcal{Q} \varphi(\mathrm{P}, \mathcal{Q})$ be a Σ_1^1 sentence of the countable admissible fragment $\mathsf{L}_{\mathbb{A}}$. If for each countable structure $\mathfrak{M}$ there are less than $2^{\aleph_0}$ relations P such that*

$$(\mathfrak{M}, P) \models \exists \mathcal{Q} \varphi(\mathrm{P}, \mathcal{Q})$$

then there is a sentence σ of $\mathsf{L}_{\mathbb{A}}$ of the form

$$\bigvee_{i \in I} \exists y_1, \ldots, \exists y_{m_i} \forall x_1, \ldots, \forall x_n [\mathrm{P}(x_1, \ldots, x_n) \leftrightarrow \psi_i(x_1, \ldots, x_n, y_1, \ldots, y_{m_i})]$$

which is a logical consequence of $\varphi(\mathrm{P}, \mathcal{Q})$, where each ψ_i contains only symbols of L not in $\mathcal{Q} \cup \{\mathrm{P}\}$.

The converse is obvious. In fact, the conclusion implies that every such P is in any admissible set containing $\mathfrak{M}$ and φ so there are $\leqslant \aleph_0$ such P.

Note that Theorem 4.3 is the special case of Theorem 4.6 where $\mathsf{L}_{\mathbb{A}}$ is $\mathsf{L}_{\omega\omega}$ and where the Q's do not occur in $\varphi(\mathrm{P}, \mathcal{Q})$.

Before attempting to prove 4.6 it is good to get some idea of what it says by applying it to prove 4.2 and strengthen 4.4.

4.7 Corollary. *Under the assumption of Theorem 4.4 there is an $\mathbf{S}' \in \mathbb{HYP}_{\mathfrak{M}}$ such that $\mathbf{S} \subseteq \mathbf{S}'$.*

Proof. Suppose $P \in \mathbf{S}$ iff $(\mathfrak{M}, P) \models \exists \mathrm{Q} \varphi_0(\mathrm{P}, \mathrm{Q})$. Let φ be the conjunction of $\varphi_0(\mathrm{P}, \mathrm{Q})$, diagram $(\mathfrak{M})$ and $\forall x \bigvee_{p \in M} (x = \bar{\mathrm{p}})$. The hypothesis of 4.6 is satisfied so let σ be as in the conclusion of 4.6, σ of the form

$$\bigvee_{i \in I} \exists y_1, \ldots, \exists y_{m_i} \forall x_1, \ldots, \forall x_n [\mathrm{P}(x_1, \ldots, x_n) \leftrightarrow \psi_i(x_1, \ldots, x_n, y_1, \ldots, y_{m_i})],$$

where each ψ_i is in the language $\mathsf{L} \cup \{\bar{\mathrm{p}} | \in M\}$. For each $i \in I$ and $q_1, \ldots, q_{m_i} \in M$ let

$$P_{i, \vec{q}} = \{(x_1, \ldots, x_n) | \mathfrak{M} \models \psi_i[x_1, \ldots, x_n, q_1, \ldots, q_m]\}.$$

Each $P_{i,\vec{q}} \in \mathbb{HYP}_{\mathfrak{M}}$ by Δ_1 Separation and, as an operation of i and $\vec{q}$, $P_{i,\vec{q}}$ is a Σ operation in $\mathbb{HYP}_{\mathfrak{M}}$ so we may form the set

$$\mathbf{S}' = \{P_{i,\vec{q}} \mid i \in I, \vec{q} \in M\} \in \mathbb{HYP}_{\mathfrak{M}}$$

by Σ Replacement and $\mathbf{S} \subseteq \mathbf{S}'$. ▯

4.8 Theorem. *Let $\mathfrak{M} = \langle M, R_1, \ldots, R_l \rangle$ be a countable recursively saturated structure (i.e. $o(\mathbb{HYP}_{\mathfrak{M}}) = \omega$). Let $\mathbf{S}$ be a second order Σ_1^1 predicate with* $\operatorname{card}(\mathbf{S}) < 2^{\aleph_0}$, *say $\mathbf{S} \subseteq \operatorname{Power}(M^n)$. There is a finite set of finitary formulas*

$$\psi_1(\vec{x}, y_1, \ldots, y_{m_1}), \ldots, \psi_k(\vec{x}, y_1, \ldots, y_{m_k})$$

of $\mathsf{L}_{\omega\omega}$ such that for each $S \in \mathbf{S}$ there is an i, $1 \leqslant i \leqslant k$, and elements $q_1, \ldots, q_{m_i}$ of $\mathfrak{M}$ so that S is defined by

$$S(\vec{x}) \text{ iff } \mathfrak{M} \models \psi_i[\vec{x}, q_1, \ldots, q_{m_i}].$$

Proof. Using 4.7 choose $\mathbf{S}'$ so that $\mathbf{S}' \subseteq \operatorname{Power}(M^n)$ and

$$\mathbf{S} \subseteq \mathbf{S}' \in \mathbb{HYP}_{\mathfrak{M}}.$$

Since $o(\mathbb{HYP}_{\mathfrak{M}}) = \omega$ we have, by II.7.3,

$$\forall S \in \mathbf{S}' \, \exists \psi \, \exists \vec{q}$$

[ψ is a formula of $\mathsf{L}_{\omega\omega}$, $\vec{q}$ is an m-tuple of elements of M (where the free variables of ψ are among $v_1, \ldots, v_{n+m}$) so that for all $x_1, \ldots, x_n \in M$:

$$\langle x_1, \ldots, x_n \rangle \in S \text{ iff } \mathfrak{M} \models \psi[x_1, \ldots, x_n, q_1, \ldots, q_m]].$$

Since L is finite we can assume $\mathsf{L}_{\omega\omega}$ is coded up on $\mathbb{HF}$. By Σ Collection in $\mathbb{HYP}_{\mathfrak{M}}$ there is a finite set Φ of formulas such that each ψ can be chosen in Φ. ▯

4.9 Example. *Let $\mathcal{N} = \langle \omega, 0, +, \cdot \rangle$ and let $\mathfrak{M}$ be a countable recursively saturated elementary extension of $\mathcal{N}$. Then there are $2^{\aleph_0}$ distinct $\mathfrak{M}_0$ such that*

(i) *$\mathfrak{M}_0 \prec \mathfrak{M}$, and*

(ii) *$\mathfrak{M}_0$ is an initial segment of $\mathfrak{M}$.*

Proof. Let

$$\mathbf{S} = \{M_0 \subseteq M \mid M_0 \text{ is the universe of an } \mathfrak{M}_0 \text{ with (i) and (ii)}\}.$$

The techniques of § 2 show that $\mathbf{S}$ is Σ_1^1 on $\mathfrak{M}$. Suppose, toward a contradiction, that $\operatorname{card}(\mathbf{S}) < 2^{\aleph_0}$. Then since $\omega \in \mathbf{S}$, there is a formula $\psi(x, q_1, \ldots, q_m)$ with parameters from $\mathfrak{M}$ such that

$$\omega = \{x \mid \mathfrak{M} \models \psi[x, q_1, \ldots, q_m]\}$$

which is a contradiction. □

Before turning to the proof of Theorem 4.6, we show how 4.8 can be used to strengthen the Chang-Makkai-Reyes Theorem (4.3). The result is interesting because of the light it sheds on the usual proofs of this theorem by means of saturated (or special) models.

4.10 Corollary. *Let $\varphi(\mathsf{P},\mathsf{Q})$ be a finitary sentence such that for each* recursively saturated *countable model $\mathfrak{M}$, there are less than $2^{\aleph_0}$ different P with*

$$(\mathfrak{M}, P) \models \exists \mathsf{Q}\, \varphi(\mathsf{P},\mathsf{Q}).$$

Then there is a finite list of finitary formulas $\psi_1(\vec{x},\vec{y}), \ldots, \psi_m(\vec{x},\vec{y})$ such that

$$\models \varphi(\mathsf{P},\mathsf{Q}) \to \bigvee_{1 \leq i \leq m} \exists \vec{y}\, \forall \vec{x}\, [\mathsf{P}(\vec{x}) \leftrightarrow \psi(\vec{x},\vec{y})].$$

Proof. Suppose that the hypothesis holds but that the conclusion falls. Let T be the theory

$$\varphi(\mathsf{P},\mathsf{Q})$$
$$\neg \exists \vec{y}\, \forall \vec{x}\, [\mathsf{P}(\vec{x}) \leftrightarrow \psi(\vec{x},\vec{y})], \quad \text{for all} \quad \psi \in \mathsf{L}_{\omega\omega}.$$

By the ordinary compactness theorem, this theory is consistent. By Theorem II.8.8, it has a countable recursively saturated model $(\mathfrak{M}, P)$. But this structure $\mathfrak{M}$ has $< 2^{\aleph_0} P'$ such that $(\mathfrak{M}, P') \models \exists \mathsf{Q}\, \varphi(\mathsf{P},\mathsf{Q})$ so, by 4.8, each of these P' (in particular the original P) is definable, contradicting the fact that $(\mathfrak{M}, P)$ is a model of T. □

4.11. Proof of 4.2 from 4.6. We must cheat a bit by quoting a result, Scott's Theorem, from Chapter VII. Let $\mathfrak{M}, P, \mathbf{S}$ be given as in 4.2 and suppose that $\operatorname{card}(\mathbf{S}) < 2^{\aleph_0}$. Let $\varphi(\mathsf{P})$ be the Scott sentence of $(\mathfrak{M}, P)$ so that for all countable structures $(\mathfrak{M}', P')$,

$$(\mathfrak{M}', P') \models \varphi(\mathsf{P}) \text{ iff } (\mathfrak{M}, P) \cong (\mathfrak{M}', P').$$

(The sentence $\varphi(\mathsf{P})$ involves only constants from $\mathsf{L} \cup \{\mathsf{P}\}$.) Thus there are, for each model $\mathfrak{M}'$, fewer than $2^{\aleph_0} P'$ such that

$$(\mathfrak{M}', P') \models \varphi(\mathsf{P}).$$

From 4.6 we get a $\psi(x_1, \ldots, x_n, y_1, \ldots, y_m)$ such that for some $q_1, \ldots, q_m \in M$

$$(\mathfrak{M}, P) \models \forall \vec{x} [\mathsf{P}(\vec{x}) \leftrightarrow \psi(x_1, \ldots, x_n, q_1, \ldots, q_m)] .$$

which yields the conclusion of 4.2. □

Having no excuse for further procrastination, we begin the proof of 4.6.

4.12. Proof of 4.6. Since 4.6 implies 4.4 we expect to use considerations similar to those used in proving 4.4, that is, the method of § III.8. The chief difference is that instead of constructing $2^{\aleph_0}$ distinct models $\mathfrak{M}$ we need one model with $2^{\aleph_0}$ distinct P such that

$$(\mathfrak{M}, P) \models \exists \mathcal{Q} \varphi(\mathsf{P}, \mathcal{Q}) .$$

This accounts for the complications in the proof below. We prove the contra-positive, so suppose $\varphi(\mathsf{P}, \mathcal{Q})$ does not have any sentence of the desired form as a logical consequence. Let us simplify matters by assuming that $\mathcal{Q}$ has only one relation symbol Q and, further, that P is unary. The proof will make it clear that these assumptions do not really matter. Let

$$\mathsf{L}^0 = \mathsf{L} - \{\mathsf{P}, \mathsf{Q}\}, \quad C = \{\mathsf{c}_n \mid n < \omega\}, \quad \mathsf{K}^0 = \mathsf{L}^0 \cup C, \quad \mathsf{K} = \mathsf{L} \cup C .$$

Call a set s of sentences of $\mathsf{K}_{\mathbb{A}}$ *special* if the following conditions are fulfilled, conditions (D1)—(D7) coming from (C1)—(C7) of III.2.2 respectively.

(D1) If $\varphi \in s$ then $\neg \varphi \notin s$.
(D2) If $\neg \varphi \in s$ then $\sim \varphi \in s$.
(D3) If $\bigwedge \Phi \in s$ then $\varphi \in s$ for all $\varphi \in \Phi$.
(D4) If $\forall v \varphi(v) \in s$ then $\varphi(\mathsf{c}) \in s$ for all $\mathsf{c} \in C$.
(D5) If $\bigvee \Phi \in s$ then $\varphi \in s$ for some $\varphi \in \Phi$.
(D6) If $\exists v \varphi(v) \in s$ then for some $\mathsf{c} \in C$, $\varphi(\mathsf{c}) \in s$.
(D7) If t is a basic term of $\mathsf{L}_{\mathbb{A}}$ and $\mathsf{c}, \mathsf{d} \in C$ then: if $(\mathsf{c} \equiv \mathsf{d}) \in s$ then $(\mathsf{d} \equiv \mathsf{c}) \in s$; if $\varphi(t)$, $(\mathsf{c} \equiv t) \in s$ then $\varphi(\mathsf{c}) \in s$; for some $\mathsf{e} \in C$, $(\mathsf{e} \equiv t) \in s$.
(D8) If $\varphi \in \mathsf{K}^0_{\mathbb{A}}$ then $\varphi \in s$ or $\neg \varphi \in s$.

In the proof of the Model Existence Theorem we first constructed a set s_ω satisfying (D1)—(D7) and then showed that any set s satisfying (D1)—(D7) gave rise to a unique canonical model $\mathfrak{M}$ by the conditions

$$\mathfrak{M} \models \mathsf{R}(\mathsf{c}_1, \ldots, \mathsf{c}_n) \text{ iff } \mathsf{R}(\mathsf{c}_1, \ldots, \mathsf{c}_n) \in s .$$

Furthermore, this model was a model of each $\varphi \in s$. We shall use these facts here. Note that if a consistency property S has the property

$$\text{(C8) if } s \in S \text{ and } \varphi \in \mathsf{K}^0_{\mathbb{A}} \text{ then } s \cup \{\varphi\} \in S \text{ or } s \cup \{\neg \varphi\} \in S$$

then the resulting s_ω will satisfy (D8) and hence will be a special set of sentences.

Now recall the notation from § III.8:

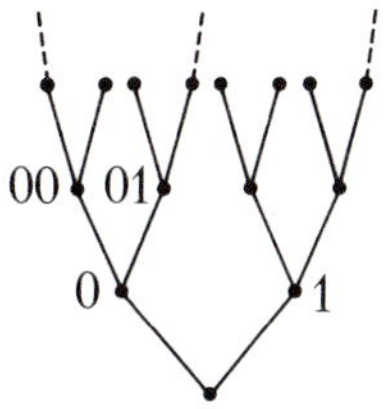

d is a typical node on the tree; $d0$ extends d by putting a 0 on right end; $d1$ a 1; and b is a typical branch.

The *level* of a node is just its length as a sequence. The plan for the proof is to attach a finite set s_d of sentences of $\mathsf{K}_{\mathbb{A}}$ to each node d of the tree in a way that insures the following conditions:

(1) *$\{\varphi(\mathsf{P},\mathsf{Q})\}$ is placed at the bottom of the tree; i. e., $s_{<>}=\{\varphi(\mathsf{P},\mathsf{Q})\}$.*

(2) *If b is any branch and $s^b=\bigcup\{s_d \mid d$ a node on $b\}$ then s^b is a special set of sentences of $\mathsf{K}_{\mathbb{A}}$.*

(3) *Any two sets s_d and $s_{d'}$ on the tree are consistent with respect to the sentences of $\mathsf{K}^0_{\mathbb{A}}$*; that is, if $\varphi\in\mathsf{K}^0_{\mathbb{A}}$ and $\varphi\in s_d$ then $(\neg\varphi)\notin s_{d'}$.

(4) *Distinct branches through the tree are inconsistent with respect to the symbol* P; that is, if b_1, b_2 split at d then there is a constant symbol c so that $\mathsf{P}(\mathsf{c})$ is in s_{d0}, but $\neg\mathsf{P}(\mathsf{c})$ is in s_{d1}.

Suppose we contrive to fulfill (1)—(4). The canonical model determined by a branch b through the tree will have the form $(\mathfrak{M}^b, P^b, Q^b)$ with $\varphi(\mathsf{P},\mathsf{Q})$ true by (1), (2) and the above remarks on special sets. Furthermore, $\mathfrak{M}^{b_1}=\mathfrak{M}^{b_2}$ for all branches b_1, b_2. For if $\mathsf{R}\in\mathsf{L}^0$ and $\mathsf{R}(\mathsf{c}_1,\mathsf{c}_2)$ holds in $\mathfrak{M}^{b_1}$ then $\mathsf{R}(\mathsf{c}_1,\mathsf{c}_2)\in s_d$ for some d on b_1 but then $\neg\mathsf{R}(\mathsf{c}_1,\mathsf{c}_2)$ is never put into any $s_{d'}$ on b_2, by (3), so $\mathsf{R}(\mathsf{c}_1,\mathsf{c}_2)$ is in some $s_{d'}$ on b_2 by (D8) so $\mathsf{R}(\mathsf{c}_1,\mathsf{c}_2)$ holds in $\mathfrak{M}^{b_2}$. Finally, if b_1, b_2 are distinct branches then $P^{b_1}\neq P^{b_2}$ by (4). In other words we have one model $\mathfrak{M}$ with $2^{\aleph_0}$ distinct P each satisfying

$$(\mathfrak{M}, P)\models\exists\mathsf{Q}\ \varphi(\mathsf{P},\mathsf{Q})$$

and so we will have proved our theorem. Satisfying (1)—(4), though, is not so trivial.

In order ultimately to satisfy condition (4), we would like to have a symbol P^b for each branch b thru the tree, but this would make our language uncountable. Instead we introduce new relation symbols P^d, Q^d for each node d on the tree.

We think of P^d as our original P with a ghostly superscript d just barely visible. Our original P, Q are $\mathsf{P}^d, \mathsf{Q}^d$ where d is the empty sequence, $d = \langle\ \rangle$. We denote this expanded language by K^g and the admissible fragment by K^g_A. As usual we consider only formulas with finite many c's and, this time, only finitely many different P^d's ans Q^d's. A finite set s of sentences of K^g_A is *g-consistent* if all the nodes occuring as ghostly superscripts in s lie on some branch (e.g., P^{010} and Q^{01010} could both occur in s but P^{010} and Q^{011} could not). If s is g-consistent then $\hat{s}$ is the result of increasing all superscripts in s to the longest one appearing in s. E.g., if 010 and 01010 are the only superscripts in s then $\hat{s}$ has all P^{010} and Q^{010} replaced by P^{01010} and Q^{01010}. We define a giant consistency machine $\mathbf{S}$ by $\{s_1, \ldots, s_n\} \in \mathbf{S}$ iff $s_1, \ldots, s_n$ are each finite, g-consistent, and $\hat{s}_1 \cup \cdots \cup \hat{s}_n$ does not imply any sentence of $\mathsf{K}^g_{\mathbb{A}}$ of the form

$$(*) \qquad \bigvee_{1 \le i \le n} \bigvee_{\psi \in \Psi_i} [\exists \vec{y}\, \forall x\, \mathsf{P}^{d_i}(x) \leftrightarrow \psi(x, \vec{y})]$$

where each $\psi \in \mathsf{L}^0_{\mathbb{A}}$ and d_i is the longest node in s_i. (Note that if $\{s_1, \ldots, s_n\} \in \mathbf{S}$ then $\hat{s}_1 \cup \cdots \cup \hat{s}_n$ is consistent which will give us (3) above.) Our hypothesis insures us that

$$(5) \quad \{\{\varphi(\mathsf{P}, \mathsf{Q})\}\} \in \mathbf{S}.$$

While $\mathbf{S}$ is not really a consistency property, it generates many of them.

(6) *If* $\{s_1, \ldots, s_n, s_{n+1}\} \in \mathbf{S}$ *then*

$$S = \{s \mid \{s_1, \ldots, s_n, s\} \in \mathbf{S}\}$$

is a consistency property satisfying (C8) *above.*

Most of the clauses are routine. Let us check (C5) and (C8).

(C5) Suppose $\bigvee \Theta \in s \in S$, but that for each $\theta \in \Theta$, $s \cup \{\theta\} \notin S$ so that

$$\{s_1, \ldots, s_n, s \cup \{\theta\}\} \notin \mathbf{S}.$$

Since s is g-consistent so is $s \cup \{\theta\}$ so the problem comes from (*). We must have, for each $\theta \in \Theta$, some σ_θ of the form (*) such that

$$\hat{s}_1 \cup \cdots \cup \hat{s}_n \cup \widehat{s \cup \{\theta\}} \vdash \sigma_\theta.$$

Now, just as in the proof of the interpolation theorem, we can assume the σ_θ is given as a function of θ, a function in our admissible set (σ_θ will be the disjunction of the σ's given by strong Σ replacement). But then $\sigma = \bigvee_{\theta \in \Theta} \sigma_\theta$ is again of the form (*), once you rearrange it a bit, and

$$\hat{s}_1 \cup \cdots \cup \hat{s}_n \cup \hat{s} \vdash \sigma,$$

a contradiction.

(C8) Suppose $\varphi(\mathsf{c}_1,\dots,\mathsf{c}_n)\in \mathsf{K}^0_{\mathbb{A}}$, that $s\in S$ but neither $s\cup\{\varphi(\mathsf{c}_1,\dots,\mathsf{c}_n)\}$ nor $s\cup\{\neg\varphi(\mathsf{c}_1,\dots,\mathsf{c}_n)\}\in S$. Then there are sentences σ_1, σ_2 of the form (*) such that

$$\hat{s}_1\cup\cdots\cup\hat{s}_n\cup\hat{s}\vdash\varphi(\mathsf{c}_1,\dots,\mathsf{c}_n)\rightarrow\sigma_1$$

$$\hat{s}_1\cup\cdots\cup\hat{s}_n\cup\hat{s}\vdash\neg\varphi(\mathsf{c}_1,\dots,\mathsf{c}_n)\rightarrow\sigma_2$$

but then

$$\hat{s}_1\cup\cdots\cup\hat{s}_n\cup\hat{s}\vdash\sigma_1\vee\sigma_2$$

and $\sigma_1\vee\sigma_2$ is equivalent to a sentence of the form (*).

We now come to the crucial step which will yield (4) above.

(7) *If* $\{s_1,\dots,s_n\}\in\mathbf{S}$, *if* d *is the longest node in* s_n, *if* $d0, d1$ *do not occur in* $s_1\cup\cdots\cup s_n$, *and if* c *is a constant symbol not in* $s_1\cup\cdots\cup s_n$ *then*

$$\{s_1,\dots,s_{n-1},s_n\cup\{\mathsf{P}^{d0}(\mathsf{c})\},s_n\cup\{\neg\mathsf{P}^{d1}(\mathsf{c})\}\}$$

is in $\mathbf{S}$.

We use the Interpolation Theorem for $\mathsf{K}_{\mathbb{A}}$ to prove (7). We invite the student to try the case $n=1$ for himself before going on. We do the case $n=2$ because it exhibits the problems that arise in general. Now, if (7) fails, the trouble cannot arise from g-consistency since

$$s_1, s_2\cup\{\mathsf{P}^{d0}(\mathsf{c})\},\quad s_2\cup\{\mathsf{P}^{d1}(\mathsf{c})\}$$

are g-consistent so it must be that there are sentences $\sigma_1,\sigma_2,\sigma_3$ where σ_i is of the form

$$\bigvee_{\psi\in\Psi_i}\exists\vec{y}\,\forall x\,[\mathsf{P}_i(x)\leftrightarrow\psi(x,\vec{y})]$$

(where P_1 is the symbol P^d in $\hat{s}_1$, P_2 is P^{d0}, P_3 is P^{d1}), such that

$$\hat{s}_1\cup\widehat{s_2\cup\{\mathsf{P}^{d0}(\mathsf{c})\}}\cup\widehat{s_2\cup\{\neg\mathsf{P}^{d1}(\mathsf{c})\}}\vdash\sigma_1\vee\sigma_2\vee\sigma_3. \tag{8}$$

We show that (8) implies $\{s_1,s_2\}\notin\mathbf{S}$ by finding a sentence σ of the form (*) such that

$$\hat{s}_1\cup\hat{s}_2\vdash\sigma.$$

Rewrite (8) as follows:

$$[s_1(\mathsf{P}_1,\mathsf{Q}_1)\wedge\neg\sigma_1(\mathsf{P}_1)\wedge s_2(\mathsf{P}^{d0},\mathsf{Q}^{d0})\wedge\neg\sigma_2(\mathsf{P}^{d0})\wedge\mathsf{P}^{d0}(\mathsf{c})]$$
$$\rightarrow[s_2(\mathsf{P}^{d1},\mathsf{Q}^{d1})\wedge\neg\sigma_3(\mathsf{P}^{d1})\rightarrow\mathsf{P}^{d1}(\mathsf{c})]$$

where $s_2(\mathsf{P}^{d0},\mathsf{Q}^{d0})$ indicates the result of replacing P^d by P^{d0} in $\hat{s}_2$. Notice that the only symbols on both sides of the implication sign are in K^0. By the Interpolation Theorem there is a $\psi(\mathsf{c},\mathsf{c}_1,\ldots,\mathsf{c}_m)$ which is an interpolant. We may write this as:

$$\hat{s}_1(\mathsf{P}_1,\mathsf{Q}_1)\wedge\hat{s}_2(\mathsf{P}^{d0},\mathsf{Q}^{d0})\wedge\mathsf{P}^{d0}(\mathsf{c})\rightarrow\sigma_1(\mathsf{P}_1)\vee\sigma_2(\mathsf{P}^{d0})\vee\psi(\mathsf{c},\mathsf{c}_1,\ldots,\mathsf{c}_m),\text{ and}$$
$$\hat{s}_2(\mathsf{P}^{d1},\mathsf{Q}^{d1})\wedge\psi(\mathsf{c},\mathsf{c}_1,\ldots,\mathsf{c}_m)\rightarrow\mathsf{P}^{d1}(\mathsf{c})\vee\sigma_3(\mathsf{P}^{d1}).$$

Now replace P^{d0}, Q^{d0} by P^d, Q^d in the top line, P^{d1}, Q^{d1} by P^d, Q^d in the second line. We obtain

$$\hat{s}_1\cup\hat{s}_2\rightarrow\sigma_1(\mathsf{P}_1)\vee\sigma_2(\mathsf{P}^d)\vee\sigma_3(\mathsf{P}^d)\vee[\mathsf{P}^d(\mathsf{c})\leftrightarrow\psi(\mathsf{c},\mathsf{c}_1,\ldots,\mathsf{c}_m)].$$

Since c does not occur in $\hat{s}_1\cup\hat{s}_2$ we get

$$\hat{s}_1\cup\hat{s}_2\vdash\sigma_1(\mathsf{P}_1)\vee\sigma_2(\mathsf{P}^d)\vee\sigma_3(\mathsf{P}^d)\vee\exists y_1,\ldots,\exists y_m\,\forall x\,[\mathsf{P}^d(x)\leftrightarrow\psi(x,y_1,\ldots,y_m)]$$

and hence $\{s_1,s_2\}\notin\mathbf{S}$.

Now we are ready to decorate our tree. List the sentences of $\mathsf{K}^g_{\mathbb{A}}$ as a sequence

$$\varphi_0,\varphi_1,\ldots,\varphi_n,\ldots$$

in such a way that any node d appearing in φ_n is of level $\leqslant n$. List the terms occuring in $\mathsf{L}_{\mathbb{A}}$:

$$t_0,t_1,\ldots,t_n,\ldots.$$

We work our way up the tree as follows. Place $\{\varphi(\mathsf{P},\mathsf{Q})\}$ at $<\ >$. Assume we have placed sets s_d at every node d of level n so that d is the longest node in s_d and the set

$$\{s_d\mid d\text{ a node at level }n\}$$

is in our consistency machine $\mathbf{S}$.

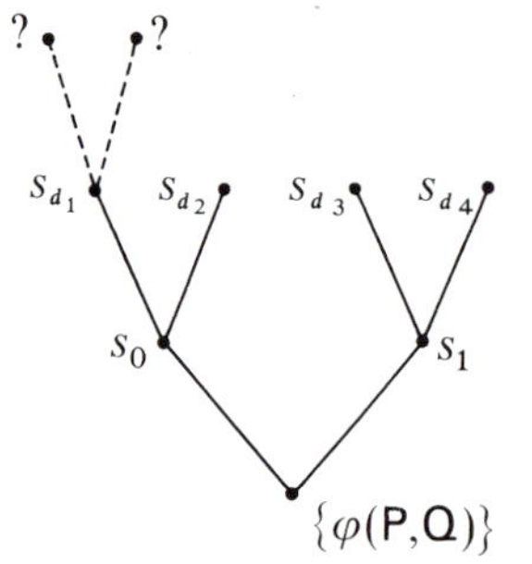

Given s_{d_1} we first take care of t_n and φ_n (if φ_n happens to be g-consistent with s_{d_1}) as in the proof of the Model Existence Theorem, using (6), giving us some

$$\{s', s_{d_2}, s_{d_3}, s_{d_4}\} \in \mathbf{S}.$$

We then apply (7) to get

$$\{s' \cup \{\mathsf{P}^{d0}(\mathsf{c})\}, s' \cup \{\neg \mathsf{P}^{d1}(\mathsf{c})\}, s_2, s_3, s_4\} \in \mathbf{S}$$

and we let $s_{d_1 0} = s' \cup \{\mathsf{P}^{d0}(\mathsf{c})\}$, $s_{d_1 1} = s' \cup \{\neg \mathsf{P}^{d1}(\mathsf{c})\}$. In this way we work our way along level $n+1$ and on up the tree. We see that any finite set of nodes on the tree is in **S**. This takes care of (3) since, otherwise, they would certainly imply a formula of the form (*). Now that there is a set at each node, let the superscripts vanish and you will discover we have satisfied (1), (2), (3) and (4), proving our theorem. □

4.13 – 4.17 Exercises

4.13. Show that Example 4.9 is not true without the assumption $o(\mathbb{HYP}_{\mathfrak{M}}) = \omega$. [Let $\mathfrak{M}$ be a minimal elementary extension of $\mathcal{N} = \langle \omega, 0, +, \cdot \rangle$].

4.14. Let $\mathfrak{M} = \langle M, 0, +, \cdot \rangle$ be a countable nonstandard model of Peano arithmetic with $o(\mathbb{HYP}_{\mathfrak{M}}) = \omega$. Show that there are $2^{\aleph_0}$ initial segments of $\mathfrak{M}$ which are models of Peano arithmetic.

4.15. Improve 4.14 to get $2^{\aleph_0}$ initial submodels of $\mathfrak{M}$ which are isomorphic to $\mathfrak{M}$. [Hint: Use a theorem of Friedman [1973] to the effect that every countable nonstandard model of Peano arithmetic is isomorphic to some initial segment of itself.]

4.16. Use 4.4 to show that if a countable abelian group G has $< 2^{\aleph_0}$ divisible subgroups then they are all in $\mathbb{HYP}_G$ and hence there are at most $\aleph_0$ of them. Give a direct group theoretic proof of this fact.

4.17. Extend Theorem 4.6 from simple sentences to Σ_1 theories. Similarly extend the applications of 4.6 given above.

4.18 Notes. The results of this section are called perfect set results because one always ends up constructing, by a tree argument, a perfect set of objects, perfect in the topological sense.

5. *Recursively Saturated Structures*

Having discovered several interesting facts about structures $\mathfrak{M}$ with $O(\mathfrak{M}) = \omega$, we take time in this section to relate this condition on $\mathbb{HYP}_{\mathfrak{M}}$ to more traditional notions.

Recall that a structure $\mathfrak{M}=\langle M,R_1,\ldots,R_l\rangle$ for L is $\aleph_0$-saturated if for every $k<\omega$ and every set $\Phi(x,v_1,\ldots,v_k)$ of formulas of $\mathsf{L}_{\omega\omega}$ with free variables among $x,v_1,\ldots,v_k$ the following infinitary sentence is true in $\mathfrak{M}$:

$$\forall v_1,\ldots,v_k\,[(\bigwedge_{\Phi_0\in S_\omega(\Phi)}\exists x\bigwedge\Phi_0(x,v_1,\ldots,v_k))\rightarrow\exists x\bigwedge\Phi(x,v_1,\ldots,v_k)]$$

where $S_\omega(\Phi)$ is the set of all finite subsets of Φ.

5.1 Definition. The structure $\mathfrak{M}=\langle M,R_1,\ldots,R_l\rangle$ for L is *recursively saturated* if the above holds for all $k<\omega$ and all recursive sets $\Phi(x,v_1,\ldots,v_k)$ of $\mathsf{L}_{\omega\omega}$.

Just as in the case of $\aleph_0$-saturated we have the following lemma.

5.2 Lemma. *Let $\mathfrak{M}$ be recursively saturated and let $\Phi(x_1,\ldots,x_n,v_1,\ldots,v_k)$ be a recursive set of formulas with free variables as indicated. The following infinitary sentence is true in $\mathfrak{M}$:*

$$\forall v_1,\ldots,v_k\,[(\bigwedge_{\Phi_0\in S_\omega(\Phi)}\exists x_1,\ldots,x_n\bigwedge\Phi_0)\rightarrow\exists x_1,\ldots,x_n\bigwedge\Phi].$$

Proof. The proof is by induction on n, the case $n=1$ being the hypothesis. It clearly suffices to prove the result for Φ satisfying the condition

$$\Phi_0\in S_\omega(\Phi)\quad\text{implies}\quad\bigwedge\Phi_0\in\Phi,$$

since we could close Φ under finite conjunctions. Let $\Psi(x_1,\ldots,x_n,v_1,\ldots,v_k)$ be the set of all formulas

$$\exists x_{n+1}\,\varphi(x_1,\ldots,x_n,x_{n+1},v_1,\ldots,v_k)$$

for $\varphi\in\Phi$. Suppose that $q_1,\ldots,q_k\in\mathfrak{M}$ are such that

$$\mathfrak{M}\models\exists x_1,\ldots,x_{n+1}\bigwedge\Phi_0(\vec{x},\vec{q})$$

for all $\Phi_0\in\Phi$. By the induction hypothesis, there are $p_1,\ldots,p_n\in\mathfrak{M}$ such that

$$\mathfrak{M}\models\bigwedge\Psi(p_1,\ldots,p_n,q_1,\ldots,q_k)$$

and hence

$$\mathfrak{M}\models\exists x_{n+1}\bigwedge\Phi_0(p_1,\ldots,p_n,x,q_1,\ldots,q_k)$$

for all $\Phi_0\in S_\omega(\Phi)$, since every such $\exists x_{n+1}\bigwedge\Phi_0$ is in Ψ. But then since $\mathfrak{M}$ is recursively saturated there is a $p_{n+1}\in\mathfrak{M}$ such that

$$\mathfrak{M}\models\Phi(p_1,\ldots,p_{n+1},q_1,\ldots,q_k).\quad\square$$

The principal link between recursively saturated structures and admissible sets is the following theorem of John Schlipf.

5.3 Theorem. *Let $\mathfrak{M}=\langle M, R_1,\ldots,R_l\rangle$ be a structure for* L. *$\mathfrak{M}$ is recursively saturated iff $O(\mathfrak{M})=\omega$.*

Proof. We prove the easy half first. Suppose that $o(\mathbb{HYP}_{\mathfrak{M}})=\omega$. Let $\Phi(v,w_1,\ldots,w_k)$ be a recursive set of formulas of $\mathsf{L}_{\omega\omega}$. We may consider Φ as a Δ_1 subset of $\mathbb{HF}$ by II.2.3. Since $\mathbb{HF}$ is Δ_1 on every admissible set, Φ is also Δ_1 on $\mathbb{HYP}_{\mathfrak{M}}$. Let $\vec{q}=q_1,\ldots,q_k\in\mathfrak{M}$ be such that

$$\mathfrak{M}\models\neg\exists v\bigwedge\Phi(v,q_1,\ldots,q_k).$$

We need to find a finite subset Φ_0 of Φ such that

$$\mathfrak{M}\models\neg\exists v\bigwedge\Phi_0(v,q_1,\ldots,q_k).$$

Now, since

$$\forall p\in M\ \exists\varphi\,[\varphi\in\Phi\wedge\mathfrak{M}\models\neg\varphi[p,\vec{q}]]$$

we have, by strong Σ Collection, a set b such that

(1) $\forall p\in M\ \exists\varphi\in b\,[\varphi\in\Phi\wedge\mathfrak{M}\models\neg\varphi[p,\vec{q}]]$

and

(2) $\forall\varphi\in b\ \exists p\in M\,[\varphi\in\Phi\wedge\mathfrak{M}\models\neg\varphi(p,\vec{q}]]$.

From (2) we see that $b\subseteq\Phi$ so let $\Phi_0=b$. Φ_0 is finite since it is in $\mathbb{HYP}_{\mathfrak{M}}$, is a set of pure sets, and has finite rank. From (1) we see that $\Phi_0(v,\vec{q})$ is not satisfiable on $\mathfrak{M}$.

To prove the other half of the theorem, let $\mathfrak{M}$ be recursively saturated. We need to prove that $\mathrm{L}(\mathfrak{M},\omega)$ is admissible; i.e., that it satisfies Δ_0 Collection. Call a set $a\in\mathrm{L}(\mathfrak{M},\omega)$ *simple* if there is a single term $\mathscr{F}(v_1,\ldots,v_{k+1})$ built up from $\mathscr{F}_1,\ldots,\mathscr{F}_N$, $\mathscr{D}$ such that each $x\in a$ is of the form

$$x=\mathscr{F}(p_1,\ldots,p_k,M)$$

for some $p_1,\ldots,p_k\in M$. Assume, for the moment, that we have established (3) and (4):

(3) *Every $a\in\mathrm{L}(\mathfrak{M},\omega)$ is the union of a finite number of simple sets;*

(4) *If $z\in\mathrm{L}(\mathfrak{M},\omega)$ and if a simple, then $\mathrm{L}(\mathfrak{M},\omega)$ satisfies*

$$\forall x\in a\ \exists y\ \varphi(x,y,z)\rightarrow\exists b\ \forall x\in a\ \exists y\in b\ \varphi(x,y,z)$$

for all Δ_0 formulas $\varphi(x,y,z)$.

Assuming this, let $\varphi(x,y,z)$ be a Δ_0 formula such that $L(\mathfrak{M},\omega)$ satisfies

$$\forall x \in a\, \exists y\, \varphi(x,y,z).$$

Write $a = a_1 \cup \cdots \cup a_m$ where each a_i is simple. Since

$$\forall x \in a_i\, \exists y\, \varphi(x,y,z)$$

holds in $L(\mathfrak{M},\omega)$ there are sets $b_1,\ldots,b_m$ in $L(\mathfrak{M},\omega)$ such that

$$\forall x \in a_i\, \exists y \in b_i\, \varphi(x,y,z).$$

But then let $b = b_1 \cup \cdots \cup b_m$. Then $b \in L(\mathfrak{M},\omega)$ and

$$\forall x \in a\, \exists y \in b\, \varphi(x,y,z)$$

so $L(\mathfrak{M},\omega)$ satisfies Δ_0 Collection.

To prove (3) note that in the proof of II.7.7 we showed that for each n there are a finite number of terms $\mathscr{F}^1,\ldots,\mathscr{F}^m$ such that each $x \in L(\mathfrak{M},n)$ is of the form

$$x = \mathscr{F}^i(\vec{p}, M)$$

for some $i \leqslant m$ and some $\vec{p} \in M$. If $a \in L(\mathfrak{M},n)$ then $a \subseteq L(\mathfrak{M},n)$. Define, by Δ_0 Separation, sets $a_1,\ldots,a_m$ by

$$a_i = \{x \in a \mid \exists \vec{p} \in M\; x = \mathscr{F}^i(\vec{p},M)\}.$$

Then $a = a_1 \cup \cdots \cup a_m$.

Finally we prove (4). Let φ be given. By II.7.7 and II.7.6 we may assume that the only parameters in φ are M and some $\vec{q} \in M$. Given the simple set a let $\mathscr{F}^0(v_1,\ldots,v_{n+1})$ be as given in the definition of simple. Let $a = \mathscr{F}^1(r_1,\ldots,r_k,M)$ for some $r_1,\ldots,r_k \in M$. Rather than prove (4) we prove its contrapositive. Let ψ be $\neg\varphi$, so that we want to verify that $L(\mathfrak{M},\omega)$ is a model of

$$\forall b\, \exists x \in a\, \forall y \in b\, \psi(x,y,\vec{q},M) \rightarrow \exists x \in a\, \forall y\, \psi(x,y,\vec{q},M).$$

Assume the hypothesis. In particular we have, for each $m < \omega$,

$$(5)_m \quad \exists x \in a\, \forall y \in L(M,m)\, \varphi(x,y,\vec{q},M)$$

which becomes

$$(6)_m \quad \exists p_1,\ldots,p_n \in M\, [\mathscr{F}^0(\vec{p},M) \in \mathscr{F}^1(\vec{r},M) \wedge \forall y \in L(M,m)\, \psi(x,y,\vec{q},M)].$$

This is a Δ_0 formula of $\vec{p},\vec{q},\vec{r}$ so, by the effective version of II.7.8, we can find a formula $\psi_m(\vec{p},\vec{q},\vec{r})$ of $L_{\omega\omega}$ equivalent to it. Note that by (5) we have

$$\mathfrak{M} \models \forall v_1,\ldots,v_n\, [\psi_m(v_1,\ldots,v_n,\vec{q},\vec{r}) \rightarrow \psi_{m'}(v_1,\ldots,v_n,\vec{q},\vec{r})],$$

whenever $m \geqslant m'$. By $(6)_m$ we see that

$$\Phi = \{\psi_m(v_1, \ldots, v_n, \vec{q}, \vec{r}) \mid m < \omega\}$$

is finitely satsfiable. Since it is clearly a recursive set (by the exercises at the end of II.7) and $\mathfrak{M}$ is recursively saturated there are $p_1, \ldots, p_n \in \mathfrak{M}$ so that

$$\mathfrak{M} \models \psi_m(\vec{p}, \vec{q}, \vec{r})$$

for all $m < \omega$. Thus for this $\vec{p}$, we have, setting $x = \mathscr{F}^0(\vec{p}, M)$, $x \in a$, and for all $m < \omega$,

$$\forall y \in \mathrm{L}(M, m)\, \psi(x, y, \vec{q}, M)$$

and hence

$$\forall y \in \mathrm{L}(M, \omega)\, \psi(x, y, \vec{q}, M)$$

as desired. □

Schlipf discovered 5.3 by generalizing the results 5.4 and 5.7 below.

5.4 Corollary. *If* $\mathfrak{M} = \langle M, R_1, \ldots, R_l \rangle$ *is* $\aleph_0$*-saturated then* $O(\mathfrak{M}) = \omega$. □

5.5 Corollary. *If* $\mathfrak{M} = \langle M, R_1, \ldots, R_l \rangle$ *is recursively saturated and* $\Phi(x, v_1, \ldots, v_k)$ *is any set of formulas of* $\mathsf{L}_{\omega\omega}$ *which is* Σ_1 *on* $\mathbb{HYP}_{\mathfrak{M}}$ *then* $\mathfrak{M}$ *satisfies:*

$$\forall v_1, \ldots, v_k \left[\left(\bigwedge\nolimits_{\Phi_0 \in S_\omega(\Phi)} \exists x \bigwedge \Phi_0 \right) \rightarrow \exists x \bigwedge \Phi \right].$$

Proof. The proof that $o(\mathbb{HYP}_{\mathfrak{M}}) = \omega$ implies $\mathfrak{M}$ is recursively saturated actually proves this stronger result. □

5.6 Corollary. *For every infinite* $\mathfrak{M} = \langle M, R_1, \ldots, R_l \rangle$ *there is a proper elementary extension* $\mathfrak{N}$ *of* $\mathfrak{M}$ *of the same cardinality such that* $\mathfrak{N}$ *is recursively saturated.*

Proof. Immediate from 5.3 and II.8.6. □

The above ω corollary shows a contrast between the notions of recursively saturated and $\aleph_0$-saturated structures since there is no countable $\aleph_0$-saturated elementary extension of $\mathcal{N} = \langle \omega, 0, +, \cdot \rangle$. Of course one could also prove 5.6 by a more standard model theoretic argument using elementary chains.

The following result shows that 5.3 can be improved for countable structures. It shows that if $\mathfrak{M}$ is countable and $o(\mathbb{HYP}_{\mathfrak{M}}) = \omega$ then $\mathfrak{M}$ is saturated for certain sets of Σ_1^1 formulas.

5.7 Theorem. *Let* $\mathfrak{M} = \langle M, R_1, \ldots, R_l \rangle$ *be a countable structure for* L *with* $O(\mathfrak{M}) = \omega$. *Let* $\mathsf{K} = \mathsf{L} \cup \{\mathsf{S}_1, \ldots, \mathsf{S}_m\}$ *and let* $\Phi(x_1, \ldots, x_n, v_1, \ldots, v_k, \mathsf{S}_1, \ldots, \mathsf{S}_m)$ *be a* *set of formulas of* $\mathsf{K}_{\omega\omega}$ *which is* Σ_1 *on* $\mathbb{HYP}_{\mathfrak{M}}$. *The following infinitary second order sentence holds in* $\mathfrak{M}$:

$$\forall v_1,\ldots,v_k[(\bigwedge_{\Phi_0\in S_\omega(\Phi)} \exists \mathsf{S}_1,\ldots,\mathsf{S}_m\, \exists x_1,\ldots,x_n \bigwedge \Phi_0(\vec{x},\vec{v},\mathsf{S}))\to \exists \mathsf{S}_1,\ldots,\mathsf{S}_m\, \exists x_1,\ldots,x_n \bigwedge \Phi].$$

Proof. We use Theorem III.5.8. Let $q_1,\ldots,q_k\in M$ be given so that

$$\mathfrak{M}\models \exists \mathsf{S}_1,\ldots,\mathsf{S}_m\, \exists x_1,\ldots,x_n \bigwedge \Phi_0(\vec{x},\mathsf{S},q_1,\ldots,q_k)$$

for all $\Phi_0\in S_\omega(\Phi)$. We can assume that $\mathsf{K}\cup\{\mathsf{c}_1,\ldots,\mathsf{c}_n,\mathsf{d}_1,\ldots,\mathsf{d}_k\}$ is coded up on $\mathbb{HF}$. Let T be the theory

$$\Phi(\mathsf{c}_1,\ldots,\mathsf{c}_n,\mathsf{d}_1,\ldots,\mathsf{d}_k,\mathsf{S}_1,\ldots,\mathsf{S}_m).$$

Introduce symbols $\overline{\mathsf{p}}$ for $p\in M$ as usual and let $T'=\{\psi\}$ be the conjunction of

$$\bigwedge \text{Diagram}(\mathfrak{M})$$
$$\forall x \bigvee_{p\in M} x=\overline{\mathsf{p}}$$
$$\mathsf{d}_1=\overline{\mathsf{q}}_1,\ldots,\mathsf{d}_k=\overline{\mathsf{q}}_k.$$

For every finite subset T_0 of T, $T_0\cup T'$ has a model, so $T\cup T'$ has a model. This model is isomorphic to some

$$(\mathfrak{M},S_1,\ldots,S_m,p_1,\ldots,p_n,q_1,\ldots,q_k)$$

with

$$(\mathfrak{M},S_1,\ldots,S_m)\models\Phi[p_1,\ldots,p_n,q_1,\ldots,q_k]. \quad \square$$

5.8—5.14 Exercises

5.8. Show that every recursively saturated structure is ω-homogeneous.

5.9. Suppose $\mathfrak{M}$ is uncountable. Show that $o(\mathbb{HYP}_{\mathfrak{M}})=\omega$ iff for all relations $T_1,\ldots,T_k$ on $\mathfrak{M}$ there is a countable recursively saturated structure $\mathfrak{N}$ with

$$(\mathfrak{N},T_1{\restriction}N,\ldots,T_k{\restriction}N)\prec(\mathfrak{M},T_1,\ldots,T_k).$$

5.10. Show that the predicate

"$\mathfrak{M}$ is recursively saturated"

is absolute (Δ_1) for models of KPU + Infinity but that the predicate

"$\mathfrak{M}$ is $\aleph_0$-saturated"

cannot be expressed by a Σ formula.

5.11 (J. Schlipf, J.-P. Ressayre). Let α be an admissible ordinal and let $\mathbb{A} = L(\alpha)$. Let L be a language with a finite number of symbols. A structure $\mathfrak{M}$ for L is α-recursively saturated iff for every set $\Phi(x, v_1, \ldots, v_k)$ of sentences of $\mathsf{L}_{\mathbb{A}}$ which is Δ_1 on A the following sentence holds in $\mathfrak{M}$:

$$\forall v_1, \ldots, v_k \left[\left(\bigwedge_{\Phi_0 \in S_{\mathbb{A}}(\Phi)} \exists x \bigwedge \Phi_0(x, v_1, \ldots, v_k) \right) \rightarrow \exists x \bigwedge \Phi(x, v_1, \ldots, v_k) \right]$$

where $S_{\mathbb{A}}(\Phi) = \{\Phi_0 \subseteq \Phi \mid \Phi_0 \in \mathbb{A}\}$.

(i) Prove that if $L(\alpha)_{\mathfrak{M}}$ is admissible then $\mathfrak{M}$ is α-recursively saturated.

(ii) Prove that $O(\mathfrak{M})$ = the least α such that α is recursively saturated. (This result, due to J. Schlipf, strenghtens a special case of a theorem of J.-P. Ressayre. Schlipf's proof uses notions from Chapters V and VI.) Makkai has translated Ressayre's result into our setting to show that for α countable, admissible and greater than ω, $L(\alpha)_{\mathfrak{M}}$ is admissible iff $\mathfrak{M}$ is α-recursively saturated and satisfies the following condition: Suppose $\varphi_{\beta,\gamma}(v_1, \ldots, v_n)$ is an α-recursive function of β, γ. Suppose further that for some $p_1, \ldots, p_n \in \mathfrak{M}$ and some $\beta_0 < \alpha$:

$$\mathfrak{M} \models \bigwedge_{\beta < \beta_0} \bigvee_{\gamma < \alpha} \varphi_{\beta,\gamma}(\vec{p}).$$

Then there is a $\gamma_0 < \alpha$ such that

$$\mathfrak{M} \models \bigwedge_{\beta < \beta_0} \bigvee_{\gamma < \gamma_0} \varphi_{\beta,\gamma}(\vec{p}).$$

5.12. Show that $\mathfrak{M}$ is $\aleph_0$-saturated iff

(i) $o(\mathbb{HYP}_{\mathfrak{M}}) = \omega$.

(ii) for every $X \subseteq \omega$, $(\mathbb{HYP}_{\mathfrak{M}}, X)$ is admissible.

5.13. In this exercise we sketch some interesting connections between recursively saturated models of Peano arithmetic and models of nonstandard analysis. To simplify matters, we identify analysis with second order arithmetic (a standard perversion among logicians). Thus we add to the first order language of number theory new second order variables $X_1, X_2, \ldots$ and a membership symbol $\in$ which can hold between first order objects and second order objects ($(x_i \in X_j)$ is a formula but $(X_j \in x_i)$ isn't). The *axiom of induction* asserts:

$$\forall X \left[\mathbf{0} \in X \wedge \forall x (x \in X \rightarrow (x+1) \in X) \rightarrow \forall x (x \in X) \right].$$

(Warning: when working in systems weaker than the one described here it is often necessary to replace this single axiom by an axiom scheme.) The *axiom of comprehension* asserts the following, for every formula $\varphi(x, y_1, \ldots, y_k)$:

$$\forall \vec{y}\, \exists x \left[\forall x (x \in X \leftrightarrow \varphi(x, y_1, \ldots, y_k)) \right].$$

By *analysis* we mean the usual axioms of Peano arithmetic plus the axiom of induction and the axiom of comprehension. (Of course there is no need to include the first order form of induction since it follows from our second order axioms.) A *model of analysis* consists of a pair $(\mathfrak{N}, \mathscr{H})$, where $\mathscr{H}$ is a collection of subsets of the first order structure $\mathfrak{N}$, which makes all the axioms of analysis true. Any such model of analysis gives rise to a model $\mathfrak{N}$ of Peano arithmetic, but not every model of arithmetic can be expanded to a model of analysis. A *model of nonstandard analysis* is a model $(\mathfrak{N}, \mathscr{H})$ of analysis with $\mathfrak{N}$ not isomorphic to the standard model of arithmetic.

i) Prove that if $(\mathfrak{N}, \mathscr{H})$ is a model of nonstandard analysis, then $\mathfrak{N}$ is recursively saturated.

ii) Let $\mathfrak{N}$ be a nonstandard countable model of arithmetic. Let

$$\mathscr{H} = \bigcap\{\mathscr{H} \mid (\mathfrak{N}, \mathscr{H}) \models \text{analysis}\}.$$

Prove that either $\mathscr{H}$ is empty or that $\mathscr{H}$ consist of exactly the definable subsets of $\mathfrak{N}$. [This is easy from (i) and Theorem 1.1.]

5.14. Prove that there are two nonisomorphic countable recursively saturated elementary extensions of $\mathcal{N} = \langle \omega, +, x \rangle$.

5.15 Notes. It is not known whether or not there is a complete theory T in an finite language such that all models of T are recursively saturated but T is not $\aleph_0$-categorical.

6. *Countable $\mathfrak{M}$-Admissible Ordinals*

Since this chapter concerns the interplay of model theory and recursion theory, it seems appropriate to discuss one of the first applications of infinitary logic to the theory of admissible ordinals.

Let $\mathcal{N} = \langle \omega, 0, +, \cdot \rangle$. Most countable admissible ordinals α (other than ω) that arise in recursion theory are of the form

$$\alpha = O((\mathcal{N}, R))$$

for some relation R on ω. The question arose: Is every countable admissible $\alpha, \alpha > \omega$, of the above form? Sacks eventually answered this in the affirmative by means of "perfect set" forcing. His proof remains unpublished since Friedman-Jensen [1968] presented a simple proof of the result by means of the Barwise Compactness Theorem. We extend this theorem as follows.

6.1 Theorem. *Let $\mathfrak{M} = \langle M, R_1, \ldots, R_l \rangle$ be a countable infinite structure and let α be a countable ordinal. The following are equivalent:*

(i) *α is $\mathfrak{M}$-admissible;*

(ii) *for some relation S on $\mathfrak{M}$,*

$$\alpha = O(\mathfrak{M}, S);$$

(iii) *for some linear ordering* $\prec$ *of* $\mathfrak{M}$,

$$\alpha = O(\mathfrak{M}, \prec)$$

and the order type of the largest well-ordered initial segment of $\prec$ *is* α.

Proof. The implications (iii)$\Rightarrow$(ii) and (ii)$\Rightarrow$(i) are obvious. To prove (i)$\Rightarrow$(iii) we borrow a fact from Section V.3:

(1) *If* $r \subseteq a \times a$ *is a linear ordering, r an element of an admissible set* $\mathbb{A}$, *and if* β *is the length of a well-ordered initial segment of r then* $\beta \leqslant o(\mathbb{A})$.

This could be proved now, but it is easier to wait for the Second Recursion Theorem. Let α be $\mathfrak{M}$-admissible. Then there is a countable admissible set $\mathbb{A} = \mathbb{A}_{\mathfrak{M}}$ above $\mathfrak{M}$ with

$$\alpha = o(\mathbb{A}_{\mathfrak{M}})$$

by II.3.3. Let K be the language L^* plus new constant symbols c, r, and $\overline{\mathsf{x}}$ for each $\mathsf{x} \in \mathbb{A}_{\mathfrak{M}}$. Let $\mathsf{K}_{\mathbb{A}}$ be the admissible fragment of $\mathsf{K}_{\infty\omega}$ given by $\mathbb{A}_{\mathfrak{M}}$. Let T be the theory which asserts:

KPU$^+$

Diagram$(\mathbb{A}_{\mathfrak{M}})$

"$\overline{\mathsf{M}}$ *is the set of all urelements*"

$\forall v\,[v \in \overline{\mathsf{a}} \rightarrow \bigvee_{x \in a} v = \overline{\mathsf{x}}]$ (for all $a \in \mathbb{A}_{\mathfrak{M}}$),

"c *is an ordinal*"

$\mathsf{c} > \overline{\beta}$ (for all $\beta < \alpha$),

"r *is a linear ordering of* $\overline{\mathsf{M}}$ *of order type* $\in \cap (\mathsf{c} \times \mathsf{c})$".

T has a model of the form

$$(\mathfrak{M}; H(\omega_1)_{\mathfrak{M}}, \in, \alpha, r)$$

for any well-ordering r of M of order type α. By III.7.5 T has a model

$$(\mathfrak{M}; B, E, c, r)$$

with $\alpha = o\,\mathcal{W}\!f(\mathfrak{M}; B, E)$. Let $\mathbb{A}'_{\mathfrak{M}} = \mathcal{W}\!f(\mathfrak{M}; B, E)$ which is an admissible set by the Truncation Lemma. Since $r \subseteq M \times M$, $r \in \mathbb{A}'_{\mathfrak{M}}$ so $\mathbb{A}'_{\mathfrak{M}}$ is actually admissible above $(\mathfrak{M}, r)$. Hence $\alpha \geqslant o(\mathbb{HYP}_{(\mathfrak{M},r)})$. But r has an initial segment of order type α (by T) so, by (1) applied to $\mathbb{HYP}_{(\mathfrak{M},r)}$, $\alpha \leqslant o(\mathbb{HYP}_{(\mathfrak{M},r)})$. We let $\prec$ be r. $\square$

6.2—6.5 Exercises

6.2. Let $(\mathfrak{M}, \prec)$ be as in 6.1(iii). Show that $\mathbb{HYP}_{(\mathfrak{M},\prec)}$ is a model of $\neg$Beta.

6.3. Prove (1) above.

6.4. Let $\mathbb{A}_{\mathfrak{M}}$ be countable, admissible above $\mathfrak{M}$ with $o(\mathbb{A}_{\mathfrak{M}})>\omega$. Find a larger admissible set $\mathbb{B}_{\mathfrak{M}}$ above $\mathfrak{M}$ with the same ordinal such that $\mathbb{B}_{\mathfrak{M}}$ is locally countable; i. e.,

$$\mathbb{B}_{\mathfrak{M}} \models \forall a \text{ (“}a\text{ is countable”).}$$

[Hint: Use the $\overline{\text{YY}}$ Compactness Theorem and Theorem II.7.5.]

6.5 (Schlipf). Prove that for every countable admissible ordinal β there is an elementary extension $\mathfrak{M}$ of $\mathcal{N}=\langle\omega,0,+,x\rangle$ such that $\beta=o(\mathbb{HYP}_{\mathfrak{M}})$. [Hint: i) Show that if $\mathfrak{M}$ is not recursively saturated and the set $\{n<\omega \mid \mathfrak{M}\models \text{“}n \text{ divides } k\text{”}\}$ codes a well-ordering of ω, and if α is the length of the well-ordering, then $o(\mathbb{HYP}_{\mathfrak{M}})>\alpha$. ii) Show that if $\mathfrak{M}$ is a model of Peano arithmetic generated by a single element k, usually written $\mathfrak{M}=\mathcal{N}[k]$, then $\mathfrak{M}$ is not recursively saturated.]

6.6 Notes. Theorem 6.1 and Exercise 6.4 are just two of many results that can be proved by either forcing arguments or by compactness arguments. See the appendix for a few references. Kunen has recently removed the hypothesis of countability from 6.5.

7. *Representability in $\mathfrak{M}$-Logic*

One of our principle results in this chapter, Theorems 3.1 and 3.3, identifies the relations on $\mathfrak{M}$ which are Σ_1 on $\mathbb{HYP}_{\mathfrak{M}}$ as the Π_1^1 relations on $\mathfrak{M}$, as long as M is countable. In Chapter VI we will search for the absolute version of this result. The results of this section will be of central importance in this search.

The reader should recall the notions of representability used to characterize the r. e. and recursive sets. The following are the infinitary analogues.

7.1 Definition. Let $\mathfrak{M}$ be an L-structure, T a set of finitary sentences of L^+ which are consistent in $\mathfrak{M}$-logic, $\varphi(v_1,\ldots,v_n)$ a finitary formula of L^+ and S an n-ary relation on $\mathfrak{M}$.

i) We say that $\varphi(v_1,\ldots,v_n)$ *strongly represents* S in T by the $\mathfrak{M}$-rule if, for all $q_1,\ldots,q_n\in\mathfrak{M}$,

$$S(q_1,\ldots,q_n) \text{ implies } T\vdash_{\mathfrak{M}}\varphi(\overline{\mathsf{q}}_1,\ldots,\overline{\mathsf{q}}_n), \text{ and}$$
$$\neg S(q_1,\ldots,q_n) \text{ implies } T\vdash_{\mathfrak{M}}\neg\varphi(\overline{\mathsf{q}}_1,\ldots,\overline{\mathsf{q}}_n);$$

whereas it *weakly represents* S in T using the $\mathfrak{M}$-rule if for all $q_1,\ldots,q_n\in\mathfrak{M}$

$$S(q_1,\ldots,q_n) \text{ iff } T\vdash_{\mathfrak{M}}\varphi(\overline{\mathsf{q}}_1,\ldots,\overline{\mathsf{q}}_n).$$

ii) We say that $\varphi(v_1,\dots,v_n)$ *invariantly defines* S in T in $\mathfrak{M}$-logic if for all $q_1,\dots,q_n\in\mathfrak{M}$

$$S(q_1,\dots,q_n) \text{ implies } T\models_{\mathfrak{M}}\varphi(\overline{\mathsf{q}}_1,\dots,\overline{\mathsf{q}}_n)$$
$$\neg S(q_1,\dots,q_n) \text{ implies } T\models_{\mathfrak{M}}\neg\varphi(\overline{\mathsf{q}}_1,\dots,\overline{\mathsf{q}}_n)$$

where as it *semi-invariantly defines* S in T in $\mathfrak{M}$-logic if for all $q_1,\dots,q_n\in\mathfrak{M}$

$$S(q_1,\dots,q_n) \text{ iff } T\models_{\mathfrak{M}}\varphi(\overline{\mathsf{q}}_1,\dots,\overline{\mathsf{q}}_n).$$

The following is an immediate consequence of the $\mathfrak{M}$-Completeness Theorem.

7.2 Proposition.

Strongly representable $\Rightarrow$ *invariantly definable*

weakly representable $\Rightarrow$ *semi-invariantly definable*

and, if $\mathfrak{M}$ *and* L^+ *are countable, the converses hold.* ▯

These are excellent examples of notions which agree in ordinary recursion theory but which diverge, yield two interesting distinct notions, in generalized recursion theory.

7.3 Theorem. *Let* $\mathfrak{M}=\langle M,R_1,\dots,R_l\rangle$ *and let* S *be a relation on* $\mathfrak{M}$.

i) *If* S *is* Σ_1 *on* $\mathbb{HYP}_{\mathfrak{M}}$ *then* S *is weakly representable in* KPU$^+$ *using the* $\mathfrak{M}$*-rule.*

ii) *If* $S\in\mathbb{HYP}_{\mathfrak{M}}$ *then* S *is strongly representable in* KPU$^+$ *using the* $\mathfrak{M}$*-rule.*

Proof. Our language L^+ for $\mathfrak{M}$-logic consists of $\mathsf{L}\cup\{\overline{\mathsf{p}}\mid p\in M\}$ as in III.3.2(ii). We prove the results for countable $\mathfrak{M}$. In Chapter VI we will show that the results are absolute. We prove (i) first. Choose $\varphi(x_1,\dots,x_n,p_1,\dots,p_k,M)$ as in II.8.8. We can rewrite this using the relation symbol $\overline{\mathsf{M}}$ in place of the single set M. Thus we have, for $q_1,\dots,q_n\in M$

$$S(q_1,\dots,q_n) \text{ iff } \mathrm{KPU}^+\models_{\mathfrak{M}}\varphi(\overline{\mathsf{q}}_1,\dots,\overline{\mathsf{q}}_n,\overline{\mathsf{p}}_1,\dots,\overline{\mathsf{p}}_k,\overline{\mathsf{M}})$$

which, by 7.2, gives the desired result.

Now we prove (ii). Let us assume S is unary to simplify notation. Using II.5.15 let $\varphi(x,p_1,\dots,p_n,M)$ be a good Σ_1 definition of S so that

$$\mathbb{HYP}_{\mathfrak{M}}\models\varphi[S,p_1,\dots,p_n,M]$$

and

$$\mathfrak{A}_{\mathfrak{M}}\models\exists!x\,\varphi(x,p_1,\dots,p_n,M)$$

for all models $\mathfrak{A}_{\mathfrak{M}}$ of KPU$^+$, and hence

$$\mathrm{KPU}^+\vdash_{\mathfrak{M}}\exists!x\,\varphi(x,\overline{\mathsf{p}}_1,\dots,\overline{\mathsf{p}}_n,\overline{\mathsf{M}})$$

by the $\mathfrak{M}$-completeness theorem. We claim that S is strongly represented by the formula $\psi(v)$ given by

$$\exists x\,[\varphi(x,\bar{\mathsf{p}}_1,\ldots,\bar{\mathsf{p}}_n,\bar{\mathsf{M}})\wedge v\in x].$$

If $S(q)$ holds then $\mathfrak{A}_{\mathfrak{M}}\models\psi(\bar{\mathsf{q}})$ for all models $\mathfrak{A}_{\mathfrak{M}}$ of KPU$^+$ so KPU$^+\vdash_{\mathfrak{M}}\psi(\bar{\mathsf{q}})$. If $\neg S(q)$ then, for any $\mathfrak{A}_{\mathfrak{M}}\models$ KPU$^+$, since $\mathfrak{A}_{\mathfrak{M}}\models\varphi(S)\wedge\exists!x\,\varphi(x)$, $\mathfrak{A}_{\mathfrak{M}}\models\neg\psi(\bar{\mathsf{q}})$ and hence, KPU$^+\vdash_{\mathfrak{M}}\neg\psi(\bar{\mathsf{q}})$. □

We now prove a strong converse to Theorem 7.3. The first time through this result the student should think of T as KPU$^+$ or some strong extension of it in L^* given by an r.e. set of axioms.

7.4 Theorem. *Let T be a set of finitary sentences of L^+ which is Σ_1 on $\mathbb{HYP}_{\mathfrak{M}}$ and is consistent in $\mathfrak{M}$-logic. Let S be a relation on $\mathfrak{M}$.*

(i) *If S is strongly representable in T using the $\mathfrak{M}$-rule then $S\in\mathbb{HYP}_{\mathfrak{M}}$.*

(ii) *If S is weakly representable in T using the $\mathfrak{M}$-rule then S is Σ_1 on $\mathbb{HYP}_{\mathfrak{M}}$.*

Proof. First note that (ii)⇒(i) since S strongly representable implies S and $\neg S$ are weakly representable so S and $\neg S$ are Σ_1 on $\mathbb{HYP}_{\mathfrak{M}}$, so S is Δ_1 and hence $S\in\mathbb{HYP}_{\mathfrak{M}}$ by Δ Separation. We prove (ii) for the case where $\mathfrak{M}$ and L^+ are countable leaving the absoluteness of 7.4 to Chapter VI. Let $\varphi(v_1,\ldots,v_n)$ weakly represent S in T. Then we see that the following are equivalent:

$$S(q_1,\ldots,q_n),$$

$$T\vdash_{\mathfrak{M}}\varphi(\bar{\mathsf{q}}_1,\ldots,\bar{\mathsf{q}}_n),$$

$$T\models_{\mathfrak{M}}\varphi(\bar{\mathsf{q}}_1,\ldots,\bar{\mathsf{q}}_n),$$

$$T\models\psi(\bar{\mathsf{q}}_1,\ldots,\bar{\mathsf{q}}_n).$$

I.e., the infinitary sentence $\psi(\bar{\mathsf{q}}_1,\ldots,\bar{\mathsf{q}}_n)$ is a logical consequence of T, where $\psi(\bar{\mathsf{q}}_1,\ldots,\bar{\mathsf{q}}_n)$ is

$$\bigwedge\text{Diagram}(\mathfrak{M})\wedge\forall v\,[\bar{\mathsf{M}}(v)\leftrightarrow\bigvee\nolimits_{p\in M}v\equiv\bar{\mathsf{p}}]\rightarrow\varphi(\bar{\mathsf{q}}_1,\ldots,\bar{\mathsf{q}}_n).$$

The sentence $\psi(\bar{\mathsf{q}}_1,\ldots,\bar{\mathsf{q}}_n)\in\mathbb{HYP}_{\mathfrak{M}}$ and the map $(q_1,\ldots,q_n)\mapsto\psi(\bar{\mathsf{q}}_1,\ldots,\bar{\mathsf{q}}_n)$ is Σ_1 definable so, by the Extended Completeness Theorem, S is Σ_1 on $\mathbb{HYP}_{\mathfrak{M}}$. □

It should be obvious from the proof of 7.4 that there was no real reason to demand that T be a set of finitary sentences. It is just that we only bothered to define $\vdash_{\mathfrak{M}}$ for finite sentences. T could have been a set of sentences each in $\mathbb{HYP}_{\mathfrak{M}}$ as long as T is Σ_1 on $\mathbb{HYP}_{\mathfrak{M}}$ and the proof would go through unchanged.

One might well ask about what happens to invariant and semi-invariant definability in the uncountable case where they no longer coincide with the representability notion. They turn out to be significant classes of predicates, ones we study in Chapter VIII.

7.5 Exercise. Let $\mathfrak{M}=\langle M, R_1, \ldots, R_l\rangle$ be a structure for L. Let L^+ be as in 7.3.

i) Assume that we have added a Σ function symbol F to L^* for the operation $F(x,y)=x\cup\{y\}$ and a constant symbol $\emptyset$ for the empty set. Show that each $x\in \mathbb{HF}_{\mathfrak{M}}$ is denoted by a closed term t_x of L^+.

ii) Show that $S\subseteq \mathbb{HF}_{\mathfrak{M}}$ is Σ_1 on $\mathbb{HYP}_{\mathfrak{M}}$ iff S is weakly representable in KPU$^+$ using the $\mathfrak{M}$-rule.

7.6 Notes. The representability approach to the hyperarithmetic sets goes back to Grzegorczyk, Mostowski and Ryll-Nardzewski [1961].

Part B

The Absolute Theory

"... the central notions of model theory are absolute, and absoluteness, unlike cardinality, is a logical concept."

G. Sacks, from
Saturated Model Theory

Chapter V

The Recursion Theory of Σ_1 Predicates on Admissible Sets

There are many equivalent definitions of the class of recursive functions on the natural numbers. Different definitions have different uses while the equivalence of all the notions provides evidence for Church's thesis, the thesis that the concept of recursive function is the most reasonable explication of our intuitive notion of effectively calculable function.

As the various definitions are lifted to domains other than the integers (e. g., admissible sets) some of the equivalences break down. This break-down provides us with a laboratory for the study of recursion theory. By studying the notions in the general setting one sees with a clearer eye the truths behind the results on the integers.

The most dramatic breakdown results in two competing notions of r.e. on admissible sets, notions which happen to coincide on countable admissible sets. We refer to these as the *syntactic* and *semantic* notions of r.e. and study the former in this chapter. The semantic notion is discussed in Chapter VIII.

1. Satisfaction and Parametrization

In view of Theorem II.2.3 (which shows that r.e. on ω is just Σ_1 on $\mathbb{HF}$) it is natural to ask oneself what properties of r.e. and recursive lift up to Σ_1 and Δ_1 on an arbitrary admissible set. Luckily, the more important results, results like Kleene's Enumeration and Second Recursion Theorem, lift to completely arbitrary admissible sets.

1.1 Definition. Let $\mathbb{A}$ be admissible and let R be a relation on $\mathbb{A}$.

(i) R is $\mathbb{A}$-*r.e.* if R is Σ_1 on $\mathbb{A}$.

(ii) R is $\mathbb{A}$-*recursive* if R is Δ_1 on $\mathbb{A}$.

(iii) R is $\mathbb{A}$-*finite* if $R \in \mathbb{A}$.

(iv) A function f with domain and range subsets of $\mathbb{A}$ is $\mathbb{A}$-*recursive* if its graph is $\mathbb{A}$-r.e.

If $\mathbb{A} = \mathrm{L}(\alpha)$ then we refer to these notions as α-*r.e.*, α-*recursive* and α-*finite*, respectively.

As in ordinary ω-recursion theory, a *total* $\mathbb{A}$-recursive function will have an $\mathbb{A}$-recursive graph.

The first result of ω-recursion theory we want to generalize is Kleene's Enumeration Theorem.

1.2 Definition. Let $\mathbf{S}$ be a collection of n-ary relations on some set X. Let $Y \subseteq X$. An $n+1$-ary relation T on X *parametrizes* $\mathbf{S}$ *(with indices from* Y*)* if $\mathbf{S}$ consists of all relations of the form

$$S_e = \{(x_1, \ldots, x_n) \mid T(e, x_1, \ldots, x_n)\}$$

as e ranges over Y.

1.3 Theorem. *Let* $\mathbb{A} = (\mathfrak{M}; A, \in, \ldots)$ *be an admissible set. There is an* $\mathbb{A}$*-r.e. relation* T_n *which parametrizes the class of n-ary* $\mathbb{A}$*-r.e. relations, with indices from* A.

To prove this theorem we make use of our earlier formalization in KPU of syntax and semantics. The proof is more important than the theorem itself.

There is a systematic ambiguity which has served us well until now. We have been using $\varphi, \psi, \ldots$ to range over formulas of our metalanguage L^* as well as over formulas of formalized languages. We must avoid this confusion in this section.

Let $\mathsf{L}^* = \mathsf{L}(\in, \ldots)$ be fixed and finite. For simplicity we assume L^* has only relation symbols. The extension to the general case is sketched in the exercises. We consider L^* here as a single sorted language with variables $x_1, x_2, \ldots$ and unary symbols U (for "urelement") and S (for "set"). Let l^* be some effective coding of L^* in $\mathbb{HF}$. For basic symbols like R we let $\ulcorner \mathsf{R} \urcorner$ be the set in $\mathbb{HF}$ which names R. For definiteness we take $v_n = \ulcorner x_n \urcorner = \langle 0, n \rangle$. To each formula φ of L^* there corresponds its formalized version $\ulcorner \varphi \urcorner$, an element of $l^*_{\omega\omega} \subseteq \mathbb{HF}$, defined by recursion equations

$$\ulcorner \varphi \wedge \psi \urcorner = \langle \bigwedge, \{\ulcorner \varphi \urcorner, \ulcorner \psi \urcorner\} \rangle$$

$$\ulcorner \exists x_n \varphi \urcorner = \langle \exists, v_n, \ulcorner \varphi \urcorner \rangle$$

and so forth.

Define, in KPU, an operation $\mathfrak{N}_a$ on sets a by: $\mathfrak{N}_a$ is a structure for l^* with universe $\mathrm{TC}(a)$ which interprets the symbols of l^* as follows:

Symbol	*Interpretation*
$\ulcorner \mathsf{U} \urcorner$	$\{p \mid p \in \mathrm{TC}(a)\}$
$\ulcorner \mathsf{S} \urcorner$	$\{b \mid b \in \mathrm{TC}(a)\}$
$\ulcorner \in \urcorner$	$\{\langle x, y \rangle \mid x, y \in \mathrm{TC}(a), x \in y\}$
$\ulcorner \mathsf{R} \urcorner$	$\{\langle x_1, \ldots, x_n \rangle \in \mathrm{TC}(a)^n \mid R(x_1, \ldots, x_n)\}$.

Clearly $\mathfrak{N}_a$ is a Σ_1 operation of a. Recall the notation $\varphi^{(a)}$ from § I.4.

1.4 Lemma. *For each formula* $\varphi(x_1,\ldots,x_n)$ *of* L^* *the following is a theorem of* KPU: *for all sets* a *and all* $x_1,\ldots,x_n \in \mathrm{TC}(a)$, *if* $s=\{\langle v_i, x_i\rangle \mid i=1,\ldots,n\}$ *then*

$$\varphi^{(\mathrm{TC}(a))}(x_1,\ldots,x_n) \quad \textit{iff} \quad \mathfrak{N}_a \models \ulcorner\varphi\urcorner[s].$$

Proof. For φ atomic this follows from the definition of $\mathfrak{N}_a$. The result follows by induction on formulas. ☐

1.5 Definition. Let $\Sigma\text{-Sat}_n$ be the following Σ_1 formula of L^* with variables $y, x_1,\ldots,x_n$:

"*y is a Σ formula of l^* with free variables among $v_1,\ldots,v_n$ and there is a transitive set a with $x_1,\ldots,x_n \in a$ such that*

$$\mathfrak{N}_a \models y[s]$$

where $s=\{\langle v_i, x_i\rangle \mid i=1,\ldots,n\}$".

That this can be expressed by a Σ_1 formula follows from the results in § III.1.

1.6 Proposition. *Let* $\varphi(x_1,\ldots,x_n)$ *be a* Σ *formula of* L^*. *The following is a theorem of* KPU: *for all* $x_1,\ldots,x_n$,

$$\varphi(x_1,\ldots,x_n) \quad \textit{iff} \quad \Sigma\text{-Sat}_n(\ulcorner\varphi(x_1,\ldots,x_n)\urcorner, x_1,\ldots,x_n).$$

Proof. Assume the axioms of KPU. The following are equivalent:

$\varphi(x_1,\ldots,x_n)$
$\exists a\,[\mathrm{Tran}(a) \wedge x_1,\ldots,x_n \in a \wedge \varphi^{(a)}(x_1,\ldots,x_n)]$
$\exists a\, \exists s\,[\mathrm{Tran}(a) \wedge x_1,\ldots,x_n \in a \wedge \mathfrak{N}_a \models \ulcorner\varphi\urcorner[s]$, *where* $s=\{\langle v_i,x_i\rangle \mid i=1,\ldots,n\}]$
$\Sigma\text{-Sat}_n(\ulcorner\varphi\urcorner, x_1,\ldots,x_n)$.

The first two lines are equivalent by Σ Reflection, the middle two by Lemma 1.4 and the last two by the definition of $\Sigma\text{-Sat}_n$. ☐

Define $T_n(e, x_1,\ldots,x_n)$ to be the Σ_1 formula:

"e is an ordered pair $\langle\psi, z\rangle$ and $\Sigma\text{-Sat}_{n+1}(\psi, x_1,\ldots,x_n,z)$".

Proof of Theorem 1.3. Since T_n is Σ_1 any predicate defined by

$$R(x_1,\ldots,x_n) \quad \text{iff} \quad T_n(e, x_1,\ldots,x_n)$$

is $\mathbb{A}$-r.e. To prove the converse, let R be an n-ary $\mathbb{A}$-r.e. predicate. By using ordered pairs it has a Σ_1 definition on $\mathbb{A}$ with exactly one parameter z, say

$$R(x_1,\ldots,x_n) \quad \text{iff} \quad \mathbb{A} \models \psi(x_1,\ldots,x_n,z).$$

Then let $e = \langle \ulcorner \psi(x_1, \ldots, x_{n+1}) \urcorner, z \rangle$ and apply 1.6. $\square$

1.7 Corollary. *Let $\mathbb{A}$ be admissible. There is an $\mathbb{A}$-r.e. set which is not $\mathbb{A}$-recursive.*

Proof. Just as for ω-recursion theory define

$$K = \{e \in \mathbb{A} \mid \mathbb{A} \models T_1(e,e)\}.$$

If $\mathbb{A} - K$ were $\mathbb{A}$-r.e. there would be an e_0 such that for all $e \in \mathbb{A}$, $e \notin K$ iff $T_1(e_0, e)$, and hence $e_0 \notin K$ iff $e_0 \in K$. $\square$

Let $\mathfrak{M} = \langle M \rangle$ be an infinite set with no additional relations. Note that if $X \subseteq M$ is $\mathbb{HYP}_{\mathfrak{M}}$-r.e. then X is $\mathbb{HYP}_{\mathfrak{M}}$-finite since by II.9.3, X or $M - X$ is finite. Thus Corollary 1.7 cannot in general be improved to get a $\mathbb{A}$-r.e. subset of $\mathfrak{M}$ which is not $\mathbb{A}$-recursive.

1.8—1.10 Exercises

1.8. Suppose $\mathsf{L}^* = \mathsf{L}(\in, \mathsf{f}, \ldots)$ has a function symbol f. Show that under the standard treatment of function symbols as relation symbols, Δ_0 formulas transform into both Σ and Π formulas (but not necessarily into Δ_0 formulas). Hence Σ_1 formulas transform into Σ formulas.

1.9. Let L^* be a finite language with function symbols. Define Σ-Sat$_n$ for L^* in such a way that 1.6 and hence 1.3 become provable.

1.10. Find an admissible set $\mathbb{A}_{\mathfrak{M}}$ such that the class of $\mathbb{A}_{\mathfrak{M}}$-r.e. subsets of $\mathfrak{M}$ cannot be parametrized by an $\mathbb{A}_{\mathfrak{M}}$-r.e. binary relation with indices from M.

2. *The Second Recursion Theorem for* KPU

The Second Recursion Theorem in ω-recursion theory is a mysterious device for implicitly defining recursive partial functions, or equivalently, r.e. predicates. The theorem is equally mysterious and equally useful in our setting.

Let $\mathsf{L}^* = \mathsf{L}(\in, \ldots)$ be a finite language (as in § 1) and let R be a new n-ary relation symbol, $n \geq 1$.

2.1 Definition. The collection of R-*positive* formulas of $\mathsf{L}^*(\mathsf{R})$ is the smallest class of formulas containing all formulas of L^*, all atomic formulas of $\mathsf{L}^*(\mathsf{R})$, and closed under

$$\wedge,\ \vee,\ \forall u \in v,\ \exists u \in v,\ \forall u,\ \exists u$$

for all variables u, v. We use the notation

$$\varphi(\mathsf{R}_+)$$

to indicate that φ is an R-positive formula.

Given a formula $\varphi(\mathsf{R})$ of $\mathsf{L}^*(\mathsf{R})$ and a formula $\psi(x_1,\dots,x_n)$ of L^* we use the notations

$$\varphi(\psi/\mathsf{R})$$

$$\varphi(\lambda x_1,\dots,x_n\, \psi(x_1,\dots,x_n))$$

more or less interchangeably to denote the formula resulting by replacing each occurrence of an atomic formula of the form $\mathsf{R}(t_1,\dots,t_n)$ in $\varphi(\mathsf{R})$ by $\psi(t_1/x_1,\dots,t_n/x_n)$ (unless some t_i is not free for x_i in ψ in which case we must first rename bound variables in ψ, but then we agreed in Chapter I not to mention such details). Thus $x_1,\dots,x_n$ do not occur free in $\varphi(\psi/\mathsf{R})$ (unless they are free in $\varphi(\mathsf{R})$), and R does not occur in $\varphi(\psi/\mathsf{R})$.

2.2 Lemma. *If $\varphi(\mathsf{R}_+)$ is a Σ formula of $\mathsf{L}^*(\mathsf{R})$ and if $\psi(x_1,\dots,x_n)$ is a Σ formula of L^* then $\varphi(\psi/\mathsf{R})$ is a Σ formula of L^*.*

Proof. By induction on the class of R-positive formulas $\varphi(\mathsf{R}_+)$. □

2.3 The Second Recursion Theorem. *Let $\varphi(\vec{x},\vec{y},\mathsf{R}_+)$ be an R-positive Σ formula where R is n-ary, $\vec{x}=x_1,\dots,x_n$ and $\vec{y}=y_1,\dots,y_k$. There is a Σ formula $\psi(\vec{x},\vec{y})$ of L^* so that the following is a theorem of* KPU: *for all parameters $\vec{y}$ and all $x_1,\dots,x_n$*

$$\psi(x_1,\dots,x_n,\vec{y}) \quad \textit{iff} \quad \varphi(x_1,\dots,x_n,\vec{y},\lambda x_1,\dots,x_n\,\psi(x_1,\dots,x_n,\vec{y})).$$

Proof. To simplify notation we assume $n=k=1$. Let $\theta(x,y,z)$ be the Σ formula

$$\varphi(x,y,\lambda x\, \Sigma\text{-}\mathrm{Sat}_3(z,x,y,z)).$$

Let $e=\ulcorner\theta(x,y,z)\urcorner\in\mathbb{HF}$ and let $\psi(x,y)$ be $\theta(x,y,e)$, or rather, the Σ formula equivalent to it obtained by replacing the constant e by a good Σ_1 definition of e. Then we have, in KPU, that the following are equivalent:

$$\psi(x,y)$$

$$\theta(x,y,e)$$

$$\varphi(x,y,\lambda x\, \Sigma\text{-}\mathrm{Sat}_3(e,x,y,e))$$

$$\varphi(x,y,\lambda x\, \theta(x,y,e))$$

$$\varphi(x,y,\lambda x\, \psi(x,y)). \quad □$$

Since any Σ formula is equivalent, in KPU, to a Σ_1 formula, we could have demanded that the ψ of 2.3 be Σ_1.

We give a simple application of the Second Recursion Theorem. In any admissible set $\mathbb{A}_{\mathfrak{M}}$, $\mathbb{HF}$ is an $\mathbb{A}_{\mathfrak{M}}$-recursive subset since

$$a \in \mathbb{HF} \text{ iff } \mathrm{sp}(a) = 0 \wedge (\mathrm{rk}(a) \text{ is a natural number}).$$

$\mathbb{HF}_{\mathfrak{M}}$, however, *is not always* Δ_1 *definable.* (The student can find an example of this in Exercise 2.6.)

2.4 Proposition. *There is a* Σ_1 *formula* $\psi(x)$ *such that in any admissible set* $\mathbb{A}_{\mathfrak{M}}$,

$$\mathbb{HF}_{\mathfrak{M}} = \{a \in \mathbb{A}_{\mathfrak{M}} \mid \mathbb{A}_{\mathfrak{M}} \models \psi[a]\}.$$

Proof. Let R be unary and let $\varphi(x, \mathsf{R}_+)$ be the Σ formula

$$(x \textit{ is a finite set}) \wedge \forall y \in x \ (\textit{if } y \textit{ is a set then } \mathsf{R}(y)).$$

Now apply the Second Recursion Theorem to get a formula ψ such that

$$\mathrm{KPU} \vdash \psi(x) \leftrightarrow (x \textit{ is a finite set} \wedge \forall y \in x \ (y \textit{ is a set} \rightarrow \psi(y)).$$

Now let $\mathbb{A}_{\mathfrak{M}}$ be admissible. A trivial proof by induction on $\in$ shows that

$$a \in \mathbb{HF}_{\mathfrak{M}} \quad \text{iff} \quad \mathbb{A}_{\mathfrak{M}} \models \psi[a]$$

for all $a \in \mathbb{A}_{\mathfrak{M}}$. □

2.5—2.6 Exercises

2.5. Show that a formula $\varphi(\mathsf{R})$ is logically equivalent to an R-positive formula iff the result of pushing negations inside φ as far as possible (using de Morgan's laws) results in a formula in which $\neg \mathsf{R}$ does not occur.

2.6. Let $\mathfrak{M}$ be a recursively saturated model of Peano arithmetic, KP or ZF. Show that $\mathbb{HF}_{\mathfrak{M}}$ is not $\mathbb{HYP}_{\mathfrak{M}}$-recursive.

3. *Recursion Along Well-founded Relations*

In this section we use the Second Recursion Theorem to give a new principle of definition by recursion along well-founded relations. This serves as a useful warm-up exercise in the use of the Second Recursion Theorem.

3.1 Theorem. *Let* $\mathbb{A}=\mathbb{A}_{\mathfrak{M}}$ *be admissible, let* p *be an* $\mathbb{A}$*-recursive function and define a binary relation* $\prec$ *by*

$$x \prec y \quad \textit{iff} \quad x \in p(y)$$

for all $y \in \operatorname{dom}(p)$.

(i) *The well-founded part of* $\prec$, $\mathscr{W}\!f(\prec)$, *is* $\mathbb{A}$*-r.e.*

(ii) *If* G *is a total* $k+2$*-ary* $\mathbb{A}$*-recursive function then there is an* $\mathbb{A}$*-recursive* F *with*

$$\operatorname{dom}(F)=(M \cup A)^k \times \mathscr{W}\!f(\prec)$$

such that

$$F(\vec{z},x)=G(\vec{z},x,\{\langle y,F(\vec{z},y)\rangle \mid y \prec x\})$$

for all $\vec{z} \in (M \cup A)^k$ *and all* $x \in \mathscr{W}\!f(\prec)$.

Proof. Recall that $\mathscr{W}\!f(\prec)$ is the largest subset B of Field$(\prec)$ such that:

$$x \prec y,\ y \in B \text{ implies } x \in B, \text{ and}$$
$$\prec \restriction B^2 \text{ is well founded}.$$

There is such a largest set by II.8.2. Note that $\operatorname{pred}(x) \subseteq \mathscr{W}\!f(\prec)$ implies $x \in \mathscr{W}\!f(\prec)$. Part (i) of the theorem follows from part (ii) but we need (i) in the proof of (ii). Besides, (i) is an easy example of the use of the Second Recursion Theorem.

Define a Σ_1 formula $\psi(x,\alpha)$ such that

(1) $$\psi(x,\alpha) \quad \text{iff} \quad \exists z(z=p(x) \wedge \forall y \in z\, \exists \beta < \alpha\, \psi(y,\beta))$$

is a theorem of KPU and hence true in $\mathbb{A}$. Since this is only our second use of the Second Recursion Theorem, perhaps we should be a bit more explicit. Let $\eta(x,z)$ define the graph of p; η may have some other parameters which remain fixed throughout (the y's of the Second Recursion Theorem). Let R be a new binary relation symbol and let $\varphi(x,\alpha,\mathsf{R}_+)$ be the Σ formula

$$\exists z[\eta(x,z) \wedge \forall y \in z\, \exists \beta \in \alpha\, \mathsf{R}(y,\beta)]$$

of $\mathsf{L}^*(\mathsf{R})$. Note that R does indeed occur positively in this formula. Now apply the Second Recursion Theorem to get ψ satisfying (1). We will never again be this explicit; rather we'll just write an equation like (1) and leave it to the reader to see that the right-hand side is of the appropriate form. Now given ψ, one proves, for $\alpha \in \mathbb{A}$,

(2) $$\mathbb{A} \models \psi(x,\alpha) \quad \text{implies} \quad x \in \mathscr{W}\!f(\prec)$$

by a simple induction on α, using (1). A little less trivial is

(3) $$x \in \mathcal{W}\!f(\prec) \quad \text{implies} \quad \mathbb{A} \models \exists \alpha\, \psi(x,\alpha)\,.$$

Assume $x \in \mathcal{W}\!f(\prec)$. Since $\mathcal{W}\!f(\prec) \subseteq \mathrm{Field}(\prec)$, $p(x)$ is defined. We may assume by induction ($\prec \restriction \mathcal{W}\!f(\prec)$ is well founded so induction over it is legitimate) that for each $y \in p(x)$

$$\mathbb{A} \models \exists \beta\, \psi(y,\beta)$$

and hence

$$\mathbb{A} \models \forall y \in p(x)\, \exists \beta\, \psi(y,\beta),$$

so by Σ Reflection there is an $\alpha \in \mathbb{A}$ with

$$\mathbb{A} \models \forall y \in p(x)\, \exists \beta < \alpha\, \psi(y,\beta)\,.$$

By (1),

$$\mathbb{A} \models \psi(x,\alpha)\,.$$

Combining (2), (3) we have

$$\mathcal{W}\!f(\prec) = \{x \in \mathbb{A} \mid \mathbb{A} \models \exists \alpha\, \psi(x,\alpha)\}$$

which makes $\mathcal{W}\!f(\prec)$ an $\mathbb{A}$-r.e. set.

To prove (ii) we use the Second Recursion Theorem again. We want to define the graph of F by a Σ_1 formula $\psi(\vec{z},x,w)$. Let us suppress the parameters since they are held fixed throughout. We want

$$\begin{aligned} \psi(x,w) \quad &\text{iff} \quad x \in \mathcal{W}\!f(\prec) \wedge F(x) = w \\ &\text{iff} \quad x \in \mathcal{W}\!f(\prec) \wedge \exists f[f = F \restriction p(x) \wedge w = G(x,f)]\,. \end{aligned}$$

The Second Recursion Theorem gives us a $\Sigma_1\, \psi$ so that

$$\begin{aligned} &\psi(x,w) \quad \text{iff} \quad x \in \mathcal{W}\!f(\prec) \wedge \exists f\,[f \textit{ is a function} \wedge \mathrm{dom}(f) = p(x) \\ &\wedge \forall y \in p(x)\, \psi(y, f(y)) \wedge w = G(x,f)] \end{aligned}$$

is true in $\mathbb{A}$ for all x, w. Using Σ Replacement one shows by induction on $\prec \restriction \mathcal{W}\!f(\prec)$ that

$$x \in \mathcal{W}\!f(\prec) \quad \text{implies} \quad \mathbb{A} \models \exists!\, w\, \psi(x,w)$$

so we may use $\psi(x,w)$ as a definition of an $\mathbb{A}$-recursive F. One then checks that F satisfies the desired equation, again by induction on $\prec \restriction \mathcal{W}\!f(\prec)$. $\square$

3.2 Definition. Let $\prec$ be a binary relation with nonempty wellfounded part. Define the *$\prec$-rank function* $\rho^{\prec}$, for $x \in \mathcal{W}f(\prec)$, by

$$\rho^{\prec}(x) = \sup\{\rho^{\prec}(y)+1 \mid y \prec x\}.$$

Define the *rank of* $\prec$, $\rho(\prec)$, by

$$\rho(\prec) = \sup\{\rho^{\prec}(x)+1 \mid x \in \mathcal{W}f(\prec)\}.$$

3.3 Corollary. *Let $\mathbb{A}$ be an admissible set with $\prec$ an element of $\mathbb{A}$ and $\alpha = o(\mathbb{A})$.*
(i) $\rho(\prec) \leqslant \alpha$.
(ii) *If $\mathcal{W}f(\prec) \in \mathbb{A}$ (for example, if $\prec$ is well founded) then $\rho(\prec) < \alpha$.*
(iii) *If $\mathcal{W}f(\prec) \in \mathbb{A}$ then $\rho(\prec) = \alpha$.*

Proof. To apply Theorem 3.1 define an $\mathbb{A}$-recursive function p by

$$p(x) = \{y \in \mathrm{Field}(\prec) \mid y \prec x\}.$$

(This is the reason we assumed $\prec \in \mathbb{A}$.) Then the definition

$$\rho^{\prec}(x) = \sup\{\rho^{\prec}(y)+1 \mid y \prec x\}$$

falls under 3.1 (ii) so always gives values in $\mathbb{A}$. This proves (i).

If $\mathcal{W}f(\prec) \in \mathbb{A}$ then we may use Σ Replacement to form

$$\sup\{\rho^{\prec}(x)+1 \mid x \in \mathcal{W}f(\prec)\}$$

in $\mathbb{A}$. This gives (ii). To prove (iii), suppose $\rho(\prec) = \beta \in \mathbb{A}$, and let us prove $\mathcal{W}f(\prec) \in \mathbb{A}$. For $\gamma < \beta$ let

$$\begin{aligned} F(\gamma) &= \textstyle\bigcup_{\xi<\gamma} F(\xi) \cup \{x \in \mathrm{Field}(\prec) \mid p(x) \subseteq \bigcup_{\xi<\gamma} F(\xi)\} \\ &= \{x \in \mathrm{Field}(\prec) \mid \rho^{\prec}(x) \leqslant \gamma\} \end{aligned}$$

be defined by Σ Recursion for $\gamma < \beta$. But then

$$\mathcal{W}f(\prec) = \textstyle\bigcup_{\gamma<\beta} F(\gamma)$$

is in $\mathbb{A}$ by Σ Replacement. □

While the most useful results of ω-recursion theory lift to an arbitrary admissible set, many of the more pleasing facts of recursion theoretic life on ω carry over only to special admissible sets. In particular, there are many results of recursion theory which use the effective well-ordering of the domain in an essential way.

3.4 Definition. Let $\mathbb{A} = (\mathfrak{M}; A, \in, \ldots)$ be an admissible set with $\alpha = o(\mathbb{A})$. $\mathbb{A}$ is *recursively listed* if there is an $\mathbb{A}$-recursive bijection of α onto $M \cup A$.

Lemma II.2.4 shows that $\mathbb{HF}$ is recursively listed. We will study the recursion theory of recursively listed admissible sets in the next section. They are related to this section by means of the following result.

3.5 Proposition. *$\mathbb{A}$ is recursively listed iff there is a total $\mathbb{A}$-recursive function p such that*

$$x \prec y \quad \textit{iff} \quad x \in p(y)$$

defines a well-ordering $\prec$ of $M \cup A$.

Proof. Suppose $e: \alpha \to \mathbb{A}$ is an $\mathbb{A}$-recursive enumeration of $M \cup A$. Note that e^{-1} is $\mathbb{A}$-recursive. Define $p(y) = \{e(\beta) \mid \beta < e^{-1}(y)\}$ and note that $x \in p(y)$ iff $e^{-1}(x) < e^{-1}(y)$.

Now suppose p is given as above. Note that, by 3.1, $\rho^{\prec}$ is an $\mathbb{A}$-recursive function. Since $\prec$ is a linear ordering, $\rho^{\prec}$ is one-one so we can let e be the inverse of $\rho^{\prec}$. By Σ Replacement, $\rho^{\prec}$ has range α so e has domain α. □

Recall the definition of $\mathrm{L}(\alpha)$ given (in KPU) in II.5.

$$\mathrm{L}(0) = 0$$

$$\mathrm{L}(\alpha+1) = \mathscr{D}(\mathrm{L}(\alpha) \cup \{\mathrm{L}(\alpha)\})$$

$$\mathrm{L}(\lambda) = \textstyle\bigcup_{\alpha<\lambda} \mathrm{L}(\alpha) \quad \text{for } \lambda \text{ a limit ordinal}$$

where

$$\mathscr{D}(b) = b \cup \{\mathscr{F}_i(x, y) \colon x, y \in b,\ 1 \leqslant i \leqslant N\}.$$

We have shown that if α is admissible then $\mathrm{L}(\alpha)$ is the smallest admissible set $\mathbb{A}$ with $o(\mathbb{A}) = \alpha$. There is a natural well-ordering of $\mathrm{L}(\alpha)$ given by putting everything in $\mathrm{L}(\beta)$ before everything in $\mathrm{L}(\delta)$ for $\beta < \delta$ and ordering the elements a of $\mathrm{L}(\beta+1) - \mathrm{L}(\beta)$ according to which $\mathscr{F}_i(x, y) = a$. To make this precise define, in KPU, a Σ_1 formula $\psi(x, y)$, which we write as $x <_{\mathrm{L}} y$, as follows. First let

$$F(x) = \text{the least } \alpha (x \in \mathrm{L}(\alpha+1))$$

$$G(x) = \begin{cases} 0 \text{ if } \ x \in \mathscr{S}(\mathrm{L}(F(x))), \\ \text{the least } i,\ 1 \leqslant i \leqslant N, \quad \text{such that} \quad x = \mathscr{F}_i(z_1, z_2) \quad \text{for some} \\ \qquad z_1, z_2 \in \mathscr{S}(\mathrm{L}(F(x))) \quad \text{otherwise}. \end{cases}$$

$\langle z_1, z_2 \rangle \lhd_\alpha \langle w_1, w_2 \rangle$ if

$$z_1 <_{\mathrm{L}} w_1 \quad \text{or}$$

$$z_1 \in \mathrm{L}(\alpha) \wedge w_1 = \mathrm{L}(\alpha) \quad \text{or}$$

$$z_1 = w_1 \quad \text{and,}$$

$$z_2 <_{\mathrm{L}} w_2 \quad \text{or}$$

$$z_2 \in \mathrm{L}(\alpha) \wedge w_2 = \mathrm{L}(\alpha).$$

Given that $<_L$ already wellorders $L(\alpha)$, $\lhd_\alpha$ wellorders the pairs z_1, z_2 from $\mathscr{S}(L(\alpha)) = L(\alpha) \cup \{L(\alpha)\}$ "lexicographically after putting $L(\alpha)$ itself at the end of the alphabet".

Now define $<_L$ by

$$\begin{aligned} x <_L y \quad \text{if} \quad x \in L \quad \text{and} \quad y \in L \quad \text{and} \quad & F(x) < F(y) \quad \text{or} \\ & F(x) = F(y) \wedge G(x) < G(y) \quad \text{or} \\ & F(x) = F(y) \wedge G(x) = G(y) \quad \text{and there is a pair} \\ & \quad z_1, z_2 \in \mathscr{S}(L(F(x))) \quad \text{such that} \quad x = \mathscr{F}_{G(x)}(z_1, z_2) \\ & \quad \text{but for all} \quad w_1, w_2 \in \mathscr{S}(L(F(x))), \quad \text{if} \quad y = \mathscr{F}_{G(x)}(w_1, w_2) \\ & \quad \text{then} \quad \langle z_1, z_2 \rangle \lhd_{F(x)} \langle w_1, w_2 \rangle . \end{aligned}$$

We could define $<_L$ explicitly, if we really had to, but for our purposes here we can be content to use any such formula given by the Second Recursion Theorem. To see that the Second Recursion Theorem applies we need only observe that, once $\lhd_\alpha$ is replaced by its definition, the right-hand side is a Σ formula and that $<_L$ occurs positively.

3.6 Lemma (of KPU). *For each α, $<_L \upharpoonright L(\alpha) \times L(\alpha)$ well orders $L(\alpha)$ in such a way that for $\beta < \gamma < \alpha$, if $x \in L(\beta)$, $y \in L(\gamma) - L(\beta)$ then $x <_L y$.*

Proof. By induction on α. $\square$

3.7 Theorem. *If α is an admissible ordinal then $L(\alpha)$ is a recursively listed admissible set.*

Proof. Since, for $x, y \in L$, $\neg(x <_L y)$ iff $x = y \vee y <_L x$, we see that $<_L$ is Δ_1 when restricted to L. Also we can define

$$\begin{aligned} p(x) &= \{y \in L(F(x)+1) \mid y <_L x\} \\ &= \{y \in L \mid y <_L x\} \end{aligned}$$

for all $x \in L(\alpha)$ so p is α-recursive, and

$$x <_L y \quad \text{iff} \quad x \in p(y)$$

so we may apply Proposition 3.5. $\square$

3.8—3.11 Exercises

3.8. An admissible set $\mathbb{A}_{\mathfrak{M}}$ is *resolvable* if there is an $\mathbb{A}_{\mathfrak{M}}$-recursive function f with $\operatorname{dom}(f) = o(\mathbb{A}_{\mathfrak{M}})$ such that $\mathbb{A}_{\mathfrak{M}} = \bigcup \operatorname{rng}(f)$.

i) Show that if $\mathbb{A}_{\mathfrak{M}}$ is resolvable then there is a function f with the above properties which also satisfies: $f(\beta)$ is always transitive and $\beta < \gamma$ implies $f(\beta) \in f(\gamma)$. Such an f is a *resolution* of $\mathbb{A}_{\mathfrak{M}}$.

ii) A well-founded relation $\prec$ is a *pre-wellordering* if for all $x, y \in \mathrm{Field}(\prec)$,

$$\rho^{\prec}(x) < \rho^{\prec}(y) \quad \text{implies} \quad x \prec y .$$

Show that an admissible set $\mathbb{A}_{\mathfrak{M}}$ is resolvable iff there is a total $\mathbb{A}_{\mathfrak{M}}$-recursive function p with $\mathbb{A}_{\mathfrak{M}} = \bigcup \mathrm{rng}(p)$ such that

$$x \prec y \quad \text{iff} \quad x \in p(y)$$

defines a pre-wellordering of $\mathbb{A}_{\mathfrak{M}}$.

3.9. Show that every admissible set of the form $\mathrm{L}(a,\alpha)$ is resolvable. In particular, $\mathbb{HYP}_{\mathfrak{M}}$ is resolvable.

3.10. Let $\mathrm{L}(a,\alpha)$ be admissible and assume that there is a well-ordering $\prec$ of a, $\prec$ an element of $\mathrm{L}(a,\alpha)$. Modify the definition of $<_{\mathrm{L}}$ to show that $\mathrm{L}(a,\alpha)$ is recursively listed. In particular, if $\mathcal{N} = \langle \omega, +, \times \rangle$ and if $\mathrm{L}(\alpha)_{\mathcal{N}}$ is admissible then it is recursively listed. Hence $\mathbb{HYP}_{\mathcal{N}}$ is recursively listed.

3.11. Let $\mathbb{A}$ be admissible, $\prec \in \mathbb{A}$, $\prec$ not well-founded but

$$\mathbb{A} \models \text{“}{\prec}\text{is well founded”} .$$

(In other words, every subset X of $\mathrm{Field}(\prec)$ which happens to be an element of $\mathbb{A}$ has a $\prec$-minimal element.) Show that $\rho(\prec) = o(\mathbb{A})$.

4. *Recursively Listed Admissible Sets*

In this section we show how the elementary parts of the theory of r.e. sets generalize from ω-recursion theory to any recursively listed admissible set.

4.1 Theorem. *Let* $\mathbb{A} = \mathbb{A}_{\mathfrak{M}}$ *be a recursively listed admissible set, with* $\alpha = o(\mathbb{A})$, *and let* B *be a nonempty subset of* $\mathbb{A}$. *The following are equivalent:*

(i) *B is $\mathbb{A}$-r.e.*

(ii) *B is the range of a total $\mathbb{A}$-recursive function.*

(iii) *B is the range of an $\mathbb{A}$-recursive function with domain α.*

Proof. We have (iii)$\Rightarrow$(ii) since there is an $\mathbb{A}$-recursive bijection e of α onto $M \cup A$. Clearly (ii)$\Rightarrow$(i) so we prove (i)$\Rightarrow$(iii). Let

$$x \in B \quad \text{iff} \quad \mathbb{A} \models \exists y\, \varphi(x, y)$$

where φ is Δ_0. Fix $x_0 \in B$. Define an $\mathbb{A}$-recursive f by

$$f(\beta) = 1^{\text{st}} e(\beta) \quad \text{if} \quad \varphi(1^{\text{st}} e(\beta), 2^{\text{nd}} e(\beta))$$
$$= x_0 \quad \text{otherwise}.$$

Then $B = \text{rng}(f)$ and $\alpha = \text{dom}(f)$. □

4.2 Reduction Theorem. *Let $\mathbb{A} = \mathbb{A}_{\mathfrak{M}}$ be a recursively listed admissible set. For any pair B, C of $\mathbb{A}$-r.e. sets there is a pair B_0, C_0 of disjoint $\mathbb{A}$-r.e. sets with $B_0 \subseteq B$, $C_0 \subseteq C$ and $B_0 \cup C_0 = B \cup C$.*

Proof. We may assume B and C are nonempty. Use 4.1 to choose $\mathbb{A}$-recursive functions F, G with domain $o(\mathbb{A})$ such that

$$B = \text{rng}(F), \qquad C = \text{rng}(G).$$

Define B_0 and C_0 by:

$$x \in B_0 \quad \text{iff} \quad \exists \beta [F(\beta) = x \wedge \forall \gamma < \beta\, G(\gamma) \neq x]$$
$$x \in C_0 \quad \text{iff} \quad \exists \gamma [G(\gamma) = x \wedge \forall \beta \leqslant \gamma\, F(\beta) \neq x]$$

Then clearly B_0 and C_0 are disjoint $\mathbb{A}$-r.e. sets with $B_0 \subseteq B$, $C_0 \subseteq C$. If $x \in B - C$ then $x \in B_0$. If $x \in C - B$ then $x \in C_0$. If $x \in B \cap C$ then let β be the least ordinal with $F(\beta) = x$, γ the least with $G(\gamma) = x$. If $\beta \leqslant \gamma$ then $x \in B_0$ but if $\beta > \gamma$ then $x \in C_0$ so $B \cup C \subseteq B_0 \cup C_0$. □

4.3 Corollary (Separation Theorem). *Let $\mathbb{A} = \mathbb{A}_{\mathfrak{M}}$ be a recursively listed admissible set. For any pair B, C of disjoint Π_1 sets on $\mathbb{A}$ there is an $\mathbb{A}$-recursive set containing B but disjoint from C.*

Proof. Apply 4.2 to $\mathbb{A} - B$, $\mathbb{A} - C$ to get disjoint sets B_0, C_0 with $B \subseteq B_0$, $C \subseteq C_0$, $B_0 \cup C_0 = A$. Then B_0 is $\mathbb{A}$-recursive. □

4.4 Uniformization Theorem. *Let $\mathbb{A} = \mathbb{A}_{\mathfrak{M}}$ be a recursively listed admissible set and let R be an $\mathbb{A}$-r.e. binary relation. There is an $\mathbb{A}$-recursive function F with*

(i) $\text{dom}(F) = \{x \mid \exists y\, R(x, y)\}$

(ii) *for* $x \in \text{dom}(F)$,

$$R(x, F(x)).$$

Proof. Let e be an $\mathbb{A}$-recursive bijection of $o(\mathbb{A})$ onto $\mathbb{A}$. Let R be given by

$$R(x, y) \quad \text{iff} \quad \exists z\, S(x, y, z)$$

where S is $\mathbb{A}$-recursive. Define F by

$$F(x)=y \quad \text{iff} \quad \exists\beta[S(x,1^{\text{st}}e(\beta),2^{\text{nd}}e(\beta))\wedge\forall\gamma<\beta\,\neg S(x,1^{\text{st}}e(\gamma),2^{\text{nd}}e(\gamma))\wedge y=1^{\text{st}}e(\beta)]\,. \quad \square$$

The passage, in 4.4, from the Σ_1 definition of R to the Σ_1 definition of F was explicitly given, so we can get the following more complicated but stronger result. For $z\in\mathbb{A}$ we let

$$W_z^2=\{(x,y)\mid T_2(z,x,y)\}$$

where T_2 is the $\mathbb{A}$-r.e. relation which parametrizes the $\mathbb{A}$-r.e. binary relations, as it was defined in § 1.

4.5 Theorem. *Let $\mathbb{A}$ be a recursively listed admissible set. There is a total $\mathbb{A}$-recursive function G such that for all $z\in\mathbb{A}$:*

(i) *$W^2_{G(z)}$ is the graph of an $\mathbb{A}$-recursive function,*
(ii) *$W^2_{G(z)}\subseteq W^2_z$, and*
(iii) *$\text{dom}(W^2_z)=\text{dom}(W^2_{G(z)})$.*

Proof. See 4.4 and remarks following it. $\square$

Using this we get the following analogue of Kleene's T-predicate for recursive partial functions.

4.6 Theorem. *Let $\mathbb{A}$ be a recursively listed admissible set. There is an $\mathbb{A}$-r.e. predicate T_2^* of three arguments which parametrizes the collection of all partial $\mathbb{A}$-recursive functions, with indices from the ordinals of $\mathbb{A}$.*

Proof. Let $e\colon o(\mathbb{A})\to\mathbb{A}$ be a recursive listing and let G be as given in 4.5. Define

$$T_2^*(\beta,x,y) \quad \text{iff} \quad T_2(G(e(\beta)),x,y)\,.$$

Then for each β,

$$f_\beta=\{\langle x,y\rangle\mid T_2^*(\beta,x,y)\}$$

is a partial function with Σ_1 graph (by 4.4i). If $f=W^2_z$ then pick β so that $e(\beta)=z$. Then since

$$W_{G(z)}=W_z$$

by 4.4, $f_\beta=f$. $\square$

4.7 Corollary. *Let $\mathbb{A}$ be a recursively listed admissible set. There are disjoint $\mathbb{A}$-r.e. sets which cannot be separated by an $\mathbb{A}$-recursive set.*

Proof. Let B, C be the disjoint $\mathbb{A}$-r.e. sets defined by

$$B = \{\beta \mid T_2^*(\beta, \beta, 0)\}$$
$$C = \{\beta \mid T_2^*(\beta, \beta, 1)\}$$

where T_2^* is given in Theorem 4.6. Suppose D were an $\mathbb{A}$-recursive set with $B \subseteq D$, $C \cap D = 0$. Let

$$\begin{aligned} g(x) &= 1 \quad \text{if} \quad x \in D \\ &= 0 \quad \text{if} \quad x \notin D \end{aligned}$$

so that g is $\mathbb{A}$-recursive. Pick β so that

$$g(x) = y \quad \text{iff} \quad T_2^*(\beta, x, y).$$

If $\beta \in D$ then $g(\beta) = 1$ so $T_2^*(\beta, \beta, 1)$ which implies $\beta \in C$, but $C \cap D = 0$. If $\beta \notin D$ then $g(\beta) = 0$ so $T_2^*(\beta, \beta, 0)$ which implies $\beta \in B$, but $B \subseteq D$. But $\beta \in D$ or $\beta \notin D$ so we have a contradiction in either case. Thus there can be no such D. □

It is an open problem to determine whether the conclusion of 4.7 holds for arbitrary admissible sets.

4.8—4.10 Exercises

4.8. Let $\mathfrak{M} = \langle M, R_1, \ldots, R_l \rangle$ be countable and suppose there is a well-ordering of M which is Δ_1^1 on $\mathfrak{M}$. Prove the following:

(i) Let B be a Π_1^1 subset of $\mathfrak{M}$. There is a function F with domain $o(\mathbb{HYP}_{\mathfrak{M}})$ such that

$$B = \bigcup_{\alpha < o(\mathbb{HYP}_{\mathfrak{M}})} F(\alpha)$$

and for each $\beta < o(\mathbb{HYP}_{\mathfrak{M}})$

$$\bigcup_{\alpha < \beta} F(\alpha)$$

is Δ_1^1 on M. [Pick an F which is $\mathbb{HYP}_{\mathfrak{M}}$ recursive.]

(ii) (Reduction) If B, C are Π_1^1 subsets of $\mathfrak{M}$ then there are disjoint Π_1^1 subsets $B_0 \subseteq B$, $C_0 \subseteq C$ with $B_0 \cup C_0 = B \cup C$.

(iii) (Separation) If B, C are disjoint Σ_1^1 subsets of $\mathfrak{M}$ then there is a Δ_1^1 set D with

$$B \subseteq C, \qquad C \cap D = 0.$$

(iv) (Uniformization) If $R \subseteq M \times M$ is Π_1^1 on $\mathfrak{M}$ there is a Π_1^1 subrelation $R_0 \subseteq R$ such that

$$\mathrm{dom}(R_0) = \mathrm{dom}(R)$$
$$x \in \mathrm{dom}(R_0) \Rightarrow \exists!\, y R_0(x, y).$$

If $\mathrm{dom}(R)=M$ then R_0 is a Δ_1^1 relation.

4.9. Show that for any admissible $\mathbb{A}_{\mathfrak{M}}$ and any $B\subseteq\mathbb{A}_{\mathfrak{M}}$, B is $\mathbb{A}_{\mathfrak{M}}$-r.e. iff $B=\mathrm{dom}(f)$ for some $\mathbb{A}_{\mathfrak{M}}$-recursive function f.

4.10. Show that if $\mathbb{A}_{\mathfrak{M}}$ is resolvable then the Reduction and Separation Theorems, 4.2 and 4.3, still hold. In particular, show that 4.8(i), (ii), (iii) hold without the hypothesis that $\mathfrak{M}$ has a Δ_1^1 well-ordering.

5. Notation Systems and Projections of Recursion Theory

An important stimulus in the earlier development of admissible ordinals was the desire to understand the analogy between Π_1^1 and r.e. sets of natural numbers. The metarecursion theory of Kreisel-Sacks [1965] explained this by developing a recursion theory on ω_1^c, the first nonrecursive ordinal, with the property that a set of natural numbers in Π_1^1 on ω iff it is ω_1^c-r.e. The theory was developed by using a notation system for the recursive ordinals to define the notions of ω_1^c-recursive, ω_1^c-r.e. and ω_1^c-finite.

The development by means of admissible sets proceeds the other way around. Instead of using known facts about Π_1^1 sets to develop a recursion theory on ω_1^c by means of a notation system, we have a recursion theory given on ω_1^c (it is the first admissible ordinal $>\omega$; see 5.11) and then transfer the results to Π_1^1 subsets of ω via a notation system.

5.1 Definition. Let $\mathbb{A}=\mathbb{A}_{\mathfrak{M}}$ be admissible.

(i) A *notation system for* $\mathbb{A}$ is a total $\mathbb{A}$-recursive function π such that if $x\neq y$ then $\pi(x)$ and $\pi(y)$ are disjoint non-empty sets. (We think of $\pi(x)$ as a set of *notations for* x.)

(ii) The *domain* of a notation system π, D_π, is defined by (!)

$$D_\pi=\bigcup_{x\in A}\pi(x)\,. \quad \text{(Thus } D_\pi \text{ is the set of all notations.)}$$

(iii) Associated with a notation system π is a function $|\cdot|_\pi$ with domain D_π and range $A\cup M$ defined by

$$|y|_\pi=x \quad \text{iff} \quad y\in\pi(x)\,.$$

(Thus, for any notation y, y is a notation for $|y|_\pi$.)

(iv) $\mathbb{A}$ is *projectible into* C if C is $\mathbb{A}$-r.e. and there is a notation system π with $D_\pi\subseteq C$.

It is best to think of the notation system as the triple $D_\pi, |\cdot|_\pi, \pi$ even though the first two can be defined in terms of the third. We require C to be $\mathbb{A}$-r.e. in (iv) only because that is the only kind of C that interests us in this context.

5.2 Lemma. *Let π be a notation system for the admissible set $\mathbb{A}$.*

(i) *π is a one-one function.*
(ii) *D_π is $\mathbb{A}$-r.e. but not $\mathbb{A}$-finite.*
(iii) *The graph of $|\cdot|_\pi$ is an $\mathbb{A}$-*recursive *relation. In particular, $|\cdot|_\pi$ is an $\mathbb{A}$-recursive function.*

Proof. The only part which is not absolutely immediate is the fact that D_π is not $\mathbb{A}$-finite. But if $D_\pi \in \mathbb{A}$ then, by Σ Replacement, the range of $|\cdot|_\pi$ would be an element of $\mathbb{A}$ whereas this range is all of $M \cup A$. □

Our plan for this section is to first exhibit some useful notation systems and then use them to transfer results.

5.3 Theorem.

(i) *For any structure $\mathfrak{M}$, $\mathbb{HYP}_{\mathfrak{M}}$ is projectible into $\mathbb{HF}_{\mathfrak{M}}$.*
(ii) *For any admissible set $\mathbb{A}$, $\mathbb{HYP}(\mathbb{A})$ is projectible into $\mathbb{A}$.*

The theorem is a simple consequence of the following lemma, an effective version of Theorem II.5.14.

5.4 Lemma. *Let L be a finite language, let $\mathfrak{M}$ be a structure for L, let $\mathsf{L}^* = \mathsf{L}(\in)$ and let $a \in \mathbb{V}_{\mathfrak{M}}$ be a transitive set with $M \subseteq a$. Let $\mathsf{L}' = \mathsf{L}^* \cup \{\overline{\mathsf{x}} \mid x \in a \cup \{a\}\}$ be the usual language with constant symbol $\overline{\mathsf{x}}$ for x. Let α be the least ordinal such that*

$$\mathbb{A}_{\mathfrak{M}} = (\mathfrak{M}; L(a, \alpha), \in)$$

is admissible and assume L' is coded up on $\mathbb{A}_{\mathfrak{M}}$ in a way that makes the syntactic operations of $\mathsf{L}'_{\omega\omega}$ all $\mathbb{A}_{\mathfrak{M}}$-recursive. There is a total $\mathbb{A}_{\mathfrak{M}}$-recursive function π such that for each $x \in \mathbb{A}_{\mathfrak{M}}$, $\pi(x)$ is a set of good Σ_1 definitions of x with parameters from $a \cup \{a\}$.

Proof. We already know, from Theorem II.5.14, that each $x \in \mathbb{A}_{\mathfrak{M}}$ *has* a good Σ_1 definition with parameters from $a \cup \{a\}$. The object here is to use the Second Recursion Theorem to show how we can go $\mathbb{A}_{\mathfrak{M}}$-recursively from x to a set $\pi(x)$ of good Σ_1 definitions of x, by reexamining the proof of II.5.14. If we look back at that proof we see that this is really pretty obvious. We write out clauses in the definition of π. In each case it is assumed that none of the earlier cases hold. We also arrange things so that v is the only free variable in any formula considered.

Case 1ne. If $x \in a \cup \{a\}$ then $\pi(x)$ is the set whose only member is the $\mathsf{L}'_{\omega\omega}$ Δ_0 formula

$$v = \overline{\mathsf{x}}.$$

Case 2wo. If $x=\beta+1$ then $\pi(x)$ is the set of formulas

$$\exists w[v=\mathscr{S}(w)\wedge\varphi(w/v)]$$

where $\varphi(v)\in\pi(\beta)$ and w is the first variable not in $\varphi(v)$.

Case 3hree. If $\pi(\beta)$ is defined then $\pi(\mathrm{L}(a,\beta))$ is the set of formulas of the form

$$\exists w[v=\mathrm{L}(\bar{\mathsf{a}},w)\wedge\varphi(w/v)]$$

where $\varphi(v)\in\pi(\beta)$. We may use "$v=\mathrm{L}(\bar{\mathsf{a}},w)$" since $\mathrm{L}(\cdot,\cdot)$ is a Σ_1 operation symbol.

Case 4our. If $x\in\mathrm{L}(a,\beta+1)-\mathscr{S}(L(a,\beta))$ then $\pi(x)$ is defined as follows. Find the least i, $1\leqslant i\leqslant N$, such that for some $y,z\in\mathrm{L}(a,\beta)\cup\{\mathrm{L}(a,\beta)\}$,

$$x=\mathscr{F}_i(y,z)\,.$$

Then $\pi(x)$ is the set of all formulas of the form

$$\exists w_1\,\exists w_2[v=\mathscr{F}_i(w_1,w_2)\wedge\varphi(w_1/v)\wedge\psi(w_2/v)]$$

where, for some $y,z\in\mathscr{S}(L(a,\beta))$, $x=\mathscr{F}_i(y,z)$ and $\varphi(v)\in\pi(y)$ and $\psi(v)\in\pi(z)$ and w_1,w_2 are the first two distinct variables not appearing anywhere in φ or ψ. The set of all such formulas exists by Σ Replacement. This clause in the definition of $\pi(x)=y$ is Σ, as can be seen by writing it out.

Case 5ive. If $\beta<\alpha$ is a limit ordinal then $\pi(\beta)$ is defined $\mathbb{A}_{\mathfrak{M}}$-effectively as follows. Find the first Δ_0 formula $\varphi(x,y,z_1,\ldots,z_n)$ of L^* (first in some effective well-ordering of $\mathbb{HF}$, say that given by II.2.4 or 3.7 of this chapter) such that for some $d,z_1,\ldots,z_n\in\mathrm{L}(a,\beta)$

(1) $$\mathrm{L}(a,\beta)\models\forall x\in d\,\exists y\,\varphi(x,y,z_1,\ldots,z_n)$$

but

(2) $$\mathrm{L}(a,\beta)\models\neg\exists b\,\forall x\in d\,\exists y\in b\,\varphi(x,y,z_1,\ldots,z_n)\,.$$

Now given φ let $\theta(\beta)(=\theta(\beta,d,z_1,\ldots,z_n))$ be formed from φ just as in the proof of II.5.14. Let $\pi(\beta)$ be the set of all formulas of the form

$$\exists w,w_1,\ldots,w_n[\theta(v,w,w_1,\ldots,w_n)\wedge\psi(w/v)\wedge\textstyle\bigwedge_{j=1}^n\sigma_j(w_j/v)]$$

such that for some $d,z_1,\ldots,z_n\in\mathrm{L}(a,\beta)$, (1) and (2) hold and $\psi\in\pi(d)$ and, for $1\leqslant j\leqslant n$, $\sigma_j(v)\in\pi(z_j)$. Again, this clause in the definition of $\pi(x)=y$ can be seen to be Σ and so, by the Second Recursion Theorem, π is an $\mathbb{A}_{\mathfrak{M}}$-recursive function. $\square$

Proof of 5.3. For (i) simply note that $\mathsf{L}'_{\omega\omega}$ can be coded up on $\mathbb{HF}_{\mathfrak{M}}$ in this case. For (ii) we can code $\mathsf{L}'_{\omega\omega}$ on $\mathbb{A}_{\mathfrak{M}}$ itself. The admissibility of $\mathbb{A}_{\mathfrak{M}}$ comes in only in that this coding can be done on $\mathbb{A}_{\mathfrak{M}}$ and is far stronger than we need. □

We will see in § VI.4 that if $\mathfrak{M}$ has a "built in pairing function" then $\mathbb{HYP}_{\mathfrak{M}}$ is projectible into $\mathfrak{M}$.

5.5 Corollary. *Let $\mathcal{N} = \langle \omega, +, \cdot \rangle$ be the structure of the natural numbers. $\mathbb{HYP}_{\mathcal{N}}$ is projectible into $\mathcal{N}$.*

Proof. The simplest proof is just to observe that in this case the coding used in the proof of 5.3(i) can be done on $\mathcal{N}$ itself. An alternate explicit proof will appear in §VI.4. □

We now give some examples of the use of notation systems. Combined with 5.5 and the results of §IV.3, the next two results show that, over $\mathcal{N}$, the Π_1^1 relations are parameterized by a Π_1^1 relation, that there are Π_1^1 sets which are not Δ_1^1, and that there are Δ_1^1 sets which are not first order definable over $\mathcal{N}$.

5.6 Theorem. *Let $\mathbb{A}$ be an admissible set which is projectible into C.*

(i) *For $n \geq 1$ there is an $(n+1)$-ary $\mathbb{A}$-r.e. relation S on C which parametrizes the class of all n-ary relations on C which are $\mathbb{A}$-r.e.*

(ii) *There is subset of C which is $\mathbb{A}$-r.e. but not $\mathbb{A}$-recursive.*

Proof. (ii) follows from (i) just as in the proof of 1.7. To prove (i) let π be a notation system for $\mathbb{A}$ with $D_\pi \subseteq C$. Let T_n be the $(n+1)$-ary relation on $\mathbb{A}$ which parametrizes the n-ary $\mathbb{A}$-r.e. relations. Define

$$S(y, x_1, \ldots, x_n) \quad \text{iff} \quad x_1, \ldots, x_n \in C, \quad y \in D_\pi \quad \text{and} \quad T_n(|y|_\pi, x_1, \ldots, x_n).$$

S is $\mathbb{A}$-r.e. since C and D_π are $\mathbb{A}$-r.e. and $|\cdot|_\pi$ is $\mathbb{A}$-recursive. Now let $R \subseteq C^n$ be $\mathbb{A}$-r.e. Pick a z such that

$$R(x_1, \ldots, x_n) \quad \text{iff} \quad T_n(z, x_1, \ldots, x_n).$$

Then for any $y \in \pi(z)$,

$$R(x_1, \ldots, x_n) \quad \text{iff} \quad S(y, x_1, \ldots, x_n). \quad \square$$

5.7 Theorem. *Let $\mathbb{A}$ be an admissible set with $o(\mathbb{A}) > \omega$. Let $\mathfrak{N} = \langle N, \ldots \rangle$ be a structure (for a language K) which is an element of $\mathbb{A}$ and suppose that $\mathbb{A}$ is projectible into N.*

(i) *There is an $\mathbb{A}$-recursive $(n+1)$-ary relation S on N which parametrizes the n-ary relations on $\mathfrak{N}$ which are first order definable over $\mathfrak{N}$ (using parameters).*

(ii) *There is a subset of $\mathfrak{N}$ which is $\mathbb{A}$-recursive but not first order definable over $\mathfrak{N}$.*

Proof. As usual (ii) follows from (i) by diagonalization. To prove (i) define

$S_0(y, x_1, \ldots, x_n)$ iff $y = \langle \varphi, s \rangle$ where $\varphi(v_1, \ldots, v_n, w_1, \ldots, w_m)$ is a formula of $\mathsf{K}_{\omega\omega}$ and s is an assignment with values in $\mathfrak{N}$, $s(v_i) = x_i$ all $i \leqslant n$, and $\mathfrak{N} \models \varphi[s]$.

S_0 is clearly Δ_1 on $\mathbb{A}$. Since $o(\mathbb{A}) > \omega$ the set X of all relevant pairs $\langle \varphi, s \rangle$ is an element of $\mathbb{A}$. Let π be the notation system for $\mathbb{A}$ with $D_\pi \subseteq N$. Define

$$S(z, x_1, \ldots, x_n) \quad \text{iff} \quad \exists y \in X[\pi(y) = z \wedge S_0(y, x_1, \ldots, x_n)].$$

Since $X \in \mathbb{A}$, the quantifier on y is bounded so S is indeed $\mathbb{A}$-recursive. It clearly parametrizes the relations definable over $\mathfrak{N}$. ☐

We now turn to a result, Theorem 5.9, which will allow us to identify $O(\mathcal{N})$. A notation system π is *univalent* if each $\pi(x)$ is a singleton, that is, if it assigns a unique notation to each $x \in \mathbb{A}$.

5.8 Proposition. (i) *Let $\mathbb{A}$ be a recursively listed admissible set projectible into C. There is a univalent notation system which projects $\mathbb{A}$ into C.*

(ii) $\mathbb{H}\mathrm{YP}_\mathcal{N}$ *has a univalent notation system which projects into $\mathcal{N}$.*

Proof. (i) If π projects $\mathbb{A}$ into C then define π_1, the univalent notation system, by

$$\pi_1(x) = \{y\} \quad \text{where } y \text{ is the first member of } \pi(x).$$

Part (ii) follows from (i) and 3.10. ☐

5.9 Theorem. *Let $\mathbb{A}$ be an admissible set which is projectible into C.*

$$\begin{aligned} o(\mathbb{A}) &= \{\rho(\prec) \mid \prec \text{ is a well-founded relation}, \prec \subseteq C^2, \prec \in \mathbb{A}\} \\ &= \{\rho(\prec) \mid \prec \text{ is a pre-wellordering}, \prec \subseteq C^2, \prec \in \mathbb{A}\}. \end{aligned}$$

If there is a univalent notation system projecting $\mathbb{A}$ into C then

$$o(\mathbb{A}) = \{\rho(\prec) \mid \prec \text{ a well-ordering}, \prec \subseteq C^2, \prec \in \mathbb{A}\}.$$

Proof. Every well-founded relation $\prec \in \mathbb{A}$ has $\rho(\prec) < o(\mathbb{A})$ by 3.3(ii) so we need only show that each $\beta \in \mathbb{A}$ is of the form $\rho(\prec)$ for some pre-wellordering $\prec \in \mathbb{A}$, $\prec \subseteq C^2$. Let π be a notation system projecting $\mathbb{A}$ into C. Let $b = \bigcup \mathrm{rng}(\pi \restriction \beta) \in \mathbb{A}$. Now $b \subseteq D_\pi \subseteq C$ and b is the set of all notations for ordinals $\gamma < \beta$. Define $\prec \subseteq b \times b$ by

$$x \prec y \quad \text{iff} \quad |x|_\pi < |y|_\pi.$$

Then $\prec$ is a pre-wellordering of b of length β and it is a well-ordering if π happens to be univalent. ☐

5.10 Corollary. $O(\mathcal{N})=\{\rho(\prec)\mid \prec$ *is a* Δ_1^1 *well-ordering,* $\prec\subseteq\mathcal{N}\times\mathcal{N}\}$
$=\omega_1^c$.

Proof. The first equality is immediate 5.9, 5.10 and §IV.3. The second follows from the first and the result from ordinary recursion theory that every Δ_1^1 well-ordering of $\mathcal{N}$ has order type some $\alpha<\omega_1^c$. □

The reader unfamiliar with the result used in the above proof can take

$$\omega_1^c=\{\rho(\prec)\mid \prec \text{ is a } \Delta_1^1 \text{ well-ordering, } \prec\subseteq\mathcal{N}\times\mathcal{N}\}$$

as the definition of ω_1^c.

5.11 Corollary. ω_1^c *is the first admissible ordinal greater than* ω.

Proof. ω_1^c is admissible by 5.10. Let α be the least admissible $>\omega$ so that $L(\alpha)$ is admissible and $\omega_1^c\geqslant\alpha$. But if $\omega_1^c>\alpha$ then α is the order type of some Δ_1^1 well-ordering $\prec$ of $\mathcal{N}$ and hence of some Δ_1^1 well-ordering $\prec$ of ω. But then $\prec\in L(\alpha)$ by §IV.3 which contradicts 3.3(ii). □

5.12—5.13 Exercises

5.12. *For any* $\mathfrak{M}=\langle M,R_1,\dots,R_l\rangle$ *show that*

$$O(\mathfrak{M})=\{\rho(\prec)\mid \prec \textit{ is a pre-wellordering, } \prec\in \mathbb{HYP}_{\mathfrak{M}},\ \prec\subseteq \mathbb{HF}_{\mathfrak{M}}^2\}.$$

5.13. Let $\mathbb{A}$ be a recursively listed admissible set. Show that there is a single-valued notation system with domain $o(\mathbb{A})$. Hence the recursion theory of $\mathbb{A}$ can be transfered to $o(\mathbb{A})$.

5.14 Notes. Notation systems are standard tools in ordinal recursion theory but don't seem to have been treated systematically before over arbitrary admissible sets. The definitions used above are stronger than those of Moschovakis [1974]. In the case where $\mathbb{A}$ is projectible into some $C\in\mathbb{A}$ (the only case of interest to Moschovakis) they are equivalent.

Corollary 5.11 is due to Kripke and Platek, but with more complicated proofs.

6. *Ordinal Recursion Theory: Projectible and Recursively Inaccessible Ordinals*

In the final sections of this chapter we return to the origins of the theory of admissible sets, recursion theory on admissible ordinals. We are thus in the domain of admissible sets without urelements.

Let τ_β be the βth admissible ordinal; that is, let

$$\tau_0 = \omega,$$

$$\tau_\beta = \text{least } \alpha[\alpha \text{ is admissible } \wedge \alpha > \tau_\gamma \text{ for all } \gamma < \beta].$$

In this section we begin looking at the sequence of admissible ordinals and the relationships between various members of it.

6.1 Definition. An admissible ordinal α is *projectible into* β (where $\beta \leqslant \alpha$) if there is a total α-recursive function mapping α one-one into β. The least β such that α is projectible into β is called the *projectum* of α and is denoted by α^*. If $\alpha^* < \alpha$ then α is said to be *projectible*; otherwise α is *nonprojectible.*

If α is admissible then $L(\alpha)$ is recursively listed so we see that α is projectible into β in the sense of 6.1 iff $L(\alpha)$ is projectible into β in the sense of 5.1 (iv). Similarly, if β is also admissible then α is projectible into β (in the sense of 6.1) iff $L(\alpha)$ is projectible into $L(\beta)$ (in the sense of 5.1(iv)).

6.2 Proposition

(i) *If* $\kappa \geqslant \omega$ *is a cardinal then* κ *is nonprojectible.*
(ii) *For any* $\beta, \tau_{\beta+1}$ *is projectible into* τ_β.
(iii) *If* τ_β *is projectible into* τ_γ *and* τ_γ *is projectible into* δ *then* τ_β *is projectible into* δ.

Proof. (i) is obvious by cardinality considerations since otherwise κ would have the same cardinality as some $\beta < \kappa$. For (ii), note that $L(\tau_{\beta+1}) = \mathbb{H}\mathrm{YP}(L(\tau_\beta))$ so $L(\tau_{\beta+1})$ is projectible into $L(\tau_\beta)$ by 5.3. Part (iii) is obvious. We simply compose projections. □

From this proposition we see that there are many projectible ordinals. We also see that $\tau_n^* = \omega$ for all $n = 0, 1, 2, \ldots$.

As we mentioned at the beginning of this chapter, one use of generalized recursion theory is as a laboratory for understanding ordinary recursion theory. One important aspect of ordinary recursion theory is the number of different versions of the notion of finite that arise. For examples, a set $B \subseteq \omega$ is finite iff any one of the following hold: B is recursive and bounded, B is R.E. and bounded, or B is bounded. By defining a set $B \subseteq L(\alpha)$ to be α-finite if $B \in L(\alpha)$ we have chosen to use the first. This means that when we meet some use of a different version of "finite" in ordinary recursion theory we may have trouble lifting this to α-recursion theory. The following theorem shows us that if α is projectible then there are going to be α-r.e. subsets of ordinals $\beta < \alpha$ which are *not* α-finite. Thus, for projectible ordinals we may expect some aspects of ordinary recursion theory to become more subtle. This is particularly true in the study of α-degrees, a subject not treated in this book.

6.3 Theorem. *Let α be admissible. The following are equivalent:*

(i) *α is nonprojectible.*

(ii) $L(\alpha) \models \Sigma_1$ Separation.

(iii) *If $\beta < \alpha$ and B is an α-*r.e.* subset of β then B is α-finite.*

Proof. We first prove (i)$\Rightarrow$(ii). Suppose $B \subseteq a \in L(\alpha)$ and B is Σ_1 definable on $L(\alpha)$. We wish to prove $B \in L(\alpha)$. Pick $\beta_0 < \alpha$ such that $a \in L(\beta_0)$; hence $B \subseteq L(\beta_0)$. The recursive listing f of $L(\alpha)$ given by $<_L$ puts everything in $L(\beta_0)$ before everything in $L(\alpha) - L(\beta_0)$. If we show that the set

$$C = \{\gamma \mid f(\gamma) \in B\}$$

is an element of $L(\alpha)$, then $B \in L(\alpha)$ by Σ Replacement. But $C \subseteq \beta_1$ for some $\beta_1 < \alpha$. Use 4.1 to pick an α-recursive function G mapping α onto C and define H by Σ Recursion as follows:

$$H(\beta) = G(\text{least } \gamma [G(\gamma) \notin \{H(\delta) \mid \delta < \beta\}]) .$$

Now H is α-recursive, one-one, and is defined on some initial segment of α. It cannot be defined for all $\beta < \alpha$, however, for this would give a projection of α into $\beta_1 < \alpha$ and α is nonprojectible. Let β_2 be the least ordinal for which H is not defined. The only reason $H(\beta_2)$ can be undefined is that

$$C = \{H(\beta) \mid \beta < \beta_2\}$$

so that $C \in L(\alpha)$ by Σ Replacement.

The implication (ii)$\Rightarrow$(iii) is trivial. We prove (iii)$\Rightarrow$(i) by contraposition. Thus, let $p: \alpha \to \beta$ be an α-recursive one-one mapping of α into β, $\beta < \alpha$, and let $B = \text{rng}(p)$. Then B is α-r.e., $B \subseteq \beta$ but B cannot be α-finite, since

$$\alpha = \{p^{-1}(x) \mid x \in B\}$$

and p^{-1} is α-recursive. ☐

6.4 Corollary. *If α is projectible into β then there is an α-r.e. subset of β which is not α-finite.* ☐

6.5 Corollary. *If α is nonprojectible then* $L(\alpha) \models$ *Beta.*

Proof. $L(\alpha) \models \Sigma_1$ Separation, and Σ_1 Separation implies Beta. ☐

6.6 Corollary. *Let κ be an uncountable cardinal. For every $\beta < \kappa$ there is a nonprojectible α between β and κ.*

Proof. $L(\kappa) \models \Sigma_1$ Separation, so apply Theorem II.3.3 with $\mathbb{A}_{\mathfrak{M}} = L(\kappa)$, $A_0 = \beta$. The resulting admissible set satisfies the axiom $V = L$ (i.e. $\forall x\, L(x)$) and so is $L(\alpha)$ for some $\alpha < \kappa$. Since $L(\alpha) \equiv L(\kappa)$, $L(\alpha) \models \Sigma_1$ Separation and hence α is nonprojectible. ☐

Now that we know there are lots of nonprojectible ordinals we can ask how big the first one is. So far, all we know is that it is bigger than τ_n for each $n<\omega$. Is it τ_ω? To shed some light on the size of the first nonprojectible we introduce the recursively inaccessible ordinals.

6.7 Definition. An admissible ordinal α is *recursively inaccessible* if α is the least upper bound of all admissibles less than α.

6.8 Theorem. *If α is nonprojectible and greater than ω then α is recursively inaccessible.*

Proof. Assume that α is admissible, $\alpha>\omega$ but that the ordinal

$$\beta=\sup\{\gamma<\alpha \mid \gamma \text{ is admissible}\}$$

is less than α. We will prove that α is projectible into β. Let $e:\alpha\to L(\alpha)$ be the recursive listing of $L(\alpha)$ given by $<_L$. Since β is a sup of admissible ordinals, $e\upharpoonright\beta$ is the canonical listing of $L(\beta)$ by ordinals $<\beta$. Thus, if $L(\alpha)$ were projectible into $L(\beta)$, then it would be projectible into β and so α would be projectible with $\alpha^*\leqslant\beta$. But $L(\alpha)$ is the smallest admissible set with $L(\beta)$ as an element, i.e. $L(\alpha)=\mathbb{HYP}(L(\beta))$ so $L(\alpha)$ is projectible into $L(\beta)$ by Lemma 5.4. ☐

If we combine Theorem 6.8 with the next result we see that the first nonprojectible is fairly large, much larger than τ_ω.

6.9 Theorem. *If τ_α is recursively inaccessible then $\tau_\alpha=\alpha$, and conversely.*

We isolate part of the proof of 6.9 which will be used again.

6.10 Lemma. *Define $G(\beta)=\tau_\beta$ for $\beta<\alpha$. Then G is a τ_α-recursive function.*

Proof. The result is literally trivial if $\alpha=0$. For $\alpha>0$ we can define G by

$$G(0)=\omega,$$

$$G(\beta)=\text{least}\,\gamma\,[L(\gamma)\models KP \wedge \gamma\notin\{G(\delta)\mid\delta<\beta\}]$$

for $\beta<\alpha$. Since KP is an ω-recursive set of axioms, it is in $L(\tau_\alpha)$ so this is a Σ Recursive definition of G. ☐

Proof of 6.9. Note first that $\tau_\alpha\geqslant\alpha$ for all α, by induction. Suppose $\tau_\alpha=\alpha$. Then for each $\beta<\tau_\alpha$, $\beta\leqslant\tau_\beta<\tau_\alpha$ so τ_α is the sup of all smaller admissibles. Now suppose τ_α is recursively inaccessible, but that $\tau_\alpha>\alpha$. Note that α is a limit ordinal, since $\tau_{\beta+1}$ can never be recursively inaccessible. Let G be as in Lemma 6.10 and observe that

$$\tau_\alpha=\sup\{G(\beta)\mid\beta<\alpha\}.$$

But this is a contradiction, for G is τ_α-recursive and hence the right-hand side of this equality is in $L(\tau_\alpha)$ by Σ Replacement. $\square$

We see, by 6.9, that none of the following are recursively inaccesible and, hence, all are projectible:

$$\tau_1, \tau_2, \ldots, \tau_\omega, \tau_{\omega+1}, \ldots, \tau_{\tau_1}, \tau_{\tau_1+1}, \ldots, \tau_{\tau_2}, \ldots, \tau_{\tau_{\tau_2}}, \quad \text{etc.}$$

What are their projectums? We will show in the next section that all are projectible into ω by showing that projectums are always admissible.

The interest in projectums stems largely from the following property which is quite useful in priority arguments involving α-degrees.

6.11 Theorem. *Let α be admissible and let α^* be its projectum. If B is α-r.e., $B \subseteq \beta$ for some $\beta < \alpha^*$, then B is α-finite.*

Proof. The proof is like the proof of (i)$\Rightarrow$(ii) in Theorem 6.3. Define an α-recursive function F by

$$F(\gamma) = \gamma^{\text{th}} \quad \text{member of } B$$

i.e.

$$F(\gamma) = G(\text{least } \delta (G(\delta) \notin \{F(\xi) \mid \xi < \gamma\}))$$

where G maps α onto B. Now, since $\beta < \alpha^*$, F cannot be a one-one mapping of α into β. Thus $F(\gamma)$ is undefined for some $\gamma < \alpha$. If γ_0 is the least such then $B = \{F(\gamma) : \gamma < \gamma_0\}$ so $B \in L(\alpha)$ by Σ Replacement. $\square$

To prove stronger facts about nonprojectible ordinals we need to use the notion of *stable* ordinal introduced in the next section.

6.12 Exercise. Let α be a limit of admissibles. Prove that $L(\alpha) \models \text{Beta}$ even if α is not admissible. This is an improvement of 6.5.

6.13 Notes. The concepts and results of this section are all due to Kripke and Platek. The student interested in the uses of the projectum in the study of α-degrees should consult Simpson's excellent survey article, Simpson [1974].

7. Ordinal Recursion Theory: Stability

Given structures $\mathfrak{A}_{\mathfrak{M}} \subseteq \mathfrak{B}_{\mathfrak{N}}$, we write $\mathfrak{A}_{\mathfrak{M}} \prec_1 \mathfrak{B}_{\mathfrak{N}}$ if for every Σ_1 formula $\varphi(v_1, \ldots, v_n)$ and every $x_1, \ldots, x_n \in \mathfrak{A}_{\mathfrak{M}}$,

$$\mathfrak{B}_{\mathfrak{N}} \models \varphi[x_1, \ldots, x_n] \quad \text{iff} \quad \mathfrak{A}_{\mathfrak{M}} \models \varphi[x_1, \ldots, x_n].$$

7.1 Definition. An ordinal α is *stable* if $L(\alpha) \prec_1 L$. The sequence of stable ordinals is defined by

$$\sigma_0 = \text{the least stable ordinal,}$$

$$\sigma_\gamma = \text{the least stable ordinal greater than each } \sigma_\beta \text{ for } \beta < \gamma .$$

The first theorem shows that there are lots of stable ordinals and that they are better behaved under sups than the admissible ordinals.

7.2 Theorem. (i) *If* $\lambda > 0$ *is a limit ordinal then* $\sigma_\lambda = \sup\{\sigma_\beta \mid \beta < \lambda\}$.
(ii) *Every uncountable cardinal is stable.*
(iii) *If* $\omega \leqslant \beta < \kappa$, *where* κ *is a cardinal, then there is a stable ordinal* $\alpha, \beta < \alpha < \kappa$.
(iv) *If* κ *is a cardinal then* $\kappa = \sigma_\kappa$.

Proof. To prove (i) let $\lambda > 0$ be a limit. Since $\sigma_\lambda \geqslant \sup\{\sigma_\beta \mid \beta < \lambda\}$ by definition, it suffices to prove that the ordinal

$$\gamma = \sup\{\sigma_\beta \mid \beta < \lambda\}$$

is stable. Let φ be a Σ_1 formula, let $x_1, \ldots, x_n \in L(\gamma)$ and suppose

$$L \models \varphi[x_1, \ldots, x_n] .$$

Pick $\beta < \lambda$ such that $x_1, \ldots, x_n \in L(\sigma_\beta)$. Then $L(\sigma_\beta) \models \varphi[x_1, \ldots, x_n]$ by stability, and then $L(\gamma) \models \varphi[x_1, \ldots, x_n]$ by persistence of Σ_1 formulas. (What we are really proving here is that the union $\mathfrak{A} = \bigcup_{\beta<\lambda} \mathfrak{A}_\beta$ of a chain of $\prec_1$-extensions $\mathfrak{A}_\beta$ is a $\prec_1$-extension of each $\mathfrak{A}_\beta$.)

Now let $\kappa > \omega$ be a cardinal and suppose that $x_1, \ldots, x_n \in L(\kappa)$ and that

$$L \models \exists y \psi[x_1, \ldots, x_n]$$

where ψ is Δ_0. We need to see that $L(\kappa)$ satisfies the same formula. But, for large enough cardinal λ,

$$H(\lambda) \models \exists \alpha \exists y \in L(\alpha) \psi[x_1, \ldots, x_n]$$

so, by II.3.5,

$$H(\kappa) \models \exists \alpha \exists y \in L(\alpha) \psi[x_1, \ldots, x_n]$$

and so there is an $\alpha < \kappa$ such that $L(\alpha) \models \exists y \psi[x_1, \ldots, x_n]$ and hence $L(\kappa) \models \exists y \psi[x_1, \ldots, x_n]$, as desired.

To prove (iii) we apply Theorem II.3.3. Note that we need only prove the result for κ regular since every singular κ is a limit of regular cardinals. Let $\alpha_0 = \beta + 1$. Given α_n apply II.3.3 to get an admissible set $\mathbb{B}$ such that

$$L(\alpha_n) \subseteq \mathbb{B},$$

$$\text{card}(\alpha_n) = \text{card}(\mathbb{B}) < \kappa,$$

$$L(\kappa) \models \varphi[\vec{x}] \quad \text{iff} \quad \mathbb{B} \models \varphi[\vec{x}]$$

for every formula φ and every $x_1, \ldots, x_n \in \mathrm{L}(\alpha_n)$. Now since $\mathbb{B} \equiv \mathrm{L}(\kappa)$, $B = \mathrm{L}(\gamma)$ for some admissible $\gamma < \kappa$ and we let α_{n+1} be this γ. Let $\alpha = \sup_{n<\omega} \alpha_n < \kappa$. We claim that α is stable; i.e., that $\mathrm{L}(\alpha) \prec_1 \mathrm{L}$. It suffices to prove that $\mathrm{L}(\alpha) \prec_1 \mathrm{L}(\kappa)$ since κ is stable. Let φ be Σ_1 and $\mathrm{L}(\kappa) \models \varphi[x_1, \ldots, x_n]$, where $x_1, \ldots, x_n \in \mathrm{L}(\alpha)$. Pick $k < \omega$ so that $x_1, \ldots, x_n \in \mathrm{L}(\alpha_k)$. Then $\mathrm{L}(\alpha_{k+1}) \models \varphi[x_1, \ldots, x_n]$ by choice of α_{k+1} and then $\mathrm{L}(\alpha) \models \varphi[x_1, \ldots, x_n]$ by persistence of Σ_1 formulas.

Part (iv) follows from (i) and (iii). In fact, if f is any continuous increasing function on the ordinals such that for all cardinals $\kappa > \omega$, $f(\alpha) < \kappa$ implies $f(\alpha+1) < \kappa$, one always has for all $\kappa > \omega$, $f(\kappa) = \kappa$. First assume κ is regular and consider the set B of β such that $f(\beta) < \kappa$. B is an initial segment of the ordinals and has no largest element so B is a limit ordinal λ. But then, by continuity,

$$f(\lambda) = \sup\{f(\beta) \mid \beta < \lambda\} \leqslant \kappa$$

but $f(\lambda) \not< \kappa$ since $\lambda \notin B$ so $f(\lambda) = \kappa$. Since κ is regular, $\lambda = \kappa$. Now for singular κ the result follows by continuity since every singular κ is the sup of regular cardinals. For if $\kappa = \sup_{\beta<\gamma} \lambda_\beta$, where the λ_β are regular, then $f(\kappa) = \sup_{\beta<\gamma} f(\lambda_\beta) = \sup_{\beta<\gamma} \lambda_\beta = \kappa$. □

There is a useful relative notion of stability.

7.3 Definition. An ordinal α is *β-stable* if $\alpha \leqslant \beta$ and

$$\mathrm{L}(\alpha) \prec_1 \mathrm{L}(\beta).$$

Since we have allowed $\alpha = \beta$ there is always at least one β-stable ordinal.

7.4 Proposition. (i) *If $\alpha \leqslant \beta \leqslant \gamma$ and α is γ-stable then α is β-stable.*

(ii) *If α is β-stable and β is γ-stable then α is γ-stable.*

(iii) *If β is stable and $\alpha < \beta$ then α is stable iff α is β-stable.*

(iv) *If B is a nonempty set of β-stable ordinals and $\alpha = \sup B$ then α is β-stable.*

Proof. These are all simple consequences of the definition and the persistence of Σ_1 formulas. □

7.5 Theorem. *If $\alpha < \beta$ and α is β-stable then α is admissible. In particular, every stable ordinal is admissible.*

Proof. Suppose $\alpha < \beta$ and $\mathrm{L}(\alpha) \prec_1 \mathrm{L}(\beta)$. Note that since the operations $\mathscr{F}_1, \ldots, \mathscr{F}_N$ all have Δ_0 graphs, and for $x, y \in \mathrm{L}(\alpha)$,

$$\mathrm{L}(\beta) \models \exists z(\mathscr{F}_i(x, y) = z),$$

we have

$$\mathrm{L}(\alpha) \models \exists z(\mathscr{F}_i(x, y) = z),$$

so L(α) is closed under the operations $\mathscr{F}_1, \ldots, \mathscr{F}_N$. Thus L($\alpha$), in addition to being transitive, is closed under pair and union and satisfies Δ_0 separation. It remains to check Δ_0 Collection. Suppose

$$L(\alpha) \models \forall x \in a \, \exists y \, \varphi(x, y, z)$$

where φ is Δ_0 and $a, z \in L(\alpha)$. Then, letting $b = L(\alpha) \in L(\beta)$, we have

$$L(\beta) \models \forall x \in a \, \exists y \in b \, \varphi(x, y, z)$$

so

$$L(\beta) \models \exists b \, \forall x \in a \, \exists y \in b \, \varphi(x, y, z)$$

and so, by $L(\alpha) \prec_1 L(\beta)$,

$$L(\alpha) \models \exists b \, \forall x \in a \, \exists y \in b \, \varphi(x, y, z) . \quad \square$$

7.6 Corollary. (i) *If β is admissible, $\alpha < \beta$ and α is β-stable then α is recursively inaccessible.*

(ii) *Every stable ordinal is recursively inaccessible.*

Proof. (i) Let $\beta = \tau_\gamma$ and $\alpha = \tau_\delta$ where $\delta \leqslant \alpha$ and $\delta < \gamma$. We need to see that $\delta = \alpha$. Suppose $\delta < \alpha$. Then $L(\beta) \models \exists x [x = \tau_\delta]$, so, by Lemma 6.10, and $L(\alpha) \prec_1 L(\beta)$, $L(\alpha) \models \exists x [x = \tau_\delta]$ (one needs to observe that no parameters occur in the definition of G in 6.10) from which we have $\tau_\delta \in L(\alpha)$, which is ridiculous since $\alpha = \tau_\delta$ and $\alpha \notin L(\alpha)$. Part (ii) follows from (i). $\square$

The definition of α is β-stable appears to be model theoretic until one reformulates it as follows: If f is a β-recursive function then whenever $x_1, \ldots, x_n \in L(\alpha)$, if $f(x_1, \ldots, x_n)$ is defined then $f(x_1, \ldots, x_n) \in L(\alpha)$. This reformulation suggests a way of generating the β-stable and the stable ordinals. First, however, a lemma.

Notice that we did not assume $\mathfrak{A}_{\mathfrak{M}} \subseteq_{\text{end}} \mathfrak{B}_{\mathfrak{N}}$ in the definition of $\mathfrak{A}_{\mathfrak{M}} \prec_1 \mathfrak{B}_{\mathfrak{N}}$. We are going to apply the notion to a case where we do not know, ahead of time, that this holds.

7.7 Lemma (Tarski Criterion for $\prec_1$). *If $\mathfrak{A}_{\mathfrak{M}} \subseteq \mathfrak{B}_{\mathfrak{N}}$ then $\mathfrak{A}_{\mathfrak{M}} \prec_1 \mathfrak{B}_{\mathfrak{N}}$ iff the following condition holds for every Δ_0 formula $\varphi(v_1, \ldots, v_n)$ and every $x_1, \ldots, x_{n-1} \in \mathfrak{A}_{\mathfrak{M}}$: if*

$$\mathfrak{B}_{\mathfrak{N}} \models \exists v_n \varphi[x_1, \ldots, x_{n-1}]$$

then there is an $x_n \in \mathfrak{A}_{\mathfrak{M}}$ such that

$$\mathfrak{B}_{\mathfrak{N}} \models \varphi[x_1, \ldots, x_{n-1}, x_n] .$$

Proof. $\mathfrak{A}_{\mathfrak{M}} \prec_1 \mathfrak{B}_{\mathfrak{N}}$ clearly implies the condition. To prove the converse, one first uses the criterion to prove

$$\mathfrak{A}_{\mathfrak{M}} \models \psi[x_1, \ldots, x_n] \quad \text{iff} \quad \mathfrak{B}_{\mathfrak{N}} \models \psi[x_1, \ldots, x_n]$$

for all Δ_0 formulas ψ and all $x_1, \ldots, x_n \in \mathfrak{A}_{\mathfrak{M}}$, by induction on ψ. The atomic cases hold by $\mathfrak{A}_{\mathfrak{M}} \subseteq \mathfrak{B}_{\mathfrak{N}}$, the propositional connectives take care of themsevles and the criterion gets us past bounded quantifiers. Now suppose $\exists v_n \varphi(v_1, \ldots, v_n)$ is a Σ_1 formula, $x_1, \ldots, x_{n-1} \in \mathfrak{A}_{\mathfrak{M}}$. If $\mathfrak{A}_{\mathfrak{M}} \models \exists v_n \varphi[x_1, \ldots, x_{n-1}]$ then there is an $x_n \in \mathfrak{A}_{\mathfrak{M}}$ such that $\mathfrak{A}_{\mathfrak{M}} \models \varphi[x_1, \ldots, x_{n-1}, x_n]$ so $\mathfrak{B}_{\mathfrak{N}} \models \varphi[x_1, \ldots, x_{n-1}, x_n]$, since φ is Δ_0, and hence

$$\mathfrak{B}_{\mathfrak{N}} \models \exists v_n \varphi[x_1, \ldots, x_{n-1}].$$

The proof of the converse first uses the criterion to pick $x_n \in \mathfrak{A}_{\mathfrak{M}}$ and then applies the result for Δ_0 formulas. □

We now come to the main theorem on the generation of stable ordinals. The proof is rather amusing since we use the Collapsing Lemma to collapse a set that is already transitive.

7.8 Theorem. *Let β be an admissible ordinal and let $0 \leqslant \gamma < \beta$. Let A be the set of those $a \in \mathrm{L}(\beta)$ for which there is a Σ_1 definition of a in $\mathrm{L}(\beta)$ using parameters $< \gamma$ (Σ_1 definable as elements in the sense of* II.5.13). *Let α be the least ordinal not in A. Then*

(i) *$A = \mathrm{L}(\alpha)$, and*
(ii) *α is the least β-stable ordinal $\geqslant \gamma$.*

Proof. It is not transparent that A is even transitive, let alone admissible. The first step in the proof is to show

(1) $\langle A, \in \cap A^2 \rangle \prec_1 \langle L(\beta), \in \rangle$.

We use the Tarski Criterion. Suppose $\mathrm{L}(\beta) \models \exists y \varphi[a_1, \ldots, a_n]$, where $a_1, \ldots, a_n \in A$. We need to find a $b \in A$ such that $\mathrm{L}(\beta) \models \varphi[a_1, \ldots, a_n, b]$.

Since each $a_i \in A$ is Σ_1 definable by a formula with parameters $< \gamma$, we may replace each a_i by its definition and assume all the parameters are ordinals $< \gamma$, except that φ now becomes Σ_1 instead of Δ_0. Write φ as $\exists z \psi(v_1, \ldots, v_m, y, z)$, so that $\mathrm{L}(\beta) \models \exists y \exists z \psi(\lambda_1, \ldots, \lambda_m, y, z)$, where $\lambda_1, \ldots, \lambda_m < \gamma$. Let $b = 1^{\mathrm{st}}(c)$ where c is the least pair $\langle y, z \rangle$ in $\mathrm{L}(\beta)$ (least under the ordering $<_{\mathrm{L}}$) such that $\mathrm{L}(\beta) \models \psi(\lambda_1, \ldots, \lambda_m, y, z)$. Then b is Σ_1 definable in $\mathrm{L}(\beta)$ with parameters $\lambda_1, \ldots, \lambda_m$ so $b \in A$ and $\mathrm{L}(\beta) \models \varphi(a_1, \ldots, a_n, b)$. Which proves (1).

Let $B = \mathrm{clpse}(A)$ so that B is transitive, and

$$c_A : \langle A, \in \cap A^2 \rangle \cong \langle B, \in \rangle.$$

Let τ be the least ordinal not in B and note that $\tau \leqslant \beta$ since there is an embedding of τ into β. We claim that

(2) $B \subseteq \mathrm{L}(\tau)$.

The predicate (of x and δ)

$$x \in L(\delta)$$

is Δ_1 in KP so we can find a Σ_1 formula equivalent to it in KP:

$$x \in L(\delta) \quad \text{iff} \quad \exists y\, \theta(x, \delta, y)$$

where θ is Δ_0. Now pick any $x \in B$. We will show that $x \in L(\delta)$ for some $\delta \leqslant \tau$. Write x as $c_A(a)$ for some $a \in A$. Since $A \subseteq L(\beta)$ there is an ordinal $\lambda < \beta$ such that $a \in L(\lambda)$. By (1), A is a model of

$$\exists \lambda, y\, \theta(a, \lambda, y).$$

Hence B is a model of

$$\exists \lambda, y\, \theta(x, \lambda, y),$$

but then x really is in $L(\delta)$ for some $\delta \in B$, proving (2).

Next we prove that

(3) $A = B$.

To prove this it suffices to prove that $c_A(a) = a$ for all $a \in A$. Since $\gamma \subseteq A$, $c_A(\lambda) = \lambda$ for all $\lambda < \gamma$. Let $a \in A$ be Σ_1 definable in $L(\beta)$ by the Σ_1 formula $\varphi(x, \lambda_1, \ldots, \lambda_n)$ where $\lambda_1, \ldots, \lambda_n < \gamma$,

$$L(\beta) \models \exists!\, x\, \varphi(x, \lambda_1, \ldots, \lambda_n),$$
$$L(\beta) \models \varphi[a, \lambda_1, \ldots, \lambda_n].$$

If we can prove that $L(\beta) \models \varphi[c_A(a), \lambda_1, \ldots, \lambda_n]$ then we will have $a = c_A(a)$. But from (1) it follows that $\langle A, \in \cap A^2 \rangle \models \varphi[a, \lambda_1, \ldots, \lambda_n]$, so $B \models \varphi[c_A(a), c_A(\lambda_1), \ldots, c_A(\lambda_n)]$. As we mentioned, $c_A(\lambda_i) = \lambda_i$ so $B \models \varphi[c_A(a), \lambda_1, \ldots, \lambda_n]$. By (2), $B \subseteq L(\beta)$ so $L(\beta) \models \varphi[c_A(a), \lambda_1, \ldots, \lambda_n]$ by persistence of φ. This proves $A = B$.

Since B is transitive and $B \prec_1 L(\beta)$, it follows that B is admissible and that $B = L(\tau)$. But of course $\tau = \alpha$ so $A = B = L(\alpha)$. Thus α is β-stable. Since $\gamma \subseteq A$, $\gamma \leqslant \alpha$. If $\gamma \leqslant \alpha' \leqslant \beta$ and α' is also β-stable then every element of A must be in $L(\alpha')$ so $\alpha \leqslant \alpha'$. Hence α is the least β-stable ordinal $\geqslant \gamma$. □

7.9 Corollary. *The stable ordinals are generated as follows.*

(i) $\sigma_0 = \{\alpha \mid \alpha$ *is* Σ_1 *definable in* L *without parameters*$\}$,
$L(\sigma_0) = \{x \in L \mid x$ *is* Σ_1 *definable in* L *without parameters*$\}$;

(ii) $\sigma_{\gamma+1} = \{\alpha \mid \alpha$ *is* Σ_1 *definable in* L *with parameters* $\leqslant \sigma_\gamma\}$,
$L(\sigma_{\gamma+1}) = \{x \in L \mid x$ *is* Σ_1 *definable in* L *with parameters* $\leqslant \sigma_\gamma\}$;

(iii) *If* λ *is a limit ordinal then*

$$\sigma_\lambda = \sup\{\sigma_\gamma \mid \gamma < \lambda\},$$
$$L(\sigma_\lambda) = \textstyle\bigcup_{\gamma < \lambda} L(\sigma_\gamma).$$

Proof. For (i) apply 7.8 with $\beta=\omega_1$, $\gamma=0$. For (ii) apply 7.8 with β equal to some cardinal $>\sigma_\gamma$ and the γ of 7.8 equal to $\sigma_\gamma+1$. Part (iii) is just a restatement of part of 7.2 included for completeness. □

We will study σ_0 in some depth in the next section and give a classical description of it. Part (i) of the next theorem will play a crucial role.

7.10 Theorem. (i) *σ_0 is projectible into ω.*
(ii) *$\sigma_{\gamma+1}$ is projectible into σ_γ.*
(iii) *If λ is a limit ordinal then σ_λ is nonprojectible.*

Proof. Let's dispose of (iii) first since it's fairly trivial. We prove that

$$L(\sigma_\lambda)\models\Sigma_1 \ \textit{Separation}$$

and then apply Theorem 6.3 to see that σ_λ is nonprojectible. Let $a\in L(\sigma_\lambda)$, let φ be Σ_1 and form the set

$$b=\{x\in a \mid L(\sigma_\lambda)\models\varphi[x]\}\,.$$

Pick $\gamma<\lambda$ large enough that a and the parameters in φ are members of $L(\sigma_\gamma)$. Then, by $L(\sigma_\gamma)\prec_1 L(\sigma_\lambda)$, we have

$$b=\{x\in a \mid L(\sigma_\gamma)\models\varphi[x]\}$$

so $b\in L(\sigma_\lambda)$ by Δ Separation.

Now for (i). The idea is that we want to assign to each $\alpha<\sigma_0$ some Σ_1 definition of α, thus projecting σ_0 into $\mathbb{HF}$. The trouble is that

$$L(\sigma_0)\models\exists!x\,\varphi(x)$$

is not a σ_0-r.e. predicate of the formula φ. To get around this we use the Uniformization Theorem, Theorem 4.4.

Recall the Σ_1 formula $\Sigma\text{-Sat}_1(z,y)$ from § 1. Let F be given by 4.4 so that F is σ_0-recursive,

$$\begin{aligned}\operatorname{dom}(F)&=\{\psi(x)\mid\psi(x) \text{ is a } \Sigma_1 \text{ formula } \wedge L(\sigma_0)\models\exists y\,\psi(y)\}\\&=\{z\mid L(\sigma_0)\models\text{“}z \text{ is } \Sigma_1\wedge\exists y\,\Sigma\text{-Sat}_1(z,y)\text{”}\}\end{aligned}$$

and for each $\psi(x)\in\operatorname{dom}(F)$, $L(\sigma_0)\models\psi(F(\psi))$. Now whenever a is Σ_1 definable there is a ψ such that

$$L(\sigma_0)\models\exists!y\,\psi(y)\wedge\psi[a]$$

so $F(\psi)=a$. We may project $L(\sigma_0)$ into $\mathbb{HF}$ by

$$\pi(a)=\text{least } \Sigma_1 \text{ formula } \psi \text{ such that } F(\psi)=a\,,$$

where by least we mean in some well-ordering of $\mathbb{HF}$ as given in II.2.4. The proof for (ii) is similar, using Σ-Sat_3 instead of Σ-Sat_1, once we observe that every $a \in \mathrm{L}(\sigma_{\gamma+1})$ is definable by a formula

$$\exists ! x\, \varphi[x, b, \sigma_\gamma]$$

with $b \in \mathrm{L}(\sigma)$ and no other parameters, by just using b to code a finite sequence of ordinals. Now apply Uniformization to get a $\sigma_{\gamma+1}$-recursive F such that

$$\mathrm{dom}(F) = \{\langle \psi(x,y,z), b\rangle \mid \psi \text{ is } \Sigma \text{ and } \mathrm{L}(\sigma_{\gamma+1}) \models \exists x\, \psi(x, b, \sigma_\gamma)\}$$

and, if $F(\psi(x,y,z), b)$ is defined then

$$\mathrm{L}(\sigma_{\gamma+1}) \models \psi(F(\psi, b), b, \sigma_\gamma)\,.$$

Then define

$$\pi(a) = \text{least pair } \langle \psi, b\rangle \text{ such that } F(\psi, b) = a\,.$$

This π projects $\mathrm{L}(\sigma_{\gamma+1})$ into $\mathrm{L}(\sigma_\gamma)$. Since $\mathrm{L}(\sigma_\gamma)$ is recursively listed, this amounts to projecting $\mathrm{L}(\sigma_{\gamma+1})$ into σ_γ. □

The use of Uniformization in 7.10 is very typical of more advanced work in L. We also use it to prove the next result.

7.11 Theorem. *Let β be an admissible ordinal whose projectum β^* is not ω. Then β^* is the limit of smaller β-stable ordinals. Hence β^* is β-stable and admissible.*

Before proving 7.11 we state some of its consequences.

7.12 Corollary. *If $\alpha > \omega$ is nonprojectible then α is the limit of smaller α-stable ordinals.* □

Next we present the result promised at the end of the last section.

7.13 Corollary. *For any admissible ordinal α, α^* is admissible and nonprojectible.*

Proof. By 7.11, α^* is admissible if $\alpha^* > \omega$. But if $\alpha^* = \omega$ it is also admissible. Nonprojectibility is obvious. □

Thus, if α is an admissible ordinal less than the first nonprojectible then $\alpha^* = \omega$. We saw that the first nonprojectible ordinal was recursively inaccessible. We can iterate this result using 7.11. We give only a sample result which shows that the first nonprojectible is much larger than the first recursively inaccessible.

7.14 Corollary. *Let ρ_β be the βth recursively inaccessible ordinal. If α is nonprojectible and $\alpha > \omega$ then $\alpha = \rho_\alpha$.*

Proof. Assume that $\alpha^* = \alpha$ but that $\alpha = \rho_\gamma$ for some $\gamma < \alpha$. Apply 7.12 to find an α-stable ordinal λ, $\gamma < \lambda < \alpha$. The predicate

$$\gamma < \alpha \quad \text{and } \gamma \text{ is recursively inaccessible}$$

is α-recursive since it holds iff

$$\gamma \text{ is admissible} \wedge \forall x < \gamma\, \exists \tau < \lambda\, (x < \tau \wedge \tau \text{ is admissible}).$$

Define $H(x) = \rho_x$ for $x < \gamma$. Then H is α-recursive and

$$\mathrm{L}(\alpha) \models \forall x < \gamma\, \exists y\, (H(x) = y)$$

so

$$\mathrm{L}(\lambda) \models \forall x < \gamma\, \exists y\, (H(x) = y)$$

since λ is α-stable. But this is ridiculous for λ itself is recursively inaccessible by 7.6, so $\lambda = H(x)$ for some $x < \gamma$. □

Some authors refer to ordinals α such that $\alpha = \rho_\alpha$ as being *recursively hyper-inaccessible.*

We now return to prove Theorem 7.11.

7.15 Lemma (Π_2-reflection). *Let* $\alpha > \omega$ *be admissible and let* $\forall x\, \exists y\, \varphi(x, y)$ *be a sentence which holds in* $\mathrm{L}(\alpha)$, *where* φ *is* Δ_0. *Then for every* $\gamma < \alpha$ *there is a* λ, $\gamma \leqslant \lambda < \alpha$ *such that*

$$\mathrm{L}(\lambda) \models \forall x\, \exists y\, \varphi(x, y).$$

Proof. Let $\gamma \leqslant \lambda_0 < \alpha$ where all parameters in φ are members of $\mathrm{L}(\lambda_0)$. Let λ_{n+1} be the least ordinal such that for all $x \in \mathrm{L}(\lambda_n)$ there is a $y \in \mathrm{L}(\lambda_{n+1})$ such that $\varphi(x, y)$. There is such a λ_{n+1} by Σ Reflection. The sequence $\langle \lambda_n : n < \omega \rangle$ is α-recursive so

$$\lambda = \sup_{n < \omega} \lambda_n$$

is less than α. □

Proof of Theorem 7.11. For several years all that was known about the projectum β^* of an admissible ordinal was that

(4) β^* is admissible or the limit of admissibles.

For suppose $\beta^* < \beta$ but that there is an admissible ordinal τ_γ such that

$$\beta \geqslant \tau_\gamma > \beta^* > \sup_{\delta < \gamma} \tau_\delta.$$

But by Theorem 6.8 (and its proof) τ_γ is projectible into $\sup_{\delta<\gamma}\tau_\delta$ and hence so is β, contradicting the definition of β^*.

For the purposes of this proof we call an ordinal γ *nice* if $<_L\upharpoonright L(\gamma)\times L(\gamma)$ has order type γ so that the function enumerating L, definable in KP, maps γ onto $L(\gamma)$. We know that every admissible ordinal is nice. The only point of proving (4) was to prove that

(5) *if* $\omega\leqslant\xi<\beta^*$ *then there is a nice limit ordinal* γ, $\xi\leqslant\gamma<\beta^*$.

If β^* is admissible, this follows by Π_2 Reflection. If β^* is the limit of admissibles we pick γ to be an admissible.

We are now ready to prove that if $\omega\leqslant\gamma<\beta^*$ then there is a β-stable α, $\gamma\leqslant\alpha<\beta^*$. By (5) it suffices to prove this for nice γ. Let α be the least β-stable ordinal $\geqslant\gamma$. Now $L(\alpha)$ is the set of $a\in L(\beta)$ definable by Σ_1 formulas with parameters $<\gamma$. Since γ is nice, though, we can code all these parameters into one so that

$$L(\alpha)=\{a\in L(\beta)\mid \text{for some } \Sigma_1 \text{ formula } \varphi(v_1,v_2) \text{ and some } \xi<\gamma,\ L(\beta)\models(\exists! v_1\varphi(v_1,\xi)\wedge\varphi(a,\xi))\}.$$

As in Theorem 7.10, let F be a β-recursive function uniformizing Σ-Sat_2. Note that Σ-Sat_2 and hence the graph of F are β-r.e. definable by Σ_1 formulas without parameters. Thus

$$L(\alpha)=\mathrm{rng}(F\upharpoonright(\mathbb{HF}\times\gamma)).$$

Since γ is nice we can identify $\mathbb{HF}\times\gamma$ with γ and apply Theorem 6.11 to see that $\mathrm{dom}(F\upharpoonright(\mathbb{HF}\times\gamma))\in L(\beta)$ since it is, essentially, a β-r.e. subset of $\gamma<\beta^*$. But if the domain of a β-recursive function is in $L(\beta)$, so is its range, so $L(\alpha)\in L(\beta)$. That is, $\alpha<\beta$. We need to see that $\alpha<\beta^*$. Suppose $\beta^*\leqslant\alpha$. The inverse of $F\upharpoonright(\mathbb{HF}\times\gamma)$ maps $L(\alpha)$ into γ so we could then project $L(\beta)$ into γ, contradicting the definition of β^*. Thus $\alpha<\beta^*$ so we have proven that β^* is the sup of smaller β-stable ordinals. Thus β^* is itself β-stable and admissible. □

7.16—7.25 Exercises

7.16. Prove that every stable ordinal is the limit of smaller nonprojectible ordinals. In particular, the first nonprojectible ordinal is less than the first stable ordinal, even though the first stable ordinal is projectible into ω.

7.17. Compute σ^*_γ. [Hint: $\sigma^*_{\omega+1}=\omega$.]

7.18 (Jensen). Show that α is admissible iff α is a limit ordinal and $L(\alpha)$ satisfies Δ_1 Separation.

7.19. Prove the converse of Lemma 7.15. That is show that a limit ordinal α is admissible and $>\omega$ iff every Π_2 sentence $\forall x\,\exists y\,\varphi(x,y,z)$ true in $L(\alpha)$ is true in $L(\beta)$ for arbitrarily large $\beta<\alpha$.

7.20. Let α be admissible, $\alpha > \omega$. An ordinal $\beta < \alpha$ is an *α-cardinal* if there is no $f \in L(\alpha)$ mapping β one-one into an ordinal $\gamma < \beta$.

(i) Show that if α^* (the projectum of α) is $< \alpha$ then α^* is an α-cardinal.

(ii) Prove that every α-cardinal $\kappa > \omega$ (if there are any) is α-stable. [Show that the proof of 7.2(ii) can be effectivized so as to hold inside $L(\alpha)$.]

(iii) Prove that if κ is an α-cardinal $> \omega$ and $\gamma < \kappa$ then there is an α-stable ordinal β, $\gamma < \beta < \kappa$. [Modify the proof of 7.11.]

7.21. Suppose α is admissible and $\omega < \alpha^* < \alpha$. Show that any α-r.e. subset of some $L(\beta)$, for $\beta < \alpha^*$, is α^*-finite, not just α-finite as stated in 6.11. [Use 7.11.]

7.22. Assume that there is an α such that $L(\alpha)$ is a model of ZF. Show that the least such α is less than σ_0.

7.23. Let $\mathbb{A}_{\mathfrak{M}}, \mathbb{B}_{\mathfrak{N}}$ be countable, admissible sets and suppose that $\mathbb{A}_{\mathfrak{M}}$ is $\mathbb{B}_{\mathfrak{N}}$-stable; i.e., that

$$\mathbb{A}_{\mathfrak{M}} \prec_1 \mathbb{B}_{\mathfrak{N}} .$$

Let T be a theory of $\mathsf{L}_{\mathbb{B}}$ which is definable over by a Σ_1 formula with parameters from $\mathbb{A}_{\mathfrak{M}}$. Show that if every $T_0 \subseteq T$ with $T_0 \in \mathbb{A}_{\mathfrak{M}}$ has a model then T has a model. [Use the Extended Completeness Theorem.]

7.24. Let $\mathbb{A}_{\mathfrak{M}}$ be countable, admissible. Show that the following are equivalent:

(i) $\mathbb{A}_{\mathfrak{M}} \prec_1 \mathbb{HYP}(\mathbb{A}_{\mathfrak{M}})$;

(ii) $\mathbb{A}_{\mathfrak{M}}$ is Π_1^1 reflecting; i.e. if $\Phi(v)$ is a Π_1^1 formula and $\mathbb{A}_{\mathfrak{M}} \models \Phi[x]$, then there is an admissible set $\mathbb{A}'_{\mathfrak{M}} \in \mathbb{A}_{\mathfrak{M}}$ such that $\mathbb{A}'_{\mathfrak{M}} \models \Phi[x]$. In particular, if $\alpha < \omega_1$ then τ_α is $\tau_{\alpha+1}$-stable iff $L(\tau_\alpha)$ is Π_1^1 reflecting. [Use the Completeness Theorem.]

7.25. An admissible ordinal α is *recursively Mahlo* if every α-recursive closed unbounded subset of α contains an admissible ordinal. (This is the "effective version" of the definition of Mahlo cardinal. See Chapter VIII.)

(i) Prove that if α is recursively Mahlo then it is recursively inaccessible, recursively hyperinaccessible, etc.

(ii) Prove that if α is the limit of smaller α-stable ordinals then α is recursively Mahlo.

(iii) Prove that if α is nonprojectible then it is recursively Mahlo.

7.26 Notes. The stability of uncountable cardinals is due to Takeuti [1960]. The concepts and other results in 7.1—7.10 are due to Kripke and Platek, independently. Theorem 7.11 (and hence 7.12, 7.13, 7.20, 7.21) are due to Kripke. The student interested in further similar results should study Jensen's theory of the fine structure of L as presented, for example, in Devlin [1973]. Exercise 7.23 appears in Barwise [1969]. Exercise 7.24 is due to Aczel-Richter [1973] and, in an absolute form, to Moschovakis [1974]. Exercise 7.25 goes back to Kripke and Platek.

Putting admissible ordinals $\alpha > \omega$ in their place.

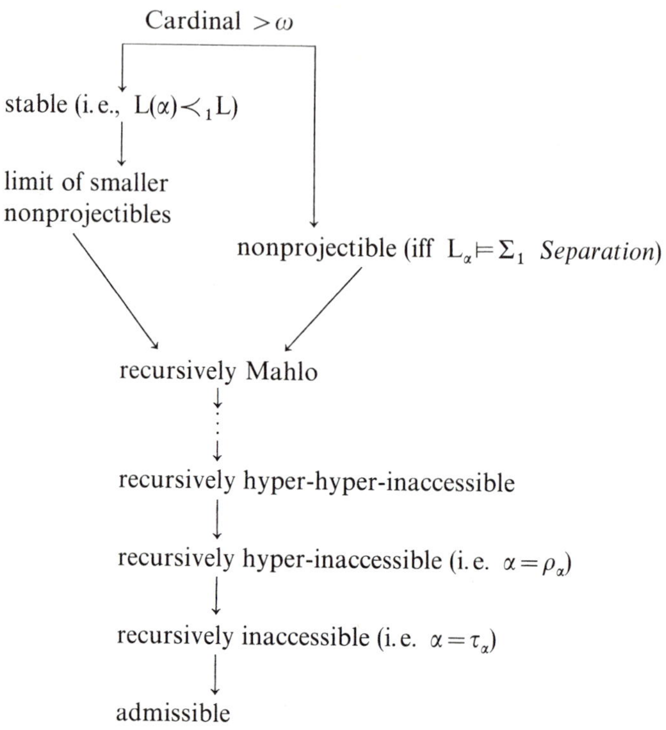

Notes:

1. No arrows are missing.
2. No arrows reverse.
3. The first stable ordinal σ_0 is projectible into ω; the $\beta+1^{\text{st}}$ stable ordinal $\sigma_{\beta+1}$ is projectible into σ_β.
4. For λ a limit, σ_λ is nonprojectible.
5. If α is projectible then its projectum α^* is admissible and nonprojectible.

8. *Shoenfield's Absoluteness Lemma and the First Stable Ordinal*

In § 5 we saw that the first admissible ordinal $\tau_1 > \omega$ is the least ordinal not the order type of a Δ_1^1 well-ordering of $\mathbb{HF}$ and that a set $X \subseteq \mathbb{HF}$ is τ_1-r.e. iff X is Π_1^1 on $\mathbb{HF}$. In this section we prove an analogous result for the first stable ordinal σ_0.

A relation R on $\mathbb{HF}$ is Σ_2^1 if it can be defined by a second order formula of the form $\exists \mathsf{S}_1 \forall \mathsf{S}_2 \varphi$, where φ is first order:

$$R(\vec{x}) \quad \text{iff} \quad \langle \mathbb{HF}, \in \rangle \models \exists \mathsf{S}_1 \forall \mathsf{S}_2 \varphi(\vec{x}, \mathsf{S}_1, \mathsf{S}_2).$$

If the complement $\mathbb{HF}^n - R$ of R is Σ_2^1, then R is said to be Π_2^1. If R is both Σ_2^1 and Π_2^1 then R is Δ_2^1.

At first glance the step from Δ_1^1 to Δ_2^1 seems a small one. We will show, however, that it is an enormous jump, taking us from τ_1 past the first recursively inaccessible, past the first nonprojectible all the way to σ_0, the first stable ordinal. The precise statement is contained in Corollary 8.3 below. The main step in the proof is the following theorem, known as the Shoenfield-Lévy Absoluteness Lemma.

8.1 Theorem. *Any Σ_1 sentence without parameters true in $\mathbb{V}$ is true in* L.

Warning: this does not say that $\mathrm{L} \prec_1 \mathbb{V}$ because parameters are not permitted. Some extensions with parameters are discussed in the exercises.

We defer the proof of 8.1 to the end of the section (Corollary 8.11) since it leads away from our chief concern.

8.2 Theorem. *Let σ_0 be the first stable ordinal and let R be a relation on* $\mathbb{HF}$.
(i) *R is Σ_2^1 on $\langle \mathbb{HF}, \in \rangle$ iff R is σ_0-r.e.*
(ii) *R is Δ_2^1 on $\langle \mathbb{HF}, \in \rangle$ iff $R \in \mathrm{L}(\sigma_0)$.*

Proof. As usual, (ii) follows from (i). We first prove the ($\Leftarrow$) half of (i). Let R be Σ_1 on $\mathrm{L}(\sigma_0)$. We know that $\mathrm{L}(\sigma_0) \prec_1 \mathrm{L}$ by the definition of σ_0 and that every $x \in \mathrm{L}(\sigma_0)$ is Σ_1 definable (as an element) in L by a formula without parameters (by 7.9). It follows that every $x \in \mathrm{L}(\sigma_0)$ is Σ_1 definable in $\mathrm{L}(\sigma_0)$ by a Σ_1 formula without parameters. Thus any parameters in a Σ_1 definition of the relation R can be eliminated so we may assume that

$$R(x) \quad \text{iff} \quad \mathrm{L}(\sigma_0) \models \exists y \, \varphi(x, y)$$

where φ is Δ_0 and contains no parameters. But then we claim that

(1) $R(x)$ iff $\exists \alpha [\alpha \text{ admissible} \wedge \mathrm{L}(\alpha) \models \exists y \, \varphi(x, y)]$.

The proof of ($\Rightarrow$) in (1) is trivial since we can let $\alpha = \sigma_0$. The other half ($\Leftarrow$) of (1) follows from $\mathrm{L}(\sigma_0) \prec_1 \mathrm{L}$ for $\mathrm{L}(\alpha) \models \exists y \, \varphi(x, y)$ implies $\mathrm{L} \models \exists y \, \varphi(x, y)$ and hence

$L(\sigma_0) \models \exists y\, \varphi(x,y)$ by stability. Using (1), it is not too difficult to rewrite R to be Σ_2^1 over $\langle \mathbb{HF}, \in \rangle$. Namely, $R(x)$ holds iff $\langle \mathbb{HF}, \in \rangle$ satisfies

(2) $\exists E, F[\langle \mathbb{HF}, E\rangle \models \mathrm{KP} + \mathrm{V} = \mathrm{L} \wedge E$ *is well founded* $\wedge F$ *is an isomorphism of* $\langle \mathbb{HF}, \in \rangle$ onto an *initial submodel of* $\langle \mathbb{HF}, E\rangle \wedge \langle \mathbb{HF}, E\rangle \models \exists y\, \varphi(F(x), y)]$

since any such $\langle \mathbb{HF}, E\rangle$ is isomorphic to $L(\alpha)$ for some admissible α. By the techniques of IV.2, everything inside the brackets is Δ_1^1 in $\in, E, F$ except the condition

E *is well founded.*

But this is Π_1^1 by the very definition:

$$\forall X[\exists z(z \in X) \to \exists z(z \in X \wedge \forall w(w \in X \to \neg w E z))].$$

Hence the whole of (2) has the form

$$\exists E, F[\text{---}]$$

where $[\text{---}]$ is Π_1^1 so (2) is Σ_2^1. (If you insist, you can always collapse the two existential second order quantifiers to one.)

To prove the other half of (i), let $R \subseteq \mathbb{HF}$ be Σ_2^1 over $\langle \mathbb{HF}, \in \rangle$, say

$$R(x) \text{ iff } \langle \mathbb{HF}, \in \rangle \models \exists \mathsf{S}_1 \forall \mathsf{S}_2\, \varphi(x, \mathsf{S}_1, \mathsf{S}_2).$$

For each relation S_1 on $\langle \mathbb{HF}, \in \rangle$, let $\sigma(S_1)$ be the infinitary sentence

$$\mathrm{Diagram}(\langle \mathbb{HF}, \in, S_1\rangle) \wedge \forall v \bigvee_{x \in \mathbb{HF}} (v = \overline{\mathsf{x}}).$$

Thus S_1 occurs in $\sigma(S_1)$.

We claim that $R(x)$ holds iff $L(\sigma_0)$ is a model of

(3) $\exists S_1 \exists P[S_1 \subseteq \mathbb{HF} \wedge P$ *is a proof of* $(\sigma(S_1) \to \varphi(\overline{\mathsf{x}}, \mathsf{S}_1, \mathsf{S}_2))]$,

which will show that R is Σ_1 over $L(\sigma_0)$. To show this, first suppose $R(x)$ holds. Then there is an S_1 such that

$$(\mathbb{HF}, \in, S_1) \models \forall \mathsf{S}_2\, \varphi(x, \mathsf{S}_1, \mathsf{S}_2)$$

and hence

$$\sigma(S_1) \to \varphi(\overline{\mathsf{x}}, \mathsf{S}_1, \mathsf{S}_2)$$

is logically valid. It is a countable infinitary sentence, so it is provable. Hence (3) holds in $\mathbb{V}$. The only parameter in (3) is x and it is Σ_1 definable, being in $\mathbb{HF}$. Thus (3) holds in L by 8.1 and hence in $L(\sigma_0)$. Thus $R(x)$ implies $L(\sigma_0) \models (3)$. To prove the converse, suppose (3) holds in $L(\sigma_0)$. Then there is an $S_1 \in L(\sigma_0)$ such that

$$\sigma(S_1) \to \varphi(\overline{\mathsf{x}}, \mathsf{S}_1, \mathsf{S}_2)$$

is provable, and hence, is logically valid. Thus,

$$(\mathbb{HF}, \in, S_1) \models \forall \mathsf{S}_2\, \varphi(x, \mathsf{S}_1, \mathsf{S}_2)$$

so $R(x)$ holds. □

8.3 Corollary. *The first stable ordinal is the least ordinal not the order type of some well-ordering which is Δ_2^1 on $\langle \mathbb{HF}, \in \rangle$.*

Proof. Every Δ_2^1 well-ordering R is in $L(\sigma_0)$ so its order type is less than σ_0, by 3.3. To prove the converse, recall that σ_0 is projectible into ω by Corollary 7.10.

Let p be some one-one σ_0-recursive map of σ_0 into ω. For $\beta < \sigma_0$ let

$$R_\beta = \{\langle p(x), p(y)\rangle \mid x < y < \beta\}$$

which is in $L(\sigma_0)$ by Σ Replacement. But then R_β is a well-ordering of order type β and R_β is Δ_2^1 by Theorem 8.2(ii). □

We can now project the recursion theory from σ_0-r.e. sets of ordinals to Σ_2^1 sets of integers using Section 5. We state some of the simplest results.

8.4 Corollary. (i) *For any Σ_2^1 subsets B, C of* $\mathbb{HF}$ *there are disjoint Σ_2^1 sets B_0, C_0 with $B_0 \subseteq B$, $C_0 \subseteq C$ and $B \cup C = B_0 \cup C_0$.*

(ii) *Any two disjoint Π_2^1 subsets of* $\mathbb{HF}$ *can be separated by a Δ_2^1 set.*

(iii) *There are disjoint Σ_2^1 subsets of* $\mathbb{HF}$ *which cannot be separated by a Δ_2^1 set.*

Proof. These are translations and projections of results we know about σ_0. □

8.5 Corollary. *Every Σ_2^1 subset of* $\mathbb{HF}$ *is constructible.*

Proof. If R is Σ_2^1 on $\mathbb{HF}$ then it is Σ_1 on $L(\sigma_0)$ and hence an element of $L(\sigma_0 + \omega)$. □

It follows, of course, that every Π_2^1 subset of $\mathbb{HF}$ is constructible, but this is as far as one can go. It is consistent with ZFC to assume there is a nonconstructible Δ_3^1 subset of $\mathbb{HF}$, where Δ_3^1 means expressible in both the forms

$$\exists \mathsf{S}_1 \forall \mathsf{S}_2 \exists \mathsf{S}_3(\text{---}),$$
$$\forall \mathsf{S}_1 \exists \mathsf{S}_2 \forall \mathsf{S}_3(\text{---}).$$

We now turn to the proof of Theorem 8.1. We need several preliminary lemmas. A finitary formula $\varphi(x_1, \ldots, x_n)$ is an $\forall\exists$-*formula* if it has the form

$$\forall y_1, \ldots, y_k \exists z_1, \ldots, z_l\, \psi(\vec{x}, \vec{y}, \vec{z})$$

where ψ is quantifier-free.

8.6 Lemma (Skolem $\forall\exists$ normal form). *Let* K *be a language and let* Ψ *be a finite set of formulas of* $\mathsf{K}_{\omega\omega}$. *There is an expansion*

$$\mathsf{L} = \mathsf{K} \cup \{\mathsf{S}_1, \ldots, \mathsf{S}_n\}$$

by a finite number of new relation symbols and an $\forall\exists$*-sentence* φ *of* $\mathsf{L}_{\omega\omega}$ *with the following properties:*

(i) *Every* K*-structure* $\mathfrak{M}$ *has a unique expansion* $\mathfrak{M}' = (\mathfrak{M}, S_1, \ldots, S_n)$ *with* $\mathfrak{M}' \models \varphi$.

(ii) *For each formula* $\psi(y_1, \ldots, y_n)$ *in* Ψ *there is a quantifier free formula* ψ_0 *of* $\mathsf{L}_{\omega\omega}$ *such that*

$$\models \varphi \rightarrow \forall \vec{y}\,[\psi(\vec{y}) \leftrightarrow \psi_0(\vec{y})]\,.$$

Proof. We may assume Ψ is closed under subformulas. Introduce, for each $\psi(y_1, \ldots, y_n) \in \Psi$ a new relation symbol $\mathsf{S}_{\psi(y_1, \ldots, y_n)}$. Let φ be the conjunction of the universal closures of the following:

$\mathsf{S}_\psi(y_1, \ldots, y_n) \leftrightarrow \psi(y_1, \ldots, y_n)$ if $\psi \in \Psi$ is atomic,
$\mathsf{S}_{\neg\psi}(y_1, \ldots, y_n) \leftrightarrow \neg \mathsf{S}_\psi(y_1, \ldots, y_n)$ if $\neg\psi \in \Psi$,
$\mathsf{S}_{\theta\wedge\psi}(y_1, \ldots, y_n) \leftrightarrow \mathsf{S}_\theta(y_1, \ldots, y_n) \wedge \mathsf{S}_\psi(y_1, \ldots, y_n)$ if $(\theta \wedge \psi) \in \Psi$,
$\mathsf{S}_{\theta\vee\psi}(y_1, \ldots, y_n) \leftrightarrow \mathsf{S}_\theta(y_1, \ldots, y_n) \vee \mathsf{S}_\psi(y_1, \ldots, y_n)$ if $(\theta \vee \psi) \in \Psi$,
$\mathsf{S}_{\exists y_m\psi}(y_1, \ldots, \hat{y}_m, \ldots, y_n) \leftrightarrow \exists y_m\, \mathsf{S}_\psi(y_1, \ldots, y_n)$ if $(\exists y_m \psi) \in \Psi$,
$\mathsf{S}_{\forall y_m\psi}(y_1, \ldots, \hat{y}_m, \ldots, y_n) \leftrightarrow \forall y_m\, \mathsf{S}_\psi(y_1, \ldots, y_m)$ if $(\forall y_m \psi) \in \Psi$.

Here we use $y_1, \ldots, \hat{y}_m, \ldots, y_n$ to denote $y_1, \ldots, y_{m-1} y_{m+1}, \ldots, y_m$ if $m \leqslant n$, to denote $y_1, \ldots, y_n$ if $m > n$. Now φ clearly has the desired properties. □

8.7 Corollary. *Let* ψ *be any sentence of* $\mathsf{K}_{\omega\omega}$. *There is an expansion* L *of* K *by a finite number of new relation symbols and an* $\forall\exists$*-sentence* ψ' *of* $\mathsf{L}_{\omega\omega}$ *such that*

(i) *Every model* $\mathfrak{M}$ *of* ψ *has a unique expansion to a model* $\mathfrak{M}'$ *of* ψ'.

(ii) *If* $\mathfrak{M}' \models \psi'$ *and* $\mathfrak{M}$ *is the reduct of* $\mathfrak{M}'$ *to a* K*-structure then* $\mathfrak{M} \models \psi$.

Proof. Let $\Psi = \{\psi\}$ and apply 8.6. Let φ be as given there and let ψ_0 be quantifier free such that

$$\models \varphi \rightarrow (\psi \leftrightarrow \psi_0)\,.$$

The desired ψ' is $(\varphi \wedge \psi_0)$, or rather, the $\forall\exists$-sentence equivalent to it after one moves the quantifiers in φ out front. □

The next lemma gives us an easy way to construct models of $\forall\exists$-sentences and accounts for our sudden preoccupation with them.

8.8 Lemma. *Let* φ *be an* $\forall\exists$*-sentence of* $\mathsf{L}_{\omega\omega}$, *say*

$$\forall x_1, \ldots, x_n \exists y_1, \ldots, y_k \psi(\vec{x}, \vec{y})$$

where ψ *is quantifer-free. Let*

$$\mathfrak{M}_0 \subseteq \mathfrak{M}_1 \subseteq \cdots \subseteq \mathfrak{M}_l \subseteq \cdots$$

be a chain of L*-structures. Suppose that for each* $l<\omega$ *and each* $x_1,\ldots,x_n\in\mathfrak{M}_l$,

$$\mathfrak{M}_{l+1}\models\exists y_1,\ldots,y_k\psi(x_1,\ldots,x_n,y_1,\ldots,y_k).$$

If $\mathfrak{M}=\bigcup_{l<\omega}\mathfrak{M}_l$, *then* $\mathfrak{M}\models\varphi$.

Proof. Trivial, since $\mathfrak{M}_{l+1}\models\exists\vec{y}\,\psi(\vec{x},\vec{y})$ implies $\mathfrak{M}\models\exists\vec{y}\psi(\vec{x},\vec{y})$. □

The next lemma contains the secret to proving a number of important results, including Theorem 8.1.

8.9 Lemma. *Let* $\langle X,\prec\rangle$ *be a non-wellfounded partially ordered structure which is constructible (i.e. is an element of* L*). There is a sequence* $\langle x_n\rangle_{n<\omega}$ *in* L *such that*

$$x_{n+1}\prec x_n$$

for all $n<\omega$.

Proof. The hypothesis is that $\langle X,\prec\rangle\in\mathrm{L}$ and that

$$\mathbb{V}\models\langle X,\prec\rangle \quad \text{is not well founded}.$$

We claim that

(4) $\mathrm{L}\models\langle X,\prec\rangle$ is not well founded.

For otherwise, since $\mathrm{L}\models Beta$, there would be a function $f\in\mathrm{L}$ such that $f(x)=\{f(y)\mid y\prec x\}$ for all $x\in X$. But then $\langle X,\prec\rangle$ really would be well founded (see Exercise I.9.9). Now since (4) holds, there is a nonempty $X_0\in\mathrm{L}$, $X_0\subseteq X$ such that

$$\forall y\in X_0\,\exists z\in X_0(z\prec y).$$

But then, using the axiom of (dependent) choice in L, there is a sequence of the desired kind. □

8.10 Theorem (of ZF). *Let* φ *be a finitary sentence in a language* L *containing* $\in$ *and some other relation symbols* $\mathsf{R}_1,\ldots,\mathsf{R}_l$. *If* φ *is true in some structure* $\mathbb{A}=\langle A,\in,R_1,\ldots,R_l\rangle$ *where* A *is transitive, then there is a transitive structure* $\mathbb{B}=\langle B,\in,R_1',\ldots,R_l'\rangle$ *which is constructible and a model of* φ.

Proof. We may assume that extensionality is a consequence of φ since it holds in $\mathbb{A}$. By 8.7 we may also assume that φ is $\forall\exists$, say

$$\forall\vec{x}\psi(\vec{x})$$

where $\psi(\vec{x})$ is

$$\exists\vec{y}\theta(\vec{x},\vec{y})$$

and θ is quantifier free. Let $\beta = \mathrm{rk}(A)$. We define a non-wellfounded structure $\langle X, \prec \rangle \in \mathrm{L}$. The set X consits of all pairs $\langle \mathfrak{B}, f \rangle$ such that $\mathfrak{B} = \langle B, E, R_1, \ldots, R_l \rangle$ is a finite structure with $B \subseteq \omega$, $f: B \to \beta$ and xEy implies $f(x) < f(y)$. We define

$$\langle \mathfrak{B}_1, f_1 \rangle \prec \langle \mathfrak{B}_0, f_0 \rangle$$

to mean that $\mathfrak{B}_0 \subseteq \mathfrak{B}_1$, $f_0 \subseteq f_1$ and for every $\vec{x} \in B_0$,

$$\mathfrak{B}_1 \models \psi[\vec{x}].$$

Now the definitions of X and $\prec$ are absolute so $\langle X, \prec \rangle \in \mathrm{L}$. We claim that

(5) $\langle X, \prec \rangle$ is not well founded.

Assuming (5) for a moment, let us finish the proof of the theorem. By Lemma 8.9 there is a sequence

$$\langle \mathfrak{B}_n, f_n \rangle_{n<\omega}$$

in L such that

$$\langle \mathfrak{B}_{n+1}, f_{n+1} \rangle \prec \langle \mathfrak{B}_n, f_n \rangle$$

for each n. Let $\mathfrak{B} = \bigcup \mathfrak{B}_n$. By Lemma 8.8, $\mathfrak{B} \models \varphi$. Let $f = \bigcup_n f_n$. Then, if $\mathfrak{B} = \langle B, E, R'_1, \ldots, R'_l \rangle$, then $f: B \to \beta$ and xEy implies $f(x) < f(y)$ so E is well founded. Now since $\mathrm{L} \models \mathrm{Beta}$, there is a transitive structure $\mathbb{B} \in \mathrm{L}$ isomorphic to $\mathfrak{B}$. This $\mathbb{B}$ satisfies the conclusion of the theorem.

Now let's go back and prove (5). Let X_0 be the set of those $(\mathfrak{B}, f) \in X$ such that there is an embedding i of $\mathfrak{B}$ into the original $\mathbb{A}$ such that

$$f(x) = \mathrm{rk}(i(x))$$

for all $x \in B$. The set X_0 is nonempty since $\langle \mathfrak{A}_0, \{\langle 0, 0 \rangle\} \rangle \in X_0$ where $\mathfrak{A}_0$ is the substructure of $\mathbb{A}$ with universe $\{0\}$. It remains to show that X_0 has no $\prec$ minimal member. Let $\langle \mathfrak{B}_0, f_0 \rangle \in X_0$ with $i_0: \mathfrak{B}_0 \to \mathbb{A}$ the associated embedding. Let $\mathfrak{A}_0 \subseteq \mathbb{A}$ be isomorphic to $\mathfrak{B}_0$ via i_0. Since

$$\mathbb{A} \models \forall \vec{x}\, \exists \vec{y}\, \theta(\vec{x}, \vec{y})$$

there is a finite structure $\mathfrak{A}_1$, $\mathfrak{A}_0 \subseteq \mathfrak{A}_1 \subseteq \mathbb{A}$, such that for all $\vec{x} \in \mathfrak{A}_0$,

$$\mathfrak{A}_1 \models \psi[\vec{x}].$$

Now choose $\mathfrak{B}_1 = \langle B_1, E_1, \ldots \rangle$ extending $\mathfrak{B}_0$ with $B_1 \subseteq \omega$ so that for some i_1 extending i_0,

$$i_1: \mathfrak{B}_1 \cong \mathfrak{A}_1.$$

Let $f_1(x)=\mathrm{rk}(i_1(x))$ for $x \in B_1$. Then $\langle \mathfrak{B}_1, f_1\rangle \prec \langle \mathfrak{B}_0, f_0\rangle$ and $\langle \mathfrak{B}_1, f_1\rangle \in X_0$. ☐

Theorem 8.1 is the informal version of the next result.

8.11 Corollary. *For any Σ sentence φ of set theory*

$$(\varphi \rightarrow \varphi^{(\mathrm{L})})$$

is a theorem of ZF.

Proof. We work in ZF. Assume φ. Then there is a transitive $\langle A, \in\rangle \models \varphi$. But then by Theorem 8.10, there is a transitive $\langle B, \in\rangle \in \mathrm{L}$ such that $\langle B, \in\rangle \models \varphi$. And $\langle B, \in\rangle \subseteq_{\text{end}} \langle \mathrm{L}, \in\rangle$ so $\langle \mathrm{L}, \in\rangle \models \varphi$; i.e., $\varphi^{(\mathrm{L})}$. ☐

Some extensions of these results are sketched in the exercises.

8.12—8.20 Exercises

8.12. Show that if

$$\langle \mathbb{HF}, \in\rangle \models \exists \mathsf{S}_1 \exists \mathsf{S}_2 \forall \mathsf{R}\, \varphi(\mathsf{S}_1, \mathsf{S}_2, \mathsf{R}, x)$$

where φ is first order, then there is an $S_1 \in \mathrm{L}(\sigma_0)$ such that

$$\langle \mathbb{HF}, \in, S_1\rangle \models \exists \mathsf{S}_2 \forall \mathsf{R}\, \varphi(\mathsf{S}_1, \mathsf{S}_2, \mathsf{R}, x).$$

This is the original version of Shoenfield's Absoluteness Lemma. A proof of it can be discovered inside the proof of Theorem 8.2.

8.13. Show that there is a Σ_2^1 well-ordering of a subset of ω of order type σ_0.

8.14. Improve 8.11 by replacing ZF by KP + Beta.

8.15. Improve Theorem 8.10 as follows. Let φ, $\mathbb{A}$ be as in 8.10, let $\beta=\mathrm{rk}(A)$. Let α be the least admissible $\tau_\gamma > \beta$ if $\mathrm{L}(\tau_\gamma) \models$ Beta, otherwise let $\alpha=\tau_{\gamma+1}$. Show that there is a transitive model $\mathbb{B}$ of φ which is an element of $\mathrm{L}(\alpha)$.

8.16. Let α be a limit of admissibles. Show that any Σ_1 sentence (without parameters) true in $\mathrm{V}(\alpha)$ is true in $\mathrm{L}(\alpha)$. [Use 8.15.]

8.17. Let T be a countable set of finitary sentences true in some transitive structure $\mathbb{A}=\langle A, \in, R_1, R_2, \ldots\rangle$. Show that there is a transitive model $\mathbb{B}$ of T which is an element of $\mathrm{L}(T)$ and is countable in $\mathrm{L}(T)$. [Hint: Modify the definition of $\langle X, \prec\rangle$ in the proof of 8.10 so that bigger structures take care of more of the sentences in T.]

8.18. Prove that the following is a theorem of ZF (by using 8.17): for each Σ formula $\varphi(v)$

$$\forall x \subseteq \mathbb{HF}\left[\varphi(x) \rightarrow \varphi(x)^{L(x)}\right].$$

8.19. Let α be the constructible $\aleph_1$, i.e. the ordinal which, in L, is the first uncountable cardinal. Prove that

$$L(\alpha) \prec_1 \mathbb{V}.$$

It is consistent with ZFC to assume α is countable. Prove that if α *is* countable and if $\beta > \alpha$ then $L(\beta) \not\prec_1 \mathbb{V}$.

8.20. Let Φ be a Σ_2^1 sentence true in some countable structure $\mathfrak{M}$. Prove that there is an $\mathfrak{M} \in L(\sigma_0)$ which satisfies Φ. If Φ is Π_1^1 you can improve this bound. How?

8.21 Notes. The original Shoenfield Absoluteness Lemma (Exercise 8.12) was proved in Shoenfield [1961]. Theorem 8.1 appears as Theorem 43 in Lévy [1965]. The proof given in this section and some of the generalizations found in the Exercises appeared in Barwise-Fisher [1970]. Exercise 8.16 is due to Jensen-Karp [1972].

Theorem 8.2 and its Corollary 8.3 are due to Kripke [1963] and Platek [1965] in the form stated here. Their content, however, goes back to Takeuti-Kino [1962].

Chapter VI
Inductive Definitions

"Let X be the smallest set containing ... and closed under ---." A definition expressed in this form is called an inductive definition. We have used this method of definition repeatedly in the previous chapters; for example, in defining the notions of Δ_0 formula, Σ formula, infinitary formula, provable using the $\mathfrak{M}$-rule, etc. In this chapter we turn method into object by studying inductive definitions in their own right. We will see that their frequent appearance is more than an accident.

1. Inductive Definitions as Monotonic Operators

Let A be an arbitrary set. An n-ary *inductive definition* on A is simply a mapping Γ from n-ary relations on A to n-ary relations on A which is *monotone increasing*; i.e. for all n-ary relations R, S on A

$$R \subseteq S \quad \text{implies} \quad \Gamma(R) \subseteq \Gamma(S).$$

If $\Gamma(R)=R$ then R is a *fixed point of* Γ.

1.1 Theorem. *Every inductive definition on A has a smallest fixed point. Indeed, there is a relation R such that:*

(i) $\Gamma(R)=R$,

(ii) *for any relation S on A, if* $\Gamma(S)\subseteq S$ *then* $R\subseteq S$.

Proof. Let $\mathbf{C}=\{S\subseteq A^n \mid \Gamma(S)\subseteq S\}$. Since $A^n\in\mathbf{C}$, $\mathbf{C}$ is non-empty. Let $R=\bigcap\mathbf{C}$. Since (ii) now holds by definition it remains to prove (i), that is, that $\Gamma(R)=R$. Let S be an arbitrary member of $\mathbf{C}$. Since $R\subseteq S$ and Γ is monotone we have $\Gamma(R)\subseteq\Gamma(S)$, but $\Gamma(S)\subseteq S$, so $\Gamma(R)\subseteq S$. Since S was an arbitrary member of $\mathbf{C}$, and $R=\bigcap\mathbf{C}$, we have $\Gamma(R)\subseteq R$. To show that $R\subseteq\Gamma(R)$ it suffices to prove that $\Gamma(R)\in\mathbf{C}$. But since $\Gamma(R)\subseteq R$ we have, by monotonicity, $\Gamma(\Gamma(R))\subseteq\Gamma(R)$ so $\Gamma(R)\in\mathbf{C}$. ☐

The proof of 1.1, while correct, tells us next to nothing about the smallest fixed point of Γ and is certainly not the way we mentally justify a typical inductive definition. Let us look at an example.

1.2 Example. Our very first use of an inductive definition was the definition of the class of Δ_0 formulas. We defined it as the smallest class containing the atomic formulas and closed under $\wedge, \vee, \neg, \forall u \in v, \exists u \in v$. How do we convince ourselves that there is such a smallest set? We simply say: start with the atomic formulas and close under (i.e., iterate) the operations $\wedge, \vee, \neg, \forall u \in v, \exists u \in v$. We can turn this process into a much more instructive proof of Theorem 1.1. (By the way, to make the class of Δ_0 formulas fall under 1.1 we let A be the class of formulas of L* and define the 1-ary Γ by

$\Gamma(U) = \{\varphi \in A \mid \varphi$ *is atomic or* $\varphi = (\psi \wedge \theta)$ *for some* $\psi, \theta \in U$ *or ... or* $\varphi = \exists u \in v \psi$ *for some* $\psi \in U\}$.)

Motivated by the above example we make the following definitions.

1.3 Definition. Let Γ be any n-ary inductive definition on a set A.

(i) The αth-*iterate* of Γ, denoted by I_Γ^α, is the n-ary relation defined by

$$I_\Gamma^\alpha = \Gamma\left(\bigcup_{\beta < \alpha} I_\Gamma^\beta\right).$$

(ii) $I_\Gamma = \bigcup_\alpha I_\Gamma^\alpha$, where the union is taken over all ordinals.

We will show that I_Γ is the smallest fixed point of Γ referred to in Theorem 1.1. We use the notation

$$I_\Gamma^{<\alpha} = \bigcup_{\beta < \alpha} I_\Gamma^\beta$$

to simplify some equations.

1.4 Lemma. *Let Γ be any n-ary inductive definition on a set A.*

(i) $I_\Gamma^0 = \Gamma(0)$,

(ii) $I_\Gamma^\alpha = \Gamma(I_\Gamma^{<\alpha})$ *for all* α,

(iii) $\alpha \leqslant \beta$ *implies* $I_\Gamma^\alpha \subseteq I_\Gamma^\beta$, *and*

(iv) $I_\Gamma^{\alpha+1} = \Gamma(I_\Gamma^\alpha)$ *for all* α.

Proof. Parts (i) and (ii) are immediate from the definitions. Part (iii) follows from monotonicity since

$$I_\Gamma^{<\alpha} = \bigcup_{\zeta < \alpha} I_\Gamma \subseteq \bigcup_{\zeta < \beta} I_\Gamma = I_\Gamma^{<\beta}$$

implies

$$I_\Gamma^\alpha = \Gamma(I_\Gamma^{<\alpha}) \subseteq \Gamma(I_\Gamma^{<\beta}) = I_\Gamma^\beta.$$

Part (iv) follows from (ii) since

$$I_\Gamma^{\alpha+1} = \Gamma\left(\bigcup_{\zeta \leqslant \alpha} I_\Gamma^\zeta\right) = \Gamma(I_\Gamma^\alpha). \quad \square$$

1.5 Theorem. *Let Γ be an n-ary inductive definition on a set A.*

(i) *There is an ordinal γ (of cardinality $\leqslant \operatorname{card}(A^n)$) such that*

$$I_\Gamma^\gamma = I_\Gamma^{<\gamma}$$

and hence

$$I_\Gamma = \bigcup_{\alpha<\gamma} I_\Gamma^\alpha .$$

(ii) *I_Γ is the smallest fixed point of Γ.*

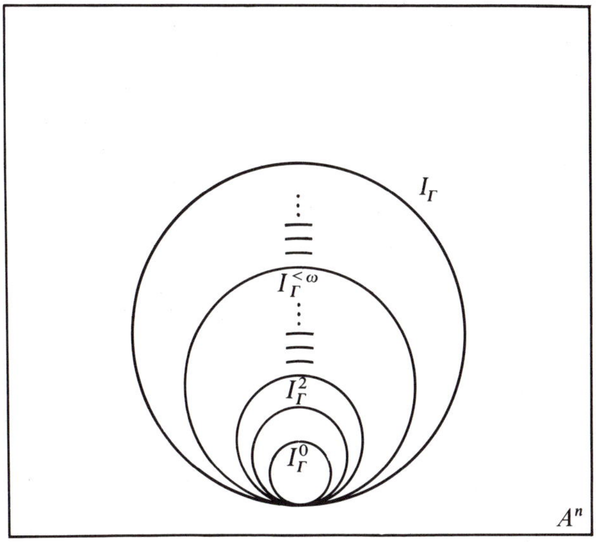

Fig. 1A. Building up the smallest fixed point I_Γ

Proof. First notice that the relations form an increasing sequence of subsets of A^n,

$$I_\Gamma^0 \subseteq I_\Gamma^1 \subseteq \cdots \subseteq I_\Gamma^\alpha \subseteq I_\Gamma^{\alpha+1} \subseteq \cdots,$$

and hence the sequence must stop strictly increasing for some γ of cardinality $\leqslant \operatorname{card}(A^n)$, i.e.,

$$I_\Gamma^\gamma = \bigcup I_\Gamma^{<\gamma} .$$

But then $I_\Gamma^\alpha = I_\Gamma^{<\gamma}$ for all $\alpha \geqslant \gamma$ so $I_\Gamma = I_\Gamma^{<\gamma}$. To prove (ii), note that

$$\Gamma(I_\Gamma) = \Gamma(I_\Gamma^{<\gamma}) = I_\Gamma^\gamma = I_\Gamma^{<\gamma} = I_\Gamma$$

by using (i) repeatedly. Hence I_Γ is a fixed point and it remains to show that I_Γ is the smallest such. Let $\Gamma(S) \subseteq S$. We prove $I_\Gamma^\alpha \subseteq S$ for all α, by induction. The induction hypothesis asserts that $I_\Gamma^\beta \subseteq S$ for all $\beta < \alpha$ so $I_\Gamma^{<\alpha} \subseteq S$. By monotonicity we have

$$I_\Gamma^\alpha = \Gamma(I_\Gamma^{<\alpha}) \subseteq \Gamma(S) \subseteq S . \quad \square$$

1.6 Definition. Given an inductive definition Γ, the least ordinal γ such that $I_\Gamma^\gamma = I_\Gamma^{<\gamma}$ is called the *closure ordinal of* Γ and is denoted by $\|\Gamma\|$.

Most of the inductive definitions we have used in the previous chapters have had closure ordinal ω so that

$$I_\Gamma = \bigcup_{n<\omega} I_\Gamma^n .$$

One of the most important however, the set of sentences provable using the $\mathfrak{M}$-rule, will in general have closure ordinal greater than ω. (In fact, this inductive definition has closure ordinal $O(\mathfrak{M})$. See Exercise 3.19.)

Our interest in this chapter is in inductive definitions which are definable over an L-structure $\mathfrak{M}$ or over an admissible set $\mathbb{A}_{\mathfrak{M}}$. In order to insure the monotonicity condition on Γ we need the notion of an R-monotone formula.

1.7 Definition. Let $\mathfrak{N}$ be a structure for some language K (usually L or L* in applications). A formula $\varphi(x_1, \ldots, x_n, \mathsf{R})$ of $\mathsf{K} \cup \{\mathsf{R}\}$ (possibly having parameters from $\mathfrak{N}$) is R-*monotone* on $\mathfrak{N}$ if for all $x_1, \ldots, x_n \in \mathfrak{N}$ and all relations $R_1 \subseteq R_2$ on $\mathfrak{N}$,

$$(\mathfrak{N}, R_1) \models \varphi(\mathsf{R})[x_1, \ldots, x_n]$$

implies

$$(\mathfrak{N}, R_2) \models \varphi(\mathsf{R})[x_1, \ldots, x_n] .$$

Recall the notion R-positive and corresponding notation $\varphi(\mathsf{R}_+)$ from V.2.1.

1.8 Lemma. *If $\varphi(x_1, \ldots, x_n, \mathsf{R}_+)$ is an* R-*positive formula of* K *then it is* R-*monotone for all* K-*structures* $\mathfrak{N}$.

Proof. Fix $\mathfrak{N}$ and prove the result by induction following the inductive definition of R-positive. □

Most inductive definitions are actually given by R-positive formulas because most inductive definitions do not really depend on the particular structure $\mathfrak{N}$ and any formula which is R-monotone for all structures $\mathfrak{N}$ is equivalent to an R-positive formula (see Exercise 1.14).

1.9 Notation and restatement of results. Let $\mathfrak{N}$ be a structure for a language K. Let R be a new n-ary relation symbol and let $\varphi(x_1, \ldots, x_n, \mathsf{R})$ be R-monotone on $\mathfrak{N}$.

(i) The n-ary *inductive definition given by* φ, denoted by Γ_φ, is defined by

$$(x_1, \ldots, x_n) \in \Gamma_\varphi(R) \quad \text{iff} \quad (\mathfrak{N}, R) \models \varphi(\mathsf{R})[x_1, \ldots, x_n] .$$

(ii) We let I_φ denote I_{Γ_φ} and similarly for I_φ^α and $I_\varphi^{<\alpha}$. Thus I_φ is an n-ary relation on $\mathfrak{N}$ satisfying

$$(\mathfrak{N}, I_\varphi) \models \forall x_1, \ldots, x_n [\varphi(x_1, \ldots, x_n, \mathsf{R}) \leftrightarrow \mathsf{R}(x_1, \ldots, x_n)] .$$

Furthermore, if R is an n-ary relation on $\mathfrak{N}$ satisfying

$$(\mathfrak{N}, R) \models \forall x_1, \ldots, x_n [\varphi(x_1, \ldots, x_n, \mathsf{R}) \to \mathsf{R}(x_1, \ldots, x_n)]$$

then $I_\varphi \subseteq R$. I_φ is called the *smallest fixed point* of the inductive definition Γ_φ and I_φ^α is called the αth *stage* of Γ_φ. It satisfies

$$I_\varphi^\alpha = \{(x_1, \ldots, x_n) \mid (\mathfrak{N}, I_\varphi^{<\alpha}) \models \varphi(\mathsf{R})[x_1, \ldots, x_n]\}.$$

1.10 Proposition. *Let $\mathfrak{N}$ be any* K*-structure and let $\varphi(x_1, \ldots, x_n, \mathsf{R})$ be* R*-monotone on $\mathfrak{N}$, where* R *is a new n-ary symbol. The fixed point I_φ is a Π_1^1 relation on $\mathfrak{N}$.*

Proof. By 1.9 we see that $(x_1, \ldots, x_n) \in I_\varphi$ iff $\forall \mathsf{R}[\Gamma_\varphi(\mathsf{R}) \subseteq \mathsf{R} \to \mathsf{R}(x_1, \ldots, x_n)]$ which becomes

$$\mathfrak{N} \models \forall \mathsf{R}[\forall y_1, \ldots, y_n(\varphi(y_1, \ldots, y_n, \mathsf{R}) \to \mathsf{R}(y_1, \ldots, y_n)) \to \mathsf{R}(x_1, \ldots, x_n)]$$

when written out in full. $\square$

Let $\mathcal{N} = \langle \omega, 0, +, \cdot \rangle$. Spector [1959] observed that Kleene's analysis of Π_1^1 relations on $\mathcal{N}$ showed that every Π_1^1 relation or could be obtained by means of an inductive definition. This result will follow from more general results in § 3. We present the classical proof, nevertheless, since it is attractive and illustrates several important points.

1.11 Theorem. *Let $\mathcal{N} = \langle \omega, 0, +, \cdot \rangle$ and let S be an n-ary Π_1^1 relation on $\mathcal{N}$. There is a formula $\varphi(x_1, \ldots, x_n, y, R_+)$ with R $n+1$-ary such that*

$$S(x_1, \ldots, x_n) \Longleftrightarrow I_\varphi(x_1, \ldots, x_n, 1)$$

for all $x_1, \ldots, x_n \in \mathcal{N}$.

Proof. We prove the result for $n=1$ and use the following normal form of Kleene for Π_1^1 sets S:

$$S(x) \quad \text{iff} \quad \forall f \, \exists n \, P(x, \bar{f}(n))$$

where the following are assumed:

P is recursive,
$\bar{f}(n)$ is a number s coding up the sequence $\langle f(0), \ldots, f(n-1) \rangle$,
$s_1 \prec s_2$ means that s_1 is a sequence (code) properly extending s_2,
$P(x, s_2)$ and $s_1 \prec s_2$ implies $P(x, s_1)$,
1 codes the empty sequence,
if s codes $\langle x_1, \ldots, x_n \rangle$ then $s \frown y$ codes $\langle x_1, \ldots, x_n, y \rangle$.

The desired inductive definition φ is given by

$$s \text{ is a sequence code and, } P(x,s) \text{ or } \forall y\, \mathsf{R}(x, s\frown y).$$

We first prove that

$$\text{(1)}\quad I_\varphi(x,s) \text{ implies } \forall f\, [\textit{if } f \textit{ extends } s \textit{ then } \exists n\, P(x, \bar{f}(n))].$$

Let R be the set of pairs (x,s) satisfying the right side of (1). Note that $P(x,s) \to R(x,s)$. It suffices to prove that $\Gamma_\varphi(R) \subseteq R$. If $(x,s) \in \Gamma_\varphi(R)$ then either $P(x,s)$ or else $\forall y\, R(x, s\frown y)$. But then $R(x,s)$ since every function extending s extends $s\frown y$ for some y.

Next we prove the converse of (1), or rather, as much of it as we need:

$$\text{(2)}\quad \forall f\, \exists n\, P(x, \bar{f}(n)) \quad \text{implies} \quad I_\varphi(x,1).$$

If $P(x,1)$ then $(x,1) \in I_\varphi^0$ so we may assume $\neg P(x,1)$. Assuming the left side of (2) consider the set S of all s such that $\neg P(x,s)$. This set is well founded (under $\prec$) since any infinite descending sequence would produce an f with $\neg P(x, \bar{f}(n))$ holding for arbitrarily large n, and hence for all n. Let us write, in this proof, $\rho(s)$ for $\rho^{\prec \restriction S}(s)$; $\rho(s)$ is defined for all $s \in S$ since S is well founded. We prove by induction on ξ that if $\rho(s) = \xi$ then $(x,s) \in I_\varphi^{\xi+1}$. (Since $1 \in S$ we then have $(x,1) \in I_\varphi^{\xi+1}$ where $\xi = \rho(1)$.) Observe that

$$\rho(s) = \sup\{\rho(s\frown y) + 1 \mid \neg P(x, s\frown y),\ y \in \omega\}.$$

Now for each y, if $P(x, s\frown y)$ then $(x, s\frown y) \in I_\varphi^0$, and if $\neg P(x, s\frown y)$ then $(x, s\frown y) \in I_\varphi^{\beta+1}$ for some $\beta < \xi$ by the induction hypothesis. In either case

$$(x, s\frown y) \in I_\varphi^\xi.$$

But then by the definition of φ,

$$(x,s) \in I_\varphi^{\xi+1}$$

as desired. Combining (1) and (2) yields the theorem. □

One of our goals in this chapter is to prove some generalizations of this result to arbitrary structures. Looking at the above theorem and its proof, we are struck by three facts.

The most prominent fact is that the proof uses a normal form for Π_1^1 predicates on $\mathcal{N}$ which has no generalization to Π_1^1 over arbitrary structures. If we can ignore this unsettling fact, however, we can go on to make two useful observations.

First, and very typical of the whole subject of inductive definitions, is that the Π_1^1 relation S was not defined as a fixed point but rather as a "section" of a fixed point:

$$S(\vec{x}) \Longleftrightarrow I_\varphi(\vec{x}, 1).$$

The proof makes it clear that the last coordinate of I_φ is where all the work is going on. It is only at the very last minute that we can set $s=1$. (To clinch matters, Feferman [1963] proves that not every Π_1^1 set over $\mathcal{N}$ is a fixed point.) This motivates the next definition.

1.12 Definition. Let K be a language, $\mathfrak{N}$ be any structure for K and let Φ be a set of formulas such that each $\varphi \in \Phi$ is of the form $\varphi(x_1, \ldots, x_n, \mathsf{R})$ for some n and some n-ary relation symbol R not in K) and is R-monotone on $\mathfrak{N}$.

(i) If $S = I_\varphi$ for some $\varphi \in \Phi$ then S is called a *Φ-fixed point.*

(ii) A relation S of m arguments is *Φ-inductive* if there is a Φ fixed point S' of $m+n$ arguments ($n \geqslant 0$) and $y_1, \ldots, y_n \in \mathfrak{N}$ such that

$$S(x_1, \ldots, x_m) \quad \text{iff} \quad S'(x_1, \ldots, x_m, y_1, \ldots, y_n)$$

for all $x_1, \ldots, x_m \in \mathfrak{N}$. S is called a *section* of S'.

Combining 1.10 and 1.11 (and the triviality that a section of a Π_1^1 relation is Π_1^1) we see that a relation S on $\mathcal{N}$ is Π_1^1 iff it is first order inductive.

A final point on the proof of Theorem 1.11. We made heavy use of coding in the proof, coding of pieces of functions by sequences and sequences by numbers, not to mention the coding which goes into the proof of the normal form theorem. In an admissible set, coding presents no trouble. In an arbitrary structure $\mathfrak{M}$, however, we may be out of luck. In this case we have two options. One is to restrict ourselves to $\mathfrak{M}$ which have built in coding machinery (this amounts to Moschovakis [1974]'s use of "acceptable" structures). The second option, more natural in our context is to replace induction on $\mathfrak{M}$ by inductions on $\mathbb{HF}_{\mathfrak{M}}$. We study both approaches in the latter parts of this chapter.

1.13—1.19 Exercises

1.13. Let K be a language with only relation symbols. One form of the Lyndon Interpolation Theorem asserts that if $\varphi, \psi \in \mathsf{K}_{\omega\omega}$, if φ or ψ is R-positive, and if

$$\models (\varphi \to \psi)$$

then there is a θ which is R-positive and has symbols common to φ and ψ such that

$$\models (\varphi \to \theta) \quad \text{and} \quad \models (\theta \to \psi).$$

Prove a generalization of this to arbitrary countable, admissible fragments $\mathsf{K}_{\mathbb{A}}$.

1.14. Prove that if $\varphi(x_1, \ldots, x_n, \mathsf{R})$ is R-monotone for all models $\mathfrak{N}$ of some theory T of $\mathsf{K}_{\omega\omega}$ (T not involving R, of course) then there is an R-positive $\psi(x_1, \ldots, x_n, \mathsf{R}_+)$ of $\mathsf{K}_{\omega\omega}$ such that

$$T \vdash \forall x_1, \ldots, x_n (\varphi \leftrightarrow \psi).$$

[Use the $\mathsf{K}_{\omega\omega}$ version of 1.13.]

1.15. Let Γ be an inductive definition, i. e. a monotonic increasing operation on n-ary relations on some set A. Show that Γ has a largest fixed point.

1.16. Let Γ be an n-ary inductive definition on A and define

$$J_\Gamma^0 = \Gamma(A^n)$$

and for $\alpha > 0$,

$$J_\Gamma^\alpha = \Gamma(\bigcap_{\beta<\alpha} J_\Gamma^\beta).$$

Let

$$J_\Gamma = \bigcap_\alpha J_\Gamma^\alpha.$$

Show that J_Γ is the largest fixed point of Γ refered to in 1.15.

1.17. Let Γ be an n-ary inductive definition on A and let Γ' be defined by

$$\Gamma'(R) = A^n - \Gamma(A^n - R).$$

Prove that Γ' is an inductive definition. Prove that, for each α,

$$x \in I_{\Gamma'}^\alpha \quad \text{iff} \quad x \notin J_\Gamma^\alpha$$

and hence that

$$I_{\Gamma'} = A^n - J_\Gamma.$$

1.18. Let Φ_1, Φ_2 be classes of formulas R-monotone on a structure $\mathfrak{N}$, closed under logical equivalence and such that

$$\varphi(x_1,\ldots,x_k,\mathsf{R}) \in \Phi_1 \quad \text{iff} \quad (\neg\varphi(x_1,\ldots,x_k,\neg\mathsf{R})) \in \Phi_2.$$

A relation S on $\mathfrak{N}$ is *Φ_1-coinductive* iff for some $\varphi \in \Phi_1$ and some parameters $y_1,\ldots,y_n \in \mathfrak{N}$

$$S(x_1,\ldots,x_m) \quad \text{iff} \quad (x_1,\ldots,x_m,y_1,\ldots,y_n) \in J_\varphi$$

for all $x_1,\ldots,x_n \in \mathfrak{N}$. Show that S is Φ_1-coinductive iff $\neg S$ is Φ_2-inductive. (Hence every coinductive relation on $\mathfrak{N}$ is Σ_1^1. You can also prove this directly.)

1.19. Let G be an abelian p-group. Define $\Gamma(H)$, for $H \subseteq G$, by

$$\Gamma(H) = \{px \mid x \in H\}.$$

Show that J_Γ is the largest divisible subgroup of G. In this case the least ordinal α such that $J_\Gamma = \bigcap_{\beta<\alpha} J_\Gamma^\beta$ is usually called the length of the group G. It plays a key role in the study of p-groups.

1.20 Notes. We have built monotonicity of Γ into our definition of "inductive definition". There are also things called "non-monotonic inductive definitions" which have interesting relationships with admissible sets. For references on these operators, we refer the reader to Richter-Aczel [1974] and Moschovakis [1975].

All the results of § 1 are standard.

2. Σ *Inductive Definitions on Admissible Sets*

Let $\Sigma(\mathsf{R}_+)$ be the collection of R-positive Σ formulas of $\mathsf{L}^*(\mathsf{R})$ and let Σ_+ be the union of the $\Sigma(\mathsf{R}_+)$ as R ranges over all relation symbols not in L^*. Applying Definition 1.12 (with $\mathsf{K}=\mathsf{L}^*$, $\mathfrak{N}=\mathfrak{A}_{\mathfrak{M}}$ and $\Phi=\Sigma_+$) we have the companion notions of Σ_+ *fixed point* and Σ_+ *inductive relation.* These notions are the primary object of study of this section. The proofs, however, give information about a wider class of relations.

Let $\mathscr{K}$ be a class of L^*-structures and let $\Sigma(\mathsf{R}\uparrow\mathscr{K})$ be the collection of Σ formulas $\varphi(x_1,\ldots,x_n,\mathsf{R})$ of $\mathsf{L}^*(\mathsf{R})$ which are monotone increasing on each structure in $\mathscr{K}$. We let $\Sigma(\uparrow\mathscr{K})$ be the union of the $\Sigma(\mathsf{R}\uparrow\mathscr{K})$ as R varies. (Read "Σ increasing on $\mathscr{K}$" for $\Sigma(\uparrow\mathscr{K})$.) Given a structure $\mathfrak{A}_{\mathfrak{M}}\in\mathscr{K}$ we have corresponding notions of $\Sigma(\uparrow\mathscr{K})$ *fixed point on* $\mathfrak{A}_{\mathfrak{M}}$ and $\Sigma(\uparrow\mathscr{K})$ *inductive relation on* $\mathfrak{A}_{\mathfrak{M}}$. If $\mathscr{K}=\{\mathfrak{A}_{\mathfrak{M}}\}$ then we write $\Sigma(\uparrow\mathfrak{A}_{\mathfrak{M}})$ for $\Sigma(\uparrow\mathscr{K})$.

Note that by Lemma 1.8, $\Sigma_+\subseteq\Sigma(\uparrow\mathscr{K})$ for all $\mathscr{K}$. If $\mathscr{K}$ is the class of all structures for L^* which are models of some theory T then Exercise 1.14 tells us that $\Sigma_+=\Sigma(\uparrow\mathscr{K})$, up to logical equivalence. In the results below, however, $\mathscr{K}$ is usually a single admissible set or a class of admissible sets.)

We have already studied the most important Σ_+ inductive definition at some length back in Chapter III. Let $\mathsf{K}_{\mathbb{A}}$ be an admissible fragment and let $Thm_{\mathbb{A}}$ be the set of theorems of $\mathsf{K}_{\mathbb{A}}$. By definition, $Thm_{\mathbb{A}}$ is the smallest set of formulas of $\mathsf{K}_{\mathbb{A}}$ containing the axioms (A 1)—(A 7) and closed under (R 1)—(R 3). This is, of course, a typical example of an inductive definition. Let Γ_0 be this inductive definition.

2.1 Proposition. *Using the notation just above we have*

(i) *Γ_0 is a Σ_+ inductive definition, and hence*

(ii) *$Thm_{\mathbb{A}}$ is a Σ_+ fixed point.*

Proof. We simply write out the definition of Γ_0 to see that it is in fact Σ_+. Let R be a new unary symbol and recall that

$$x\in\Gamma_0(R) \quad \text{iff} \quad x\in\mathsf{K}_{\mathbb{A}}\wedge[(\mathrm{A})\vee(\mathrm{R}\,1)\vee(\mathrm{R}\,2)\vee(\mathrm{R}\,3)]$$

where we have used

(A) "x is an instance of (A 1)—(A 7)".

(R 1) $\exists y\,[y\in R\wedge(y\rightarrow x)\in R]$.

(R 2) "x is of the form $(\psi\rightarrow\forall v\,\theta(v))$ where v is not free in ψ and $(\psi\rightarrow\theta(v))\in R$".

(R 3) "x is of the form $(\psi\rightarrow\bigwedge\Phi)$ and, for each $\varphi\in\Phi$, $(\psi\rightarrow\varphi)\in R$".

We can rewrite this schematically in the form

$$x \in \Gamma_0(R) \quad \text{iff} \quad \Delta_1 \wedge [\Delta_1 \vee \Sigma_1(\mathsf{R}_+) \vee \Delta_1(\mathsf{R}_+) \vee \Delta(\mathsf{R}_+)],$$

so Γ_0 is indeed a $\Sigma(\mathsf{R}_+)$ inductive definition. □

Now one of the primary aims of § III.5 was to prove that $Thm_{\mathbb{A}}$ was in fact Σ_1 *definable* on $\mathbb{A}$. In this section we use this fact to prove that every Σ_+ inductive relation on an admissible set is Σ_1 on that admissible set. For $\mathbb{A}$ countable, even more is true.

2.2 Theorem. *Let $\mathbb{A}$ be a countable admissible set. Every $\Sigma(\uparrow\mathbb{A})$ inductive relation on $\mathbb{A}$ is Σ_1 on $\mathbb{A}$.*

Proof. It clearly suffices to prove that every $\Sigma(\uparrow\mathbb{A})$ fixed point on $\mathbb{A}$ is Σ_1 since the Σ_1 relations are closed under sections. Let $\varphi(x_1,\ldots,x_n,\mathsf{R})\in\Sigma(\uparrow\mathbb{A})$. The proof goes back to the Extended Completeness Theorem for countable admissible fragments and, hence, to our analysis of Γ_0 carried out in § III.5. Let K be the formalized version of $\mathsf{L}^*(\mathsf{R})\cup\{\overline{\mathsf{x}}\mid x\in\mathbb{A}\}$ and let $\mathsf{K}_{\mathbb{A}}$ be the fragment given by $\mathbb{A}$ $(=\mathbb{A}_{\mathfrak{M}})$. Let T be the $\mathsf{K}_{\mathbb{A}}$ theory:

$$\text{Diagram}(\mathbb{A}),$$

$$\forall v\,[v\in\overline{\mathsf{a}}\leftrightarrow \textstyle\bigvee_{x\in a} v=\overline{\mathsf{x}}] \quad \text{for all} \quad a\in\mathbb{A},$$

$$\forall v_1,\ldots,v_n\,[\varphi(v_1,\ldots,v_n,\mathsf{R})\to \mathsf{R}(v_1,\ldots,v_n)].$$

We claim that

(1) $(x_1,\ldots,x_n)\in I_\varphi$ iff $T\models \mathsf{R}(\overline{\mathsf{x}}_1,\ldots,\overline{\mathsf{x}}_n)$

from which the conclusion follows by the Extended Completeness Theorem. The $(\Leftarrow)$ half of (1) follows from the observation that

$$(\mathbb{A},I_\varphi)\models T$$

when R is interpreted by I_φ. To prove $(\Rightarrow)$ suppose that $(\mathfrak{B}_{\mathfrak{N}},R)$ is an arbitrary model of T. We need to prove that whenever $(x_1,\ldots,x_n)\in I_\varphi$, we have

$$(\mathfrak{B}_{\mathfrak{N}},R)\models \mathsf{R}(\overline{\mathsf{x}}_1,\ldots,\overline{\mathsf{x}}).$$

If we let $R_0=R\restriction\mathbb{A}_{\mathfrak{M}}$ then we note that (up to isomorphism)

$$(\mathbb{A}_{\mathfrak{M}},R_0)\subseteq_{\text{end}}(\mathfrak{B}_{\mathfrak{N}},R)$$

so what we need to prove is that $I_\varphi\subseteq R_0$. This will follow (from 1.5 (ii)) if we prove that $\Gamma_\varphi(R_0)\subseteq R_0$); i. e., that

(2) $(\mathbb{A}_{\mathfrak{M}},R_0)\models\forall y_1,\ldots,y_n\,[\varphi(y_1,\ldots,y_n,\mathsf{R})\to \mathsf{R}(y_1,\ldots,y_n)]$.

So suppose that $y_1,\ldots,y_n \in \mathbb{A}_{\mathfrak{M}}$ and

$$(\mathbb{A}_{\mathfrak{M}}, R_0) \models \varphi(y_1,\ldots,y_n,\mathsf{R}).$$

Since φ is a Σ formula and $(\mathfrak{B}_{\mathfrak{N}}, R)$ is an end extension of $(\mathbb{A}_{\mathfrak{M}}, R_0)$ we have

$$(\mathfrak{B}_{\mathfrak{N}}, R) \models \varphi(y_1,\ldots,y_n,\mathsf{R}),$$

and so, by the last axiom of T, $R(y_1,\ldots,y_n)$ holds, and hence $R_0(y_1,\ldots,y_n)$. This establishes (2) and hence the theorem. □

Let $\varphi(x_1,\ldots,x_n,v_1,\ldots,v_k,\mathsf{R})$ be a fixed Σ formula of $\mathsf{L}^*(\mathsf{R})$. The following remarks are intended to lift much of Theorem 2.2 to arbitrary admissible sets by means of the Absoluteness Principle.

2.3 Remark. The Σ_1 formula defining I_φ in Theorem 2.2 is independent of $\mathbb{A}$ except for the parameters occuring in φ. More fully, let

$$I_\varphi(\mathbb{A}, y_1,\ldots,y_k)$$

denote the smallest fixed point defined on $\mathbb{A}$ by Γ_φ when $v_1=y_1,\ldots,y_k=y_k$ (provided $\varphi(x_1,\ldots,x_n,y_1,\ldots,y_k,\mathsf{R}) \in \Sigma(\mathsf{R}\!\upharpoonright\!\mathbb{A})$). *There is a* Σ_1 *formula* $\psi(x_1,\ldots,x_n,v_1,\ldots,v_k)$ *of* L^* *such that for all countable, admissible* $\mathbb{A}$ *and all* $y_1,\ldots,y_k \in \mathbb{A}$,

(3) *if* $\varphi(x_1,\ldots,x_n,y_1,\ldots,y_k,\mathsf{R})$ *is* R-*monotone on* $\mathbb{A}$ *then for all* $x_1,\ldots,x_n \in \mathbb{A}$,

$$(x_1,\ldots,x_n) \in I_\varphi(\mathbb{A}, y_1,\ldots,y_k) \quad \textit{iff} \quad \mathbb{A} \models \psi(x_1,\ldots,x_n,y_1,\ldots,y_k).$$

Proof. Let ψ be the formula which expresses

$$\exists p\ [p \textit{ is a proof of } \sigma \rightarrow \mathsf{R}(\overline{\mathsf{x}}_1,\ldots,\overline{\mathsf{x}}_n) \textit{ where } \sigma \textit{ is a conjunction of members of } T],$$

where T is as in the proof of 2.2, and examine the proof of Theorem 2.2. □

2.4 Remark. The operation $I_\varphi^\alpha(\mathbb{A}, y_1,\ldots,y_k)$, is a Σ operation of $\mathbb{A}$, $y_1,\ldots,y_k$, since it is defined by Σ Recursion on α. In ZF we proved the existence of an α (depending on $\mathbb{A}, y_1,\ldots,y_k$) such that

$$I_\varphi^\alpha(\mathbb{A}, y_1,\ldots,y_k) = I_\varphi^{<\alpha}(\mathbb{A}, y_1,\ldots,y_k).$$

(This step takes us outside KPU since it requires some form of Σ_1 Separation.) Thus, in ZF, *the predicate*

$$(x_1,\ldots,x_n) \in I_\varphi(\mathbb{A}, y_1,\ldots,y_k)$$

is a Δ_1 *predicate of* $\mathbb{A}$, $x_1,\ldots,x_n,y_1,\ldots,y_k$. It is expressed by the Σ_1 formula

$$\exists \alpha\ [(x_1,\ldots,x_n) \in I_\varphi^\alpha(\mathbb{A}, y_1,\ldots,y_k)]$$

and the Π_1 formula

$$\forall\alpha\,[I_\varphi^\alpha(\mathbb{A}, y_1, \ldots, y_k) = I_\varphi^{<\alpha}(\mathbb{A}, y_1, \ldots, y_k) \rightarrow (x_1, \ldots, x_n) \in I_\varphi^\alpha(\mathbb{A}, y_1, \ldots, y_k)]\,.$$

(The characterization of $I_\varphi(\mathbb{A}, y_1, \ldots, y_k)$ as smallest fixed point of Γ_φ gives another Π_1 definition.)

2.5 Remark. *The conclusion of line* (3) *above is a* Δ_1 *predicate of* $\mathbb{A}$, $y_1, \ldots, y_k$. The hypothesis, however, is a Π_1 predicate of $\mathbb{A}$, $y_1, \ldots, y_k$ which makes (3) of the form $\Pi_1 \rightarrow \Delta_1$ and hence a Σ_1 predicate of $\mathbb{A}$, $y_1, \ldots, y_k$. To apply the Absoluteness Principle we would need the result to be Π_1.

We are now ready to lift Theorem 2.2 to the uncountable. We give two proofs because each contains information not available in the other (see the two corollaries 2.7 and 2.8).

2.6 Gandy's Theorem. *Let* $\mathbb{A}$ *be any admissible set. Every* Σ_+ *inductive relation on* $\mathbb{A}$ *is* Σ_1 *on* $\mathbb{A}$.

First Proof of Theorem 2.6. Fix $\varphi(x_1, \ldots, x_n, v_1, \ldots, v_k, \mathsf{R}) \in \Sigma(\mathsf{R}_+)$. Since φ is R-positive it is R-monotone for all structures for L^* and hence for all admissible sets. The troublesome hypothesis of line (3) is thus superfluous and we see that we have proved for all countable $\mathbb{A}$:

if $\mathbb{A}$ *is admissible then for all* $y_1, \ldots, y_k \in \mathbb{A}$ *and all* $x_1, \ldots, x_n \in \mathbb{A}$

$$(x_1, \ldots, x_n) \in I_\varphi(\mathbb{A}, y_1, \ldots, y_k) \leftrightarrow \mathbb{A} \models \psi(x_1, \ldots, x_n, y_1, \ldots, y_k)\,.$$

The displayed part is Δ_1 so by the Lévy Absoluteness Principle, the result holds for all $\mathbb{A}$. □

2.7 Corollary. *Let* $\mathscr{K}$ *be a class of admissible sets which is* Σ_1 *definable in* ZFC. *Then for any* $\mathbb{A} \in \mathscr{K}$, *every* $\Sigma(\uparrow\mathscr{K})$ *inductive relation on* $\mathbb{A}$ *is* Σ_1 *on* $\mathbb{A}$.

Proof. The hypothesis asserts that there is a Σ_1 formula $\theta(x)$ without parameters such that

$$\mathbb{A} \in \mathscr{K} \quad \text{iff} \quad \theta(\mathbb{A}),$$

$$\mathrm{ZFC} \vdash \theta(\mathbb{A}) \rightarrow \mathbb{A} \quad \text{is admissible.}$$

Replace "$\mathbb{A}$ is admissible" by "$\theta(\mathbb{A})$" in the above proof. □

For example, the $\mathscr{K}$ in 2.6 might be the class of all admissible sets or the class of $L(\alpha)$ where α is recursively inaccessible or nonprojectible.

Second Proof of Theorem 2.6. This proof is more traditional in that it uses the Second Recursion Theorem. For simplicity we let $n=1$ and we suppress parameters $y_1,\ldots,y_k$ entirely since they are held constant in this proof. To simplify notation, whenever S is a relation on $\mathbb{A}$ and $\varphi(x,\mathsf{R})\in\Sigma(\mathsf{R}_+)$, we write

$$\mathbb{A}\models\varphi(x,S)$$

instead of the more accurate

$$(\mathbb{A},S)\models\varphi(\mathsf{R})\,[x].$$

Now let $\varphi(x,\mathsf{R})\in\Sigma(\mathsf{R}_+)$. Use the Second Recursion Theorem to define a Σ_1 formula ψ of L^* such that

$$\mathrm{KPU}\vdash\psi(x,\beta)\leftrightarrow\varphi(x,\exists\gamma<\beta\,\psi(\cdot\,,x)).$$

(More precisely,

$$\mathrm{KPU}\vdash\psi(x,\beta)\leftrightarrow\varphi(x,\lambda y\,\exists\gamma<\beta\,\psi(y,\gamma)).$$

To fit thus into Second Recursion Theorem, first let S be a new binary symbol and let $\varphi'(x,\beta,\mathsf{S})$ be $\varphi(x,\exists\gamma<\beta\,\mathsf{S}(\cdot\,,\gamma))$ and then apply the Second Recursion Theorem.) We claim that

(4) for $\beta<o(\mathbb{A})$

$$x\in I_\varphi^\beta \quad\text{iff}\quad \mathbb{A}\models\psi(x,\beta).$$

The proof proceeds by induction on β. The induction hypothesis gives us, for $\gamma<\beta$,

$$I_\varphi^\gamma=\{x\mid\mathbb{A}\models\psi(x,\gamma)\}$$

so, taking unions,

$$I_\varphi^{<\beta}=\{x\mid\mathbb{A}\models\exists\gamma<\beta\,\psi(x,\gamma)\}.$$

Then for any $x\in\mathbb{A}$ we have

$$\begin{aligned} x\in I_\varphi^\beta \quad&\text{iff}\quad x\in\Gamma_\varphi(I_\varphi^{<\beta})\\ &\text{iff}\quad \mathbb{A}\models\varphi(x,I_\varphi^{<\beta})\\ &\text{iff}\quad \mathbb{A}\models\varphi(x,\exists\gamma<\beta\,\psi(\cdot\,,\gamma))\\ &\text{iff}\quad \mathbb{A}\models\psi(x,\beta).\end{aligned}$$

Let $\alpha=o(\mathbb{A})$. From (4) we obtain

(5) $I_\varphi^{<\alpha}=\{x\mid\mathbb{A}\models\exists\beta\,\psi(x,\beta)\}.$

Now we claim that

(6) $\Gamma_\varphi(I_\varphi^{<\alpha}) = I_\varphi^{<\alpha}$.

It suffices to prove $\Gamma_\varphi(I_\varphi^{<\alpha}) \subseteq I_\varphi^{<\alpha}$, so suppose $x \in \Gamma_\varphi(I_\varphi^{<\alpha})$, i.e., that

$$\mathbb{A} \models \varphi(x, I_\varphi^{<\alpha}).$$

By (5) this becomes

$$\mathbb{A} \models \varphi(x, \exists\beta\, \psi(\cdot, \beta)).$$

By the Σ Reflection Theorem and Lemma V.2.2 there is a $\delta < \alpha$ such that

$$\mathbb{A} \models \varphi(x, \exists\beta < \delta\, \psi(\cdot, \beta))$$

which, by (4), is equivalent to

$$\mathbb{A} \models \varphi(x, I_\varphi^{<\delta}).$$

Thus $x \in \Gamma_\varphi(I_\varphi^{<\delta}) = I_\varphi^\delta$. But $I_\varphi^\delta \subseteq I_\varphi^{<\alpha}$ so $x \in I_\varphi^{<\alpha}$ as desired. But (6) immediately implies that $I_\varphi = I_\varphi^{<\alpha}$, so $\|\Gamma_\varphi\| \leqslant \alpha$ and

$$I_\varphi = \{x \mid \mathbb{A} \models \exists\beta\, \psi(x, \beta)\},$$

which proves that I_φ is Σ_1 on $\mathbb{A}$. □

2.8 Corollary (Second half of Gandy's Theorem). *Let $\mathbb{A}$ be admissible and let $\varphi(x_1, \ldots, x_n, \mathsf{R}_+)$ be a Σ formula with parameters from $\mathbb{A}$. Let $\alpha = o(\mathbb{A})$.*
(i) $\|\Gamma_\varphi\| \leqslant \alpha$.
(ii) *For all β, I_φ^β is Σ_1 on $\mathbb{A}$.*

Proof. Part (i) was explicitly mentioned in the second proof of 2.6. For (ii) we have the result for $\beta \geqslant \alpha$ by 2.6 and for $\beta < \alpha$ by line (4) above. □

The results mentioned in 2.8 also hold for arbitrary R-monotone $\varphi(x, \mathsf{R})$ if the admissible set $\mathbb{A}$ is countable. The proof of this, however, must await a stronger reflection principle, the $s-\Pi_1^1$ Reflection Principle.

For sets of the form $\mathrm{L}(\alpha)$ the conclusions of 2.6 and 2.8 are actually equivalent to the hypothesis of admissibility. This will follow from Theorem 3.17 in the next section.

2.9—2.11 Exercises

2.9. Let $\mathfrak{A}_\mathfrak{M}$ be a nonstandard model of KPU. Show that $\mathscr{W}\!\!\mathscr{f}(\mathfrak{A}_\mathfrak{M})$ is a Σ_+ fixed point which is not first order definable over $\mathfrak{A}_\mathfrak{M}$. What is the length of the inductive definition?

2.10 (Stavi). Show that there are pure transitive sets which are not admissible but such that every Σ_+ inductive relation is Σ_1. [Hint: Let $\mathbb{A} = L(\tau_1) \cap V(\alpha)$ for suitably nice $\alpha, \omega < \alpha < \tau_1$.]

2.11. Let $\varphi(x_1, \ldots, x_n, \mathsf{R}_+)$ be a Π formula and let J_φ be the largest fixed point of Γ_φ on an admissible set $\mathbb{A}$. Show that J_φ is Π_1^0.

2.12 Notes. The fact that, over an admissible set $\mathbb{A}$, a Σ_+ inductive definition Γ_φ has a Σ_1 fixed point and closure ordinal $\|\Gamma_\varphi\| \leqslant o(\mathbb{A})$ is usually called Gandy's Theorem. He proved this theorem in lectures at the UCLA Logic year in 1968 by adapting the proof-theoretic approach used to prove the Barwise Completeness Theorem. A similar approach is taken in Gandy [1974]. We have given two new proofs for this theorem, one which shows that the result can be derived from the Barwise Completeness Theorem, the other a much more standard recursion theoretic approach using the Second Recursion Theorem.

The recursion theoretic approach to Gandy's Theorem suggests an alternate approach to the material in this book. One could prove Gandy's Theorem (by means of the Second Recursion Theorem) and then quote it to prove that the set $Thm_{\mathbb{A}}$ of theorems of an admissible fragment $\mathsf{K}_{\mathbb{A}}$ is Σ_1 on $\mathbb{A}$. This would suffice for many applications of the Completeness Theorem, but not all. Some applications actually need the notion of $\mathsf{K}_{\mathbb{A}}$-*proof* used in § III.5, since there is important information coded inside the proof.

The approach taken here also has the advantage of stressing the interplay of all branches of mathematical logic, which is one of the attractive features of admissible set theory.

3. *First Order Positive Inductive Definitions and* $\mathbb{HYP}_{\mathfrak{M}}$

We have seen various ways in which $\mathbb{HYP}_{\mathfrak{M}}$ is a mini-universe of set theory above $\mathfrak{M}$. For countable $\mathfrak{M}$, we have seen that the relations on $\mathfrak{M}$ which are elements of $\mathbb{HYP}_{\mathfrak{M}}$ are exactly the Δ_1^1 relations. This characterization breaks down for uncountable $\mathfrak{M}$ (see Exercise VII.1.16) so we are left with two problems in the general case:

To characterize the relations on $\mathfrak{M}$ which are elements of $\mathbb{HYP}_{\mathfrak{M}}$, and
To characterize the Δ_1^1 relations on $\mathfrak{M}$ in terms of $\mathbb{HYP}_{\mathfrak{M}}$.

The first of these two problems is solved by Theorem 3.6 below. The second problem is solved at the end of § VIII.2.

3.1 Definition (Moschovakis). Let K be a language and let Φ be the set of all finitary formulas of the form $\varphi(\mathsf{R}_+)$, for any new relation symbol R. Let $\mathfrak{N}$ be a structure for K and let S be a relation on $\mathfrak{N}$.

(i) S is a *(first order positive) fixed point* on $\mathfrak{N}$ if S is a Φ-fixed point (in the sense of 1.12) on $\mathfrak{N}$.

(ii) S is *inductive* on $\mathfrak{N}$ of S is Φ-inductive on $\mathfrak{N}$.

(iii) S is *coinductive* on $\mathfrak{N}$ if $\neg S$ is Φ-inductive on $\mathfrak{N}$.

(iv) S is *hyperelementary* on $\mathfrak{N}$ if S is inductive and coinductive on $\mathfrak{N}$.

(For more intuition into the notion of coinductive, the student should do Exercises 1.15—1.18.)

The theorems of this section are suggested by the following classical result.

3.2 Theorem. *Let $\mathcal{N}=\langle\omega,0,+,\cdot\rangle$ and let S be a relation on $\mathcal{N}$.*

(i) *S is Π_1^1 on $\mathcal{N}$ iff S is inductive on $\mathcal{N}$.*

(ii) *S is Δ_1^1 on $\mathcal{N}$ iff S is hyperelementary on $\mathcal{N}$.*

Proof. We proved (i) in 1.10 and 1.11; (ii) is immediate from (i). □

Thus we see that for relations on $\mathcal{N}$,

$$\Sigma_1 \text{ on } \mathbb{HYP}_{\mathcal{N}} = \text{inductive on } \mathcal{N},$$

$$\text{element of } \mathbb{HYP}_{\mathcal{N}} = \text{hyperelementary on } \mathcal{N}.$$

We would like to generalize these equations from $\mathcal{N}$ to an arbitrary structure $\mathfrak{M}$. We would like to, but we can't because the generalization works only for $\mathfrak{M}$ which have some built in coding machinery. We discuss just how much coding is needed in the next section. For now we simply state one special case where all goes smoothly, and then take a different tack.

3.3 Theorem. *Let $\mathbb{A}$ be an admissible set and let S be a relation on $\mathbb{A}$.*

(i) *S is Σ_1 on $\mathbb{HYP}(\mathbb{A})$ iff S is inductive on $\mathbb{A}$.*

(ii) *S is an element of $\mathbb{HYP}(\mathbb{A})$ iff S is hyperelementary on $\mathbb{A}$.*

Proof. We merely sketch a proof since this result is a special case of Theorem 3.8 and the results of the next section. The proof sketched here is more direct. As usual, (ii) follows trivially from (i). We first show that if S is inductive on $\mathbb{A}$ then S is Σ_1 on $\mathbb{HYP}(\mathbb{A})$. It clearly suffices to prove the result for the case where S is a fixed point I_φ of some first order positive inductive definition Γ_φ. Since φ is first order over $\mathbb{A}$ it is Δ_0 in $\mathbb{HYP}(\mathbb{A})$ so Γ_φ is, in particular, a Σ_+ inductive definition over $\mathbb{HYP}(\mathbb{A})$, hence by Gandy's Theorem, I_φ is Σ_1 on $\mathbb{HYP}(\mathbb{A})$. (A more direct proof which works here but not in 3.8 is to observe that I_φ^β is a $\mathbb{HYP}(\mathbb{A})$-recursive function of β, for $\beta<o(\mathbb{HYP}(\mathbb{A}))$, and use Σ Reflection to prove that $\|\Gamma_\varphi\|\leqslant o(\mathbb{HYP}(\mathbb{A}))$. This would give the following Σ_1 definition of S:

$$S(\vec{x}) \quad \text{iff} \quad \mathbb{HYP}(\mathbb{A})\models\exists\beta\,(\vec{x}\in I_\varphi^\beta).)$$

To prove the other half, suppose $S \subseteq \mathbb{A}$ is Σ_1 on $\mathbb{HYP}(\mathbb{A})$. By Theorem IV.7.3 (or, more precisely, Corollary 3.14 below) S is weakly representable in KPU′ using the $\mathbb{A}$-rule, where KPU′ is the theory

$$\begin{aligned}&\text{KPU},\\&\text{diagram}(\mathbb{A}),\\&\overline{\mathsf{x}} \in \overline{\mathsf{A}} \quad (\text{all } x \in \mathbb{A}),\\&\exists a\, \forall v\, [v \in a \leftrightarrow \overline{\mathsf{A}}(v)].\end{aligned}$$

But the set $C_{\mathbb{A}}(\text{KPU}')$ of consequences of KPU′ using the $\mathbb{A}$-rule is clearly an inductive subset of $\mathbb{A}$. Thus we have

$$S(\vec{x}) \quad \text{iff} \quad f(\vec{x}) \in C_{\mathbb{A}}(\text{KPU}')$$

for some $\mathbb{A}$-recursive function f. An easy exercise (Exercise 3.20) establishes that S is inductive on $\mathbb{A}$. □

We have been deliberately sketchy in the above proof to give the student a feel for the main idea. This must be gone into in more detail to prove Theorem 3.8 below, the main result of this section. First, though, let's draw some easy corollaries of Theorem 3.3.

3.4 Corollary. *If $\mathbb{A}$ is a countable admissible set then*

$$\Pi_1^1 \textit{ on } \mathbb{A} = \textit{inductive on } \mathbb{A},$$

$$\Delta_1^1 \textit{ on } \mathbb{A} = \textit{hyperelementary on } \mathbb{A}.$$

Proof. This is an immediate consequence of Theorem 3.3 and the results of § IV.3. □

3.5 Lemma. *Let $\mathbb{A}$ be admissible.*

(i) *There is an $(n+1)$-ary inductive relation on $\mathbb{A}$ which parametrizes the class of n-ary inductive relations on $\mathbb{A}$.*

(ii) *There is an inductive subset of $\mathbb{A}$ which is not hyperelementary.*

Proof. By V.5.3, $\mathbb{HYP}(\mathbb{A})$ is projectible into $\mathbb{A}$. Thus the lemma is just a restatement using 3.3. □

Using these results we can show just exactly how one gets from one admissible ordinal τ_α to the next admissible ordinal $\tau_{\alpha+1}$. Namely

$$\tau_{\alpha+1} = \sup\{\|\Gamma_\varphi\| : \Gamma_\varphi \text{ is a first order positive inductive definition over } \mathrm{L}(\tau_\alpha)\},$$

and this sup is actually obtained. This is a special case of the following result.

3.6 Corollary. *Let $\mathbb{A}$ be admissible and let $\alpha = o(\mathbb{HYP}(\mathbb{A}))$. Then α is equal to the* sup *of all $\|\Gamma_\varphi\|$ where Γ_φ is a first order positive inductive definition over $\mathbb{A}$, and this* sup *is actually attained.*

Proof. We know that any first order positive inductive definition Γ_φ over A is Σ_+ over $\mathbb{HYP}(\mathbb{A})$ (in fact "Δ_{0+}") so $\|\Gamma_\varphi\| \leq \alpha$ by the second half of Gandy's Theorem. To show that α is such an ordinal $\|\Gamma_\varphi\|$, use 3.5(ii) to choose an inductive subset $S \subseteq \mathbb{A}$ which is not hyperelementary. Then S is a section of some fixed point I_φ. Clearly I_φ is not hyperelementary either. We claim that $\|\Gamma_\varphi\| = \alpha$. As mentioned in the proof of Theorem 3.3, I_φ^β is a $\mathbb{HYP}(\mathbb{A})$ recursive function of β, for $\beta < \alpha$. Hence $I_\varphi^\beta \in \mathbb{HYP}(\mathbb{A})$ for all $\beta < \alpha$. But then, if $\|\Gamma_\varphi\| = \beta < \alpha$, $I_\varphi = I_\varphi^\beta \in \mathbb{H}YP(\mathbb{A})$ which makes I_φ hyperelementary, a contradiction. □

As we'll see in the next section, the hypothesis that $\mathbb{A}$ is admissible is far too strong for the above results. All we really need is a reasonable amount of coding apparatus.

What we are really after, though, is a characterization of the relations on $\mathfrak{M}$ in $\mathbb{HYP}_{\mathfrak{M}}$ which works for *all* structures $\mathfrak{M}$, not just those with built in coding machinery. The best way around this is to slightly strengthen the notion of inductive definition, so that one can do the coding needed in the inductive definition itself.

3.7 Definition. Let Φ be the set of extended first order formulas $\varphi(\mathsf{R}_+)$ of $\mathsf{L}^*(\mathsf{R})$ as defined in II.2.7, p. 50. Let $\mathfrak{M}$ be a structure for L and let S be a relation on $\mathfrak{M}$ (or even $\mathbb{HF}_{\mathfrak{M}}$).

(i) S is *extended inductive* (written *inductive**) on $\mathfrak{M}$ iff S is Φ inductive on $\mathbb{HF}_{\mathfrak{M}}$.

(ii) S is *extended hyperelementary* (written *hyperelementary**) on $\mathfrak{M}$ iff S and $\neg S$ are inductive* on $\mathfrak{M}$.

Our second, and principal, generalization of Theorem 3.2 is the following result.

3.8 Theorem. *Let $\mathfrak{M} = \langle M, R_1, \ldots, R_l \rangle$ be a structure for* L *and let S be a relation on $\mathfrak{M}$ (or even on $\mathbb{HF}_{\mathfrak{M}}$).*

(i) *S is Σ_1 on $\mathbb{HYP}_{\mathfrak{M}}$ iff S is inductive* on $\mathfrak{M}$.*

(ii) *S is Δ_1 on $\mathbb{HYP}_{\mathfrak{M}}$ iff S is hyperelementary* on $\mathfrak{M}$.*

Its corollaries are analogous to those of 3.3.

3.9 Corollary. *Let $\mathfrak{M} = \langle M, R_1, \ldots, R_l \rangle$ be a countable structure for* L.

(i) *Π_1^1 on $\mathfrak{M}$ = inductive* on $\mathfrak{M}$.*

(ii) *Δ_1^1 on $\mathfrak{M}$ = hyperelementary* on $\mathfrak{M}$.* □

3.10 Lemma. *Let $\mathfrak{M} = \langle M, R_1, \ldots, R_l \rangle$ be a structure for* L.

(i) *There is an $(n+1)$-ary inductive* relation on $\mathbb{HF}_{\mathfrak{M}}$ which parameterizes the class of n-ary inductive* relations on $\mathbb{HF}_{\mathfrak{M}}$.*

(ii) *There is an inductive* relation on $\mathbb{HF}_{\mathfrak{M}}$ which is not hyperelementary*.*

Proof. $\mathbb{HYP}_{\mathfrak{M}}$ is projectible into $\mathbb{HF}_{\mathfrak{M}}$, again by V.5.3, so the results follows from V.5.6 and 3.8. ☐

We use these corollaries to get the most intelligible description yet of $O(\mathfrak{M})$ (and hence of $\mathbb{HYP}_{\mathfrak{M}}$ since $\mathbb{HYP}_{\mathfrak{M}} = \mathrm{L}(\alpha)_{\mathfrak{M}}$ where $\alpha = O(\mathfrak{M})$).

3.11 Theorem. *If* $\mathfrak{M} = \langle M, R_1, \ldots, R_l \rangle$ *is a structure for* L *then*

$$O(\mathfrak{M}) = \sup\{\|\Gamma_\varphi\| \mid \Gamma_\varphi \text{ is an extended inductive definition over } \mathfrak{M}\}$$

and this sup *is actually attained.*

Proof. The proof of 3.11 is exactly like the proof of 3.6 when $O(\mathfrak{M}) > \omega$ for then $\mathbb{HF}_{\mathfrak{M}} \in \mathbb{HYP}_{\mathfrak{M}}$. Suppose $\mathbb{HYP}_{\mathfrak{M}}$ has ordinal ω. Let Γ_φ be an extended first order inductive definition on $\mathfrak{M}$. As we will see in the proof of Theorem 3.8, Γ_φ is Σ_+ on $\mathbb{HYP}_{\mathfrak{M}}$, so $\|\Gamma_\varphi\| \leqslant \omega$ by the second half of Gandy's Theorem. It is simple to give an example of extended first order inductive definitions of length ω, e. g.,

$$x \in \Gamma(R) \quad \text{iff} \quad \text{“}x \text{ is a natural number} \wedge \forall y < x\, R(y)\text{”}$$

defines ω in $\mathbb{HF}_{\mathfrak{M}}$ with

$$I_\Gamma^n = \{0, \ldots, n\}$$

so $\|\Gamma\| = \omega$. ☐

It is worthwhile digressing to compare 3.8 with the following consequence of 3.3, just to make sure the student is not confusing two distinct things.

3.12 Corollary. *Let* $\mathfrak{M} = \langle M, R_1, \ldots, R_l \rangle$ *be a structure for* L *which is not recursively saturated. Let S be a relation on* $\mathfrak{M}$ *(or even* $\mathbb{HF}_{\mathfrak{M}}$*).*

(i) *S is* Σ_1 *on* $\mathbb{HYP}_{\mathfrak{M}}$ *iff S is inductive on* $\mathbb{HF}_{\mathfrak{M}}$.

(ii) $S \in \mathbb{HYP}_{\mathfrak{M}}$ *iff S is hyperelementary on* $\mathbb{HF}_{\mathfrak{M}}$.

Proof. Since $\mathfrak{M}$ is not recursively saturated, $o(\mathbb{HYP}_{\mathfrak{M}}) > \omega$ so $\mathbb{HF}_{\mathfrak{M}} \in \mathbb{HYP}_{\mathfrak{M}}$. But then $\mathbb{HYP}(\mathbb{HF}_{\mathfrak{M}}) = \mathbb{HYP}_{\mathfrak{M}}$ since $\mathbb{HYP}(\mathbb{HF}_M)$ is the smallest admissible set with $\mathbb{HF}_{\mathfrak{M}}$ as an element. Thus 3.10 is a special case of 3.3. ☐

The student must be clear about the difference between inductive* definitions on $\mathfrak{M}$ and inductive definitions on $\mathbb{HF}_{\mathfrak{M}}$. The latter are, in general, much more powerful since they allow unbounded universal quantification over sets in $\mathbb{HF}_{\mathfrak{M}}$ in addition to the unbounded existential allowed by inductive* definitions.

We have already done most of the work for proving Theorem 3.8 back in § III.3, the section on $\mathfrak{M}$-logic and the $\mathfrak{M}$-rule.

In the discussion below we let $\mathfrak{M} = \langle M, R_1, \ldots, R_l \rangle$ be a fixed L-structure and we let L^+ be an expansion of L with a new unary symbol $\overline{\mathsf{M}}$ and symbols $\overline{\mathsf{p}}$

for each $p \in \mathfrak{M}$, just as in our discussion of $\mathfrak{M}$-logic in § III.3. We assume that L^+ is coded up in an effective way on $\mathbb{HF}_{\mathfrak{M}}$.

3.13 Proposition. *Let T be a set of sentences of $\mathsf{L}^+_{\omega\omega}$ which is Σ_1 on $\mathbb{HF}_{\mathfrak{M}}$. Let $C_{\mathfrak{M}}(T)$ be the set of formulas of $\mathsf{L}^+_{\omega\omega}$ which are provable from T using the $\mathfrak{M}$-rule. Then $C_{\mathfrak{M}}(T)$ is inductive*.*

Proof. We simply write out the original definition Γ of $C_{\mathfrak{M}}(T)$ given in III.3.4 and observe that it has the correct form. Let R be a new unary symbol and define Γ by

$$x \in \Gamma(R) \quad \text{if} \quad x \in \mathsf{L}^+_{\omega\omega} \wedge [(1) \vee \cdots \vee (5)]$$

where (1)...(5) are given below.

(1) (Logical Axioms) "x is an axiom of first order logic";
(2) (Nonlogical Axioms) $x \in T$;
(3) (Modus Ponens) $\exists y\, [y \in R \wedge (y \to x) \in R]$;
(4) (Generalization) "x is of the form $(\psi \to \forall v\, \theta(v))$ where v is not free in ψ and $(\psi \to \theta(v)) \in R$";
(5) ($\mathfrak{M}$-rule) "x is of the form $\forall v_0\, [\overline{\mathsf{M}}(v_0) \to \theta(v_0)]$ and for all $p \in M$, $\theta(\overline{\mathsf{p}}/v_0) \in R$".

Clearly Γ defines $C_{\mathfrak{M}}(T)$, i.e., $C_{\mathfrak{M}}(T) = I_\Gamma$ so that $C_{\mathfrak{M}}(T)$ is actually a fixed point. Γ is definable over $\mathbb{HF}_{\mathfrak{M}}$ by an R-positive formula; the only unbounded universal quantifier is in (5) and it is a quantifier over M. □

The reader may remember that we left a couple of proofs unfinished in § IV.7, the section on representability using the $\mathfrak{M}$-rule. We proved IV.7.3 and IV.7.4 in the countable case but left the absoluteness of those results until later. Proposition 3.13 allows us to finish these proofs.

3.14 Corollary. *Assume the notation of Proposition* 3.13.

(i) *$x \in C_{\mathfrak{M}}(T)$ is a Δ_1 predicate of x, T and $\mathfrak{M}$, Δ_1 in the theory* ZF.

(ii) *Consequently, the proofs given in* § IV.7 *of* IV.7.3 *and* IV.7.4 *for the countable case, together with Lévy's Absoluteness Principle, yield the general results.*

Proof. Part (i) is a consequence of Remark 2.4. For (ii), the proofs of IV.7.3 and IV.7.4 are quite similar. Since IV.7.3 is the more important for us here (we apply it in the next proof) let us treat it in some detail. Again 7.3 (i) and 7.3 (ii) are similar so we prove (i). Suppose, as in the proof of (i), that $\varphi(x_1,\ldots,x_n,p_1,\ldots,p_k,M)$ is a Σ_1 formula with the property that for all $q_1,\ldots,q_n \in M$

$$\mathbb{HYP}_{\mathfrak{M}} \models \varphi(q_1,\ldots,q_n,\vec{p},M) \quad \text{iff} \quad \mathrm{KPU}^+ \models_{\mathfrak{M}} \varphi(\overline{\mathsf{q}}_1,\ldots,\overline{\mathsf{q}}_n,\overline{\mathsf{p}},\overline{\mathsf{M}}).$$

Now, if $\mathfrak{M}$ is countable we use the $\mathfrak{M}$-completeness theorem to write

$$\mathbb{HYP}_{\mathfrak{M}} \models \varphi(q_1,\ldots,q_n,\vec{p},M) \quad \text{iff} \quad \mathrm{KPU}^+ \vdash_{\mathfrak{M}} \varphi(\overline{\mathsf{q}}_1,\ldots,\overline{\mathsf{q}}_n,\overline{\mathsf{p}},\overline{\mathsf{M}}).$$

I. e., we have for all *countable* $\mathfrak{M}$ and all $q_1,\ldots,q_n \in M$:

$$\mathbb{HYP}_{\mathfrak{M}} \models \varphi(q_1,\ldots,q_n,\vec{p},M) \quad \text{iff} \quad \varphi(\bar{\mathsf{q}}_1,\ldots,\bar{\mathsf{q}}_n,\bar{\mathsf{p}},\bar{\mathsf{M}}) \in C_{\mathfrak{M}}(\mathrm{KPU}^+).$$

We claim that this is a Δ_1 predicate of $\mathfrak{M}$, Δ_1 in ZF. The right hand side of the iff is Δ_1 by (i), and the left hand side is Δ_1 since satisfaction is Δ_1 and since $\mathbb{HYP}_{\mathfrak{M}}$ is a Σ_1 operation of $\mathfrak{M}$ by the argument given in IV.3.5. By Lévy Absoluteness, the result holds for *all* $\mathfrak{M}$. □

Theorem 3.8 will follow from Proposition 3.13 given the next lemma. It is a special case of the Combination Lemma of Moschovakis [1974].

3.15 Lemma. *Let* $U \subseteq \mathbb{HF}_{\mathfrak{M}}$ *be inductive*, let* $f\colon \mathbb{HF}_{\mathfrak{M}}^n \to \mathbb{HF}_{\mathfrak{M}}$ *be* Σ_1 *on* $\mathbb{HF}_{\mathfrak{M}}$ *and let P be defined by*

$$P(x_1,\ldots,x_n) \quad \textit{iff} \quad f(x_1,\ldots,x_n) \in U.$$

Then P is inductive on* $\mathfrak{M}$.

Proof. Suppose U is a section of the fixed point I_φ where $\varphi(v_1,v_2,\mathsf{R}_+)$ is extended first order positive on $\mathfrak{M}$, say

$$U(y) \leftrightarrow ((y,z_0) \in I_\varphi).$$

We define an $n+3$-ary inductive* definition Γ_ψ so that a section of I_ψ^ξ (with $i=0$) imitates I_φ^α and the section with $i=1$ takes care of f. Define $\psi(i,x_1,\ldots,x_n,v_1,v_2,\mathsf{S}_+)$, where S is $n+3$-ary, by the following, where $t_1,\ldots,t_n,z_1,z_2$ are arbitrary but fixed elements of $\mathbb{HF}_{\mathfrak{M}}$:

$$i=0 \wedge \vec{x}=\vec{t} \wedge \varphi(v_1,v_2,\lambda w_1 w_2\, \mathsf{S}(0,t_1,\ldots,t_n,w_1,w_2)/\mathsf{R}), \quad \text{or}$$
$$i=1 \wedge v_1,v_2=z_1,z_2 \wedge \mathsf{S}(0,t_1,\ldots,t_n,f(x_1,\ldots,x_n),z_0).$$

A simple proof by induction shows that

$$(6) \quad I_\varphi^\alpha(v_1,v_2) \quad \text{iff} \quad I_\psi^\alpha(0,t_1,\ldots,t_n,v_1,v_2)$$

so that

$$U(y) \quad \text{iff} \quad I_\psi(0,t_1,\ldots,t_n,y,z_0).$$

Another proof by induction, using (6), shows that

$$(f(x_1,\ldots,x_n),z_0) \in I_\varphi^\alpha \quad \text{iff} \quad (1,x_1,\ldots,x_n,z_1,z_2) \in I_\psi^{\alpha+1}.$$

Thus

$$P(x_1,\ldots,x_n) \quad \text{iff} \quad (1,x_1,\ldots,x_n,z_1,z_2) \in I_\psi$$

so P is a section of I_ψ. The only universal quantifiers in ψ are those in φ so ψ is extended first order positive. □

We now return to prove the main theorem of this section, Theorem 3.8.

3.16 Proof of Theorem 3.8. (i) Let Γ_φ be an extended first order inductive definition over $\mathfrak{M}$. Since $\mathbb{HF}_{\mathfrak{M}}$ is a Σ_1 subset of $\mathbb{HYP}_{\mathfrak{M}}$, relativizing the unbounded (existential) set quantifiers in Γ_φ to $\mathbb{HF}_{\mathfrak{M}}$ and relativizing the unbounded quantifiers over $\mathfrak{M}$ to the set M turns Γ_φ into a Σ_+ inductive definition over $\mathbb{HYP}_{\mathfrak{M}}$ and hence Γ_φ has a Σ_1 fixed point I_φ, by Gandy's Theorem.

To prove the other half, let us consider a relation S on $\mathfrak{M}$ which is Σ_1 on $\mathbb{HYP}_{\mathfrak{M}}$. By Theorem IV.7.3, S is weakly representable in KPU$^+$ using the $\mathfrak{M}$-rule. Thus there is a formula $\varphi(v_1,\ldots,v_n)$ of L* such that for all $x_1,\ldots,x_n \in M$,

$$S(x_1,\ldots,x_n) \quad \text{iff} \quad \varphi(\overline{\mathsf{x}}_1,\ldots,\overline{\mathsf{x}}_n) \in C_{\mathfrak{M}}(\mathrm{KPU}^+).$$

Now, by 3.13, $C_{\mathfrak{M}}(\mathrm{KPU}^+)$ is inductive* over $\mathfrak{M}$. Let $f(x_1,\ldots,x_n) = \varphi(\overline{\mathsf{x}}_1/v_1,\ldots,\overline{\mathsf{x}}_n/v_n)$. Then

$$S(x_1,\ldots,x_n) \quad \text{iff} \quad f(x_1,\ldots,x_n) \in C_{\mathfrak{M}}(\mathrm{KPU}^+)$$

so S is inductive* by Lemma 3.15. The same proof works if $S \subseteq \mathbb{HF}_{\mathfrak{M}}$ except that Exercise IV.7.5 replaces Theorem IV.7.3. Part (ii) follows from (i) as usual. □

The final results of this section show that for nonadmissible sets of the form $\mathrm{L}(\alpha)_{\mathfrak{M}}$ (for example), Σ_+ inductive definitions are just as strong as arbitrary first order inductive definitions, and that they are just as long. The results thus yield partial converses to the results of § 2 by showing how necessary the assumption of admissibility was for those results.

3.17 Theorem. *Let $M \subseteq a$ where a is transitive in $\mathbb{V}_M$ and let β be any limit ordinal such that*

$$\mathbb{A}_{\mathfrak{M}} = (\mathfrak{M}; \mathrm{L}(a,\beta) \cap \mathbb{V}_{\mathfrak{M}}, \in)$$

is not admissible.

(i) *A relation S on $\mathbb{A}_{\mathfrak{M}}$ is Σ_1 on $\mathbb{HYP}(\mathbb{A}_{\mathfrak{M}})$ iff S is Σ_+ inductive on $\mathbb{A}_{\mathfrak{M}}$.*

(ii) *The ordinal $o(\mathbb{HYP}(\mathbb{A}_{\mathfrak{M}}))$ is equal to*

$$\sup\{\|\Gamma_\varphi\| \mid \Gamma_\varphi \text{ is a } \Sigma_+ \text{ inductive definition on } \mathbb{A}_{\mathfrak{M}}\}$$

and the sup *is actually attained.*

3.18 Corollary. *Let $M \subseteq a$ where a is transitive in $\mathbb{V}_{\mathfrak{M}}$ and let β be any limit ordinal. Let*

$$\mathbb{A}_{\mathfrak{M}} = (\mathfrak{M}; \mathrm{L}(a,\beta) \cap \mathbb{V}_{\mathfrak{M}}, \in).$$

The following are equivalent, where $\alpha = o(\mathbb{A}_{\mathfrak{M}})$.

(i) $\mathbb{A}_{\mathfrak{M}}$ *is admissible.*

(ii) *Every* Σ_+ *inductive set on* $\mathbb{A}_{\mathfrak{M}}$ *is* Σ_1 *on* $\mathbb{A}_{\mathfrak{M}}$.

(iii) *For every* Σ_+ *inductive definition* Γ_φ *on* $\mathbb{A}_{\mathfrak{M}}$, $\|\Gamma_\varphi\| \leqslant \alpha$.

Proof. By the results of the previous section, (i)$\Rightarrow$(ii) and (i)$\Rightarrow$(iii). To prove (ii)$\Rightarrow$(i), suppose $\mathbb{A}_{\mathfrak{M}}$ is not admissible. Let S be a subset of $\mathbb{A}_{\mathfrak{M}}$ which is Σ_1 on $\mathbb{HYP}(\mathbb{A}_{\mathfrak{M}})$ but not $\mathbb{HYP}(\mathbb{A}_{\mathfrak{M}})$-finite; such an S exists since $\mathbb{HYP}(\mathbb{A}_{\mathfrak{M}})$ is projectible into $\mathbb{A}_{\mathfrak{M}}$. But then S is Σ_+ inductive on $\mathbb{A}_{\mathfrak{M}}$ by 3.17. S cannot be Σ_1 on $\mathbb{A}_{\mathfrak{M}}$ for then it would be Δ_0 on $\mathbb{HYP}(\mathbb{A}_{\mathfrak{M}})$, hence in $\mathbb{HYP}(\mathbb{A}_{\mathfrak{M}})$. Thus $\neg$(i)$\Rightarrow\neg$(ii). For the same reason, the length $\|\Gamma_\varphi\|$ of an inductive definition of S could not be $\leqslant \alpha$ so $\neg$(i)$\Rightarrow\neg$(iii). $\square$

The proof of Theorem 3.17 uses ideas similar to those used in the proofs of Theorem 3.3 and 3.8. We leave a few of the details to the student.

Proof of Theorem 3.17. We prove (i) assuming $\mathbb{A}_{\mathfrak{M}}$ is countable, leaving the extension (via Lévy's Absoluteness Principle) to the student. The ($\Leftarrow$) half of (i) is obvious, so let S be a relation on $\mathbb{A}_{\mathfrak{M}}$ which is Σ_1 on $\mathbb{HYP}(\mathbb{A}_{\mathfrak{M}})$. Every $x \in \mathbb{HYP}(\mathbb{A}_{\mathfrak{M}})$ has a good Σ_1 definition with parameters from $\mathrm{L}(a,\beta) \cup \{\mathrm{L}(a,\beta)\}$ by II.5.14. Since $\mathbb{A}_{\mathfrak{M}}$ is not admissible, β and hence $\mathrm{L}(a,\beta)$ also have Σ_1 definitions on $\mathbb{HYP}(\mathbb{A}_{\mathfrak{M}})$ with parameters from $\mathrm{L}(a,\beta)$ by the last step in the proof of II.5.14. Thus every $x \in \mathbb{HYP}(\mathbb{A}_{\mathfrak{M}})$ has a Σ_1 definition with parameters from $\mathrm{L}(a,\beta)$. But then S has a Σ_1 definition (as a subset, now, not an element) with parameters from $\mathrm{L}(a,\beta)$ since the other parameters can be defined away. Thus suppose that for all $x \in \mathrm{L}(a,\beta)$

$$S(x) \quad \text{iff} \quad \mathbb{HYP}(\mathbb{A}_{\mathfrak{M}}) \models \varphi(x,y)$$

where $y \in \mathrm{L}(a,\beta)$ and φ is Σ_1. By the Truncation Lemma $S(x)$ is equivalent to

(7) for all $\mathfrak{B}_{\mathfrak{M}} \supseteq_{\text{end}} \mathbb{A}_{\mathfrak{M}}$, if $\mathfrak{B}_{\mathfrak{M}} \models \mathrm{KPU}$ then $\mathfrak{B}_{\mathfrak{M}} \models \varphi(x,y)$.

Since β is a limit, $\mathrm{L}(a,\beta)$ is closed under pairs, union and Δ_0 Separation so we may code up $\mathsf{K} = \mathsf{L}^* \cup \{\overline{\mathsf{x}} \mid x \in \mathbb{A}_{\mathfrak{M}}\}$ on $\mathbb{A}_{\mathfrak{M}}$. Let $\mathsf{K}_{\mathbb{A}}$ be the (nonadmissible) fragment of $\mathsf{K}_{\infty\omega}$ given by $\mathbb{A}_{\mathfrak{M}}$. Let $T \subseteq \mathsf{K}_{\mathbb{A}}$ the the theory

$$\mathrm{KPU}$$

$$\mathrm{Diagram}(\mathbb{A}_{\mathfrak{M}})$$

$$\forall v\,[v \in \overline{\mathsf{a}} \leftrightarrow \textstyle\bigvee_{x \in a} v = \overline{\mathsf{x}}], \quad \text{for all} \quad a \in \mathbb{A}_{\mathfrak{M}},$$

$$\forall p\,[p \in \overline{\mathsf{M}}].$$

Every model of T is isomorphic to some $\mathfrak{B}_{\mathfrak{M}} \supseteq_{\text{end}} \mathbb{A}_{\mathfrak{M}}$ so (7) is equivalent to

$$T \models \varphi(\overline{\mathsf{x}},\overline{\mathsf{y}}).$$

By Theorem III.4.5 (really III.4.6) this is equivalent to saying that $\varphi(\mathsf{x},\mathsf{y})$ is in the smallest set of sentences of $\mathsf{K}_{\mathbb{A}}$ containing T and (A 1)—(A 7) which is closed under (R 1)—(R 3). This clearly amounts to a Σ_+ inductive definition Γ_φ such that

$$R(x) \quad \text{iff} \quad \varphi(\mathsf{x},\mathsf{y}) \in I_\Gamma .$$

Therefore R is Σ_+ inductive by Exercise 3.21.

To prove (ii) we need only find a Σ_+ inductive definition on $\mathbb{A}_{\mathfrak{M}}$ with length $o(\mathbb{H}\mathrm{YP}(\mathbb{A}_{\mathfrak{M}}))$. Let $R \subseteq \mathbb{A}_{\mathfrak{M}}$ be $\mathbb{H}\mathrm{YP}(\mathbb{A}_{\mathfrak{M}})$-r.e. but not an element of $\mathbb{H}\mathrm{YP}(\mathbb{A}_{\mathfrak{M}})$. There is such an R since $\mathbb{H}\mathrm{YP}(\mathbb{A}_{\mathfrak{M}})$ is projectible into $\mathbb{A}_{\mathfrak{M}}$ by V.5.4. Then R is a section of I_φ, where Γ_φ is some Σ_+ inductive definition. But now the argument used earlier, in the proof of 3.6 for example, shows that $\|\Gamma_\varphi\| = o(\mathbb{H}\mathrm{YP}(\mathbb{A}_{\mathfrak{M}}))$. □

3.19—3.22 Exercises

3.19. Let $C_{\mathfrak{M}}(\mathrm{KPU}^+) = I_\varphi$, where Γ_φ is extended inductive, by 3.13. Show that $O(\mathfrak{M}) = \|\Gamma_\varphi\|$. Thus, for example, $O(\mathfrak{M})$ is just the least ordinal not assigned to a proof using the $\mathfrak{M}$-rule, under the usual assignment of ordinals to proofs.

3.20. Let $\mathfrak{A}$ be a structure, let U be inductive on $\mathfrak{A}$ and let $f: A^n \to A$ be first order definable. Modify the proof of 3.15 to show that

$$P(\vec{x}) \quad \text{iff} \quad U(f(\vec{x}))$$

defines an inductive relation on $\mathfrak{A}$.

3.21. Let $\mathfrak{A}$ be a structure, let $U \subseteq \mathfrak{A}$ be Σ_+ inductive on $\mathfrak{A}$, let $f: A^n \to A$ have a Σ_1 graph and define P by

$$P(x_1, \ldots, x_n) \quad \text{iff} \quad f(x_1, \ldots, x_n) \in U .$$

Show that P is Σ_+ inductive on $\mathfrak{A}$. [Mimic the proof of 3.15.]

3.22. Give the absoluteness argument for lifting Theorem 3.17 from the countable to the uncountable.

3.23 Notes. The main results of this section are from Barwise-Gandy-Moschovakis [1971], at least in the case of pure admissible sets. Theorem 3.17 and its corollaries are new here.

4. Coding $\mathbb{H}\mathrm{F}_{\mathfrak{M}}$ *on* $\mathfrak{M}$

A *pairing function* on a set M is simply a one-one function mapping $M \times M$ into M. An n-ary function f on a structure $\mathfrak{M}$ is inductive (or hyperelementary) if its graph is an $(n+1)$-ary inductive (or hyperelementary, respectively) relation on $\mathfrak{M}$. In this section we show how to code $\mathbb{H}\mathrm{F}_{\mathfrak{M}}$ on $\mathfrak{M}$ using an inductive pairing function on $\mathfrak{M}$. Our goal is to prove the following theorem.

4.1 Theorem. *Let $\mathfrak{M}=\langle M,R_1,\ldots,R_l\rangle$ be a structure with an inductive pairing function. The inductive and inductive* relations on $\mathfrak{M}$ coincide.*

We give the applications of this theorem (and a couple of related results obtained along the way) in the next section by showing how a great many results on inductive relations on $\mathfrak{M}$ can be obtained in a simple fashion by projecting the recursion theory of $\mathbb{H}\mathrm{YP}_{\mathfrak{M}}$. In so doing, we tie up the theory of admissible sets with the theory of inductive relations as developed in Moschovakis [1974]. Since our aim in these sections is to relate our theory to Moschovakis' theory, we feel only mildly apologetic for using without proof two results (4.2 and 4.3 below) from Chapter 1 of Moschovakis [1974]. The proofs are sketched in Exercises 4.17 and 4.18.

A relation P on $\mathfrak{M}$ is defined from Q by *hyperelementary substitution* if there are hyperelementary functions $f_1,\ldots,f_k$ so that

$$P(x_1,\ldots,x_n) \quad \text{iff} \quad Q(f_1(x_1,\ldots,x_n),\ldots,f_k(x_1,\ldots,x_n))$$

for all $x_1,\ldots,x_n \in \mathfrak{M}$.

4.2 Theorem. *The inductive relations on $\mathfrak{M}$ contain all first order relations and are closed under $\wedge$, $\vee$, $\exists$, $\forall$ and hyperelementary substitution. Hence, the hyperelementary relations on $\mathfrak{M}$ contain all first order relations on $\mathfrak{M}$ and are closed under $\neg$, $\wedge$, $\vee$, $\exists$, $\forall$ and hyperelementary substitution.*

Proof. This result follows easily from 4.3. See Theorem 1D.1 of Moschovakis [1974] or Exercise 4.18. □

The inductive relations on $\mathfrak{M}$ are closed under induction in a sense made precise by 4.3.

4.3 Theorem. *Let $S_1,\ldots,S_k$ be relations on $\mathfrak{M}$ and consider an inductive definition Γ_φ over the expanded structure $(\mathfrak{M},S_1,\ldots,S_k)$, where φ is of the form $\varphi(x_1,\ldots,x_n,\mathsf{R}_+,\mathsf{S}_1,\ldots,\mathsf{S}_k)$ in $\mathsf{L}\cup\{\mathsf{R},\mathsf{S}_1,\ldots,\mathsf{S}_k\}$.*

(i) *If $S_1,\ldots,S_k$ are hyperelementary on $\mathfrak{M}$ then the fixed point I_φ defined on $(\mathfrak{M},S_1,\ldots,S_k)$ is inductive on the original structure $\mathfrak{M}$.*

(ii) *If $S_1,\ldots,S_k$ are inductive on $\mathfrak{M}$ then the conclusion of* (i) *still holds provided φ is S_i-positive for $i=1,\ldots,k$.*

(iii) *In either case* (i) *or* (ii), *I_φ is a section of a fixed point I_ψ for some $\psi(x_1,\ldots,x_m,\mathsf{R}_+)\in\mathsf{L}\cup\{\mathsf{R}\}$ with $\|\Gamma_\psi\|\geqslant\|\Gamma_\varphi\|$.*

Proof. See Theorem 1C.3 of Moschovakis [1974] or Exercise 4.17. □

There is one simple consequence of 4.2 that deserves mention. If f is an inductive function on $\mathfrak{M}$ and if its domain D is hyperelementary (e. g., if f is total) then f is hyperelementary, since

$$f(x_1,\ldots,x_n)\neq y \quad \text{iff} \quad (x_1,\ldots,x_n\notin D \vee \exists z\,[f(x_1,\ldots,x_n)=z\wedge z\neq y].$$

Thus, if $\mathfrak{M}$ has an inductive pairing function p, p is actually hyperelementary since p is total.

The plan for the proof of Theorem 4.1 is simple. Fix an inductive pairing function p on $\mathfrak{M}$. We are going to use p to assign notations to the elements of $\mathbb{HF}_M$. The set T of notations will be inductive on $\mathfrak{M}$ but not, in general, hyperelementary. An extended first order formula of the form

$$\exists a \in \mathbb{HF}_{\mathfrak{M}}(\ldots)$$

will translate into

$$\exists x\,(x \in T \wedge \cdots)$$

which will keep us within the class of inductive relations since the inductive set T occurs positively. On the other hand, a quantifier of the form

$$\forall a \in \mathbb{HF}_{\mathfrak{M}}(\ldots)$$

would translate into

$$\forall x\,(x \notin T \vee \cdots)$$

which is not permitted since T occurs negatively. The only complications in the proof are caused by the following two facts. Since $\{p,q\}=\{q,p\}$ we are not going to be able to have unique notations for the elements of $\mathbb{HF}_{\mathfrak{M}}$. Secondly, we must find some way to handle bounded universal quantifiers in a positive way. (This accounts for the relation $\check{\mathscr{E}}$ used below and most of the other complications.)

The notation system used is based upon the fact that $\mathbb{HF}_{\mathfrak{M}}$ is the closure of $M \cup \{0\}$ under the operation

$$S(x,y) = x \cup \{y\}.$$

Define a hierarchy $\mathrm{HF}_{\mathfrak{M}}^{(n)}$ as follows:

$$\mathrm{HF}_{\mathfrak{M}}^{(0)} = \{0\},$$

$$\mathrm{HF}_{\mathfrak{M}}^{(n+1)} = \mathrm{HF}_{\mathfrak{M}}^{(n)} \cup \{a \cup \{x\} : a, x \in \mathrm{HF}_{\mathfrak{M}}^{(n)} \cup M\}.$$

This hierarchy grows more slowly than the $\mathrm{HF}_{\mathfrak{M}}(n)$ hierarchy used in § II.2 but it eventually gets the job done.

4.4 Lemma. $\mathbb{HF}_{\mathfrak{M}} = \bigcup_{n<\omega} \mathrm{HF}_{\mathfrak{M}}^{(n)}$.

Proof. Suppose there were some set $a \in \mathbb{HF}_{\mathfrak{M}}$ which did not appear at any stage of our new hierarchy. Among such sets a choose one of least rank and, among those of least rank, choose one of smallest cardinality. Since $0 \in \mathrm{HF}_{\mathfrak{M}}^{(0)}$, a is non-empty so we may write

$$a = \{x_1, \ldots, x_{k+1}\}.$$

Let $a_0=\{x_1,\dots,x_k\}$. Since $\mathrm{rk}(a_0)\leqslant\mathrm{rk}(a)$ and $\mathrm{card}(a_0)<\mathrm{card}(a)$, a_0 is formed in our new hierarchy, by choice of a. Since $\mathrm{rk}(x_{k+1})<\mathrm{rk}(a)$, x_{k+1} is also formed. Pick n so that both a_0 and x_{k+1} are in $\mathrm{HF}_{\mathfrak{M}}^{(n)}$. Then $a=S(a_0,x_{k+1})$ is in $\mathrm{HF}_{\mathfrak{M}}^{(n+1)}$. □

Let M be an infinite set with pairing function $p\colon M\times M\to M$. Let x_0, x_1, x_2 be distinct elements of M. We use the following notational conventions.

$$\emptyset \quad \text{for} \quad p(x_0,x_0),$$

$$\dot{x} \quad \text{for} \quad p(x_1,x),$$

$$x\,\dot{\delta}\,y \quad \text{for} \quad p(x_2,p(x,y)).$$

4.5 Lemma. *The functions f_1, f_2 defined below are one-one, they have disjoint ranges and $\emptyset$ is in the range of neither. They are $\mathbb{H}\mathrm{F}_{(\mathfrak{M},p)}$-recursive and hyperelementary on $(\mathfrak{M},p)$:*

$$f_1(x)=\dot{x} \qquad f_2(x,y)=x\,\dot{\delta}\,y.$$

Proof. This is immediate since p is one-one and x_0, x_1, x_2 are distinct. □

We use these functions to define two sets of closed terms: the ur-terms denote elements of M; the set-terms denote hereditarily finite sets over M.

4.6 Definition. (i) For each $x\in M$, $\dot{x}$ is an ur-term and $\dot{x}$ denotes x, written

$$|\dot{x}|=x.$$

The set of ur-terms is called T_u.

(ii) The set T_s of set-terms and the function $|\cdot|$ mapping T_s onto $\mathbb{H}\mathrm{F}_M$ are defined inductively:

a) $\emptyset$ is in T_s and $\emptyset$ is a notation for 0, i.e.,

$$|\emptyset|=0.$$

b) If x is in T_s and y is in $T_u\cup T_s$ and if $|y|\notin|x|$ then $x\,\dot{\delta}\,y$ is in T_s and

$$|x\,\dot{\delta}\,y|=|x|\cup\{|y|\}.$$

(iii) The set T of all notations is $T_u\cup T_s$.

We require $|y|\notin|x|$ to keep the set of notations of each $a\in\mathbb{H}\mathrm{F}_M$ finite.

The definition of T_s is an inductive definition, not over $(\mathfrak{M},p)$ but rather over $\mathbb{H}\mathrm{F}_{(\mathfrak{M},p)}$. One of our tasks is to show that T_s is actually inductive over $(\mathfrak{M},p)$ after all.

Note that by Lemma 4.4, every $a \in \mathbb{HF}_M$ is $|x|$ for some $x \in T_s$. Define the following relations on M:

$$x \mathscr{E} y \quad \text{iff} \quad x, y \in T \text{ and } |x| \in |y|;$$

$$x \breve{\mathscr{E}} y \quad \text{iff} \quad y \in T \text{ and if } x \in T \text{ then } |x| \notin |y|;$$

$$x \approx y \quad \text{iff} \quad x, y \in T \text{ and } |x| = |y|;$$

$$x \breve{\approx} y \quad \text{iff} \quad y \in T \text{ and if } x \in T \text{ then } |x| \neq |y|.$$

4.7 Main Lemma. *The sets T_s, T and the relations $\mathscr{E}$, $\breve{\mathscr{E}}$, $\approx$, and $\breve{\approx}$ are all inductive on (M,p). The set T_u is definable on (M,p).*

Proof. It is clear that T_u is definable on (M,p) since

$$y \in T_u \quad \text{iff} \quad \exists x\,(y = \dot{x}).$$

We will give an informal simultaneous inductive definition of the six other relations as well as two auxillary relations R and $\breve{R}$. First, however, let N be the smallest subset of M containing $\emptyset$ and closed under

$$\text{if} \quad x \in N \quad \text{then} \quad (x \,\delta\, x) \in N.$$

Thus N is inductive on (M,p) and N contains a unique notation for each natural number. We will confuse a natural number with its notation in this proof. Define

$$R(n,x) \quad \text{iff} \quad n \in N \text{ and } x \in T_s \text{ and } |x| \in \mathrm{HF}_M^{(n)};$$

$$\breve{R}(n,x) \quad \text{iff} \quad n \in N \text{ and if } x \in T_s \text{ then } |x| \notin \mathrm{HF}_M^{(n)}.$$

The following clauses constitute a simultaneous inductive definition of all the above relations. It should be pretty obvious to the reader how one could turn this into one giant inductive definition over (M,p) and then extract the given relations as sections. (If he needs help, the student can consult the Simultaneous Induction Lemma on p. 12 of Moschovakis [1974].)

(1) $x \in T_s$ iff $x = \emptyset$ or there is a $y \in T_s$ and a $z \in T_u \cup T_s$ such that $z \breve{\mathscr{E}} y$ and x is $y \,\delta\, z$.

(2) $x \in T$ iff $x \in T_u$ or $x \in T_s$.

(3) $x \mathscr{E} y$ iff $y \in T_s$ and y is of the form $u \,\delta\, v$ and $x \mathscr{E} u$ or $x \approx v$.

(4) $x \breve{\mathscr{E}} y$ iff $y \in T$ and y is $\emptyset$ or $y \in T_u$ or y is of the form $u \,\delta\, v$ and $x \breve{\mathscr{E}} u$ and $x \breve{\approx} v$.

(5) $x \approx y$ iff $x, y \in T$ and $x = y$ or $x, y \in T_s$ and for every z $(z \breve{\mathscr{E}} x \vee z \mathscr{E} y)$ and $(z \breve{\mathscr{E}} y \vee z \mathscr{E} x)$.

(6) $R(0,x)$ iff $x = \emptyset$;

$R(n+1,x)$ iff $x\in T_s$ and $R(n,x)$ or else x is of the form $y\,\check{\sigma}\,z$ where $R(n,y)$ and $(z\in T_u \vee R(n,z))$.

(7) $\check{R}(0,x)$ iff $x\neq\emptyset$;

$\check{R}(n+1,x)$ iff $\check{R}(n,x)$ and either x is not of the form $u\,\check{\sigma}\,v$ (for all u, v) or else x is of the form $u\,\check{\sigma}\,v$ but one of the following holds:

$$v\mathscr{E}u,\qquad \check{R}(n,u),\qquad \check{R}(n,v).$$

(8) $x\check{\approx}y$ iff there is an $n\in N$ such that $\check{R}(n,x)$ but $R(n,y)$ or there is an $n\in N$ such that $R(n,x)$ and $R(n,y)$ (in which case x is in T_s) and there is a z such that

$$((z\mathscr{E}x\wedge z\check{\mathscr{E}}y)\vee(z\mathscr{E}y\wedge z\check{\mathscr{E}}x)).$$

It takes a bit of checking to see that in each case the induction is pushed back, but this checking is best done on scratch paper. □

The relations $R, \check{R}$ used above are needed only to prove the Main Lemma. They should not be confused with other relations R used later on.

We are now ready to fill in the outline of the proof of Theorem 4.1. For simplicity of notation let us suppose our language L has only one binary symbol Q. Let R be a new relation symbol for use in inductive definitions. We consider $\mathsf{L}^*(\mathsf{R})=\mathsf{L}(\in,\mathsf{R})$ as a single sorted language with unary symbols U (for urelements) and S (for sets) with bounded quantification as a primitive. We let K be a new language with atomic symbols

$$\mathsf{Q},\ \mathsf{U},\ \mathsf{S},\ \mathsf{R},\ \mathscr{E},\ \check{\mathscr{E}},\ \approx,\ \check{\approx}.$$

We define a mapping $\hat{}$ from $\mathsf{L}^*(\mathsf{R})$ into K as follows: given $\varphi\in\mathsf{L}^*(\mathsf{R})$, first push the negations inside as far as possible so that the only negative subformulas in φ are negated atomic. Replace each positive occurrence of $x\in y$ by $x\mathscr{E}y$, each occurrence of $\neg(x\in y)$ by $x\check{\mathscr{E}}y$, each positive occurrence of $x=y$ by $x\approx y$, each occurrence of $\neg(x=y)$ by $x\check{\approx}y$, each bounded quantifer

$$\forall x\in y(\ldots)\quad\text{by}\quad \forall x\,(x\check{\mathscr{E}}y\vee\ldots),$$
$$\exists x\in y(\ldots)\quad\text{by}\quad \exists x\,(x\mathscr{E}y\wedge\ldots).$$

Thus, in $\hat{\varphi}$, all occurrences of $\mathscr{E}, \check{\mathscr{E}}, \approx, \check{\approx}$ are positive. If φ is extended first order then S also occurs positively in $\hat{\varphi}$ since it only appears in the contexts

$$\exists x\,(\mathsf{S}(x)\wedge\ldots)$$

and

$$\exists x\,((\mathsf{U}(x)\vee\mathsf{S}(x))\wedge\ldots).$$

Let M be the infinite set with pairing function p used above. Let Q be any binary relation on M. Define $\tilde{Q}$ on T_u by

$$\tilde{Q}(\dot{p},\dot{q})\quad\text{iff}\quad Q(p,q)$$

for all $p,q\in M$ so that map $t\to|t|$ gives an isomorphism of $(T_u,\tilde{Q})$ onto $\mathfrak{M}=(M,Q)$. We let $\tilde{\mathfrak{M}}$ be the structure for K with universe M and with interpretations given by

symbol:	U	S	Q	$\mathscr{E}$	$\breve{\mathscr{E}}$	$\approx$	$\breve{\approx}$
interpretation:	T_u	T_s	$\tilde{Q}$	$\mathscr{E}$	$\breve{\mathscr{E}}$	$\approx$	$\breve{\approx}$

Thus U, Q are interpreted by (hyper)elementary relations; the other symbols (which will occur positively in $\hat{\varphi}$ whenever φ is extended first order) are interpreted by inductive relations so things are set up to apply Theorem 4.3 (i), (ii).

Given an n-ary relation R on $\mathbb{HF}_{\mathfrak{M}}$ we define $\tilde{R}$ on T by

$$\tilde{R}(t_1,\dots,t_n) \quad \text{iff} \quad R(|t_1|,\dots,|t_n|), \quad \text{for} \quad t_1,\dots,t_n\in T.$$

4.8 Lemma. *For any formula* $\varphi(v_1,\dots,v_k,\mathsf{R})\in\mathsf{L}^*(\mathsf{R})$, *any relation* R *on* $\mathbb{HF}_{\mathfrak{M}}$, *and any* $t_1,\dots,t_k\in T$ *we have*

$$(\mathbb{HF}_{\mathfrak{M}},R)\models\varphi[|t_1|,\dots,|t_k|] \quad \textit{iff} \quad (\tilde{\mathfrak{M}},\tilde{R})\models\hat{\varphi}[t_1,\dots,t_k].$$

Proof. By induction on formulas $\varphi\in\mathsf{L}^*(\mathsf{R})$. For atomic and negated atomic formulas, it follows by the definitions. The induction step is immediate since every $x\in\mathbb{HF}_{\mathfrak{M}}$ is denoted by some term t. □

4.9 Lemma. *Let* $\varphi(x_1,\dots,x_n,\mathsf{R}_+)\in\mathsf{L}^*(\mathsf{R})$. *For each* α *and each* $t_1,\dots,t_n\in T$ *we have*

$$(|t_1|,\dots,|t_n|)\in I_\varphi^\alpha \quad \textit{iff} \quad (t_1,\dots,t_n)\in I_{\hat{\varphi}}^\alpha,$$

where the induction on the left is over $\mathbb{HF}_{\mathfrak{M}}$, *that on the right over* $\tilde{\mathfrak{M}}$.

Proof. By induction, of course. The induction hypothesis asserts that

$$(|t_1|,\dots,|t_n|)\in I_\varphi^{<\alpha} \quad \text{iff} \quad (t_1,\dots,t_n)\in I_{\hat{\varphi}}^{<\alpha},$$

i.e., that $\widetilde{(I_\varphi^{<\alpha})}=I_{\hat{\varphi}}^{<\alpha}$. But then

$$\begin{aligned}(|t_1|,\dots,|t_n|)\in I_\varphi^\alpha \quad &\text{iff} \quad (\mathbb{HF}_{\mathfrak{M}},I_\varphi^{<\alpha})\models\varphi(|t_1|,\dots,|t_n|,\mathsf{R}_+)\\ &\text{iff} \quad (\tilde{\mathfrak{M}},I_{\hat{\varphi}}^{<\alpha})\models\hat{\varphi}(t_1,\dots,t_n,\mathsf{R}_+) \quad \text{(by 4.8)}\\ &\text{iff} \quad (t_1,\dots,t_n)\in I_{\hat{\varphi}}^\alpha. \quad \square\end{aligned}$$

We are now ready to prove Theorem 4.1. The following result comes out of the proof.

4.10 Corollary. *Let* $\mathfrak{M}$ *be a structure for* L *with an inductive pairing function. If* Γ_φ *is an extended first order inductive definition over* $\mathfrak{M}$ *then there is a first order inductive definition* Γ_ψ *over* $\mathfrak{M}$ *with* $\|\Gamma_\psi\|\geq\|\Gamma_\varphi\|$.

Proof of Theorem 4.1 and Corollary 4.10. Let $\mathfrak{M}=\langle M, Q\rangle$ be an L-structure and let p be an inductive, hence hyperelementary, pairing function on $\mathfrak{M}$. By 4.2 (i), $\mathfrak{M}$ and the expanded structure $(\mathfrak{M}, p)$ have exactly the same inductive and hyperelementary relations. Thus T_u, $\tilde{Q}$ are hyperelementary on $\mathfrak{M}$, and T_s, $\mathscr{E}$, $\check{\mathscr{E}}$, $\approx$, and $\check{\approx}$ are inductive on $\mathfrak{M}$. Let $S \subseteq M^n$ be inductive*. Choose an extended first order inductive definition Γ_φ and parameters $y_1, \ldots, y_k \in M \cup \mathbb{HF}_{\mathfrak{M}}$ such that

$$S(x_1, \ldots, x_n) \quad \text{iff} \quad (x_1, \ldots, x_n, \vec{y}) \in I_\varphi .$$

Now consider the inductive definition $\Gamma_{\hat{\varphi}}$ over $\tilde{\mathfrak{M}}$. By the above lemma $\|\Gamma_\varphi\| = \|\Gamma_{\hat{\varphi}}\|$ and, for any $t_1, \ldots, t_{n+k} \in T$,

$$(t_1, \ldots, t_{n+k}) \in I_{\hat{\varphi}} \quad \text{iff} \quad (|t_1|, \ldots, |t_{n+k}|) \in I_\varphi .$$

By Theorem 4.3 (ii) and the remarks above about the relations T_s, $\mathscr{E}$, $\check{\mathscr{E}}$, $\approx$ and $\check{\approx}$ all occuring positively in $\hat{\varphi}$, $I_{\hat{\varphi}}$ is inductive over the original $\mathfrak{M}$. Choose $t_1, \ldots, t_k$ with $|t_1| = y_1, \ldots, |t_k| = y_k$. Then, for all $x_1, \ldots, x_n \in M$,

$$S(x_1, \ldots, x_n) \quad \text{iff} \quad (\dot{x}_1, \ldots, \dot{x}_n, t_1, \ldots, t_k) \in I_{\hat{\varphi}}$$

so S is obtained from the inductive set $I_{\hat{\varphi}}$ by hyperelementary substitution and, hence, is inductive. By 4.3 (iii) there is an inductive definition Γ_ψ over $\mathfrak{M}$ with $\|\Gamma_\psi\| \geqslant \|\Gamma_{\hat{\varphi}}\| = \|\Gamma_\varphi\|$, so this also proves the corollary. □

The notation system we have been using can be seen to be a notation system in the precise sense of § V.5. This follows from the next lemma. We assume the notation from above.

4.11 Lemma. *Define a function π on $\mathbb{HF}_{(M,p)}$ by*

$$\pi(x) = \{y \in T \mid |y| = x\}.$$

Then π is a total $\mathbb{HF}_{(\mathfrak{M},p)}$-recursive function.

Proof. Given a set a of cardinality $\geqslant 1$, we call a pair (a_0, x) a *splitting* of a if $a = a_0 \cup \{x\}$ but $x \notin a_0$. Let

$$\mathrm{Spl}(a) = \{(a_0, x) \mid (a_0, x) \text{ is a splitting of } a\}$$

for all $a \in \mathbb{HF}_M$. It is a simple matter to check that Spl is $\mathbb{HF}_{\mathfrak{M}}$-recursive. We first define π more explicitly and then discuss the method used to see that the definition is $\mathbb{HF}_{(\mathfrak{M},p)}$-recursive. The definition of π parallels the proof of 4.4.

$$\pi(p) = \{\dot{p}\} \quad \text{for} \quad p \in M,$$

$$\pi(0) = \{\emptyset\}.$$

For nonempty sets a, $\pi(a)$ is defined by a double induction, first on $\mathrm{rk}(a)$ and, among sets of the same rank, on $\mathrm{card}(a)$. So suppose $\pi(x)$ is defined for all $x \in a$ and all $x \subseteq a$ with $\mathrm{card}(x) < \mathrm{card}(a)$. If $a = \{x_1, \ldots, x_n\}$ with $n \geqslant 1$ then we look at any splitting (a_0, x) of a. Now $\pi(a_0)$, $\pi(x)$ are defined and, for $t_0 \in \pi(a_0)$, $|t_0| = a_0$ and for $t_1 \in \pi(x)$, $|t_1| = x$ so $|t_0 \,\delta\, t_1| = a_0 \cup \{x\} = a$. Thus we may define

$$\pi(a) = \{t_0 \,\delta\, t_1 : \text{for some } (a_0, x) \in \mathrm{Spl}(a),\ t_0 \in \pi(a_0) \text{ and } t_1 \in \pi(x)\}.$$

With this definition π is clearly $\mathbb{HF}_{(\mathfrak{M}, p)}$-recursive by the Second Recursion Theorem. $\square$

4.12 Theorem. *Let $\mathfrak{M} = \langle M, R_1, \ldots, R_l \rangle$ be a structure for L.*

(i) *If $\mathfrak{M}$ has an $\mathbb{HF}_{\mathfrak{M}}$-recursive pairing function then $\mathbb{HF}_{\mathfrak{M}}$ is projectible into $\mathfrak{M}$.*

(ii) *If $\mathfrak{M}$ has a $\mathbb{HYP}_{\mathfrak{M}}$-recursive pairing function then $\mathbb{HYP}_{\mathfrak{M}}$ is projectible into $\mathfrak{M}$.*

Proof. (i) The sets in $\mathbb{HF}_{\mathfrak{M}}$ depend only on M, not on the whole structure $\mathfrak{M}$, so if we add a pairing function p to $\mathfrak{M}$, $\mathbb{HF}_{(\mathfrak{M}, p)}$ has the same sets as $\mathbb{HF}_{\mathfrak{M}}$. By Lemma 4.11, $\mathbb{HF}_{(\mathfrak{M}, p)}$ is projectible into $\mathfrak{M}$; i.e., there is an $\mathbb{HF}_{(\mathfrak{M}, p)}$ recursive notation system π with $D_\pi \subseteq M$. But then, if p is $\mathbb{HF}_{\mathfrak{M}}$-recursive, π is also $\mathbb{HF}_M$-recursive. The proof of (ii) is similar. Let p be a $\mathbb{HYP}_{\mathfrak{M}}$-recursive pairing function so that $\mathbb{HYP}_{\mathfrak{M}}$ and $\mathbb{HYP}_{(\mathfrak{M}, p)}$ have the same universe of sets. By V.5.3 we have a notation system π_0 for $\mathbb{HYP}_{\mathfrak{M}}$ with $D_\pi \subseteq \mathbb{HF}_{\mathfrak{M}}$. By 4.11, there is a $\mathbb{HYP}_{(M, p)}$-recursive map π_1 on $\mathbb{HF}_{\mathfrak{M}}$ with $\pi_1(x) \subseteq M$, $\pi_1(x) \cap \pi_1(y) = 0$ for $x \neq y$. Let π be defined by

$$\pi(x) = \bigcup \{\pi_1(y) \mid y \in \pi_0(x)\}.$$

Then π is a notation system for $\mathbb{HYP}_{\mathfrak{M}}$ with $D_\pi \subseteq M$. $\square$

The following special case of 4.12 (ii) will be of great use to us in the next section.

4.13 Corollary. *Let $\mathfrak{M} = \langle M, R_1, \ldots, R_l \rangle$ be a structure for L with an inductive pairing function. Then $\mathbb{HYP}_{\mathfrak{M}}$ is projectible into $\mathfrak{M}$.*

Proof. If p is an inductive pairing function on $\mathfrak{M}$ then it is hyperelementary and hence an element of $\mathbb{HYP}_{\mathfrak{M}}$. Thus 4.12 (ii) applies. $\square$

4.14—4.18 Exercises

4.14. Let $\mathfrak{M} = \langle M, \sim \rangle$ where $\sim$ is an equivalence relation on M which exactly one equivalence class of each finite cardinality. Define

$$x < y \quad \text{iff} \quad \mathrm{card}(x/\sim) < \mathrm{card}(y/\sim).$$

(i) Prove that $<$ is Σ_1 on $\mathbb{H}\mathrm{F}_{\mathfrak{M}}$ and hence is extended inductive on $\mathfrak{M}$.
(ii) (Kunen). Prove that $<$ is not inductive on $\mathfrak{M}$.
(iii) Prove that $o(\mathbb{H}\mathrm{YP}_{\mathfrak{M}})>\omega$.

4.15. This exercise introduces the Moschovakis [1974] notions of *acceptable* and *almost acceptable* structures. A *coding scheme* $\mathscr{C}$ for a structure $\mathfrak{M}$ consists of:
(a) a subset $N^{\mathscr{C}}$ of M and a linear ordering $<^{\mathscr{C}}$ of $N^{\mathscr{C}}$ such that

$$\langle N^{\mathscr{C}}, <^{\mathscr{C}}\rangle \cong \langle \omega, <\rangle, \quad \text{and}$$

(b) an injection $\langle\ \rangle^{\mathscr{C}}$ of the set of all finite sequences from M into M.
Given a fixed coding scheme $\mathscr{C}$ we use $\dot{0}, \dot{1}, \dot{2}, \ldots$ to indicate the appropriate members of $N^{\mathscr{C}}$ as ordered by $<^{\mathscr{C}}$. Associated with a coding scheme $\mathscr{C}$ there are some natural relations and functions.

$$Seq^{\mathscr{C}}(x) \text{ iff } x=\langle\ \rangle^{\mathscr{C}} \text{ or } x=\langle x_1,\ldots,x_n\rangle^{\mathscr{C}} \text{ for some } n \text{ and some } x_1,\ldots,x_n.$$

$$\begin{aligned} lh^{\mathscr{C}}(x) &= \dot{0} \quad \text{if} \quad \neg Seq^{\mathscr{C}}(x) \\ &= \dot{n} \quad \text{if} \quad Seq^{\mathscr{C}}(x) \text{ and } x=\langle x_1,\ldots,x_n\rangle^{\mathscr{C}}. \end{aligned}$$

$$\begin{aligned} q^{\mathscr{C}}(x,\dot{m}) &= x_m \text{ if for some } x_1,\ldots,x_n,\ x=\langle x_1,\ldots,x_n\rangle^{\mathscr{C}} \text{ and } 1\leqslant m\leqslant n \\ &= \dot{0} \text{ otherwise.} \end{aligned}$$

A structure $\mathfrak{M}$ is *almost acceptable* (or *acceptable*) if M has a coding scheme $\mathscr{C}$ with all of $N^{\mathscr{C}}$, $<^{\mathscr{C}}$, $Seq^{\mathscr{C}}$, $lh^{\mathscr{C}}$, $q^{\mathscr{C}}$ hyperelementary (or first order, resp.).
(i) Show that every almost acceptable structure has an inductive pairing function.
(ii) Let $\mathfrak{M}$ be a structure with an inductive pairing function. Show that M is almost acceptable iff M is not recursively saturated. [It is easy to see that if $\mathfrak{M}$ is almost acceptable then $o(\mathbb{H}\mathrm{YP}_{\mathfrak{M}})>\omega$. To prove the converse use Corollary 4.10.]

4.16. Show that all models of Peano arithmetic, KPU and ZF have definable pairing functions, even the recursively saturated ones.

4.17. Let $\mathfrak{M}=\langle M, R_1,\ldots,R_l\rangle$ be an infinite structure and let Γ_ψ be an inductive definition over $\mathfrak{M}$, say $\psi=\psi(u_1,\ldots,u_4,\mathsf{S}_+)$. Now let $\mathfrak{M}'=(\mathfrak{M},S)$ where S is defined by:

$$S(x_1,x_2) \quad \text{iff} \quad (x_1,x_2,a_1,a_2)\in I_\psi.$$

Let $\varphi(v_1,\ldots,v_3,\mathsf{S}_+,\mathsf{T}_+)\in \mathsf{L}\cup\{\mathsf{S},\mathsf{T}\}$, where S is binary (to denote S) and T is 3-ary (to be used in an induction) and let Γ_φ be the natural inductive definition over $(\mathfrak{M},S)$ given by φ. We are going to outline the proof from Moschovakis [1974] that I_φ is inductive over the original structure $\mathfrak{M}$, thus proving Theorem 4.3.

Let $0, 1, \bar{u}_1, \ldots, \bar{u}_4, \bar{v}_1, \ldots, \bar{v}_3$ be constants from M with $0 \neq 1$. Let Q be a new 8-ary $(8 = 1+4+3)$ relation symbol and define $\theta(i, u_1, \ldots, u_4, v_1, \ldots, v_3, \mathsf{Q}_+)$ by

$$[i=0 \wedge \psi(u_1,\ldots,u_4, \mathsf{Q}(0, \cdot,\cdot,\cdot,\cdot,\bar{v}_1,\bar{v}_2,\bar{v}_3)/\mathsf{R}] \vee$$
$$[i=1 \wedge \varphi(v_1, v_2, v_3, \mathsf{Q}(0, \cdot,\cdot, a_1, a_2, \bar{v}_1, \bar{v}_2, \bar{v}_3)/\mathsf{S}, \mathsf{Q}(1, \bar{u}_1, \bar{u}_2, \bar{u}_3, \bar{u}_4, \cdot,\cdot,\cdot)/\mathsf{T}].$$

Consider the induction definition Γ_θ over $\mathfrak{M}$.

(i) Prove that for each α,

$$(u_1,\ldots,u_4) \in I_\psi^\alpha \quad \text{iff} \quad (0, u_1, \ldots, u_4, \bar{v}_1, \ldots, \bar{v}_3) \in I_\theta^\alpha$$

and hence

$$(u_1,\ldots,u_4) \in I_\psi \quad \text{iff} \quad (0, u_1, \ldots, u_4, \bar{v}_1, \ldots, \bar{v}_3) \in I_\theta .$$

(ii) Prove that if $(1, \bar{u}_1, \ldots, \bar{u}_4, v_1, \ldots, v_3) \in I_\theta^\alpha$ then $(v_1, \ldots, v_3) \in I_\varphi^\alpha$.

(iii) Prove that if $(v_1, \ldots, v_3) \in I_\varphi^\alpha$ then for some β, $(1, \bar{u}_1, \ldots, \bar{u}_4, v_1, \ldots, v_3 \in I_\theta^\beta$, by induction on α, using (i).

(iv) Use (ii), (iii) to conclude that I_φ is a section of I_θ and hence is inductive on $\mathfrak{M}$.

(v) Show that $\|\Gamma_\theta\| \geqslant \|\Gamma_\varphi\|$.

(vi) Prove Theorem 4.3.

4.18 Use Theorem 4.3 to prove Theorem 4.2 [For example, show that if S_1, S_2 are inductive on $\mathfrak{M}$ then $S_1 \cup S_2$ is inductive on $(\mathfrak{M}, S_1, S_2)$ with an inductive definition in which S_1, S_2 occur positively.]

4.19 Notes. The fact that an inductive pairing function suffices for coding $\mathbb{HF}_{\mathfrak{M}}$ on $\mathfrak{M}$ goes back, indirectly, to Aczel [1970]. The proof of Theorem 4.1 given above owes much to ideas of Aczel and Nyberg.

5. Inductive Relations on Structures with Pairing

Inductive and coinductive definitions appear in most branches of mathematics. Spector [1961] was the first to focus attention on them as objects worthy of study in their own right, but then only over the structure $\mathcal{N}$ of the natural numbers. The development over an absolutely arbitrary structure $\mathfrak{M}$ was not carried out until Moschovakis [1974] produced his attractive and coherent picture. Our object in this section is to view portions of Moschovakis' picture as projections of $\mathbb{HYP}_{\mathfrak{M}}$.

Let us summarize the results at our disposal.

5.1 Theorem. *Let $\mathfrak{M} = \langle M, R_1, \ldots, R_l \rangle$ be a structure with an inductively definable pairing function. Let S be a relation on $\mathfrak{M}$.*

(i) *S is inductive on $\mathfrak{M}$ iff S is Σ_1 on* $\mathbb{HYP}_{\mathfrak{M}}$.
(ii) *S is hyperelementary on $\mathfrak{M}$ iff* $S \in \mathbb{HYP}_{\mathfrak{M}}$.
(iii) *$O(\mathfrak{M})$ is equal to*

$$\sup\{\|\Gamma_\varphi\| \mid \Gamma_\varphi \textit{ is first order positive inductive on } \mathfrak{M}\}$$

and this sup *is attained.*
(iv) $\mathbb{HYP}_{\mathfrak{M}}$ *is projectible into* $\mathfrak{M}$.

Proof. Part (i) follows from Theorems 3.8 and 4.1; (ii) follows from (i). Part (iii) follows from Theorem 3.11 and Corollary 4.10. Part (iv) is Theorem 4.12 (ii). ∎

We want to use this theorem to obtain some of the results in Moschovakis [1974]. In order to facilitate comparison we use the same names for theorems as in Moschovakis, even when our theorem is a little more or a little less general.

5.2 Corollary (The Abstract Kleene Theorem). *If $\mathfrak{M}=\langle M, R_1, \ldots, R_l\rangle$ is a countable structure with an inductively definable pairing function then the Π_1^1 relations coincide with the inductive relations on $\mathfrak{M}$.*

Proof. Both classes of relations coincide with the class of relations on $\mathfrak{M}$ which are Σ_1 on $\mathbb{HYP}_{\mathfrak{M}}$ by 5.1 and § IV.3. ∎

Notice that this result makes no reference to admissible sets; it is only in the proof that they appear. The same remark applies to many of the results below. In order to make this more obvious we use Moschovakis' notation

$$\kappa^{\mathfrak{M}} = \sup\{\|\Gamma_\varphi\| \mid \Gamma_\varphi \text{ is a first order positive inductive definition over } M\}.$$

Thus $\kappa^{\mathfrak{M}} = O(\mathfrak{M})$ if $\mathfrak{M}$ has an inductive pairing function. In this section $\mathfrak{M}$ always denotes a structure $\langle M, R_1, \ldots, R_l\rangle$ for the language L.

5.3 Proposition (The Closure Theorem). *Let $\mathfrak{M}$ have an inductive pairing function and let $\varphi(x_1, \ldots, x_n, \mathsf{R}_+)$ define Γ_φ over $\mathfrak{M}$.*
(i) *For each $\alpha < \kappa^{\mathfrak{M}}$, I_φ^α is hyperelementary on $\mathfrak{M}$.*
(ii) *I_φ is hyperelementary iff $\|\Gamma_\varphi\| < \kappa^{\mathfrak{M}}$.*

Proof. I_φ^α is a $\mathbb{HYP}_{\mathfrak{M}}$-recursive function of α, for $\alpha \in \mathbb{HYP}_{\mathfrak{M}}$. Hence each $I_\varphi^\alpha \in \mathbb{HYP}_{\mathfrak{M}}$ for $\alpha \in \mathbb{HYP}_{\mathfrak{M}}$ and is thus hyperelementary by 5.1 (ii). This proves (i) and the ($\Leftarrow$) half of (ii). Consider the map ρ_φ defined on I_φ by

$$\rho_\varphi(x) = \text{least } \beta(x \in I_\varphi^\beta).$$

This is clearly $\mathbb{HYP}_{\mathfrak{M}}$-recursive. If $I_\varphi \in \mathbb{HYP}_{\mathfrak{M}}$ then, by Σ Replacement

$$\|\Gamma_\varphi\| = \sup\{\rho_\varphi(x) \mid x \in I_\varphi\}$$

exists in $\mathbb{HYP}_{\mathfrak{M}}$ and is thus less than $\kappa^{\mathfrak{M}}$. ∎

One of the awkward points in the theory of inductive definitions (when not done in the context of admissible sets) is that one needs to deal with ordinals but the ordinals are not in your structure. To get around this difficulty, Moschovakis introduces the concept of an inductive norm. A norm on a set S is simply a mapping ρ of S onto some ordinal λ. We use

$$\rho: S \twoheadrightarrow \lambda$$

to indicate that ρ is a norm mapping S onto λ. Given $\rho: S \twoheadrightarrow \lambda$, define

$$x \leqslant_\rho y \quad \text{iff} \quad x \in S \wedge (y \notin S \vee \rho(x) \leqslant \rho(y)),$$
$$x <_\rho y \quad \text{iff} \quad x \in S \wedge (y \notin S \vee \rho(x) < \rho(y)).$$

A *norm* $\rho: S \twoheadrightarrow \lambda$ *is inductive* on $\mathfrak{M}$ if the relations $\leqslant_\rho$ and $<_\rho$ are inductive on $\mathfrak{M}$. Notice that if $\rho: S \twoheadrightarrow \lambda$ is inductive then S is inductive since $S(x)$ iff $x \leqslant_\rho x$.

The most natural inductive norms are those on fixed point I_φ defined by

$$\rho_\varphi(x) = \text{least}\, \beta (x \in I_\varphi^\beta).$$

(To see that this norm $\rho = \rho_\varphi$ is inductive observe that

$$x \leqslant_\rho y \quad \text{iff} \quad x \in I_\varphi \wedge y \notin I_\varphi^{<\rho(x)},$$
$$x <_\rho y \quad \text{iff} \quad x \in I_\varphi \wedge y \notin I_\varphi^{\rho(x)}$$

and the relations on the right are clearly Σ_1 on $\mathbb{HYP}_\mathfrak{M}$, hence inductive on $\mathfrak{M}$.)

One of the most useful lemmas on inductive definitions is the Prewellordering Theorem which asserts that every inductive set has an inductive norm. In terms of admissible sets, this is a consequence of the fact that $\mathbb{HYP}_\mathfrak{M}$ is resolvable, in fact

$$\mathbb{HYP}_\mathfrak{M} = L(\alpha)_\mathfrak{M}$$

where $\alpha = O(\mathfrak{M})$. Most of the consequences of the Prewellordering Theorem in Moschovakis [1974] are actually obtained more easily from this equation. See for example, Exercise 5.19 for the Reduction and Separation Theorems.

5.4 The Prewellordering Theorem. *Let $\mathfrak{M}$ have an inductively definable pairing function. Every inductive relation S on $\mathfrak{M}$ has an inductive norm.*

Proof. Let S be Σ_1 on $\mathbb{HYP}_\mathfrak{M}$, say

$$S(x) \quad \text{iff} \quad L(\alpha)_\mathfrak{M} \models \exists z\, \varphi(x,z)$$

where φ is Δ_0 and $\alpha = o(\mathbb{HYP}_\mathfrak{M}) = \kappa^\mathfrak{M}$. Let R be the $\mathbb{HYP}_\mathfrak{M}$-recursive predicate given by

$$R(\beta, x) \quad \text{iff} \quad \exists z \in L(\beta)_\mathfrak{M}\, \varphi(x,z)$$

so

$$S(x) \quad \text{iff} \quad \exists \beta\, R(\beta, x).$$

Now the map f on S defined by

$$f(x) = \text{least}\, \beta\ R(\beta, x)$$

is not onto an ordinal so it is not a norm. Define p on S by

$$p(x) = \{y \in M \mid \exists \gamma < f(x) R(\gamma, y)\}.$$

Now

$$y < x \quad \text{iff} \quad y \in p(x)$$

is a well-founded relation so its associated rank function $\rho = \rho^{<}$ is a norm. We claim it is inductive on $\mathfrak{M}$. To see this observe that

$$y <_\rho x \quad \text{iff} \quad y \in S \quad \text{and} \quad \forall \beta \leqslant f(y) \neg R(\beta, x),$$

$$y \leqslant_\rho x \quad \text{iff} \quad y \in S \quad \text{and} \quad x \notin p(y)$$

so both relations are Σ_1 on $\mathbb{HYP}_{\mathfrak{M}}$, hence inductive on $\mathfrak{M}$. ☐

The Closure Theorem shows that every fixed point I_φ is the uniform limit of hyperelementary sets, the I_φ^β. The Prewellordering Theorem allows us to extend this from fixed points to arbitrary inductive sets. If $\rho: S \twoheadrightarrow \lambda$ then ρ endows S with stages S_ρ^β in a natural way:

$$S_\rho^\beta = \{x \in S \mid \rho(x) \leqslant \beta\}.$$

The Boundedness Theorem, Corollary 5.6, is the natural generalization of the Closure Theorem.

5.5 Theorem. *Let $\mathfrak{M}$ be a structure with an inductive pairing function. Let $\rho: S \twoheadrightarrow \lambda$ be an inductive norm on a relation S.*

(i) *$\lambda \leqslant o(\mathbb{HYP}_{\mathfrak{M}})$ and ρ is $\mathbb{HYP}_{\mathfrak{M}}$-recursive.*

(ii) *For each $\alpha < o(\mathbb{HYP}_{\mathfrak{M}})$, $S_\rho^\alpha \in \mathbb{HYP}_{\mathfrak{M}}$ and, as a function of α, S_ρ^α is a $\mathbb{HYP}_{\mathfrak{M}}$-recursive function.*

Proof. Define a function p with domain S by

$$p(x) = \{y \in M \mid \rho(y) < \rho(x)\}.$$

For $x \in S$,

$$p(x) = \{y \in M \mid y <_\rho x\} = \{y \in M \mid \neg(x \leqslant_\rho y)\}$$

so $p(x) \in \mathbb{HYP}_{\mathfrak{M}}$ by Δ_1 Separation. Further, p is $\mathbb{HYP}_{\mathfrak{M}}$-recursive since its graph is Σ_1 definable:

$$\rho(x) = z \quad \text{iff} \quad x \in S \wedge \forall y \in z\, (y <_\rho x) \wedge \forall y \in M - z\, (x \leqslant_\rho y).$$

Now we may apply V.3.1 to p. Define

$$y \prec x \quad \text{iff} \quad y \in p(x);$$

and note that $\prec$ is well founded since $y \prec x$ implies $\rho(y) < \rho(x)$. But then ρ is $\mathbb{HYP}_{\mathfrak{M}}$-recursive by V.3.1 since

$$\rho(x) = \sup\{\rho(y)+1 \mid y \prec x\}.$$

This proves (i). To prove (ii) first define $Q(\beta, x)$ by

$$Q(\beta, x) \quad \text{iff} \quad \beta < \lambda \quad \text{and} \quad \rho(x) \leqslant \beta.$$

We claim Q is $\mathbb{HYP}_{\mathfrak{M}}$-recursive. The clause $\beta < \lambda$ causes no trouble since either $\lambda = o(\mathbb{HYP}_{\mathfrak{M}})$ in which case the clause is redundant or else $\lambda < o(\mathbb{HYP}_{\mathfrak{M}})$ in which case "$\beta < \lambda$" is Δ_0. But for $\beta < \lambda$

$$Q(\beta, x) \quad \text{iff} \quad \exists \gamma\, [\rho(x) = \gamma \wedge \gamma \leqslant_\rho \beta],$$

$$\neg Q(\beta, x) \quad \text{iff} \quad \exists y\, [\rho(y) = \beta \wedge y <_\rho x]$$

so Q is Δ_1 on $\mathbb{HYP}_{\mathfrak{M}}$. But

$$S_\rho^\beta = \{x \in M \mid Q(\beta, x)\}$$

so $S_\rho^\beta \in \mathbb{HYP}_M$ by Δ_1 Separation. The graph $z = S_\rho^\beta$ is Σ_1 since it is equivalent to

$$\forall x \in z\, Q(x, \beta) \wedge \forall x \in M\, [Q(x, \beta) \rightarrow x \in z]$$

so (ii) holds. ☐

5.6 Corollary (The Boundedness Theorem). *Let $\mathfrak{M}$ be a structure with an inductive pairing function. Let $\rho: S \twoheadrightarrow \lambda$ be any inductive norm.*

(i) $\lambda \leqslant \kappa^{\mathfrak{M}}$.

(ii) *For each $\alpha < \kappa^{\mathfrak{M}}$, S_ρ^α is hyperelementary.*

(iii) *S is hyperelementary iff $\lambda < \kappa^{\mathfrak{M}}$.*

Proof. The only part left to prove, after Theorem 5.5, is that if S is hyperelementary then every inductive norm $\rho: S \twoheadrightarrow \lambda$ has $\lambda < \kappa^{\mathfrak{M}}$. This follows by Σ Replacement since

$$\lambda = \sup\{\rho(x) \mid x \in S\}$$

and ρ is $\mathbb{HYP}_{\mathfrak{M}}$-recursive. ☐

The next result, the Covering Theorem, is one of the most useful consequences of the Boundedness Theorem. We state only the special case that we need in the Exercises.

5.7 Corollary (The Covering Theorem). *Let $\mathfrak{M}$ be a structure with an inductive pairing function. Let S be an inductive subset of $\mathfrak{M}$ and let $T \subseteq S$ be coinductive on $\mathfrak{M}$. Let $\rho: S \twoheadrightarrow \lambda$ be any inductive norm on S. Then T is a subset of one of the hyperelementary resolvents S_ρ^β for $\beta < \kappa^{\mathfrak{M}}$.*

Proof. Suppose that the conclusion failed. Then we could write

$$M - S = \{x \in M \mid \forall y \in M (y \in T \rightarrow y <_\rho x)\}$$

which makes $M - S$ a Σ_1 subset of $\mathbb{HYP}_{\mathfrak{M}}$ and hence $S \in \mathbb{HYP}_{\mathfrak{M}}$ since S is also Σ_1 on $\mathbb{HYP}_{\mathfrak{M}}$. But then $S = S_\rho^\lambda$ and $\lambda < \kappa^{\mathfrak{M}}$ by 5.6, so T is, after all, a subset of the hyperelementary resolvent S_ρ^λ. ☐

We now return to more familiar matters.

5.8 Theorem. *Let $\mathfrak{M}$ be a structure with an inductive pairing function. For each $n \geqslant 1$ there is an inductive relation of $n+1$ arguments that parametrizes the class of n-ary inductive relations.*

Proof. In view of 5.1 (iv), this is just a restatement of V.5.6. ☐

As always, we have the following corollary, to be compared with 5.13 below.

5.9 Corollary. *If $\mathfrak{M}$ is a structure with an inductive pairing function, then not every inductive relation is hyperelementary.* ☐

Some further uses of $\mathbb{HYP}_{\mathfrak{M}}$ in the study of inductive relations are sketched in the exercises, see especially 5.19, 5.23 and 5.24.

We can get an excellent feeling for the inductive, coinductive and hyperelementary relations on a structure by returning to infinitary logic.

Let α be an admissible ordinal, let $\mathbb{A} = L(\alpha)$ and let $\mathsf{L}_\mathbb{A}$ be the admissible fragment of $\mathsf{L}_{\infty\omega}$ given by $\mathbb{A}$. We refer to the elements of $\mathsf{L}_\mathbb{A}$ as the *α-finite formulas.*

Let $\mathfrak{M}$ be a structure for L. A relation S on $\mathfrak{M}$ is *defined by an α-finite formula* if there is an α-finite $\varphi(x_1,\ldots,x_n,y_1,\ldots,y_k)$ and there are $q_1,\ldots,q_k \in \mathfrak{M}$ such that

(1) $\qquad S(x_1,\ldots,x_n)$ iff $\mathfrak{M} \models \varphi[x_1,\ldots,x_n,q_1,\ldots,q_k]$

for all $x_1,\ldots,x_n \in \mathfrak{M}$. S is *defined by an α-recursive n-type* if there is an α-recursive set $\Phi(x_1,\ldots,x_n,y_1,\ldots,y_k)$ of α-finite formulas and there are $q_1,\ldots,q_k \in \mathfrak{M}$ such that

(2) $\qquad S(x_1,\ldots,x_n)$ iff $\mathfrak{M} \models \bigwedge_{\varphi\in\Phi} \varphi[x_1,\ldots,x_n,q_1,\ldots,q_k]$

for all $x_1,\ldots,x_n \in \mathfrak{M}$. Replace the infinite conjunction in (2) by an infinite disjunction

(3) $\qquad S(x_1,\ldots,x_n)$ iff $\mathfrak{M} \models \bigvee_{\varphi\in\Phi} \varphi[x_1,\ldots,x_n,q_1,\ldots,q_k]$

and we say that S is *defined by an α-recursive n-cotype.* Notice that S is defined by an α-recursive type iff $\neg S$ is defined by an α-recursive cotype. The student should compare 5.10 with Theorem II.7.3. (Another version holds without the pairing function assumption; see Exercise 5.29.)

5.10 Theorem. *Let $\mathfrak{M}$ be a structure for* L *with an inductive pairing function and let $\alpha = O(\mathfrak{M})$.*

(i) *A relation S on $\mathfrak{M}$ is hyperelementary on $\mathfrak{M}$ iff S is defined by an α-finite formula.*

(ii) *A relation S on $\mathfrak{M}$ is inductive on $\mathfrak{M}$ iff S is defined by an α-recursive cotype; S is coinductive on $\mathfrak{M}$ iff S is defined by an α-recursive type.*

Proof. We first prove the ($\Leftarrow$) parts of (i) and (ii). Since $\kappa^{\mathfrak{M}} = o(\mathbb{HYP}_{\mathfrak{M}})$, $\mathrm{L}(\alpha) \subseteq \mathbb{HYP}_{\mathfrak{M}}$, so every α-finite formula is in $\mathbb{HYP}_{\mathfrak{M}}$. Thus any relation defined by an α-finite formula is in $\mathbb{HYP}_{\mathfrak{M}}$ by Δ_1 Separation and, hence, is hyperelementary. It suffices to prove either half of (ii) so suppose that Φ is an α-recursive (or even α-r.e.) set of α-finite formulas and S is defined by (3) above. Then $S(x_1, \ldots, x_n)$ iff the following is true in $\mathbb{HYP}_{\mathfrak{M}}$:

$$\exists \psi\, [\psi \in \Phi \wedge \mathfrak{M} \models \psi[x_1, \ldots, x_n, q_1, \ldots, q_k]].$$

This makes S a Σ_1 set on $\mathbb{HYP}_{\mathfrak{M}}$ so S is inductive on $\mathfrak{M}$ by 5.1.

We now prove the ($\Rightarrow$) parts of (ii) and (i). Suppose S is inductive on $\mathfrak{M}$, say

$$S(x) \quad \text{iff} \quad (x, q_0) \in I_\varphi$$

where $\varphi(v_1, v_2, q, \mathsf{R}_+)$ has R binary and has an extra parameter q. Since $\alpha = \kappa^{\mathfrak{M}}$,

$$I_\varphi = \bigcup_{\beta < \alpha} I_\varphi^\beta.$$

We define formulas ψ_β by recursion on β as follows, where $\theta(\mathfrak{f}/\mathsf{R})$ denotes the result of replacing $\mathsf{R}(t_1, t_2)$ by $t_1 \neq t_1 \wedge t_2 \neq t_2$:

$$\psi_0(v_1, v_2, v_3) \quad \text{is} \quad \varphi(v_1, v_2, v_3, \mathfrak{f}/\mathsf{R}),$$

$$\psi_\beta(v_1, v_2, v_3) \quad \text{is} \quad \varphi(v_1, v_2, v_3, \textstyle\bigvee_{\gamma < \beta} \psi\, (\cdot, \cdot, v_3)/\mathsf{R}).$$

A simple proof by induction shows that

$$(x, y) \in I_\varphi^\beta \quad \text{iff} \quad \mathfrak{M} \models \psi_\beta(x, y, q)$$

and, hence,

$$(x, y) \in I_\varphi \quad \text{iff} \quad \mathfrak{M} \models \textstyle\bigvee_{\beta < \alpha} \psi_\beta(x, y, q).$$

Then we have

$$S(x) \quad \text{iff} \quad \mathfrak{M} \models \textstyle\bigvee_{\beta < \alpha} \psi_\beta(x, q_0, q).$$

Thus it remains to check that the set

$$\Phi = \{\psi_\beta \mid \beta < \alpha\}$$

is an α-recursive set. The function $f(\beta)=\psi_\beta$ is clearly definable by Σ Recursion in $L(\alpha)$ so Φ is at least α-r.e. Define a measure of complexity of formulas, say $c(\theta)$, by recursion as follows:

$$c(\theta)=1 \quad \text{if} \quad \theta \text{ is atomic},$$

$$c(\theta)=c(\psi)+1 \quad \text{if} \quad \theta \text{ is } \neg\psi,\ \exists v\,\psi \text{ or } \forall v\,\psi,$$

$$c(\theta)=\sup\{c(\psi)+1 \mid \psi\in\Theta\} \quad \text{if} \quad \theta \text{ is } \bigwedge\Theta \text{ or } \bigvee\Theta.$$

Then $c(\psi_\beta)\geqslant\beta$ so

$$\theta\in\Phi \quad \text{iff} \quad \exists\beta\leqslant c(\theta)\,[\theta=\psi_\beta]$$

which shows that Φ is α-recursive. This finishes the proof of (ii), but what happens if S is actually hyperelementary? Then $S\in\mathbb{HYP}_{\mathfrak{M}}$ and we can define a function $g\in\mathbb{HYP}_{\mathfrak{M}}$ with $\mathrm{dom}(g)=S$ by

$$g(x)=\text{least}\,\beta\,(\mathfrak{M}\models\psi_\beta(x,q_0,q)).$$

Let $\gamma=\sup(\mathrm{rng}(g))$. Then $\gamma<\alpha$ by Σ Replacement in $\mathbb{HYP}_{\mathfrak{M}}$. Then

$$S(x) \quad \text{iff} \quad \mathfrak{M}\models\bigvee_{\beta\leqslant\gamma}\psi_\beta(x,q_0,q)$$

so S is defined by an α-finite formula. □

The converses of Theorem 5.10 (i), (ii) also hold. We prove the converse of (i) and leave the other as Exercise 5.22. First a lemma.

5.11 Lemma. *Let $\mathfrak{M}$ be an* L*-structure with an inductive pairing function, let* $\mathsf{L}_{\mathbb{A}}$ *be an admissible fragment which is an element of* $\mathbb{HYP}_{\mathfrak{M}}$, *and let*

$$\mathbf{S}^n=\{S\subseteq M^n \mid \textit{for some } \varphi\in\mathsf{L}_{\mathbb{A}}, \textit{ and some } q_1,\dots,q_k\in M,$$
$$\mathfrak{M}\models\varphi[x_1,\dots,x_n,q_1,\dots,q_k] \textit{ iff } S(x_1,\dots,x_n)$$
$$\textit{for all } x_1,\dots,x_n\in M\}.$$

(i) *The collection* $\mathbf{S}^n$ *can be parametrized by an* $n+1$*-are hyperelementary relation, with indices from* M.

(ii) *There is a hyperelementary set which is not in* $\mathbf{S}^1$.

Proof. (ii) follows from (i) by the usual diagonalization argument. The proof of (i) is a routine modification of Theorem V.5.7 since $\mathbb{HYP}_{\mathfrak{M}}$ is projectible into $\mathfrak{M}$. □

5.12 Theorem. *Let $\mathfrak{M}$ be a structure for* L *with an inductive pairing function and let α be an admissible ordinal. If the hyperelementary relations on $\mathfrak{M}$ consist of exactly the relations definable by α-finite formulas, then $\alpha = \kappa^{\mathfrak{M}}$.*

Proof. Lemma 5.11 shows us that if every hyperelementary relation is definable by an α-finite formula then $\kappa^{\mathfrak{M}} \leqslant \alpha$. We now show that if every relation definable by an α-finite formula is hyperelementary, then $\alpha \leqslant \kappa^{\mathfrak{M}}$. Suppose, to prove the contrapositive, that $\alpha > \kappa^{\mathfrak{M}}$ and let S be any inductive relation which is not hyperelementary. By 5.10, S is definable by a $\kappa^{\mathfrak{M}}$-recursive cotype. But then, S is definable by an α-finite formula since $\alpha > \kappa^{\mathfrak{M}}$, so not every relation definable by an α-finite formula is hyperelementary. $\square$

It is interesting to compare the following corollary of 5.10 and 5.12 with a result in Moschovakis [1974].

5.13 Corollary. *Let $\mathfrak{M}$ be a structure with an inductive pairing function. The following conditions on $\mathfrak{M}$ are equivalent:*

(i) *$\mathfrak{M}$ is recursively saturated.*

(ii) *Every hyperelementary relation is first-order definable.*

(iii) $\kappa^{\mathfrak{M}} = \omega$.

Proof. Since $\kappa^{\mathfrak{M}} = o(\mathrm{HYP}_{\mathfrak{M}})$, we proved (i) $\Longleftrightarrow$ (ii) back in § IV.5. We have the implication (iii) $\Rightarrow$ (ii) by 5.10 or by II.7.3. By 5.12 we have (ii) $\Rightarrow$ (iii). $\square$

Moschovakis assumes that his structures are acceptable (see Exercise 4.15), a stronger condition than having an inductive pairing function. Corollary 5 B.3 of Moschovakis [1974] asserts that if $\mathfrak{M}$ is acceptable then there is a hyperelementary relation that is not first order definable. Since an acceptable structure $\mathfrak{M}$ always has $\kappa^{\mathfrak{M}} > \omega$ (by 4.1), this follows from 5.13. But 5.13 also shows us that the restriction to acceptable structures rules out many of the most interesting structures, model theoretically interesting at any rate.

The general version of 5.13 reads as follows.

5.14 Corollary. *Let $\mathfrak{M}$ have an inductive pairing function and let α be an admissible ordinal. The following are equivalent:*

(i) *$\mathfrak{M}$ is α-recursively saturated and not β-recursively saturated for any admissible $\beta < \alpha$.*

(ii) *The hyperelementary relations are just those definable by α-finite formulas.*

(iii) $\kappa^{\mathfrak{M}} = \alpha$.

Proof. We have (ii) $\Longleftrightarrow$ (iii) by the theorems above and (i) $\Longleftrightarrow$ (iii) by Exercise IV.5.11 and the equality $\kappa^{\mathfrak{M}} = o(\mathrm{HYP}_{\mathfrak{M}})$. $\square$

Let $\mathfrak{M}$ have an inductive pairing function and let $\alpha = \kappa^{\mathfrak{M}}$. By 5.14 we see that the hyperelementary relations on $\mathfrak{M}$ are just the relations explicitly definable by α-finite formulas. One could imagine stronger notions of inductive and hyperelementary where one allowed an α-finite or even a $\mathrm{HYP}_{\mathfrak{M}}$-finite formula

$\varphi(x_1,\ldots,x_n,\mathsf{R}_+)$ to define an inductive operation Γ_φ. Refer to these notions, for the time being, as *α-inductive*, *α-hyperelementary*, $\mathbb{HYP}_\mathfrak{M}$*-inductive* and $\mathbb{HYP}_\mathfrak{M}$*-hyperelementary*. The next result shows that the notion of inductive on $\mathfrak{M}$ is "stable" in that it coincides with α-inductive and $\mathbb{HYP}_\mathfrak{M}$-inductive.

5.15 Theorem. *Let $\mathfrak{M}$ have an inductive pairing function and let $\alpha=\kappa^\mathfrak{M}$.*

(i) *The inductive, α-inductive and $\mathbb{HYP}_\mathfrak{M}$-inductive relations on $\mathfrak{M}$ all coincide.*

(ii) *Hence, the hyperelementary, α-hyperelementary and $\mathbb{HYP}_\mathfrak{M}$-hyperelementary relation on $\mathfrak{M}$ all coincide with the relations explicitly definable by α-finite formulas.*

Proof. It suffices to prove that if $\varphi(x_1,\ldots,x_n,\mathsf{R}_+)$ is a formula of $\mathsf{L}_\mathbb{A}$, where $\mathbb{A}=\mathbb{HYP}_\mathfrak{M}$, then I_φ is inductive on $\mathfrak{M}$. The proof uses the ideas from the two halves of 5.10 (ii). First note that I_φ^β is a $\mathbb{HYP}_\mathfrak{M}$-recursive function of β, for $\beta<\alpha$, since it is defined by Σ Recursion in $\mathbb{HYP}_\mathfrak{M}$. As before, the Σ Reflection theorem shows that $\|\Gamma_\varphi\|\leqslant\alpha$. Now define the formulas ψ_β as in the proof of 5.10:

$$\psi_0(x_1,\ldots,x_n)=\varphi(x_1,\ldots,x_n,\mathfrak{f}/\mathsf{R}),$$

$$\psi_\beta(x_1,\ldots,x_n)=\varphi(x_1,\ldots,x_n,\textstyle\bigvee_{\gamma<\beta}\psi\,(\ldots)/\mathsf{R})$$

so that $(x_1,\ldots,x_n)\in I_\varphi^\beta$ iff $\mathfrak{M}\models\psi_\beta[x_1,\ldots,x_n]$. Thus

$$(x_1,\ldots,x_n)\in I_\varphi \quad \text{iff} \quad \mathfrak{M}\models\textstyle\bigvee_{\beta<\alpha}\psi_\beta[x_1,\ldots,x_n].$$

But the set of $\mathbb{HYP}_\mathfrak{M}$-finite formulas $\{\psi_\beta\mid\beta<\alpha\}$ is α-r.e. (actually α-recursive) so I_φ is Σ_1 on $\mathbb{HYP}_\mathfrak{M}$ and hence inductive on $\mathfrak{M}$ by 5.1 (i). ☐

5.16—5.30 Exercises

5.16. Show that each of the following structures has a definable pairing function.

(i) $\mathcal{N}=\langle\omega,0,+,\cdot\rangle$.

(ii) Any model of Peano arithmetic.

(iii) Any model of ZF, KP or KPU.

(iv) $\mathrm{L}(a,\lambda)$ for any limit ordinal λ.

(v) $\mathcal{R}=\langle\omega^\omega\cup\omega,\omega,0,+,\cdot,\mathrm{App}\rangle$, where ω^ω is the set of all functions mapping ω into ω and

$$\mathrm{App}(f,n,m) \quad \text{iff} \quad f(m)=n.$$

5.17. Show that no nonstandard model of Peano arithmetic is acceptable. Show that some nonstandard models of Peano arithmetic are almost acceptable and that some are not. [Show that if $(\mathfrak{N},\mathcal{X})$ is a model of nonstandard analysis then $\mathfrak{N}$ is *not* almost acceptable.]

5.18 (Moschovakis [1974]). Let $\mathfrak{M}=\langle\alpha,<\rangle$ where α is any ordinal $\geqslant\omega$. Show that $\mathfrak{M}$ has an inductive pairing function. This is not easy. First assume $\alpha=\Sigma_{\beta<\alpha}(\beta\cdot2+1)$.

5.19 (Moschovakis [1974]). Let $\mathfrak{M}$ be a structure with an inductive pairing function. Prove the following results using Theorem 5.1.

(i) $\kappa^{\mathfrak{M}} = \sup\{\rho(\prec) | \prec$ is a hyperelementary pre-wellordering of $\mathfrak{M}\}$.

(ii) If $\mathfrak{M}$ has a hyperelementary well-ordering then

$$\kappa^{\mathfrak{M}} = \sup\{\rho(\prec) | \prec \text{ is a hyperelementary well-ordering of } \mathfrak{M}\}.$$

(iii) (Reduction). Let B, C be inductive on $\mathfrak{M}$. Show that there are disjoint inductive sets $B_0 \subseteq B$, $C_0 \subseteq C$ such that $B_0 \cup C_0 = B \cup C$. [See V.4.10.]

(iv) (Separation). Let B, C be disjoint coinductive subsets of $\mathfrak{M}$. Show that there is a hyperelementary set D containing B which is disjoint from C. [Use (iii).]

(v) (Hyperelementary Selection Theorem). Let $S(x,y)$ be an inductive relation on $\mathfrak{M}$. Show that there are inductive relations S_0, S_1 such that

$$S_0 \subseteq S,$$

$$\mathrm{dom}(S_0) = \mathrm{dom}(S),$$

$$x \in \mathrm{dom}(S) \Rightarrow \forall y\, (S_0(x,y) \leftrightarrow \neg S_1(x,y)).$$

5.20. We give an application of the covering theorem; in fact, the original version of it due to Spector. We use the notation from Rogers [1967]. Let

$$W = \{e | \varphi^2 \text{ is the characteristic function of a well-ordering } <_e\}.$$

Let $\rho(e)$ = the order type of $<_e$, for $e \in W$.

(i) Show that W is Π^1_1 on $\mathcal{N}$.

(ii) Show that ρ is an inductive norm,

$$\rho\colon W \twoheadrightarrow \omega_1^c.$$

(iii) Let B be a Σ^1_1 set of natural numbers, $B \subseteq W$. Show that $\sup\{\rho(e) | e \in B\} < \omega_1^c$.

5.21. Show that 5.10 (ii) remain true if "α-recursive type" is replaced by any of the following:

(i) α-r.e. type,

(ii) $\mathbb{H}\mathrm{YP}_{\mathfrak{M}}$-recursive type,

(iii) $\mathbb{H}\mathrm{YP}_{\mathfrak{M}}$-r.e. type.

5.22. Let $\mathfrak{M}$ be a structure with an inductive pairing function and let α be an admissible ordinal. Suppose that the inductive relations on $\mathfrak{M}$ are exactly the relations defined by an α-recursive cotype. Show that $\alpha = \kappa^{\mathfrak{M}}$.

5.23. Let $\mathfrak{M}$ have an inductive pairing function. Let S, T be inductive relations which are not hyperelementary.

(i) Show that $T \in \mathbb{H}\mathrm{YP}_{(\mathfrak{M},S)}$, and hence that $\mathbb{H}\mathrm{YP}_{(\mathfrak{M},S)}$ and $\mathbb{H}\mathrm{YP}_{(\mathfrak{M},T)}$ have the same universe of sets. [Show that $o(\mathbb{H}\mathrm{YP}_{(\mathfrak{M},S)}) > o(\mathbb{H}\mathrm{YP}_{\mathfrak{M}})$ and then use 5.10 (ii).]

(ii) (Moschovakis [1974]). Show that the two expanded structures $(\mathfrak{M}, S)$ and $(\mathfrak{M}, T)$ have the same inductive and hyperelementary relations.

5.24 (Moschovakis [1974]). Let $\mathfrak{M}$ be a structure with an inductive pairing function and let S be an inductive relation on $\mathfrak{M}$ which is not hyperelementary. Show that for any relation T on $\mathfrak{M}$,

$$S \text{ is hyperelementary on } (\mathfrak{M}, T) \quad \text{iff} \quad \kappa^{(\mathfrak{M},T)} > \kappa^{\mathfrak{M}}.$$

5.25. Show that Theorem 5.15 is not true without the hypothesis that $\mathfrak{M}$ has an inductive pairing function. [Use the $\mathfrak{M}$ of Exercise 4.14.]

5.26. Our proof of the Abstract Kleene Theorem, Corollary 5.2, is a bit round about. Prove it directly from the $\mathfrak{M}$-completeness theorem and Proposition 3.13. (This proof, by the way, establishes the second order version given in Moschovakis [1974] without change.)

5.27. Let $\mathfrak{M}$ be a structure for L with an inductive pairing function.

(i) Show that $C_{\mathfrak{M}}(\mathrm{KPU}^+)$, in the notation of Proposition 3.13, is inductive but not hyperelementary.

(ii) Show that $\kappa^{\mathfrak{M}}$ = closure ordinal of the inductive definition of "provable from KPU^+ by the $\mathfrak{M}$-rule".

(iii) Show that $C_{\mathfrak{M}}(\mathrm{KPU}^+)$ can be used to parametrize the inductive relations on $\mathfrak{M}$. [Use the closure of the inductive relations under hyperelementary substitution and some hyperelementary coding of formulas.]

5.28. The following definition, due to Nyberg, will be useful in Exercise VIII.9.16 and in Theorem VIII.9.5. A structure $\mathfrak{M} = \langle M, R_1, \ldots, R_k \rangle$ is a *uniform Kleene structure* if for every Π_1^1 formula $\Phi(x, \mathsf{S}_+)$ in some extra relation symbols S there is a first order $\varphi(x, y, \mathsf{R}_+, \mathsf{S}_+)$ and a $y \in M$ such that for all x and all S

$$(\mathfrak{M}, S) \models \Phi(x, \mathsf{S}_+)$$

if and only if

$$(x, y) \in I_\varphi(\mathfrak{M}, S),$$

where the R in φ is used for the induction over the structure $(\mathfrak{M}, S)$. Prove that every countable structure with an inductive pairing function is a uniform Kleene structure. Let α be any ordinal of cofinality ω. Show that $\langle \mathrm{V}(\alpha), \in \rangle$ is a uniform Kleene structure. (This last is due to Chang-Moschovakis [1970].)

5.29 (Makkai and Schlipf, independently). Improve Theorem 5.10 as follows: Let $\mathfrak{M}$ be a structure for L and let $\alpha = O(\mathfrak{M})$. Let S be a relation on $\mathfrak{M}$. Show that:

(i) $S \in \mathbb{HYP}_{\mathfrak{M}}$ iff S is defined by an α-finite formula;

(ii) S is Σ_1 on $\mathbb{HYP}_{\mathfrak{M}}$ iff S is defined by an α-recursive cotype. [Hint: Use the fact that every $a \in \mathbb{HYP}_{\mathfrak{M}}$ is of the form $\mathscr{F}(p_1, \ldots, p_n, M, L(\lambda_1)_{\mathfrak{M}}, \ldots, L(\lambda_k)_{\mathfrak{M}})$ for some limit ordinals $\lambda_1, \ldots, \lambda_k$ and a substitutable function $\mathscr{F}$.]

5.30 (Moschovakis [1974]). Let $\mathfrak{M}$ *not* have an inductive pairing function. Prove that $\kappa^{\mathfrak{M}}$ is admissible or the limit of admissibles. It is an open problem to find an $\mathfrak{M}$ where $\kappa^{\mathfrak{M}}$ is not admissible.

5.31 Notes. Some of the results discussed above hold without the pairing function assumption. For example, all of 5.3 through 5.6 are proved directly in Moschovakis [1974]. On the other hand, some of the results are false without the pairing function (like 5.2, 5.8—5.12) and those that do hold are much harder to prove without the admissible set machinery. For structures without an inductive pairing function we are left with two distinct approaches, inductive definitions and $\mathbb{HYP}_{\mathfrak{M}}$ (equivalently, inductive* definitions). Only time will tell which is the most fruitful tool for definability theory.

6. Recursive Open Games

An *open game formula* is an infinitary expression $\mathscr{G}(\vec{x})$ of the form

$$\forall y_1\, \exists z_1\, \forall y_2\, \exists y_2 \ldots \forall y_n\, \exists z_n \ldots \bigvee\nolimits_{n<\omega} \varphi_n(\vec{x}, y_1, z_1, \ldots, y_n, z_n)$$

where each φ_n is a formula of $\mathsf{L}_{\infty\omega}$. Note that $\mathscr{G}(\vec{x})$ itself is *not* a formula of $\mathsf{L}_{\infty\omega}$ due to the infinite string of quantifiers out front. If $\{\varphi_n | n<\omega\}$ is a recursive set of finitary formulas then $\mathscr{G}(\vec{x})$ is called a *recursive open game formula.*

For our study, the most important result on game formulas goes back to Svenonius [1965] where he proves that, for countable $\mathfrak{M}$, the Π_1^1 predicates are exactly those defined by recursive open game formulas (Theorem 6.8 below). This result went largely unnoticed until the formulas were rediscovered by Moschovakis [1971]. He established that for acceptable $\mathfrak{M}$ (of any cardinality), it is the inductive relations on $\mathfrak{M}$ which are definable by recursive open game formulas (Corollary 6.11 below). Thus, from our point of view, Moschovakis was proving the "absolute version" of the Svenonius theorem.

Before going into these results in detail, let's step back to examine the concept of "absolute version" with some detachment.

We have been using ZFC as a convenient informal metatheory and hence may construe all our results as statements about the universe $\mathbb{V}$ of sets. By a *class* C on $\mathbb{V}$ we mean a definable class,

$$x \in C \quad \text{iff} \quad \mathbb{V} \models \varphi[x]$$

for some formula $\varphi(v)$ of set theory. A predicate P on $\mathbb{V}$ is, by definition, given by

$$P(\vec{x}) \quad \text{iff} \quad \mathbb{V} \models \psi[\vec{x}]$$

for some formula $\psi(v_1, \ldots, v_n)$.

6.1 Definition. Let C be a class defined by a Σ_1 formula without parameters and let P be some predicate. A relation P^{abs} is an *absolute version of* P *on* C if the following conditions hold:

(i) P^{abs} is absolute on C (that is, there are Σ and Π formulas $\psi_1(v_1,\ldots,v_n)$, $\psi_2(v_1,\ldots,v_n)$ such that for all $\vec{x}\in C$

$$P^{\text{abs}}(\vec{x}) \quad \text{iff} \quad \mathbb{V}\models\psi_1[\vec{x}]$$
$$\text{iff} \quad \mathbb{V}\models\psi_2[\vec{x}]).$$

(ii) P and P^{abs} agree on $C\cap H(\omega_1)$ (that is, for all $x_1,\ldots,x_n\in C\cap H(\omega_1)$,

$$P(\vec{x}) \quad \text{iff} \quad P^{\text{abs}}(\vec{x})).$$

While not every predicate has an absolute version, at least there can be at most one absolute version.

6.2 Metatheorem. *Let C be a Σ_1 definable class, let P be some predicate and let P_1, P_2 be absolute versions of P on C. Then for all $\vec{x}\in C$,*

$$P_1(\vec{x}) \quad \text{iff} \quad P_2(\vec{x}).$$

Proof. This is just a special case of the Lévy Absoluteness Principle, one we have used several times in special cases. The hypothesis can be written

$$\forall \vec{x}\in H(\omega_1)\,[\vec{x}\in C\rightarrow(P_1(\vec{x})\leftrightarrow P_2(\vec{x}))].$$

The part within brackets is equivalent to a Π formula so the conclusion follows from the Lévy Absoluteness Principle. ☐

6.3 Example. Let C be the class of pairs $(\mathfrak{M},S)$ where $\mathfrak{M}$ is a structure. Let $P(\mathfrak{M},S)$ assert that S is Π_1^1 on $\mathfrak{M}$. Let $P^{\text{abs}}(\mathfrak{M},S)$ assert that S is Σ_1 on $\mathbb{HYP}_{\mathfrak{M}}$. Then we have shown that P and P^{abs} agree on countable structures and that P^{abs} is absolute. For other examples, see Table 5 on page 254.

The distinction between P^{abs} and P is the distinction between Part B and Part C of this book.

In this section we apply these general considerations as follows. We first prove that for all countable $\mathfrak{M}=\langle M,R_1,\ldots,R_l\rangle$, a relation S on $\mathfrak{M}$ is Π_1^1 iff it is defined by a recursive open game formula. Next we show that the notion "S is definable on $\mathfrak{M}$ by a recursive open game formula" is absolute. It will then follow that *for any* $\mathfrak{M}$,

S is Σ_1 on $\mathbb{HYP}_{\mathfrak{M}}$ iff S is definable by a recursive open game formula

and hence, by Theorem 5.1, that *if $\mathfrak{M}$ has an inductive pairing function,*

S is inductive on $\mathfrak{M}$ iff S is definable by a recursive open game formula.

(For $\mathfrak{M}$ without an inductive pairing function, we must replace inductive by inductive*.)

The first question to settle is the very meaning of an infinite string of quantifiers. Given a relation $R(y_1, z_1, \ldots, y_n, z_n, \ldots)$ of infinite sequences from $\mathfrak{M}$, what is to be meant by

$$\mathfrak{M} \models \forall y_1 \, \exists z_1 \ldots \forall y_n \, \exists z_n \ldots R(y_1, z_1, \ldots)?$$

The sensible interpretation is by means of Skolem functions. The above is defined to mean

$$\exists F_1, F_2, \ldots [(\mathfrak{M}, F_1, \ldots, F_n, \ldots) \models \forall y_1 \, \forall y_2 \ldots R(y_1, F_1(y_1), y_2, F_2(y_1, y_2), \ldots)].$$

For ease in presenting informal proofs it is convenient to rephrase this in terms of an infinite two person game, one played by players $\forall$ and $\exists$. The players take turns choosing elements $a_1, b_1, a_2, b_2, \ldots$ from $\mathfrak{M}$. Player $\exists$ wins if $R(a_1, b_1, a_2, b_2, \ldots)$; otherwise $\forall$ wins. Then

$$\mathfrak{M} \models \forall y_1 \, \exists z_1 \ldots R(y_1, z_1, \ldots)$$

is equivalent to:

Player $\exists$ has a winning strategy in the above game.

Formally, of course, a strategy for $\exists$ simply consists of a set $\{F_1, F_2, \ldots\}$ of Skolem functions such that

$$(\mathfrak{M}, F_1, \ldots) \models \forall y_1 \, \forall y_2 \, R(y_1, F_1(y_1), y_2, F_2(y_1, y_2), \ldots).$$

For games which begin with a play by $\exists$,

$$\exists y_1 \, \forall z_1 \ldots R(y_1, z_1, \ldots),$$

we use the convention that a function of 0 arguments is simply an element of $\mathfrak{M}$.

We have already defined the notion of an open game formula $\mathscr{G}(\vec{x})$

$$\forall y_1 \, \exists z_1 \ldots \bigvee_n \varphi_n(\vec{x}, y_1, z_1, \ldots, y_n, z_n).$$

The important part here is the infinite disjunction, not the fact that it begins with $\forall$ (we could always add a superfluous $\forall$ if it started with $\exists$) nor the fact the quantifiers exactly alternate one for one (again we could introduce superfluous quantifiers if necessary). The reason this is referred to as an "open" game formula is that in any given play

$$a_1, b_1, a_2, b_2, \ldots$$

of the game, if $\exists$ wins then he wins at some finite stage n and thus it wouldn't matter what he played after stage n. (That is, there is a whole neighborhood of winning plays for $\exists$ in the suitable product topology.)

The dual of an open game formula is a *closed game formula*, one of the form

$$\forall y_1 \exists z_1 \forall y_2 \exists z_2 \ldots \bigwedge_n \varphi_n(\vec{x}, y_1, z_1, \ldots, y_n, z_n).$$

In a closed game, $\exists$ must remain eternally diligent if he is to win.

6.4 Examples. (i) The simplest example of an important recursive open game sentence is given by

$$\forall y_1 \forall y_2 \ldots \bigvee_{n<\omega} \neg(y_{n+1} E y_n).$$

This sentence holds in $\langle \mathfrak{M}, E\rangle$ iff E is well founded. This is a rather boring game for $\exists$ since he never gets to play. Once $\forall$ has played a sequence $a_1, a_2, \ldots, \exists$ wins if it is not a descending sequence. Hence, $\exists$ has a winning strategy iff there are no infinite descending sequences.

(ii) The Kleene normal form for Π_1^1 relations on $\mathcal{N} = \langle \omega, 0, +, \cdot \rangle$,

$$S(x) \quad \text{iff} \quad \forall f\, \exists n\, R(\bar{f}(n), x),$$

can be considered as a reduction of Π_1^1 relations to recursive open game formulas, namely $S(x)$ iff

$$\forall y_1 \forall y_2 \ldots \bigvee_n \exists s\, [s \text{ codes } \langle y_1, \ldots, y_n\rangle \wedge R(s, x)].$$

(iii) On arbitrary countable structures we must use game formulas in which both players get to play if we are to characterize Π_1^1 relations. Suppose M is countable and let $\mathfrak{M} = \langle M, R, S\rangle$ where R, S are binary. Then $\mathfrak{M}$ is a model of

$$\forall y_1 \exists z_1 \forall y_2 \exists z_2 \ldots \bigwedge_{n,m<\omega} R(y_n, y_m) \leftrightarrow S(z_n, z_m)$$

iff $\langle M, R\rangle \cong \langle M, S\rangle$. Here we have expressed a Σ_1^1 sentence by a recursive closed game sentence.

Given a game formula $\mathscr{G}(\vec{x})$ we write

$$\mathfrak{M} \models \neg \mathscr{G}(\vec{x})$$

as shorthand for

$$\text{not} \quad (\mathfrak{M} \models \mathscr{G}(\vec{x})).$$

In general one must resist certain impulses generated by experience with finite strings of quantifiers. There is no reason to suppose that

$$\mathfrak{M} \models \neg \forall y_1 \exists z_1 \ldots R(y_1, z_1, \ldots)$$

implies

$$\mathfrak{M} \models \exists y_1 \forall z_1 \ldots \neg R(y_1, z_1, \ldots).$$

That is, just because $\exists$ has no winning strategy in the first game is no reason to suppose he does have a winning strategy in the second game. One can find R's for which this fails. For open and closed games, however, this tempting maneuver is perfectly acceptable, as Theorem 6.5 shows. We shall use the idea from this proof a couple of times later on.

6.5 Gale-Stewart Theorem. *For all* $\mathfrak{M}$ *and* $\vec{x}$,

$$\mathfrak{M} \models \neg \forall y_1 \exists z_1 \ldots \bigvee_n \varphi_n(\vec{x}, y_1, z_1, \ldots, y_n, z_n)$$

iff

$$\mathfrak{M} \models \exists y_1 \forall z_1 \ldots \bigwedge_n \neg \varphi_n(\vec{x}, y_1, z_1, \ldots, y_n, z_n).$$

Proof. Let game I be the game given by

$$\mathfrak{M} \models \forall y_1 \exists z_1 \ldots \bigvee_n \varphi_n(y_1, z_1, \ldots, y_n, z_n)$$

(we are suppressing the $\vec{x}$ since they play no role) and let game II be given by

$$\mathfrak{M} \models \exists y_1 \forall z_1 \ldots \bigwedge_n \neg \varphi_n(y_1, z_1, \ldots, y_n, z_n).$$

It is clear that $\exists$ cannot have a winning strategy in both games, for then $\forall$ could use $\exists$'s strategy from game II to defeat him in game I. Thus we have the ($\Leftarrow$) half of the theorem. (This part does not use the openness hypothesis.) Now suppose $\exists$ has no strategy in game I. We show that $\forall$ has a winning strategy in I which of course amounts to a winning strategy for $\exists$ in II. Now since $\exists$ has no strategy in I there must be a fixed a_1 such that $\exists$ still has no strategy in the game

$$\mathfrak{M} \models \exists z_1 \forall y_2 \exists z_2 \ldots \bigvee_n \varphi_n(a_1, z_1, \ldots, y_n, z_n).$$

Why? Because if each a_1 gave rise to a strategy $s(a_1)$ for $\exists$ then he would have had a winning strategy at the start; namely

answer $\forall$'s play of a_1 by using $s(a_1)$.

Thus $\forall$'s first play is to play an a_1 such that

$$\mathfrak{M} \models \neg \exists z_1 \forall y_2 \exists z_2 \ldots \bigvee_n \varphi_n(a_1, z_1, \ldots, y_n, z_n).$$

Now after $\exists$ makes some play $z_1 = b_1$, $\forall$ again plays an a_2 so that $\exists$ still has no winning strategy; i. e.

$$\mathfrak{M} \models \neg \exists z_2 \forall y_3 \exists y_3 \ldots \bigvee_n \varphi_n(a_1, b_1, a_2, z_2, \ldots, y_n, z_n).$$

The same reasoning as above shows that such an a_2 exists. Now $\forall$ keeps on playing at the mth play some a_m so that

$$\mathfrak{M} \models \neg \exists z_m \, \forall y_{m+1} \, \exists z_{m+1} \ldots \bigvee_n \varphi_n(a_1, b_1, \ldots, a_m, z_m, \ldots, z_n),$$

and, in particular

$$\mathfrak{M} \models \bigwedge_{k<m} \neg \varphi_k(a_1, b_1, \ldots, a_k, b_k).$$

Then, at the conclusion of play we have

$$\mathfrak{M} \models \bigwedge_{k<\omega} \neg \varphi_k(a_1, b_1, \ldots, a_k, b_k),$$

a win for $\forall$ in game I. We have thus defined a winning strategy for $\forall$ in game I. □

6.6 Corollary. *For all* $\mathfrak{M}, \vec{x}$,

$$\mathfrak{M} \models \neg \forall y_1 \, \exists z_1 \ldots \bigwedge_n \varphi_n(\vec{x}, y_1, z_1, \ldots, y_n, z_n)$$

iff

$$\mathfrak{M} \models \exists y_1 \, \forall z_1 \ldots \bigvee_n \neg \varphi_n(\vec{x}, y_1, z_1, \ldots, y_n, z_n).$$

Proof. The following are equivalent:

$$\mathfrak{M} \models \exists y_1 \, \forall z_1 \ldots \bigvee_n \neg \varphi_n(\vec{x}, y_1, z_1, \ldots, y_n, z_n)$$
$$\text{not } [\mathfrak{M} \models \neg \exists y_1 \, \forall z_1 \ldots \bigvee_n \neg \varphi_n(\vec{x}, y_1, z_1, \ldots, y_n, z_n)] \qquad \text{by 6.5}$$
$$\text{not } [\mathfrak{M} \models \forall y_1 \, \exists z_1 \ldots \bigwedge_n \neg\neg \varphi_n(\vec{x}, y_1, z_1, \ldots, y_n, z_n)]$$
$$\mathfrak{M} \models \neg \forall y_1 \, \exists z_1 \ldots \bigwedge \varphi_n(\vec{x}, y_1, z_1, \ldots, y_n, z_n). \quad \square$$

A simple application of the Gale-Stewart Theorem is to show that recursive open game formulas define Π_1^1 sets. We'll improve this later by improving the Gale-Stewart Theorem.

6.7 Corollary. *Let* $\mathscr{G}(\vec{x})$ *be a recursive open game formula of* L. *There is a* Π_1^1 *formula* $\Theta(\vec{x})$ *such that for all infinite* L*-structures* $\mathfrak{M}$ *and all* $x_1, \ldots, x_k \in \mathfrak{M}$,

$$\mathfrak{M} \models \mathscr{G}(\vec{x}) \quad \textit{iff} \quad \mathfrak{M} \models \Theta(\vec{x}).$$

Proof. Let $\mathscr{G}(\vec{x})$ be

$$\forall y_1 \, \exists z_1 \ldots \bigvee_n \varphi_n(\vec{x}, y_1, \ldots, z_n).$$

To prove the corollary it suffices, by the Gale-Stewart Theorem, to find a Σ_1^1 formula equivalent to

$$\exists y_1 \, \forall z_1 \ldots \bigwedge_n \neg \varphi_n(\vec{x}, y_1, \ldots, z_n).$$

This expression is equivalent to

$$\exists F\,[F \text{ is a function with } \mathrm{dom}(F) = \text{all finite sequences from } M \wedge \text{ for all } n \text{ and all } y_1,\dots,y_n \in M,\ \neg\varphi_n(\vec{x}, y_1, F(\langle y_1\rangle),\dots,y_n, F(\langle y_1,\dots,y_n\rangle))].$$

This is co-extended Σ_1^1 by Proposition IV.2.11 and hence is Σ_1^1 by Proposition IV.2.8. To see that the same Σ_1^1 formula works in all structures one simply notices that the proofs in § IV.2 were uniform. □

We now come to the theorem of Svenonius referred to above, a partial converse to 6.7.

6.8 Svenonius' Theorem. *For every Π_1^1 formula $\Theta(\vec{x})$ of* L *there is an recursive open game formula $\mathcal{G}(\vec{x})$ of* L *such that for all countable structures $\mathfrak{M}$ and all $x_1,\dots,x_k \in \mathfrak{M}$,*

$$\mathfrak{M} \models \mathcal{G}(\vec{x}) \quad \textit{iff} \quad \mathfrak{M} \models \Theta(\vec{x}).$$

Proof. It suffices, by the addition of constant symbols for the variables $x_1,\dots,x_n$, to prove the theorem for Π_1^1 sentences. We actually prove the dual, that every Σ_1^1 sentence is defined by some recursive closed game sentence in all countable structures. By the Skolem Lemma of V.8.7, any Σ_1^1 sentence is equivalent to one of the form

$$\exists \mathsf{S}_1,\dots,\mathsf{S}_m\, \forall y_1,\dots,y_l\, \exists z_1,\dots,z_k\, \varphi(\vec{y},\vec{z},\vec{\mathsf{S}})$$

where φ is quantifier free with no function symbols. We prove the special case

$$\exists \mathsf{S}\, \forall y_1\, y_2\, \exists z_1\, z_2\, \varphi(y_1, y_2, z_1, z_2, \mathsf{S}),$$

the general case being only notationally more complicated. We need the following fact.

(1) *For each quantifier free formula $\theta(\vec{v},\mathsf{S})$ there is another quantifier free formula $\theta^0(\vec{v})$ such that*

$$\theta^0(\vec{v}) \leftrightarrow \exists \mathsf{S}\, \theta(\vec{v},\mathsf{S})$$

is valid. Moreover, one can find θ^0 effectively from θ.

To prove (1), first write $\theta(\vec{v},\mathsf{S})$ as a disjunction

$$\theta_1(\vec{v},\mathsf{S}) \vee \cdots \vee \theta_p(\vec{v},\mathsf{S})$$

where each θ_i is a conjunction of atomic and negated atomic formulas. Since $\exists$ commutes with $\bigvee$ it suffices to prove (1) for formulas which are conjunctions of atomic and negated atomic formulas. So suppose we have to get rid of the $\exists\mathsf{S}$ from $\exists \mathsf{S}\, \theta(\vec{v},\mathsf{S})$ where $\theta(\vec{v},S)$ is

$$[\psi_1(\vec{v},\mathsf{S}) \wedge \cdots \wedge \psi_q(\vec{v},\mathsf{S})]$$

and each ψ_i is atomic or negated atomic. This just amounts to propositional logic. First *remove all equalities* like $(x=y)$ and make up for them by replacing x by y and y by x everywhere they occur (see examples below). Next we simply inspect the new list of formulas to see if it is consistent in propositional logic. If it is, θ^0 consists of the conjunction of all the formulas in the original list that don't mention S; if it isn't consistent, θ^0 consists of some false formula like $(x\neq x)$. We give three examples.

Example 1. Suppose $\theta(\vec{v},\mathsf{S})$ consists of

$$\mathsf{R}(x,z),\quad \mathsf{S}(x),\quad (x=y),\quad \neg\mathsf{S}(y).$$

The new list consists of

$$\mathsf{R}(x,z),\quad \mathsf{R}(y,z),\quad \mathsf{S}(x),\quad \mathsf{S}(y),\quad \neg\mathsf{S}(y),\quad \neg\mathsf{S}(x).$$

This is not consistent so there can be no such S.

Example 2. Suppose $\theta(\vec{v},\mathsf{S})$ consists of

$$\mathsf{R}(x,z),\quad \mathsf{S}(x),\quad (x\neq y),\quad (y=z).$$

The new list consists of

$$\mathsf{R}(x,z),\quad \mathsf{R}(x,y),\quad \mathsf{S}(x),\quad (x\neq y),\quad (x\neq z).$$

This is consistent so there will be such an S iff

$$\mathsf{R}(x,z)\wedge(x\neq y)\wedge(y=z).$$

Example 3. Suppose $\varphi(v,\mathsf{S})$ consists of

$$\mathsf{S}(x),\quad (x=y),\quad (y=z),\quad (x\neq z).$$

The new list will contain $(y\neq y)$ which is not consistent; there is no such S.

These examples should convince the student that the procedure decribed above actually works. It is obviously effective. This proves (1).

Now, using (1), let $\psi_n(y_{11},y_{12},z_{11},z_{12},y_{21},y_{22},z_{21},z_{22},\ldots,y_{n1},y_{n2},z_{n1},z_{n2})$ be a quantifier free formula equivalent to

$$\exists\mathsf{S}\bigwedge\nolimits_{1\leqslant m\leqslant n}\varphi(y_{m1},y_{m2},z_{m1},z_{m2},\mathsf{S})$$

and let the closed sentence $\mathscr{G}$ be

$$\forall y_{11},y_{12}\,\exists z_{11},z_{12}\,\forall y_{21},y_{22}\,\exists z_{21},z_{22}\cdots\bigwedge\nolimits_n\psi_n(y_{11},\ldots,z_{m2}).$$

First we prove that:

(2) *For any model* $\mathfrak{M}$, *if* $\mathfrak{M} \models \exists \mathsf{S} \forall y_1, y_2 \exists z_1, z_2 \varphi$, *then* $\mathfrak{M} \models \mathscr{G}$.

For suppose $(\mathfrak{M}, S) \models \forall y_1 \forall y_2 \exists z_1 \exists z_2 \varphi(y_1, y_2, z_1, z_2, \mathsf{S})$. Let $\exists$ play with the strategy:

if $\forall$ plays a_1, a_2 at stage n, then choose b_1, b_2 so that $(\mathfrak{M}, S) \models \varphi(a_1, a_2, b_1, b_2, \mathsf{S})$. This clearly presents $\exists$ with a win.

To conclude the proof we need only prove

(3) *If* $\mathfrak{M}$ *is countable and* $\mathfrak{M} \models \mathscr{G}$ *then there is a relation* S *on* $\mathfrak{M}$ *so that*

$$(\mathfrak{M}, S) \models \forall y_1, y_2 \exists z_1, z_2 \varphi(y_1, y_2, z_1, z_2, \mathsf{S}).$$

Suppose $\mathfrak{M} \models \mathscr{G}$ so that player $\exists$ has a winning strategy. Since $\mathfrak{M}$ is countable, so is M^2, so enumerate M^2, $M^2 = \{\langle a_{n1}, a_{n2} \rangle \mid n < \omega\}$. Let $\forall$ play $y_{ni} = a_{ni}$ and let $\exists$ play $z_{ni} = b_{ni} \in M$ using his winning strategy. Thus, we end up with

$$\mathfrak{M} \models \exists \mathsf{S} \bigwedge_{1 \leqslant m \leqslant n} \varphi(a_{m1}, a_{m2}, b_{m1}, b_{m2}, \mathsf{S})$$

for each $n < \omega$. Then, by the ordinary Compactness Theorem for propositional logic

$$\mathrm{Diagram}(\mathfrak{M}) \cup \{\varphi(a_{m1}, a_{m2}, b_{m1}, b_{m2}, \mathsf{S}) \mid m < \omega\}$$

is consistent. Thus there really is an S such that

$$(\mathfrak{M}, S) \models \varphi(a_{m1}, a_{m2}, b_{m1}, b_{m2}, \mathsf{S})$$

for each m, since φ is quantifier free. Thus, since every pair is $\langle a_{m1}, a_{m2} \rangle$ for some m,

$$(\mathfrak{M}, S) \models \forall y_1, y_2 \exists z_1, z_2 \varphi(y_1, y_2, z_1, z_2, \mathsf{S}).$$

This proves (3).

The proof of the theorem is complete except that $\mathscr{G}$ is not quite in the form demanded of a recursive closed game formula. But trivial modifications with superfluous quantifiers, renaming variables and renaming the subformulas obviously puts it in the desired form. □

We have carried out half our task by showing Π_1^1 is the same as "defined by a recursive open game formula" for countable structures. It remains to show that it is absolute. We prove more than this in the next two results.

The next theorem can be viewed as an effective version of the main theorem of Keisler [1965]. The proof is rather different.

Given a recursive open game formula $\mathscr{G}(\vec{x})$, say,

$$\forall y_1 \exists z_1 \forall y_2 \exists z_2 \ldots \bigvee_n \varphi_n(\vec{x}, y_1, z_1, \ldots, y_n, z_n)$$

we define its *finite approximations* $\delta_m(\vec{x})$ by:

$$\delta_m \text{ is } \quad \forall y_1 \exists z_1 \ldots \forall y_m \exists z_m \bigvee_{n \leqslant m} \varphi_n(\vec{x}, y_1, z_1, \ldots, y_n, z_n).$$

It is obvious, from a gamesmanship point of view, that

$$\forall \vec{x}[\delta_m(\vec{x}) \rightarrow \mathscr{G}(\vec{x})]$$

is true in all structures.

6.9 Theorem. *Let $\mathfrak{M}$ be recursively saturated. Then, using the notation of the previous paragraph,*

$$\mathfrak{M} \models \forall \vec{x}[\mathscr{G}(\vec{x}) \leftrightarrow \bigvee_{m<\omega} \delta_m(\vec{x})].$$

Proof. We already have the trivial implication $(\leftarrow)$. To prove the contrapositive of the other direction we imitate the proof of the Gale-Stewart theorem. We assume

$$\mathfrak{M} \models \bigwedge_{m<\omega} \neg \delta_m(\vec{x})$$

and exhibit a winning strategy for $\forall$ in the game

$$\forall y_1 \exists z_1 \forall y_2 \exists z_2 \ldots \bigvee_n \varphi_n(\vec{x}, y_1, z_1, \ldots, y_n, z_n).$$

We claim that there is an a_1 such that, for each $m<\omega$,

$$\mathfrak{M} \models \neg[\exists z_1 \forall y_2 \exists z_2 \ldots \forall y_m \exists z_m \bigvee_{n \leqslant m} \varphi_n(\vec{x}, a_1, z_1, y_2, z_2, \ldots, y_n, z_n)].$$

Why? Suppose that for every $a_1 \in M$ there is an m such that

$$\mathfrak{M} \models \exists z_1 \ldots \bigvee_{n \leqslant m} \varphi_n(\vec{x}, a_1, \ldots).$$

Now this all holds in $\mathbb{HYP}_{\mathfrak{M}}$, which has ordinal ω, so, by Σ Reflection there is a $k<\omega$ such that m can always be chosen less than k. (Here we are using the fact that φ_n is a recursive function of n, so is Σ_1 in $\mathbb{HYP}_{\mathfrak{M}}$.) But then

$$\mathfrak{M} \models \delta_k(\vec{x}),$$

contrary to assumption. Thus there is such an a_1 and we let $\forall$ play it. Let $\exists$ play $z_1 = b_1$. We claim that there is an a_2 such that, for all $m<\omega$,

$$\mathfrak{M} \models \neg[\exists z_2 \forall y_3 \exists z_3 \ldots \forall y_m \exists z_m \bigvee_{n \leqslant m} \varphi_n(\vec{x}, a_1, b_1, a_2, z_2, \ldots, y_n, z_n)].$$

The reasoning is just as for a_1. If $\forall$ continues in this way, do what $\exists$ will, a sequence $a_1 b_1 a_2 b_2 \ldots$ will be generated which satisfies

$$\mathfrak{M} \models \neg \varphi_n(a_1, b_1, \ldots, a_n, b_n)$$

for each n. Hence we have described a winning strategy for $\forall$. $\square$

Now, if $\mathfrak{M}$ is a structure with $\alpha = o(\mathbb{HYP}_{\mathfrak{M}})$ one would hope to show that, on $\mathfrak{M}$, $\mathcal{G}(x)$ is equivalent to the disjunction of its α-finite approximations:

$$\mathfrak{M} \models \forall \vec{x}[\mathcal{G}(\vec{x}) \leftrightarrow \bigvee_{\beta<\alpha} \delta_\beta(\vec{x})].$$

This turns out to be true once one has the correct definition of the δ_β's.

Let $\mathcal{G}(\vec{x})$ be a recursive open game formula, say

$$\forall y_1 \exists z_1 \forall y_2 \exists z_2 \ldots \bigvee_n \varphi_n(\vec{x}, y_1, z_1, \ldots, y_n, z_n).$$

Define formulas $\delta_\beta^n(x, y_1, z_1, \ldots, y_n z_n)$

$$\delta_0^n(\vec{x}, y_1, \ldots, z_n) \text{ is } \bigvee_{m \leqslant n} \varphi_m(\vec{x}, y_1, z_1, \ldots, y_m z_m),$$

$$\delta_{\beta+1}^n(\vec{x}, y_1, \ldots, z_n) \text{ is } \forall y_{n+1} \exists z_{n+1} \delta_\beta^{n+1}(\vec{x}, y_1, z_1, \ldots, y_{n+1}, z_{n+1}),$$

$$\delta_\lambda^n(\vec{x}, y_1, \ldots, z_n) \text{ is } \bigvee_{\beta<\lambda} \delta_\beta^n \text{ if } \lambda \text{ is a limit ordinal}.$$

Let $\delta_\beta(\vec{x})$ be $\delta_\beta^0(\vec{x})$. Note that δ_n, for $n<\omega$, has the same meaning as it did in Theorem 6.9. Also note that δ_β is an α-recursive function of $\beta<\alpha$, whenever α is an admissible ordinal.

6.10 Theorem. *Let $\alpha = o(\mathbb{HYP}_{\mathfrak{M}})$. Then, using the notation of the previous paragraph,*

$$\mathfrak{M} \models \forall \vec{x}[\mathcal{G}(\vec{x}) \leftrightarrow \bigvee_{\beta<\alpha} \delta_\beta(\vec{x})].$$

Proof. To prove the easy half ($\leftarrow$) one first proves by a straightforward induction on β that

$$\delta_\beta^n(\vec{x}, y_1, z_1, \ldots, y_n, z_n) \rightarrow \forall y_{n+1} \exists z_{n+1} \ldots \bigvee_m \varphi_m(\vec{x}, y_1, z_1, \ldots, y_m, z_m)$$

for all n. For $n=0$ this gives the desired result. The proof of the other half is so similar to the proof of Theorem 6.9 (a special case of 6.10) that we leave it to the student. ☐

6.11 Corollary. *For any structure $\mathfrak{M} = \langle M, R_1, \ldots, R_l \rangle$ and any relation S on $\mathfrak{M}$, the following are equivalent:*

(i) *S is definable by a recursive open game formula on $\mathfrak{M}$.*

(ii) *S is inductive* on $\mathfrak{M}$.*

(iii) *S is Σ_1 on $\mathbb{HYP}_{\mathfrak{M}}$.*

If $\mathfrak{M}$ has an inductive pairing function, these are also equivalent to

(iv) *S is inductive on $\mathfrak{M}$.*

Proof. It follows from 6.10 that

"S is definable on $\mathfrak{M}$ by a recursive open game formula"

is absolute so the theorem follows from Theorem 6.2. We present a slightly more direct proof which shows a bit more uniformity.

We see immediately that (i)⇒(iii), from Theorem 6.9, since

$$\exists\beta<\alpha[\mathfrak{M}\models\delta_\beta(\vec{x})]$$

is Σ_1 on $\mathbb{HYP}_{\mathfrak{M}}$. It thus suffices to prove (ii)⇒(i). Let $\varphi(\vec{x},\mathsf{R}_+)$ be any extended first order formula. Write $I_\varphi(\mathfrak{M})$ for the fixed point defined on $\mathfrak{M}$ by Γ_φ. We prove that there is a fixed recursive open game formula $\mathscr{G}(\vec{x})$ such that

(4) for all $\mathfrak{M}$, $\vec{x}\in I_\varphi(\mathfrak{M})$ iff $\mathfrak{M}\models\mathscr{G}(\vec{x})$

Now $I_\varphi(\mathfrak{M})$ is extended Π_1^1 on $\mathfrak{M}$, hence Π_1^1 on $\mathfrak{M}$ by Proposition IV.2.8, and the same Π_1^1 formula $\Phi(\vec{x})$ defines $I_\varphi(\mathfrak{M})$ for all $\mathfrak{M}$;

(5) for all $\mathfrak{M}$, $\vec{x}\in I_\varphi(\mathfrak{M})$ iff $\mathfrak{M}\models\Phi(\vec{x})$.

Now use Theorem 6.8 to choose $\mathscr{G}(\vec{x})$ such that

(6) for all countable $\mathfrak{M}$, $\mathfrak{M}\models\Phi(\vec{x})$ iff $\mathfrak{M}\models\mathscr{G}(\vec{x})$.

Now, combining lines (5) and (6) we have

for all countable $\mathfrak{M}[\vec{x}\in I_\varphi(\mathfrak{M})$ iff $\mathfrak{M}\models\mathscr{G}(\vec{x})]$

and the part in brackets is absolute. Hence, by Lévy Absoluteness, we have (4). ☐

6.12 Exercise. The Interpolation Theorem for $\mathsf{L}_{\omega\omega}$ can be stated as follows. Let $\Phi(x_1,\ldots,x_n)$ be a finitary Σ_1^1 formula of $\mathsf{L}_{\omega\omega}$ and let $\Psi(x_1,\ldots,x_n)$ be a finitary Π_1^1 formula of $\mathsf{L}_{\omega\omega}$. If every L-structure $\mathfrak{M}$ is a model of

(*) $\forall x_1,\ldots,x_n[\Phi(\vec{x})\rightarrow\Psi(\vec{x})]$

then there is a first order formula $\theta(\vec{x})$ such that every L-structure $\mathfrak{M}$ is a model of

(**) $\forall x_1,\ldots,x_n[[\Phi(\vec{x})\rightarrow\theta(\vec{x})]\wedge[\theta(\vec{x})\rightarrow\Psi(\vec{x})]]$.

We can turn this into a local result as follows.

(i) Let $\mathfrak{M}$ be a recursively saturated countable model of (*). Show that there is a $\theta(\vec{x})$ such that $\mathfrak{M}$ is a model of (**). [This is easy from Exercise V.4.8. A more direct proof goes via Svenonius' Theorem and the Approximation Theorem 6.9. Of course one could also cheat and apply the Interpolation Theorem for $\mathsf{L}_{\mathbb{A}}$ with $\mathbb{A}=\mathbb{HYP}_{\mathfrak{M}}$.]

(ii) Prove the interpolation theorem for $\mathsf{L}_{\omega\omega}$ directly from (i).

6.13 Notes. The student would profit from a comparison of our treatment with that in Moschovakis [1971], [1974]. His proof [1971] makes it clear where the approximations δ_β originate. The model theoretic interest of the Moschovakis-Svenonius results was brought out by the important paper Vaught [1973]. The student is urged to read this and Makkai [1973] in the same volume. This section (VI.6) of the book is included partly to make these papers more accessible.

Table 5. Absolute versions of some nonabsolute notions

Primitive notion P	*Absolute version* P^{abs}	*Relevant class C of objects*
1. S is Π_1^1 on $\mathfrak{M}$	S is Σ_1 on $\mathbb{HYP}_{\mathfrak{M}}$	all structures $\mathfrak{M}=\langle M, R_1, \ldots, R_l\rangle$ and relations S on $\mathfrak{M}$
2. S is Π_1^1 on $\mathfrak{M}$	S is inductive* on $\mathfrak{M}$	same as (1)
3. S is Π_1^1 on $\mathfrak{M}$	S is inductive on $\mathfrak{M}$	$\mathfrak{M}$, S as in (1) when $\mathfrak{M}$ has an inductive pairing function
4. S is Π_1^1 on $\mathfrak{M}$	S is defined by an open recursive game	same as (1)
5. $\models \varphi$	$\vdash \varphi$	all sentences of $\mathsf{L}_{\infty\omega}$
6. $\mathfrak{M} \cong \mathfrak{N}$	$\mathfrak{M} \cong_p \mathfrak{N}$ (cf. § VII.5)	all structures $\mathfrak{M}$, $\mathfrak{N}$
7. $\mathfrak{M} \cong \mathfrak{N}$	$\mathfrak{M} \equiv \mathfrak{N}(L_{\infty\omega})$ (cf. § VII.5)	same as (6)
8. S is strict Π_1^1 on $\mathbb{A}$	S is Σ_1 on $\mathbb{A}$ (cf. § VIII.3)	all admissible sets $\mathbb{A}$ and relations S on $\mathbb{A}$
9. $\mathfrak{M}$ is rigid (cf. § VII.7)	every element of $\mathfrak{M}$ is definable by a formula of $\mathsf{L}_{\infty\omega} \cap \mathbb{HYP}_{\mathfrak{M}}$ without parameters	all L-structures $\mathfrak{M}=\langle M, R_1, \ldots, R_l\rangle$

Part C

Towards a General Theory

"The *sensible* practical man realizes that the questions which he dismisses may be the key to a theory. Further, since he doesn't have a good theoretical analysis of familiar matters, sometimes not even the concepts needed to frame one, he will not be surprised if a novel situation turns out to be genuinely problematic."

G. Kreisel
Observations on Popular Discussions of Foundations

Chapter VII
More about $\mathsf{L}_{\infty\omega}$

In this chapter we resume the discussion of $\mathsf{L}_{\infty\omega}$ where we left it in Chapter III. This time, however, we do not restrict our attention to countable fragments but develop the beginning of a general theory. In this way we can gain insight into the countable case by seeing what principles are involved in the general case.

The most useful result, both for model-theoretic applications and for applications to generalized recursion theory, is the Weak Model Existence Theorem of § 2. Its model theoretic applications are discussed in §§ 3 and 4. The applications to definability theory can be found in Chapter VIII.

§§ 5, 6 and 7 are concerned with Scott sentences of $\mathsf{L}_{\infty\omega}$ and their approximations. These sections are independent of most of the rest of the book but they do illustrate the importance of $\mathsf{L}_{\infty\omega}$ and some uses of admissible sets in studying them.

1. Some Definitions and Examples

Once the hypothesis of countability is removed, all the major theorems of Chapter III fail dramatically. This section consists largely of "counter" examples to these statements. It also contains a number of definitions which will be important in our study.

1.1 Definition. An admissible set $\mathbb{A}$ is Σ_1 *compact* if for each admissible fragment of the form $\mathsf{L}_{\mathbb{A}}$ and each Σ_1 theory T of $\mathsf{L}_{\mathbb{A}}$, if every subset T_0 of T which is a element of $\mathbb{A}$ has a model, then T has a model.

The Compactness Theorem of § III.5 states that every countable, admissible set is Σ_1 compact.

1.2 Definition. An admissible set $\mathbb{A}$ is *self-definable* if for some language L containing the language of $\mathbb{A}$ there is a Σ_1 theory T of $\mathsf{L}_{\mathbb{A}}$ such that

(i) some expansion $(\mathbb{A}, \ldots)$ of $\mathbb{A}$ to an L-structure is a model of T.

(ii) if $(\mathfrak{B}, \ldots)$ is any model of T then $\mathfrak{B} \cong \mathbb{A}$.

If T can be chosen to be a single sentence of $\mathsf{L}_{\mathbb{A}}$ then $\mathbb{A}$ is called *strongly self-definable.*

We obtain a host of counter-examples to Σ_1 compactness by means of 1.3 and 1.4. The first is a trivial exercise in compactness.

1.3 Proposition. *If $\mathbb{A}$ is Σ_1 compact then $\mathbb{A}$ is not self-definable.* □

1.4 Proposition. *For all $\alpha \geqslant 0$, $H(\aleph_{\alpha+1})$ is self-definable.*

Proof. Let $A = H(\aleph_{\alpha+1})$ and let T be the theory consisting of the following sentences:

$$\mathrm{KP},$$

$$\forall x(x \in \bar{\mathsf{a}} \leftrightarrow \textstyle\bigvee_{b \in a} x = \bar{\mathsf{b}}) \quad \text{for all} \quad a \in A,$$

$$\forall x \, \exists \beta \, \exists f [\beta \leqslant \bar{\omega}_\alpha \wedge f \text{ maps TC}(x) \text{ one-one onto } \beta].$$

With the obvious interpretation of the constant symbols, A is a model of T. Suppose $\langle B, E \rangle$ is some other model of T. The infinitary sentences of T insure that we can assume

$$\langle A, \in \rangle \subseteq_{\text{end}} \langle B, E \rangle .$$

Let $x \in B$ and suppose $y \in B$ is such that

$$\langle B, E \rangle \models \text{"TC}(x) = y\text{"} .$$

Pick $\beta \leqslant \aleph_\alpha$ such that

$$\langle B, E \rangle \models \exists f [f \text{ maps } y \text{ one-one onto } \beta]$$

by the last axiom of T. Then there is some $F \subseteq \beta \times \beta$ such that $\langle B, E \rangle$ is a model of $\langle y, E \restriction y \rangle \cong \langle \beta, F \rangle$ and hence "$\langle \beta, F \rangle$ is well founded" is true in $\langle B, E \rangle$. The crucial step in the proof is to verify that

(1) $\langle \beta, F \rangle$ really is well founded.

Suppose that $\langle \beta, F \rangle$ is not well founded and let $X \subseteq \beta$ have no F-minimal member. But $\text{card}(X) < \aleph_{\alpha+1}$, so $X \in A \subseteq B$, and hence $\langle B, E \rangle$ is a model of "X has no F-minimal element", which is a contradiction. Thus (1) is established. But then the transitive set isomorphic to $\langle \beta, F \rangle$ is, on the one hand, $\langle y, E \restriction y \rangle$ and, on the other, in $H(\aleph_{\alpha+1})$. Thus $y \in H(\aleph_{\alpha+1})$ so $x \in H(\aleph_{\alpha+1})$. In other words $\langle A, \in \rangle = \langle B, E \rangle$. □

A strengthening of 1.4 is given in Exercise 1.12.

If we had wanted only to prove that $H(\aleph_{\alpha+1})$ is not Σ_1 compact, we could have come up with much simpler examples. A good example does more than just refute (the function of a *counterexample*), it makes almost explicit some of the ideas needed for understanding and generalizing existing results. Most of the examples in this section are good examples.

To understand the above example, the student should consider what happens to the proof of 1.4 if we replace $\langle H(\aleph_{\alpha+1}),\in\rangle$ by some countable, transitive set $\langle A,\in\rangle$ elementarily equivalent to it. Something must go wrong since A is Σ_1 compact. If he works through the proof he will see that the only step that fails is the proof of (1). This suggests the following proposition.

1.5 Proposition. *Let $\mathbb{A}$ be admissible.*

(i) *If $\mathbb{A}$ is self-definable then there is a Σ_1 theory $T(<)$ of $\mathsf{L}_{\mathbb{A}}$ which pins down ordinals greater than those in $\mathbb{A}$.*

(ii) *If $\mathbb{A}$ is strongly self-definable then there is a single sentence $\varphi(<)$ of $\mathsf{L}_{\mathbb{A}}$ which pins down ordinals greater than those in $\mathbb{A}$.*

Proof. We prove (i); the proof of (ii) is the same. Let T_0 be a theory which self-defines $\mathbb{A}$ and let $T = T_0 +$ "$< \, = \in\restriction$ ordinals". Then every model $\mathfrak{M}$ of T has $<^{\mathfrak{M}}$ of order type $o(\mathbb{A})$. □

Thus, self-definable admissible sets show that the theorems of § III.7 on the ordinals pinned down by Σ_1 theories of $\mathsf{L}_{\mathbb{A}}$ cannot go through in general; for example, they fail when $\mathbb{A} = H(\aleph_1)$. To get an example where a single sentence pins down large ordinals, we need some strongly self-definable admissible sets.

A set $\mathbb{A}$ is *essentially uncountable* if every countable subset $X \subseteq \mathbb{A}$ is an element of $\mathbb{A}$.

1.6 Proposition. *Let $\mathbb{A}$ be an essentially uncountable admissible set and let $\mathbb{B} = \mathbb{HYP}(\mathbb{A})$. Then $\mathbb{B}$ is strongly self-definable.*

Proof. Let ψ be the conjunction of the following:

$$\bigwedge_{a\in A\cup\{A\}} \forall v[v\in \bar{\mathsf{a}} \leftrightarrow \bigvee_{x\in a} v = \bar{\mathsf{x}}],$$

$$\bigwedge \mathrm{KPU},$$

$$\forall v\, \exists\alpha[x \in \mathrm{L}(\bar{\mathsf{A}},\alpha)],$$

$$\forall\alpha \bigvee_{\varphi\in KP} \neg\varphi^{\mathrm{L}(\bar{\mathsf{A}},\alpha)},$$

$$\forall\alpha\, \exists r[r \subset \bar{\mathsf{A}} \wedge r \text{ is a pre-wellordering of type } \alpha].$$

Since $\mathbb{HYP}(\mathbb{A})$ is projectible into $\mathbb{A}$, $\mathbb{HYP}(\mathbb{A})$ is a model of the last conjunct and hence of ψ. The well founded models of the first four conjuncts are isomorphic to $\mathbb{HYP}(\mathbb{A})$ so it remains to see that all models of ψ are well founded. Using the rank function we see that if $\langle B',E\rangle$ is a non-wellfounded model of ψ then there is a descending sequence of ordinals in $\langle B',E\rangle$ so it suffices to see that the ordinals of $\langle B',E\rangle$ are wellfounded. Let $a \in B'$ be an "ordinal" of $\langle B',E\rangle$. Apply the last conjunct of ψ to get an $r \subseteq A$ such that

$$\langle B',E\rangle \models \text{"}r \text{ has order type } a\text{"}.$$

We need to see that r really is well ordered. Suppose

$$\ldots r x_{n+1} r x_n r \ldots r x_1$$

is an r-descending sequence. Let $b=\{x_n \mid n<\omega\}$. Since b is a countable subset of A, $b\in A$. But then $b\in B'$ and b has no r-minimal element, contradicting

$$\langle B', E\rangle \models \text{"}r \text{ is well ordered"}. \quad \square$$

For example, if $\mathrm{cf}(\kappa)>\omega$ then $\mathbb{A}=\mathbb{HYP}(H(\kappa))$ is strongly self-definable. Hence $\mathsf{L}_{\mathbb{A}}$ is not Σ_1 compact and there is a single sentence of $\mathsf{L}_{\mathbb{A}}$ which pins down $o(\mathbb{A})$.

Our next examples have to do with attempts to generalize the Completeness and Extended Completeness Theorems of § III.5 to arbitrary admissible fragments.

1.7 Definition. Let $\mathbb{A}$ be an admissible set.

(i) $\mathbb{A}$ is *validity admissible* if the set of valid infinitary sentence of $\mathbb{A}$ is Σ_1 on $\mathbb{A}$.

(ii) $\mathbb{A}$ is Σ_1 *complete* if, for every Σ_1 theory T of $\mathsf{L}_{\mathbb{A}}$, the set

$$\mathrm{Cn}(T)=\{\varphi\in \mathsf{L}_{\mathbb{A}} \mid T\models\varphi\}$$

is Σ_1 on $\mathbb{A}$.

Don't forget, in reading 1.7, that the extra relations which may be part of $\mathbb{A}$ count in the definition of Σ_1. It is also important to notice that Σ_1 completeness implies validity admissibility.

1.8 Proposition. *Let $\mathbb{A}$ be admissible.*

(i) *If $\mathbb{A}$ is self-definable then $\mathbb{A}$ is not Σ_1 complete.*

(ii) *If $\mathbb{A}$ is a strongly self-definable pure admissible set then $\mathbb{A}$ is not even validity admissible.*

Proof. Recall, from § V.1, that there is a Π_1 subset of $\mathbb{A}$ which is not Σ_1. Hence, there is certainly a Π_1^1 subset of $\mathbb{A}$ which is not Σ_1. Thus the result follows from the following lemma. $\square$

1.9 Lemma. *Let $\mathbb{A}$ be admissible, let T be the theory which self-defines $\mathbb{A}$ in* 1.8 *and let $X\subseteq\mathbb{A}$ be Π_1^1 on $\mathbb{A}$. There is an $\mathbb{A}$-recursive function f such that for every $x\in\mathbb{A}$ we have $x\in X$* iff *$f(x)\in\mathrm{Cn}(T)$.*

Proof. Suppose

$$x\in X \quad \text{iff} \quad \mathbb{A}\models\forall \mathsf{R}\,\varphi(\mathsf{R},x),$$

where R is a symbol not in the language of T. In case (i) of 1.8 we may assume that T contains the diagram of $\mathbb{A}$. Then $x\in X$ iff $\varphi(\mathsf{R},\overline{\mathsf{x}})\in\mathrm{Cn}(T)$.

In case (ii) we settle the question "$x\in X$?" by checking whether the conjunction of T and the diagram of $\mathrm{TC}(\{x\})$ implies $\varphi(\mathsf{R},\overline{\mathsf{x}})$. $\square$

1.10 Corollary. *If A is pure and strongly self-definable then there are valid sentences of L_A which are not provable by the axioms and rules of Chapter III.*

Proof. The set of provable sentences is a Σ_1 set. $\square$

Thus, $H(\aleph_{\alpha+1})$ is never Σ_1 complete, even if $\alpha=0$, and $\mathbb{HYP}(H(\aleph_{\alpha+1}))$ is never validity admissible.

We conclude this section with a counterexample to the interpolation theorem. It has a rather different flavor and will not be used in the following sections.

1.11 Proposition. *Let $\mathbb{A}$ be an admissible set with an uncountable element and $o(\mathbb{A})>\omega$. The interpolation theorem fails for L_A.*

Proof. Let $\varphi(<)$ characterize $\langle\omega,<\rangle$ up to isomorphism and let ψ be

$$\bigwedge_{\substack{x,y\in a\\ x\neq y}} \overline{\mathsf{x}}\neq\overline{\mathsf{y}}$$

where $a\in A$ is uncountable. (All we reed about ψ is that it has only uncountable models and has no symbols in common with φ.) Then $\varphi,\psi\in\mathbb{A}$ and $\models\varphi\rightarrow\neg\psi$. If the interpolation theorem held for $\mathsf{L}_{\mathbb{A}}$ then there would be a sentence θ involving only equality such that $\models\varphi\rightarrow\theta$ and $\models\psi\rightarrow\neg\theta$. Thus θ is true in all countable infinite structures since such structures can always be turned into models of φ. Similarly, $\neg\theta$ is true in all structures of power $\geqslant\text{card}(a)$. But this contradicts:

(2) A sentence $\theta\in\mathsf{L}_{\infty\omega}$ involving only equality is true in all infinite structures or in none.

The proof of (2) is easy, given some notation and results of § 5, which we assume. Let $\mathfrak{M}=\langle M,=\rangle$, $\mathfrak{N}=\langle N,=\rangle$ be infinite. Let I be the set of all finite one-one maps from $M_0\subseteq M$ onto $N_0\subseteq N$. Then

$$I:\mathfrak{M}\cong_p\mathfrak{N}$$

so $\mathfrak{M}\equiv\mathfrak{N}(\mathsf{L}_{\infty\omega})$. Thus $\mathfrak{M}\models\theta$ iff $\mathfrak{N}\models\theta$. $\square$

1.12—1.17 Exercises

1.12. Suppose $0<\alpha<\aleph_\alpha$ and $\text{card}(\mathfrak{M})<\aleph_\alpha$. Show that $H(\aleph_\alpha)_{\mathfrak{M}}$ is self-definable. This includes 1.4 and $H(\aleph_\omega)$ as special cases.

1.13. A sentence $\varphi(<)$ (or theory $T(<)$) *pins down α exactly* if φ has models and every model $\mathfrak{M}$ of φ has $<^{\mathfrak{M}}$ of order type exactly α.

(i) Prove that if $\mathbb{A}$ is self-definable (strongly self-definable) then there is a Σ_1 theory T of $\mathsf{L}_{\mathbb{A}}$ (sentence φ of $\mathsf{L}_{\mathbb{A}}$) which pins down $o(\mathbb{A})$ exactly.

(ii) Let $\mathbb{A}$ be a resolvable admissible set and let T be a Σ_1 theory of $\mathsf{L}_{\mathbb{A}}$ which pins down $o(\mathbb{A})$ exactly. Show that $\mathbb{A}$ is self-definable.

1.14. Let $\mathbb{A} = \mathbb{HYP}(H(\aleph_{\alpha+1}))$. Show that there is a sentence of $L_{\mathbb{A}}$ which pins down $\aleph_{\alpha+2}$.

1.15. Show that the results of § IV.1 fail in the uncountable case.

1.16. Show that if $\mathbb{A}$ is essentially uncountable then every inductive relation on $\mathbb{A}$ is Δ_1^1. Conclude that not every Π_1^1 relation on $\mathbb{A}$ is inductive on $\mathbb{A}$, for $\mathbb{A}$ essentially uncountable.

1.17. Improve 1.8 (ii) by allowing $\mathbb{A}_{\mathfrak{M}}$ admissible above $\mathfrak{M}$.

1.18 Notes. Counterexamples to compactness go back to Hanf [1964] and earlier unpublished work of Tarski. Karp [1967] showed that, for $\mathrm{cf}(\alpha) > \omega$, the set $H(\aleph_\alpha)$ is not validity admissible. The results on pinning down large ordinals (1.14 for example) are due to Chang [1968]. The counterexample to interpolation is due to Malitz [1971]. We have tried to unify the various examples by centering them on the notion of self-definable, admissible set. Our notion is suggested by, and equivalent to, that of Kunen [1968].

Kreisel [1968] has observed that the counterexample to interpolation has the defect that it might disappear by some reasonable strengthening of the logic $L_{\mathbb{A}}$ or $L_{\infty\omega}$. The other examples of this section do not have this defect. The situation with compactness, say, could only get worse if we were to increase the expression power of the logic by introducing some new quantifier or connective. Rather than strengthen $L_{\mathbb{A}}$ we must look for strengthenings of the notion of admissibility which coincides with the old notion in the countable case. This is taken up in Chapter VIII.

2. *A Weak Completeness Theorem for Arbitrary Fragments*

The model theory of second-order logic is totally unmanageable and seems destined to remain so. Infinitary logic is an attempt to dent second-order logic by studying logics which have greater expressive power than $L_{\omega\omega}$ but still have a workable model theory. The examples of § 1 show that uncountable fragments behave more like second-order logic than do countable fragments. This makes the problem of developing a theory which handles arbitrary admissible fragments very intriguing.

In spite of, or because of, the "counter"-examples, the model theory of arbitrary admissible fragments is becoming a rich subject. In this section we present some basic tools for studying these logics. In particular, we prove an analogue of the Extended Completeness Theorem of § III.5. Recall our line of attack on the problem of completeness in Chapter III:

(1) We defined the notion: *validity property* for $L_{\mathbb{A}}$.

(2) We proved that if $L_{\mathbb{A}}$ is countable then a sentence $\varphi \in L_{\mathbb{A}}$ is valid iff φ is in every validity property.

(3) We showed that if $\mathsf{L}_{\mathbb{A}}$ is an admissible fragment then the intersection of all validity properties is a validity property which is $\mathbb{A}$-r.e., that is, Σ_1 on $\mathbb{A}$.

When we drop the assumption that $\mathsf{L}_{\mathbb{A}}$ is countable step (2) breaks down. In general, a sentence may be true in all models without being in every validity property (i.e., without being a theorem of $\mathsf{L}_{\mathbb{A}}$) as Corollary 1.10 shows. In this section we attack the problem of completeness as follows:

(1′) We define a stronger notion: *supervalidity property* for $\mathsf{L}_{\mathbb{A}}$.

(2′) We prove that a sentence $\varphi \in \mathsf{L}_{\mathbb{A}}$ is valid iff φ is in every supervalidity property.

(3′) In Chapter VIII we will introduce a semantic notion of r.e., called strict Π_1^1, and show that the intersection of all supervalidity properties for $\mathsf{L}_{\mathbb{A}}$ is a strict Π_1^1 set. When $\mathbb{A}$ is countable the notion of strict Π_1^1 reduces to Σ_1 on $\mathbb{A}$.

It is convenient in this part of the theory to work with sufficiently rich fragments, so-called Skolem fragments with constants.

2.1 Definition. Let $\mathsf{L}_{\mathbb{A}}$ be a fragment of $\mathsf{L}_{\infty\omega}$ and let C be a (possibly empty) set of constant symbols of L such that every formula of $\mathsf{L}_{\mathbb{A}}$ contains at most a finite number of constants from C.

(i) $\mathsf{L}_{\mathbb{A}}$ is a *Skolem fragment with constants* C if there is a one-one function which assigns to each formula of $\mathsf{L}_{\mathbb{A}}$ of the form

$\exists x\, \varphi(x, y_1, \ldots, y_n)$, where

φ contains no constants from C and

$y_1, \ldots, y_n$ are not bound in φ

an n-ary function symbol

$$\mathsf{F}_{\exists x \varphi}$$

of L not occuring in φ; it is called the *Skolem function symbol* for $\exists x\, \varphi(x, y_1, \ldots, y_n)$. If $C = 0$ we just say that $\mathsf{L}_{\mathbb{A}}$ is a *Skolem fragment.*

(ii) Let $\mathsf{L}_{\mathbb{A}}$ be a Skolem fragment with constants C. The *Skolem theory for* $\mathsf{L}_{\mathbb{A}}$, denoted by T_{Skolem}, consists of all sentences of $\mathsf{L}_{\mathbb{A}}$ of the form

$$\forall y_1, \ldots, y_n [\exists x\, \varphi(x, y_1, \ldots, y_n) \to \varphi(\mathsf{F}_{\exists x \varphi}(y_1, \ldots, y_n), y_1, \ldots, y_n)]$$

for all formulas $\exists x\, \varphi(x, y_1, \ldots, y_n)$ as in (i). An L-structure $\mathfrak{M}$ is a *Skolem structure for* $\mathsf{L}_{\mathbb{A}}$ if

$$\mathfrak{M} \models T_{\text{Skolem}}.$$

The extra freedom permitted by the set C of constant symbols is crucial for many applications. For now we can barely hint at their use by the following lemmas.

2.2 Lemma. *Let $\mathsf{L}_{\mathbb{A}}$ be a Skolem fragment with constants C and let $\mathscr{D}$ be any validity property for $\mathsf{L}_{\mathbb{A}}$ with $T_{\text{Skolem}} \subseteq \mathscr{D}$. Then for any formula*

$$\exists x\, \varphi(x, y_1, \ldots, y_n, \mathsf{c}_1, \ldots, \mathsf{c}_k)$$

of $\mathsf{L}_{\mathbb{A}}$ the sentence

$$\forall y_1, \ldots, y_n [\exists x\, \varphi(x, \vec{y}, \vec{\mathsf{c}}) \rightarrow \varphi(\mathsf{F}(\vec{y}, \vec{c}), \vec{y}, \vec{\mathsf{c}})]$$

is in $\mathscr{D}$, where F is the Skolem function symbol for

$$\exists x\, \varphi(x, y_1, \ldots, y_n, y_{n+1}, \ldots, y_{n+k}).$$

Proof. By the definition of T_{Skolem},

$$\forall y_1, \ldots, y_{n+k} [\exists x\, \varphi(x, y_1, \ldots, y_{n+k}) \rightarrow \varphi(\mathsf{F}(y_1, \ldots, y_{n+k}), y_1, \ldots, y_{n+k})]$$

is in $T_{\text{Skolem}} \subseteq \mathscr{D}$. Using the axioms for $\forall$ and modus ponens shows that the desired sentence is in $\mathscr{D}$. □

If $\mathsf{L}_{\mathbb{A}}$ is a fragment and C is a set of new constant symbols we use $\mathsf{L}_{\mathbb{A}}(C)$ to denote the fragment which consists of all substitution instances of formulas in $\mathsf{L}_{\mathbb{A}}$ by means of a *finite* number of constants from C. If $C = \{\mathsf{c}_1, \ldots, \mathsf{c}_n\}$ we sometimes use $\mathsf{L}_{\mathbb{A}}(\mathsf{c}_1, \ldots, \mathsf{c}_n)$ for $\mathsf{L}_{\mathbb{A}}(C)$.

2.3 Lemma. *Let $\mathsf{L}_{\mathbb{A}}$ be a Skolem fragment with constants C_0 and let C be a set of new constant symbols. Then $\mathsf{L}_{\mathbb{A}}(C)$ is a Skolem fragment with constants $C_0 \cup C$.*

Proof. Immediate from the definition. □

The next result shows us that we lose nothing (we gain a lot) by restricting ourselves to Skolem fragments and Skolem structures as far as the existence of models is concerned.

2.4 Proposition. *Let $\mathsf{L}_{\mathbb{A}}$ be a fragment of $\mathsf{L}_{\infty\omega}$. There is an expansion L' of L by new function symbols with the following properties:*

(i) *Let $\mathsf{L}'_{\mathbb{A}}$ be the set of formulas which result from a formula of $\mathsf{L}_{\mathbb{A}}$ by substituting a finite number of terms from L'. Then $\mathsf{L}'_{\mathbb{A}}$ is a Skolem fragment. Furthermore,* $\text{card}(\mathsf{L}'_{\mathbb{A}}) = \text{card}(\mathsf{L}_{\mathbb{A}})$ *and every Skolem function symbol is in $\mathsf{L}' - \mathsf{L}$.*

(ii) *Every L-structure $\mathfrak{M}$ has an expansion $\mathfrak{M}' = (\mathfrak{M}, \ldots)$ to a Skolem structure for $\mathsf{L}'_{\mathbb{A}}$.*

(iii) *If $\mathsf{L}_{\mathbb{A}}$ is an admissible fragment then we can define L' so that $\mathsf{L}'_{\mathbb{A}}$ is Δ_1 on $\mathbb{A}$ and such that the symbol $\mathsf{F}_{\exists x\varphi}$ is an $\mathbb{A}$-recursive function of $\exists x\, \varphi(x, y_1, \ldots, y_m)$. In particular, T_{Skolem} is then an $\mathbb{A}$-recursive set of sentences of $\mathsf{L}'_{\mathbb{A}}$.*

Proof. Let $\mathsf{L}^0 = \mathsf{L}$, $\mathsf{L}^0_{\mathbb{A}} = \mathsf{L}_{\mathbb{A}}$. For each formula

$$\exists x\, \varphi(x, y_1, \ldots, y_m)$$

of $\mathsf{L}^n_{\mathbb{A}}$ in which $y_1, \ldots, y_m$ are not bound, add a new function symbol

$$\mathsf{F}_{\exists x \varphi}$$

to L^n and let $\mathsf{L}^{n+1}_{\mathbb{A}}$ be the resulting fragment. Let $\mathsf{L}' = \bigcup_n \mathsf{L}^n$ so that $\mathsf{L}'_{\mathbb{A}} = \bigcup_n \mathsf{L}^n_{\mathbb{A}}$. Part (ii) is obvious from thus construction. (See Lecture 13 of Keisler [1971] for more details, if necessary.) Part (iii) is obvious if we just code up $\mathsf{F}_{\exists x \varphi}$ by something like $\langle 17, \exists x \varphi \rangle$. □

We now come to the notion of supervalidity property.

2.5 Definition. Let $\mathsf{L}_{\mathbb{A}}$ be a Skolem fragment with constants C. A validity property $\mathscr{D}$ for $\mathsf{L}_{\mathbb{A}}$ is a *supervalidity property* (s. v. p.) for $\mathsf{L}_{\mathbb{A}}$ (more precisely, for $(\mathsf{L}_{\mathbb{A}}, C)$) if $T_{\text{Skolem}} \subseteq \mathscr{D}$ and the following $\bigvee$*-rule* holds.

$\bigvee$*-Rule*: If $\bigvee \Phi$ is a SENTENCE of $\mathsf{L}_{\mathbb{A}}$ and $\bigvee \Phi \in \mathscr{D}$ then there is some $\varphi \in \Phi$ such that $\varphi \in \mathscr{D}$.

The $\bigvee$-rule causes supervalidity properties to behave in quite a different manner than ordinary validity properties. For example, it prevents the intersection of all supervalidity properties for $\mathsf{L}_{\mathbb{A}}$ from being an s. v. p. The next lemma shows just how strong the $\bigvee$-rule is.

2.6 Lemma. Let $\mathsf{L}_{\mathbb{A}}$ be a Skolem fragment with constants and let $\mathscr{D}$ be a validity property for $\mathsf{L}_{\mathbb{A}}$ with $T_{\text{Skolem}} \subseteq \mathscr{D}$. Then $\mathscr{D}$ is an s. v. p. iff $\mathscr{D}$ is complete, that is, iff for each sentence $\psi \in \mathsf{L}_{\mathbb{A}}$

$$\psi \in \mathscr{D} \quad \text{or} \quad (\neg \psi) \in \mathscr{D} .$$

Proof. Assume $\mathscr{D}$ is an s. v. p. Since all axioms of $\mathsf{L}_{\mathbb{A}}$ are in $\mathscr{D}$,

$$(\psi \vee \neg \psi) \in \mathscr{D}$$

so the conclusion follows by the $\bigvee$-rule. Now assume $\mathscr{D}$ is complete, $\bigvee \Phi$ a sentence of $\mathsf{L}_{\mathbb{A}}$, $\bigvee \Phi \in \mathscr{D}$. If for each $\varphi \in \Phi$, $\varphi \notin \mathscr{D}$, then, for each $\varphi \in \Phi$, $\neg \varphi \in \mathscr{D}$; so, by the $\bigwedge$-rule $R3$,

$$\bigwedge \{ \neg \varphi \mid \varphi \in \Phi \} \in \mathscr{D} .$$

But this sentence is just $\sim \bigvee \Phi$. Since $\mathscr{D}$ is a validity property it cannot have both $\bigvee \Phi$ and $\sim \bigvee \Phi$ as members, so $\varphi \in \mathscr{D}$ for some $\varphi \in \Phi$. □

Note that if $\mathscr{D}$ is an s. v. p. for $\mathsf{L}_{\mathbb{A}}$ and $\varphi(v_1, \ldots, v_n) \in \mathsf{L}_{\mathbb{A}}$ then

$$\varphi(v_1, \ldots, v_n) \in \mathscr{D} \quad \text{iff} \quad \forall v_1, \ldots, v_n \, \varphi(v_1, \ldots, v_n) \in \mathscr{D}$$

so that $\mathscr{D}$ is determined by its sentences. We say that an L-structure $\mathfrak{M}$ is a *model of* $\mathscr{D}$ if $\mathfrak{M}$ is a model of all sentences in $\mathscr{D}$.

2.7 Definition. Let $\mathfrak{M}$ be a Skolem structure for the Skolem fragment $L_{\mathbb{A}}$ (with constants). The supervalidity property given by $\mathfrak{M}$, denoted by $\mathscr{D}_{\mathfrak{M}}$, is the set of all $\varphi(v_1, \ldots, v_n) \in L_{\mathbb{A}}$ such that

$$\mathfrak{M} \models \forall v_1, \ldots, v_n \, \varphi(v_1, \ldots, v_n).$$

In the notation of III.4.2, $\mathscr{D}_{\mathfrak{M}} = \Gamma_{\mathfrak{M}}$. It is clear that $\mathscr{D}_{\mathfrak{M}}$ is an s.v.p. for L_A. If a sentence $\varphi \in L_A$ is in all supervalidity properties then it is in all $\mathscr{D}_{\mathfrak{M}}$; hence it is true in all Skolem structures for $L_{\mathbb{A}}$. This gives the trivial half of the next theorem.

2.8 Theorem (Weak Completeness Theorem for Arbitrary Skolem Fragments). *Let $L_{\mathbb{A}}$ be a Skolem fragment with constants C.*

(i) *A sentence φ of $L_{\mathbb{A}}$ is true in all Skolem structures for $L_{\mathbb{A}}$ iff φ is in every supervalidity property.*

(ii) *Let T be a theory of $L_{\mathbb{A}}$, φ a sentence of $L_{\mathbb{A}}$. Then φ is true in every Skolem structure $\mathfrak{M}$ which is a model of T iff φ is in every s.v.p. $\mathscr{D}$ with $T \subseteq \mathscr{D}$.*

Proof. (i) is the special case of (ii) where $T = \emptyset$. The proof of ($\Leftarrow$) in (ii) is immediate by the remarks following Definition 2.7. Most of the work for proving ($\Rightarrow$) was done back in the proof of the model existence theorem. We break its proof up in two lemmas to make this clear and because we need one of the lemmas (2.9) later.

Compare the next lemma with the definition of consistency property on p. 85.

2.9 Lemma (Weak Model Existence Theorem). *Let L have at least one constant symbol and let $L_{\mathbb{A}}$ be any fragment of $L_{\infty\omega}$. Any set S of sentences of $L_{\mathbb{A}}$ which satisfies the following rules has a model.*

Consistency rule: If φ is atomic and $\varphi \in S$ then $(\neg\varphi) \notin S$.
$\neg$-rule: If $(\neg\varphi) \in S$ then $(\sim\varphi) \in S$.
$\bigwedge$-rule: If $\bigwedge\Phi \in S$ then for all $\varphi \in \Phi$, $\varphi \in S$.
$\forall$-rule: If $(\forall v \varphi(v)) \in S$ then for each closed term t of L, $\varphi(t/v) \in S$.
$\bigvee$-rule: If $\bigvee\Phi \in S$ then for some $\varphi \in \Phi$, $\varphi \in S$.
$\exists$-rule: If $(\exists v \varphi(v)) \in S$ then for some closed term t of L, $\varphi(t(v)) \in S$.
Equality rules: For all closed terms t_1, t_2 of L:

$$\text{if } (t_1 \equiv t_2) \in S \quad \text{then } (t_2 \equiv t_1) \in S, \text{ and}$$
$$\text{if } \varphi(t_1), (t_1 \equiv t_2) \in S \text{ then } \varphi(t_2) \in S.$$

Proof. The proof of the Model Existence Theorem was in two stages. We first showed how to construct a set s_ω of sentences having the above properties (plus some others involving constants from C) and then showed how to construct a model from such a set. The second stage of that proof constitutes the proof of this lemma. □

2.10 Lemma (Alternate form of Weak Completeness Theorem). *Let $\mathsf{L}_{\mathbb{A}}$ be a Skolem fragment with constants C. Let $\mathscr{D}$ be an s.v.p. for $(\mathsf{L}_{\mathbb{A}}, C)$ and let S be the set of sentences in $\mathscr{D}$. Then S is true in some Skolem structure for L_A; i.e., $\mathscr{D}$ has a model.*

Proof. Since $T_{\text{Skolem}} \subseteq \mathscr{D}$, any model of S will be a Skolem structure for $\mathsf{L}_{\mathbb{A}}$. We need only prove that S satisfies the rules of Lemma 2.9. Since $\mathscr{D}$ contains the axioms (A 1)—(A 7) and is closed under (R 1)—(R 3), these are all routine except for the $\bigvee$ and $\exists$ rules. The $\bigvee$-rule for S follows from the $\bigvee$-rule for $\mathscr{D}$. To check the $\exists$-rule, suppose

$$\exists x\, \varphi(x, \mathsf{c}_1, \ldots, \mathsf{c}_n) \in S\,.$$

By Lemma 2.2,

$$[\exists x\, \varphi(x, \mathsf{c}_1, \ldots, \mathsf{c}_n) \rightarrow \varphi(\mathsf{F}(\mathsf{c}_1, \ldots, \mathsf{c}_n), \mathsf{c}_1, \ldots, \mathsf{c}_n)] \in S$$

for the appropriate function symbol F. Thus,

$$\varphi(\mathsf{F}(\mathsf{c}_1, \ldots, \mathsf{c}_n), \mathsf{c}_1, \ldots, \mathsf{c}_n) \in S$$

as demanded by the $\exists$-rule. ☐

Proof of Theorem 2.8 (ii) ($\Rightarrow$). Suppose $T \cup T_{\text{Skolem}} \models \varphi$. We need to see that if $\mathscr{D}$ is an s.v.p. with $T \subseteq \mathscr{D}$ then $\varphi \in \mathscr{D}$. If not, then $\neg\varphi \in \mathscr{D}$ by Lemma 2.6. Then, appying Lemma 2.10 we would get a Skolem model of $T \cup \{\neg\varphi\}$, a contradiction. ☐

We conclude this section with a result which allows us to construct interesting supervalidity properties and hence, by Weak Completeness, interesting models. It often gives us the effect of the ordinary Compactness Theorem for $\mathsf{L}_{\omega\omega}$. Given a Skolem fragment $\mathsf{L}_{\mathbb{A}}$ with constants C_0 and a Skolem fragment $\mathsf{K}_{\mathbb{B}}$ with constants C_1 we write

$$(\mathsf{L}_{\mathbb{A}}, C_0) \subseteq (\mathsf{K}_{\mathbb{B}}, C_1)$$

if $\mathsf{L}_{\mathbb{A}} \subseteq \mathsf{K}_{\mathbb{B}}$, $C_0 \subseteq C_1$, and if $\mathsf{F}_{\exists x \varphi}$ is the Skolem function symbol assigned to $\exists x\, \varphi(x, y_1, \ldots, y_n)$ by $\mathsf{L}_{\mathbb{A}}$, then it is also the one assigned to $\exists x\, \varphi(x, y_1, \ldots, y_n)$ by $\mathsf{K}_{\mathbb{B}}$.

2.11 Union of Chain Lemma. *Let I be a lineary ordered index set and suppose that, for each $i \in I$, $\mathsf{L}_{\mathbb{A}}^{(i)}$ is a Skolem fragment with constants C_i and $\mathscr{D}_i$ is a supervalidity property for $(\mathsf{L}_{\mathbb{A}}^{(i)}, C_i)$. Suppose, further, that for all $i, j \in I$, with $i < j$,*

$$(\mathsf{L}_{\mathbb{A}}^{(i)}, C_i) \subseteq (\mathsf{L}_{\mathbb{A}}^{(j)}, C_j) \quad \text{and} \quad \mathscr{D}_i \subseteq \mathscr{D}_j\,.$$

Let $\mathsf{K}_{\mathbb{B}} = \bigcup_{i \in I} \mathsf{L}_{\mathbb{A}}^{(i)}$, $C_\infty = \bigcup_{i \in I} C_i$, $\mathscr{D}_\infty = \bigcup_{i \in I} \mathscr{D}_i$. Then $\mathsf{K}_{\mathbb{B}}$ is a Skolem fragment with constants C_∞, and $\mathscr{D}_\infty$ is a supervalidity property for $(\mathsf{K}_{\mathbb{B}}, C_\infty)$.

Proof. Simple checking of the definition shows that $\mathsf{K}_{\mathbb{B}}$ is a Skolem fragment with constants C_∞. The Skolem theory for $(\mathsf{K}_{\mathbb{B}}, C_\infty)$ is the union of the Skolem theories for the various $(\mathsf{L}_{\mathbb{A}}^{(i)}, C_i)$ so the Skolem theory for $(\mathsf{K}_{\mathbb{B}}, C_\infty)$ is contained in $\mathscr{D}_\infty$. Similarly, the axioms (A 1)—(A 7) for $\mathsf{K}_{\mathbb{B}}$ are all in $\mathscr{D}_\infty$. It is a trivial matter to check (R 1), (R 2) and the $\bigvee$-rule. This time it is the $\bigwedge$-rule which requires a moment's thought. Suppose $\bigwedge\Phi\in\mathsf{K}_{\mathbb{B}}$ and that, for each $\varphi\in\Phi$, $\varphi\in\mathscr{D}_\infty$. We need to check that $\bigwedge\Phi\in\mathscr{D}_\infty$. Choose i so that $\bigwedge\Phi\in\mathsf{L}_{\mathbb{A}}^{(i)}$. We claim that, for each $\varphi\in\Phi$, $\varphi\in\mathscr{D}_i$ (so that $\bigwedge\Phi\in\mathscr{D}_i\subseteq\mathscr{D}_\infty$). Otherwise, suppose $\varphi=\varphi(v_1,\ldots,v_n)\in\Phi$ but that $\varphi\notin\mathscr{D}_i$. Then

$$\forall v_1,\ldots,v_n\,\varphi(v_1,\ldots,v_n)\notin\mathscr{D}_i.$$

By completeness (Lemma 2.6),

$$\neg\forall v_1,\ldots,v_n\,\varphi(v_1,\ldots,v_n)\in\mathscr{D}_i.$$

But $\varphi(v_1,\ldots,v_n)\in\mathscr{D}_\infty$ so for some $j>i$, $\varphi(v_1,\ldots,v_n)\in\mathscr{D}_j$. Hence

$$\forall v_1,\ldots,v_n\,\varphi(v_1,\ldots,v_n)\in\mathscr{D}_j.$$

But since $\mathscr{D}_i\subseteq\mathscr{D}_j$, this contradicts the consistency requirement for the validity property $\mathscr{D}_j$. □

All known applications of 2.11 follow from the following very special case. It exhibits the role of constants in our notion of Skolem fragment.

2.12 Union of Chain Lemma (Special form). *Let $\mathsf{L}_{\mathbb{A}}$ be a Skolem fragment. Let $C=\{\mathsf{c}_n|0<n<\omega\}$ be a countable set of new constant symbols. Suppose that for each n, $\mathscr{D}_n$ is an s.v.p. for $\mathsf{L}_{\mathbb{A}}(\mathsf{c}_1,\ldots,\mathsf{c}_n)$ and that $\mathscr{D}_n\subseteq\mathscr{D}_m$ for $n\leqslant m$. Let $\mathscr{D}_\infty=\bigcup_n\mathscr{D}_n$. Then $\mathscr{D}_\infty$ is an s.v.p. for $\mathsf{L}_{\mathbb{A}}(C)$.*

Proof. $(\mathsf{L}_{\mathbb{A}}(\mathsf{c}_1,\ldots,\mathsf{c}_n),\{\mathsf{c}_1,\ldots,\mathsf{c}_n\})\subseteq(\mathsf{L}_{\mathbb{A}}(\mathsf{c}_1,\ldots,\mathsf{c}_m),\{\mathsf{c}_1,\ldots,\mathsf{c}_m\})$ for $n\leqslant m$ so the result follows at once from 2.11. □

Applications of the results of this section appear in the next two sections as well as in Chapter VIII.

2.13—2.16 Exercises

2.13. Let $\mathsf{L}_{\mathbb{A}}$ be a fragment if $\mathsf{L}_{\infty\omega}$ and let $\mathfrak{M}$, $\mathfrak{N}$ be L-structures. $\mathfrak{M}$ is an $\mathsf{L}_{\mathbb{A}}$-*elementary substructure* of $\mathfrak{N}$, written

$$\mathfrak{M}\prec\mathfrak{N}\;(\mathsf{L}_{\mathbb{A}})$$

if $\mathfrak{M}\subseteq\mathfrak{N}$ and for every $\varphi(v_1,\ldots,v_n)\in\mathsf{L}_{\mathbb{A}}$ and every $a_1,\ldots,a_n\in\mathfrak{M}$

$$\mathfrak{M}\models\varphi[a_1,\ldots,a_n]\quad\text{iff}\quad\mathfrak{N}\models\varphi[a_1,\ldots,a_n].$$

(i) Prove that if $\mathfrak{M} \subseteq \mathfrak{N}$ then $\mathfrak{M} \prec \mathfrak{N}\ (\mathsf{L}_{\mathbb{A}})$ iff for every formula $\exists x\, \varphi(x, y_1, \ldots, y_n) \in \mathsf{L}_{\mathbb{A}}$ and every $a_1, \ldots, a_n \in \mathfrak{M}$, if

$$\mathfrak{N} \models \exists x\, \varphi(x, a_1, \ldots, a_n)$$

then there is a $b \in \mathfrak{M}$ such that

$$\mathfrak{N} \models \varphi(b, a_1, \ldots, a_n).$$

(ii) Prove that if

$$\mathfrak{M}_\alpha \prec \mathfrak{M}_\beta\ (\mathsf{L}_{\mathbb{A}})$$

for $\alpha < \beta < \gamma$ and $\mathfrak{M} = \bigcup_{\beta < \gamma} \mathfrak{M}_\beta$, then

$$\mathfrak{M}_\beta \prec \mathfrak{M}\ (\mathsf{L}_{\mathbb{A}})$$

for all $\beta < \gamma$.

2.14. Let $\mathsf{L}_{\mathbb{A}}$ be a Skolem fragment with constants and let $\mathfrak{M}$, $\mathfrak{N}$ be Skolem structures for $\mathsf{L}_{\mathbb{A}}$. Show that if $\mathfrak{M} \subseteq \mathfrak{N}$ then $\mathfrak{M} \prec \mathfrak{N}\ (\mathsf{L}_{\mathbb{A}})$. [Use 2.13 (i).]

2.15 (Downward Lowenheim-Skolem-Tarski Theorem). Let $\mathsf{L}_{\mathbb{A}}$ be a fragment of $\mathsf{L}_{\infty\omega}$ and let $\kappa \geqslant \mathrm{card}(\mathsf{L}_{\mathbb{A}})$. Let $\mathfrak{M}$ be an L-structure, $X \subseteq \mathfrak{M}$, $\kappa \leqslant \mathrm{card}(\mathfrak{M})$, $\mathrm{card}(X) \leqslant \kappa$. Prove that there is an $\mathfrak{N}$ with

$$\mathfrak{N} \prec \mathfrak{M}\ (\mathsf{L}_{\mathbb{A}}), \quad \mathrm{card}(\mathfrak{N}) = \kappa, \quad \text{and} \quad X \subseteq \mathfrak{N}.$$

[By 2.4 you may assume $\mathsf{L}_{\mathbb{A}}$ is a Skolem fragment and that $\mathfrak{M}$ is a Skolem structure for $\mathsf{L}_{\mathbb{A}}$.]

2.16. If $\mathfrak{M}$ is an L-structure and $X \subseteq M$ then $\mathrm{Hull}_{\mathfrak{M}}(X)$ is the smallest substructure of $\mathfrak{M}$ containing X.

(i) Prove

$$\mathrm{card}\,(\mathrm{Hull}_{\mathfrak{M}}(X)) = \max\{\aleph_0, \mathrm{card}(\mathsf{L}), \mathrm{card}(X)\}.$$

(ii) Prove that if $\mathfrak{M}$ is a Skolem structure for a Skolem fragment $\mathsf{L}_{\mathbb{A}}$ and $X \subseteq \mathfrak{M}$ then

$$\mathrm{Hull}_{\mathfrak{M}}(X) \prec \mathfrak{M}(\mathsf{L}_{\mathbb{A}}).$$

2.17 Notes. The essential content of the Weak Completeness Theorem is as old as the Henkin [1949] proof of the completeness theorem for $\mathsf{L}_{\omega\omega}$. As we have tried to suggest in 2.9, it is implicit in the Model Existence Theorem. Only recently, however, has it become clear that the result is useful enough to deserve to be called a Weak Completeness Theorem. (The perjorative "weak" is there for the same reason as in § III.4; there is no nice notion of provability to go along with it.)

The first explicit statement of the Weak Completeness Theorem appears as Lemma 1.5 in Barwise-Kunen [1971], where it was used to attack the model theory of uncountable fragment.

Our treatment of Skolem fragments is a modification of that contained in Lecture 13 of Keisler [1971]. In particular, the exercises are proven there (in the countable case).

3. Pinning Down Ordinals: the General Case

Several of the examples in § 1 hinge on our ability to pin down ordinals larger then $o(\mathbb{A})$ by a Σ_1 theory of $L_{\mathbb{A}}$, for certain uncountable admissible sets $\mathbb{A}$. We will see, in fact, that a good deal of the model theory of uncountable, admissible fragments revolves about this question of pinning down ordinals. For this reason we choose it as the first application of the Weak Completeness Theorem.

The proof of the next theorem proves more than we state. In fact, it will allow us to compute exactly the ordinals pinned down by theories, once we develop some recursion theoretic machinery in the next chapter. For now we content ourselves with a crude statement of the result.

3.1 Theorem. *Let $T=T(<,\ldots)$ be a set of sentences of $L_{\infty\omega}$. If T pins down ordinals then there is a ξ such that all ordinals pinned down by T are less than ξ.*

Proof. We may assume that T has models since otherwise $\xi=0$ will do. We may also assume that if T pins down α and $\beta<\alpha$ the T pins down β, by a remark in § III.7. By 2.4 we may assume that $T\subseteq L_{\mathbb{A}}$ where $L_{\mathbb{A}}$ is a Skolem fragment and that $T_{\text{Skolem}}\subseteq T$. We want to set things up to apply the special form of 2.12, the Union of Chain Lemma, so let $C=\{c_n \mid 0<n<\omega\}$ be a set of new constant symbols. Let $\mathfrak{S}_n$ be the set of all supervalidity properties $\mathscr{D}$ for $L_{\mathbb{A}}(c_1,\ldots,c_n)$ (this is just $L_{\mathbb{A}}$ if $n=0$) such that

$$T\subseteq\mathscr{D} \quad\text{and}\quad (c_2<c_1)\in\mathscr{D},\ldots,(c_n<c_{n-1})\in\mathscr{D}.$$

(For $n=0,1$, none of the sentences involving the c_i occur.) Since T has a model $\mathfrak{M}$, the s.v.p. $\mathscr{D}_{\mathfrak{M}}$ given by $\mathfrak{M}$ is in $\mathfrak{S}_0$, so $\mathfrak{S}_0\neq\emptyset$. Let

$$\mathfrak{S}=\bigcup\nolimits_{0\leqslant n<\omega}\mathfrak{S}_n$$

and put an ordering $\prec$ on $\mathfrak{S}$ by

$$\mathscr{D}'\prec\mathscr{D}$$

if $\mathscr{D}\subseteq\mathscr{D}'$ and the (unique) n such that $\mathscr{D}'\in\mathfrak{S}_n$ is greater than the unique m such that $\mathscr{D}\in\mathfrak{S}_m$. (Note that for $\mathscr{D}\in\mathfrak{S}$, we can tell which n has $\mathscr{D}\in\mathfrak{S}_n$ by just seeing what the largest n is such that $(c_n=c_n)\in\mathscr{D}$.) We claim that

(1) $\langle\mathfrak{S},\prec\rangle$ is well founded.

For suppose

$$\mathscr{D}_0 \succ \mathscr{D}_1 \succ \mathscr{D}_2 \succ \cdots.$$

Let $\mathscr{D}_\infty = \bigcup_n \mathscr{D}_n$. By the union of chain lemma, $\mathscr{D}_\infty$ is an s.v.p. and hence, by the Weak Completeness Theorem, there is a model

$$(\mathfrak{M}, a_1, a_2, \ldots)$$

of $\mathscr{D}_\infty$, where a_n is the interpretation of c_n. But then $\mathfrak{M} \models T$, and $a_{n+1} < a_n$ for all $n < \omega$ which contradicts the hypothesis that T pins down ordinals. This proves (1).

Using (1) it is easy to get an upper bound for the ordinals pinned down by T. By (1), each $\mathscr{D} \in \mathfrak{S}$ has an ordinal rank $\rho(\mathscr{D})$,

$$\rho(\mathscr{D}) = \sup\{\rho(\mathscr{D}') + 1 \mid \mathscr{D}' \in \mathfrak{S}, \mathscr{D}' \prec \mathscr{D}\},$$

and $\langle \mathfrak{S}, \prec \rangle$ has a rank

$$\xi = \sup\{\rho(\mathscr{D}) + 1 \mid \mathscr{D} \in \mathfrak{S}\}.$$

We will prove that

(2) if $\mathscr{D} \in \mathfrak{S}_n$ and $(\mathfrak{M}, a_1, \ldots, a_n) \models \mathscr{D}$ then the $<^{\mathfrak{M}}$ predecessors of a_n have order type $\leqslant \rho(\mathscr{D})$ when $n > 0$; if $n = 0$ then $<^{\mathfrak{M}}$ has order type $\leqslant \rho(\mathscr{D})$.

Since every $\mathfrak{M} \models T$ is a model of $\mathscr{D}_{\mathfrak{M}} \in S_0$, and $\rho(\mathscr{D}_{\mathfrak{M}}) < \xi$, (2) gives us:

(3) *every model $\mathfrak{M}$ of T has $<^{\mathfrak{M}}$ of order type less than ξ,*

which proves the theorem. We prove (2) by induction on $\rho(\mathscr{D})$. Suppose $\mathscr{D} \in \mathfrak{S}_n$, $\alpha = \rho(\mathscr{D})$, $(\mathfrak{M}, a_1, \ldots, a_n) \models \mathscr{D}$ but that the predecessors of a_n have order type $> \alpha$. (The case $n = 0$ is essentially the same.) Let a_{n+1} be the αth member of the field of $<^{\mathfrak{M}}$ as ordered by $<^{\mathfrak{M}}$ and let $\mathscr{D}'$ be the s.v.p. given by

$$\mathfrak{M}' = (\mathfrak{M}, a_1, \ldots, a_{n+1}).$$

Then $\mathscr{D}' \in \mathfrak{S}_{n+1}$, and $\mathscr{D} \subseteq \mathscr{D}'$ so $\mathscr{D}' \prec \mathscr{D}$ and hence $\rho(\mathscr{D}') < \alpha$. But $\mathfrak{M}'$ is a model of $\mathscr{D}'$ with the precedessors of a_{n+1} of order type $\alpha > \rho(\mathscr{D}')$, contradicting the inductive hypothesis. □

Without Theorem 3.1 we could not be sure that the next definition made sense.

3.2 Definition. Let $\mathbb{A}$ be an admissible set.

(i) $h(\mathbb{A})$ is the least ordinal not pinned down by some sentence $\varphi(<, \ldots)$ in some admissible fragment $\mathsf{L}_{\mathbb{A}}$.

(ii) $h_\Sigma(\mathbb{A})$ is the least ordinal not pinned down by some Σ_1 theory of some admissible fragment $\mathsf{L}_{\mathbb{A}}$.

In the next chapter we will determine exact recursion-theoretic descriptions of $h_\Sigma(\mathbb{A})$ and, in most cases, of $h(\mathbb{A})$.

Let us collect together remarks made at various places.

3.3 Proposition. *Let $\mathbb{A}$ be admissible.*

(i) *$h_\Sigma(\mathbb{A})$ is the* sup *of the ordinals pinned down by Σ_1 theories of $\mathsf{L}_{\mathbb{A}}$; similarly, $h(\mathbb{A})$ is the* sup *of the ordinals pinned down by single sentences of $L_{\mathbb{A}}$.*

(ii) $h_\Sigma(\mathbb{A}) \geqslant h(\mathbb{A}) \geqslant o(\mathbb{A})$.

(iii) *If $\mathbb{A}$ is countable then*

$$h_\Sigma(\mathbb{A}) = h(\mathbb{A}) = o(\mathbb{A}).$$

(iv) *If $\mathbb{A}$ is Σ_1 compact then*

$$h_\Sigma(\mathbb{A}) = h(\mathbb{A}).$$

Proof. Only (iv) needs proving. Suppose $\mathbb{A}$ is Σ_1 compact but that $h_\Sigma(\mathbb{A}) > h(\mathbb{A})$. Let $T(<)$ be a Σ_1 theory which pins down some $\beta \geqslant h(\mathbb{A})$. Add new constant symbols $\mathsf{c}_1, \ldots, \mathsf{c}_n, \ldots$ and let T' be T plus the axioms

$$\mathsf{c}_{n+1} < \mathsf{c}_n \quad (\text{all } n < \omega).$$

Since $\beta \geqslant h(\mathbb{A})$, every $\mathbb{A}$-finite subset of T has a model which is not well founded so every $\mathbb{A}$-finite subset of T' has a model. Thus, by Σ_1 compactness, T' has a model, a contradiction. □

$H(\omega_1)$ is an example of a set $\mathbb{A}$ with $h_\Sigma(\mathbb{A}) > h(\mathbb{A}) = o(\mathbb{A})$. $\mathbb{H}\mathrm{YP}(H(\omega_1))$ is an example with $h_\Sigma(\mathbb{A}) = h(\mathbb{A}) > o(\mathbb{A})$.

The next theorem is extremely useful in computations which involve $h_\Sigma(A)$ and $h(A)$.

3.4 Theorem. *Let $\mathbb{A}$ be admissible and let $F\colon \mathrm{Ord}^n \to \mathrm{Ord}$ be an n-ary function on ordinals which is Σ_1 definable in* KPU.

(i) $\alpha_1, \ldots, \alpha_n < h_\Sigma(\mathbb{A})$ *implies* $F(\alpha_1, \ldots, \alpha_n) < h_\Sigma(\mathbb{A})$.

(ii) $\alpha_1, \ldots, \alpha_n < h(\mathbb{A})$ *implies* $F(\alpha_1, \ldots, \alpha_n) < h(\mathbb{A})$.

Proof. We first prove (i) in case $n = 2$. The case for $n \neq 2$ is similar. Let

$$F\colon \mathrm{Ord} \times \mathrm{Ord} \to \mathrm{Ord}$$

be Σ_1 definable in KPU, hence in the stronger KP, say by the Σ_1 formula $\sigma(x_1, x_2, y)$:

(4) $\mathrm{KP} \vdash \forall x_1 x_2 \, \exists! y \, \sigma(x_1, x_2, y)$,

(5) for all $\alpha_1, \alpha_2 \; \mathbb{V} \models \sigma(\alpha_1, \alpha_2, F(\alpha_1, \alpha_2))$.

Suppose $\alpha_1, \alpha_2 < h_\Sigma(\mathbb{A})$ and let $\beta = F(\alpha_1, \alpha_2)$. We need to prove that $\beta < h_\Sigma(\mathbb{A})$. Let $T_1(<_1, \mathsf{R}_1)$, $T_2(<_2, \mathsf{R}_2)$ be Σ_1 theories which pin down α_1, α_2 respectively, the case with more relation symbols being similar. We will define a Σ_1 theory $T(<)$ which pins down β, but first let us exhibit its intended model $\mathfrak{M}$, the one with $<^{\mathfrak{M}}$ of type β. Let κ be a regular cardinal, $\alpha_1, \alpha_2 < \kappa$, so that $\beta < \kappa$. Let

$$\mathfrak{M}_1 = \langle M_1, <_1, R_1 \rangle \models T_1, \quad <_1 \text{ of order type } \alpha_1,$$

$$\mathfrak{M}_2 = \langle M_2, <_2, R_2 \rangle \models T_2, \quad <_2 \text{ of order type } \alpha_2.$$

By the downward Löwenheim-Skolem Theorem (Exercise 2.15) (and the fact that isomorphic models satisfy the same sentences) we may assume

$$\alpha_i \subseteq M_i \subseteq \kappa \quad \text{and} \quad <_i = \in \restriction \alpha_i .$$

Now let

$$\mathfrak{M} = \langle H(<), \in, <, M_1, <_1, R_1, M_2, <_2, R_2, \alpha_1, \alpha_2, \beta \rangle$$

where $< = \in \restriction \beta$ and α_1, α_2 and β are treated as elements, not as subsets. Then $\mathfrak{M}$ is clearly a model of the following set of sentences, where U_i is interpreted as M_i, c_i is interpreted as α_i and d as β.

$$\varphi^{(\mathsf{U}_1)} \quad \text{for all} \quad \varphi \in T_1,$$

$$\varphi^{(\mathsf{U}_2)} \quad \text{for all} \quad \varphi \in T_2,$$

$$\text{KP},$$

$$\mathsf{c}_1, \mathsf{c}_2, \mathsf{d} \text{ are ordinals},$$

$$\text{“} <_1 = \in \restriction \mathsf{c}_1 \text{”},$$

$$\text{“} <_2 = \in \restriction \mathsf{c}_2 \text{”},$$

$$\text{“} < \; = \in \restriction \mathsf{d} \text{”},$$

$$\sigma(\mathsf{c}_1, \mathsf{c}_2, \mathsf{d}).$$

If we call the above set of sentences $T(<, \ldots)$, then $\mathfrak{M}$ is a model of T with $<^{\mathfrak{M}}$ of order type β. We need to prove that every model $\mathfrak{M}$ of T has $<^{\mathfrak{M}}$ well ordered. Thus, let

$$\mathfrak{M} = \langle M, E, <, U_1, <_1, R_1, U_2, <_2, R_2, a_1, a_2, d \rangle$$

be any model of T. Identify the well-founded part of $\langle M, E \rangle$ with an admissible set $\langle B, \in \rangle$ by the Truncation Lemma. Since, for $i = 1, 2$

$$\langle U_i, <_i, R_i \rangle \models T_i,$$

$<_i$ is well ordered, so a_1 and a_2 are real ordinals and $a_1, a_2 \in B$. By (4),

$$\langle B, \in \rangle \models \exists ! y \, \sigma(a_1, a_2, y).$$

By (5), and the persistence of Σ formulas,

$$\langle B,\in\rangle \models \sigma(a_1,a_2,F(a_1,a_2))$$

and, since

$$\langle B,\in\rangle \subseteq_{\text{end}} \langle M,E\rangle,$$

we have by persistence,

$$\langle M,E\rangle \models \sigma(a_1,a_2,F(a_1,a_2))$$

so that $b=F(a_1,a_2)$. Since $b=F(a_1,a_2)\in B$, and $<\,=\in\restriction b$, $<$ is a real well-ordering. This proves (i).

The proof of (ii) is exactly the same when $o(\mathbb{A})>\omega$, since then we may form $\bigwedge$KP and the rest as a single sentence of $L_{\mathbb{A}}$. If $o(\mathbb{A})=\omega$ we must replace KP by a single sentence θ of ZF-Power (and hence true in $H(\kappa)$ since κ is regular) strong enough to insure that the standard part of any model of θ is an admissible set. We leave this to the student. ☐

All we will actually need of Theorem 3.4 is the following special case.

3.5 Corollary. *Let $\mathbb{A}$ be admissible. Then $h_\Sigma(\mathbb{A})$ and $h(\mathbb{A})$ are closed under ordinal successor, ordinal addition, multiplication, and exponentiation.*

Proof. We have shown that all these functions are Σ_1 definable in KP. ☐

The final result of this section seems almost obvious, but it needs proof.

3.6 Theorem. *Let $\mathbb{A}$ be admissible.*

(i) *If T is a Σ_1 theory of $L_{\mathbb{A}}$ which pins down ordinals then there is a $\xi<h_\Sigma(\mathbb{A})$ such that every ordinal pinned down by T is less than ξ.*

(ii) *If φ is a sentence of $L_{\mathbb{A}}$ which pins down ordinals then there is a $\xi<h(\mathbb{A})$ which is greater than all ordinals pinned down by φ.*

Proof. This is a typical example of a proof in soft model theory since the proof works for any logic. We prove (ii). We may assume that the sentence $\varphi(<)$ pins down an initial segment $\{\beta\,|\,\beta<\xi\}=\xi$ of ordinals. We show that some other sentence $\psi(\prec,\ldots)$ pins down ξ. As before, before writing down ψ, we describe its intended model $\mathfrak{M}$, the one with $\prec^{\mathfrak{M}}$ of type ξ. To simplify matters we assume $\varphi=\varphi(<,\mathsf{R})$, where R is binary, contains no other symbols. For each $\beta<\xi$, let

$$\mathfrak{M}_\beta=\langle M_\beta,<_\beta,R_\beta\rangle,\quad \mathfrak{M}\models\varphi \quad\text{and}\quad <_\beta \text{ have order type } \beta.$$

Since isomorphic structures satisfy the same sentences, we can rearrange $\mathfrak{M}_\beta$ a bit and assume $\beta\subseteq\mathfrak{M}_\beta$ and $<_\beta=\in\restriction\beta$.

Define $\mathfrak{M} = \langle M, U, \prec, N, S_1, S_2 \rangle$ where

$$M = \bigcup_{\beta < \xi} M_\beta, \qquad U = \xi \subseteq M,$$

$$N(\beta, x) \quad \text{iff} \quad \beta < \xi \wedge x \in M_\beta,$$

$$\beta \prec \gamma \quad \text{iff} \quad \beta \in \gamma \in \xi,$$

$$S_1(\beta, y, z) \quad \text{iff} \quad \beta < \xi \wedge y <_\beta z,$$

$$S_2(\beta, y, z) \quad \text{iff} \quad \beta < \xi \wedge R_\beta(y, z).$$

Thus $\mathfrak{M}$ is a structure where $\prec^{\mathfrak{M}}$ has order type ξ. Let $\psi(\prec, \ldots)$ be the sentence described as follows. Let $\theta(x, \mathsf{N}, \mathsf{S}_1, \mathsf{S}_2)$ result from $\varphi(<, \mathsf{R})$ by replacing

$$y < z \quad \text{by} \quad \mathsf{S}_1(x, y, z),$$

$$\mathsf{R}(y, z) \quad \text{by} \quad \mathsf{S}_2(x, y, z),$$

$$\forall y\,(\ldots) \quad \text{by} \quad \forall y\,(\mathsf{N}(x, y) \rightarrow \cdots), \quad \text{and}$$

$$\exists y\,(\ldots) \quad \text{by} \quad \exists y\,(\mathsf{N}(x, y) \wedge \cdots)$$

taking care to avoid clashes of variables. Let ψ be the conjunction of:

(6) $\forall x\,[\mathsf{U}(x) \rightarrow \theta(x, \mathsf{N}, \mathsf{S}_1, \mathsf{S}_2)]$;

(7) "U is linearly ordered by $\prec$";

(8) $\forall x\,[\mathsf{U}(x) \rightarrow \forall y, z(y \prec z \prec x \leftrightarrow \mathsf{S}_1(x, y, z))]$.

It is clear that $\mathfrak{M}$ is a model of ψ since (6) just asserts that each $\mathfrak{M}_\beta$ is a model of θ. We need to show that any other model

$$\mathfrak{M} = \langle M, U, \prec, N, S_1, S_2 \rangle$$

of ψ has $\prec$ well ordered. To do this it suffices to prove that for any $x \in U$, the $\prec$ predecessors of x are well ordered. Let

$$\mathfrak{M}_x = \langle M_x, <_x, R_x \rangle$$

where $M_x = \{y \mid N(x, y)\}$, $y <_x z$ iff $S_1(x, y, z)$ and $R_x(y, z)$ iff $S_2(x, y, z)$. By (6), $\mathfrak{M}_x \models \varphi$, so $<_x$ is a well-ordering and $<_x$ agrees with $\prec$ on the predecessors of x. Thus ψ does pin down ordinals, ξ among them. □

3.7—3.8 Exercises

3.7. Let $\mathbb{A}$ be admissible, $o(\mathbb{A}) = \omega$, where $\mathbb{A}$ is Σ_1 compact. Show that

$$h(\mathbb{A}) = \omega.$$

3.8. Let $\mathbb{A}$ be Σ_1 compact and suppose that $\alpha = o(\mathbb{A}) > \omega$ is such that for some $x \in \mathbb{A}$,

$$\alpha = \text{least}\, \beta\ (\mathrm{L}(x, \beta) \text{ is admissible}).$$

Prove that

$$h_\Sigma(\mathbb{A}) = o(\mathbb{A}).$$

3.9 Notes. Theorem 3.1 is due to Lopez-Escobar [1966]. His proof, however, was by way of Hanf numbers and gave no clue as to the exact description of $h(\mathbb{A})$ or $h_\Sigma(\mathbb{A})$, even for $\mathbb{A} = H(\aleph_{\alpha+1})$. The proof given here is taken from Barwise-Kunen [1971]. Theorem 3.4 is also taken from there.

There are, by the way, admissible sets which are Σ_1 compact but such that $h_\Sigma(\mathbb{A}) > o(\mathbb{A})$. This follows from Theorem VIII.8.3. It is known that $h(\mathbb{A})$ need not be admissible. It is not known whether $h_\Sigma(\mathbb{A})$ is always admissible, though it seems unlikely.

4. *Indiscernibles and Upward Löwenheim-Skolem Theorems*

In this section we show how to use the Weak Completeness Theorem and the ordinal $h_\Sigma(A)$ to tackle some model theoretic problems for L_A. The material in this section is not used elsewhere in this book.

The simplest result to state is the following theorem, stated in terms of the Beth sequence. Given a cardinal κ, define the cardinal $\beth_\alpha(\kappa)$ by induction on α:

$$\beth_0(\kappa) = \kappa,$$
$$\beth_{\alpha+1}(\kappa) = 2^{\beth_\alpha(\kappa)},$$
$$\beth_\lambda(\kappa) = \sup_{\alpha<\lambda} \beth_\alpha(\kappa).$$

We write $\beth_\alpha$ for $\beth_\alpha(0)$, but warn the reader that some authors use $\beth_\alpha$ for $\beth_\alpha(\aleph_0)$. With our definition, $\beth_\alpha = \mathrm{card}(V_\alpha)$.

4.1 Theorem. *Let $\mathbb{A}$ be an admissible set, let $\kappa = \mathrm{card}(\mathbb{A})$ and $\alpha = h_\Sigma(\mathbb{A})$. Let T be a Σ_1 theory of $L_{\mathbb{A}}$. If, for each $\beta < \alpha$, T has a model of power $\geqslant \beth_\beta(\kappa)$, then for any $\lambda \geqslant \kappa$, T has a model of power λ.*

The proof of 4.1 is given in 4.13 below. Actually the proof of this theorem is no more complicated for uncountable L_A; it is just that for countable $\mathbb{A}$ we know that $h_\Sigma(\mathbb{A}) = o(\mathbb{A})$. Thus 4.1 gives us the following corollary.

4.2 Corollary. *Let $\mathbb{A}$ be a countable, admissible set and let T be a Σ_1 theory of $L_{\mathbb{A}}$. If, for each $\beta < \alpha = o(\mathbb{A})$, T has a model of power $\geqslant \beth_\beta(\aleph_0)$, then T has a model of each infinite power.* ☐

If $\mathbb{A}_{\mathfrak{M}}$ is not $\mathbb{HF}_{\mathfrak{M}}$ then it is easy to show that for each $\beta \in A_{\mathfrak{M}}$ there is a sentence of $\mathsf{L}_{\mathbb{A}}$ which has a model of power $\beth_\beta(\aleph_0)$ but none larger (see Exercise 4.18), so 4.2 is best possible for $\mathbb{A}_{\mathfrak{M}} \neq \mathbb{HF}_{\mathfrak{M}}$. For $\mathbb{A}_{\mathfrak{M}} = \mathbb{HF}_{\mathfrak{M}}$, $\mathsf{L}_{\mathbb{A}} = \mathsf{L}_{\omega\omega}$ so we know a better result.

For applications, there are more useful upward Löwenheim-Skolem Theorems in terms of two cardinal models.

Assume our language L has a unary symbol U. A model $\mathfrak{M} = \langle M, U, \ldots \rangle$ for L is a *model of type* (κ, λ) if

$$\operatorname{card}(M) = \kappa,$$

$$\operatorname{card}(U) = \lambda.$$

A set T of sentences of $\mathsf{L}_{\infty\omega}$ is said to *admit* (κ, λ) if T has a model $\mathfrak{M}$ of type (κ, λ).

4.3 Theorem. *Let $\mathsf{L}_{\mathbb{A}}$ be an admissible fragment, let $\kappa = \operatorname{card}(\mathbb{A})$, $\alpha = h_\Sigma(A)$. Let T be a Σ_1 theory of $\mathsf{L}_{\mathbb{A}}$. If for each $\beta < \alpha$ there is a $\lambda \geqslant \kappa$ such that T admits $(\beth_\beta(\lambda), \lambda)$, then T admits (δ, κ) for all cardinals $\delta \geqslant \kappa$.*

Theorem 4.1 is an easy consequence of 4.3 by adding a new symbol U to L without mentioning it in the theory T of 4.1. On the other hand, a direct proof of 4.1 is a bit simpler than the proof of 4.3, and since the student may be interested in 4.1, we will also give a direct proof of it.

4.4 Corollary. *Let T be a Σ_1 theory of a countable admissible fragment $\mathsf{L}_{\mathbb{A}}$. Suppose that for each $\beta < \alpha = o(\mathbb{A})$, there is a $\lambda \geqslant \omega$ such that T admits $(\beth_\beta(\lambda), \lambda)$. Then T admits (λ, ω) for all $\lambda \geqslant \omega$.*

Proof. Immediate from 4.3 since $h_\Sigma(\mathbb{A}) = o(\mathbb{A})$. ▯

4.5 Corollary (Morley's Two Cardinal Theorem). *Let T be a countable theory of $\mathsf{L}_{\omega_1\omega}$. Suppose that for each $\alpha < \omega_1$ there is a $\lambda \geqslant \omega$ such that T admits $(\beth_\alpha(\lambda), \lambda)$. Then T admits (λ, ω) for all $\lambda \geqslant \omega$.*

Proof. Immediate from 4.4 by putting T in some countable admissible fragment. ▯

The reader of Keisler [1971] will have discovered many applications of Corollary 4.5. Some of these have routine generalizations using 4.3.

Two-cardinal models are extremely natural when one is working with models of set theory of urelements. How many times have we written a typical model of KPU as a single sorted structure

$$\mathfrak{A}_{\mathcal{M}} = \langle A \cup M, M, \ldots \rangle?$$

In fact, we can use such models to prove that Theorem 4.3 is an optimal result of its type, except for trivial generalizations using downward Löwenheim-Skolem arguments.

4.6 Example. *Let* $\mathbb{A}$ *be an admissible set with* $\kappa = \operatorname{card}(\mathbb{A})$, $\alpha = h_\Sigma(\mathbb{A})$. *For any* $\beta < \alpha$ *one can find a* Σ_1 *theory* $T = T(\mathsf{U}, \ldots)$ *of* $\mathsf{L}_{\mathbb{A}}$ *and a* ξ, $\beta < \xi < h_\Sigma(\mathbb{A})$ *such that*

(i) T *has a model of type* $(\beth_\beta(\kappa), \kappa)$.

(ii) *If* $\mathfrak{M}$ *is a model of* T *of type* (λ, δ) *then* $\lambda \leqslant \beth_\xi(\delta)$. *In particular,* T *has no model of type* $(\beth_\alpha(\kappa), \kappa)$.

Proof. Let $T_0 = T_0(<)$ be a Σ_1 theory of $\mathsf{L}_{\mathbb{A}}$ which pins down β. Let $\xi < h_\Sigma(\mathbb{A})$ be greater than all ordinals pinned down by T_0, by Theorem 3.6. Before describing T we describe its intended model, the one of type $(\beth_\beta(\kappa), \kappa)$. Let M be a set of urelements of power κ. Let

$$\mathfrak{M}_0 = (M_0, <, \ldots)$$

be a model of T_0 where $<$ has order type β. By the Downward Lowenheim-Skolem theorem we may assume $\operatorname{card}(\mathfrak{M}_0) \leqslant \max(\kappa, \operatorname{card}(\beta))$ so we may as well assume $M_0 \subseteq M \cup \beta$. Now let

$$\mathfrak{M} = (M \cup \mathbb{V}_{\mathfrak{M}}(\beta), M, \in, F, M_0, <, \ldots)$$

where, by definition,

$$F_0(a) = \text{rank of } a \text{ in } \mathbb{V}_M,$$

$$F(a) \;= \text{the } F_0(a)\text{-th member of } <.$$

The theory T is defined as follows. For each $x \in A$ let c_x be a constant symbol, so there are κ of them. T consists of

$\mathsf{c}_x \neq \mathsf{c}_y$ for all $x, y \in A$, $x \neq y$,

$\mathsf{U}(\mathsf{c}_x)$ for all $x \in A$,

Extensionality (as in KPU),

$\forall x\, \forall y\, [x \in y \rightarrow \mathsf{F}(x) < \mathsf{F}(y)]$,

$\varphi^{(\mathsf{U}_0)}$ for all $\varphi \in T_0$.

Here U and U_0 are new unary symbols. The theory T clearly holds in M. On the other hand, if $\mathfrak{M} = \langle A, U, E, F, U_0, <, \ldots \rangle$ is another model of T then $\langle U_0, <, \ldots \rangle \models T_0$, so $<$ is well ordered of order type $< \xi$. But then F insures that E is well founded and of rank $< \xi$ so $\langle A, U, E \rangle$ is isomorphic to a submodel of $\mathbb{V}_U(\xi)$ and hence has $\operatorname{card}(A) \leqslant \beth_\xi(\operatorname{card}(U))$. ∎

We now turn to the tools for the proofs of these theorems. Anyone familiar with the model theory of $\mathsf{L}_{\omega\omega}$ is aware of the importance of the Ehrenfeucht-Mostowski method of indiscernibles. It plays an even more important role in the model theory of $\mathsf{L}_{\infty\omega}$.

4.7 Definition. Let $\mathsf{L}_{\mathbb{A}}$ be a fragment of $\mathsf{L}_{\infty\omega}$, $\mathfrak{M}$ be an L-structure and let $\langle X, <\rangle$ be a linearly ordered set with $X \subseteq \mathfrak{M}$. We say that $\langle X, <\rangle$ is a *set of indiscernibles* (for $\mathsf{L}_{\mathbb{A}}$ in $\mathfrak{M}$) if for every n and any two increasing n-tuples from $\langle X, <\rangle$,

$$x_1 < \cdots < x_n, \qquad y_1 < \cdots < y_n$$

we have

$$(\mathfrak{M}, x_1, \ldots, x_n) \equiv (\mathfrak{M}, y_1, \ldots, y_n) \quad (\mathsf{L}_{\mathbb{A}}),$$

i.e. the n-tuples $\langle x_1, \ldots, x_n\rangle$, $\langle y_1, \ldots, y_n\rangle$ satisfy the same formulas $\varphi(v_1, \ldots, v_n)$ of $\mathsf{L}_{\mathbb{A}}$ in $\mathfrak{M}$. If $\mathfrak{M} = \langle M, U, \ldots\rangle$ then we say that $\langle X, <\rangle$ is a *set of indiscernibles over* U if, for every finite set $u_1, \ldots, u_m \in U$ and all increasing n-typles from $\langle X, <\rangle$

$$x_1 < \cdots < x_n, \qquad y_1 < \cdots < y_n$$

we have

$$(\mathfrak{M}, u_1, \ldots, u_m, x_1, \ldots, x_n) \equiv (\mathfrak{M}, u_1, \ldots, u_m, y_1, \ldots, y_n) \quad (\mathsf{L}_{\mathbb{A}}).$$

The $<$ relation on X need not be definable on $\mathfrak{M}$ in the above definition.

The latter notion is really a special case of the first, for let $\mathfrak{M} = \langle M, U, \ldots\rangle$ be a structure for $\mathsf{L}_{\mathbb{A}}$, let $C = \{\mathsf{c}_u \mid u \in U\}$ be a set of new constant symbols, and let $\mathfrak{M}' = (\mathfrak{M}, u)_{u \in U}$ be the canonical expansion of $\mathfrak{M}$ to a model for $\mathsf{L}_{\mathbb{A}}(C)$. (The language $\mathsf{L}_{\mathbb{A}}(C)$ is defined in § 2.) Then $\langle X, <\rangle$ is a set of indiscernibles over U for $\mathsf{L}_{\mathbb{A}}$ in $\mathfrak{M}$ iff $\langle X, <\rangle$ is a set of indiscernibles for $\mathsf{L}_{\mathbb{A}}(C)$ in $\mathfrak{M}'$.

Indiscernibles help us build large models, and hence prove our theorems by means of the following Stretching Theorem.

4.8 Stretching Theorem. *Let $\mathsf{L}_{\mathbb{A}}$ be a Skolem fragment with constants and let $\mathfrak{M}$ be a Skolem structure for $\mathsf{L}_{\mathbb{A}}$. Let $\langle X, <\rangle$ be an infinite set of indiscernibles for $\mathsf{L}_{\mathbb{A}}$. For any infinite linearly ordered set $\langle Y, <\rangle$ there is a Skolem structure $\mathfrak{N}$ for $\mathsf{L}_{\mathbb{A}}$ such that:*

(i) *$\langle Y, <\rangle$ is a set of indiscernibles for $\mathsf{L}_{\mathbb{A}}$ in $\mathfrak{N}$;*

(ii) *If $x_1 < \cdots < x_n$ in $\langle X, <\rangle$ and $\langle y_1 < \cdots < y_n\rangle$ in $\langle Y, <\rangle$*

$$\textit{then} \quad (\mathfrak{M}, x_1, \ldots, x_n) \equiv (\mathfrak{N}, y_1, \ldots, y_n) \quad (\mathsf{L}_{\mathbb{A}});$$

(iii) *In particular,* $\operatorname{card}(\mathfrak{N}) \geq \operatorname{card}(Y)$ *and* $\mathfrak{M} \equiv \mathfrak{N}$ $(\mathsf{L}_{\mathbb{A}})$.

Proof. Part (iii) is just part (ii) with $n = 0$. Since the distinguished constants of $\mathsf{L}_{\mathbb{A}}$ do not play any role in this proof we simply assume $\mathsf{L}_{\mathbb{A}}$ is a Skolem fragment. Let

$$C = \{\mathsf{c}_y \mid y \in Y\}$$

be a set of new constant symbols and form $\mathsf{L}_{\mathbb{A}}(C)$ as described in § 2. Then $\mathsf{L}_{\mathbb{A}}(C)$ is a Skolem fragment with constants C. We define a set $\mathscr{D}$ of formulas of $\mathsf{L}_{\mathbb{A}}(C)$

as follows. Any formula of $\mathsf{L}_{\mathbb{A}}$ can be written in the form

$$(1)\quad \varphi(v_1,\ldots,v_n,\mathsf{c}_{y_1}/v_{n+1},\ldots,\mathsf{c}_{y_m}/v_{n+m})$$

where

$$y_1<\cdots<y_m \quad \text{in} \quad \langle Y,<\rangle.$$

Put the formula (1) into $\mathscr{D}$ just in case

$$(2)\quad (\mathfrak{M},x_1,\ldots,x_m)\models\forall v_1,\ldots,v_n\,\varphi(\vec{v},\mathsf{c}_{y_1},\ldots,\mathsf{c}_{y_m})$$

for some increasing sequence

$$x_1<\cdots<x_m \quad \text{in} \quad \langle X,<\rangle,$$

where x_i interprets c_{y_i}, of course. We claim that

(3) $\mathscr{D}$ is a supervalidity property for $\mathsf{L}_{\mathbb{A}}(C)$.

If (1) is a logical axiom, then (2) certainly holds, so $(1)\in\mathscr{D}$. We need to see that if $\varphi(\vec{v},\vec{\mathsf{c}})\in\mathscr{D}$ then $(\neg\varphi(\vec{v},\vec{\mathsf{c}}))\notin\mathscr{D}$. If not, then we would have

$$(\mathfrak{M},x_1,\ldots,x_m)\models\forall v_1,\ldots,v_n\,\varphi(\vec{v},\mathsf{c}_1,\ldots,\mathsf{c}_m),$$
$$(\mathfrak{M},x_1',\ldots,x_m')\models\forall v_1,\ldots,v_n\,\neg\varphi(\vec{v},\mathsf{c}_1,\ldots,\mathsf{c}_m)$$

where $x_1<\cdots<x_m$, $x_1'<\cdots<x_m'$ in $\langle X,<\rangle$. But this contradicts the indiscernibility of $\langle X,<\rangle$. The other clauses are equally trivial. We check the $\bigvee$-rule and leave the other three to the student. Suppose $\psi(\mathsf{c}_1,\ldots,\mathsf{c}_m)=\bigvee\Phi$ is a sentence of $\mathsf{L}_{\mathbb{A}}(C)$ and $\psi(\mathsf{c}_1,\ldots,\mathsf{c}_m)\in\mathscr{D}$. Then

$$(\mathfrak{M},x_1,\ldots,x_m)\models\bigvee\Phi$$

so, for some $\varphi\in\Phi$,

$$(\mathfrak{M},x_1,\ldots,x_m)\models\varphi$$

so $\varphi\in\mathscr{D}$. Thus $\mathscr{D}$ is a supervalidity property.

(4) If $\varphi(v_1,\ldots,v_n)\in\mathsf{L}$, $y_1<\cdots<y_n$, $y_1'<\cdots<y_n'$ in $\langle Y,<\rangle$ then the following $\mathsf{L}_{\mathbb{A}}(C)$ sentence is in $\mathscr{D}$:

$$(*)\qquad \varphi(\mathsf{c}_{y_1},\ldots,\mathsf{c}_{y_n})\leftrightarrow\varphi(\mathsf{c}_{y_1'},\ldots,\mathsf{c}_{y_n'}).$$

To see what is going on here, suppose φ is $\varphi(v_1,v_2,v_3)$ and that

$$y_1<y_2<y_3 \quad \text{and} \quad y_1'<y_2'<y_3'.$$

To see that the sentence (*) in question is in $\mathscr{D}$ we must first arrange these elements of $\langle Y, < \rangle$ in order. Suppose, for example, that

$$y_1 < y_1' < y_2' = y_2 < y_3 .$$

Thus there are only five elements in this case. Let $\psi(v_1, \ldots, v_5)$ be

$$\varphi(v_1, v_4, v_5) \leftrightarrow \varphi(v_2, v_3, v_4) .$$

The definition of $\mathscr{D}$ says that (*) is in $\mathscr{D}$ iff

$$\mathfrak{M} \models \psi[x_1, x_2, x_3, x_4, x_5]$$

whenever $x_1 < x_2 < \cdots < x_5$. That is, just in case

$$\mathfrak{M} \models \varphi[x_1, x_4, x_5] \quad \text{iff} \quad \mathfrak{M} \models \varphi[x_2, x_3, x_4]$$

whenever $x_1 < \cdots < x_5$. This is obvious from the indiscernibility of $\langle X, < \rangle$, so this proves (a typical example of) (4). Apply the Weak Completeness Theorem to get a model $(\mathfrak{N}, a_y)_{y \in Y}$ of $\mathscr{D}$. Since $(\mathsf{c}_y \neq \mathsf{c}_{y'}) \in \mathscr{D}$ for $y \neq y'$, we can identify a_y with y. Then $\mathfrak{N}$ has properties (i), (ii) of the theorem. ☐

Using the Stretching Theorem we can reduce our theorems to proving the existence of models with indiscernibles, as in the next lemma.

4.9 Lemma. *Let $\mathsf{L}_{\mathbb{A}}$ be a Skolem fragment with constants and let T be a theory of $\mathsf{L}_{\mathbb{A}}$, $T_{\text{Skolem}} \subseteq T$. Let $\kappa = \text{card}(\mathsf{L}_{\mathbb{A}})$.*

(i) *If T has a model with an infinite set of indiscernibles for $\mathsf{L}_{\mathbb{A}}$ then T has a model of any power $\geqslant \kappa$.*

(ii) *If $T = T(\mathsf{U}, \ldots)$ has a model $\mathfrak{M} = \langle M, U, \ldots \rangle$ with $\langle X, < \rangle$ an infinite set of indiscernibles over U for $\mathsf{L}_{\mathbb{A}}$ then T admits $(\lambda, \text{card}(U))$ for all $\lambda \geqslant \kappa + \text{card}(U)$.*

Proof. (i) is immediate from 4.8 (iii) and the Downward Löwenheim-Skolem Theorem for $\mathsf{L}_{\mathbb{A}}$. To prove (ii) let $\mathfrak{M}$ have $\langle X, < \rangle$ an infinite set of indiscernibles over U. Let

$$C = \{\mathsf{c}_u \mid u \in U\},$$
$$\mathfrak{M}' = (\mathfrak{M}, u)_{u \in U}$$

be as usual. Thus, $\langle X, < \rangle$ is a set of indiscernibles for $\mathsf{L}_{\mathbb{A}}(C)$ in $\mathfrak{M}'$. Given $\lambda \geqslant \kappa$, let $\langle Y, < \rangle$ be a linearly ordered set of power λ and let

$$\mathfrak{N}' = (\mathfrak{N}, u)_{u \in U}$$

be as given by 4.8, the Stretching Theorem. By Exercise 2.16, we may assume

$$\mathfrak{N}' = \text{Hull}_{\mathfrak{N}'}(Y),$$

since this Hull also has properties (i), (ii) of 4.8. Write $\mathfrak{N}$ as $\mathfrak{N}=\langle N, U', \ldots\rangle$. We claim that $U=U'$. For suppose $a\in U'$. Then

$$a=t(y_1,\ldots,y_n,u_1,\ldots,u_m)$$

for some term t of $\mathsf{L}_{\mathbb{A}}$, some $u_1,\ldots,u_m\in U$ and some $y_1<\cdots<y_n$ in $\langle Y,<\rangle$. But, then,

$$\mathfrak{N}\models \mathsf{U}(t(y_1,\ldots,y_n,u_1,\ldots,u_m))$$

so, by (ii) of 4.8,

$$\mathfrak{M}\models \mathsf{U}(t(x_1,\ldots,x_n,u_1,\ldots,u_m))$$

whenever $x_1<\cdots<x_n$ in $\langle X,<\rangle$. Pick such a sequence of x's. Then there is a $u\in U$ such that

$$\mathfrak{M}\models u=t(x_1,\ldots,x_n,u_1,\ldots,u_m)$$

and, hence by (ii) of 4.8,

$$\mathfrak{N}\models u=t(y_1,\ldots,y_n,u_1,\ldots,u_m)$$

so

$$\mathfrak{N}\models u=a.$$

In other words, every member of U' is one of the original members of U. Thus, $\operatorname{card}(U')=\operatorname{card}(U)$ but

$$\begin{aligned}\operatorname{card}(\mathfrak{N})&=\operatorname{card}(\mathsf{L}_{\mathbb{A}}(C))+\operatorname{card}(Y)\\&=\kappa+\operatorname{card}(C)+\lambda\\&=\kappa+\operatorname{card}(U)+\lambda\\&=\lambda.\quad\square\end{aligned}$$

To construct a model with an infinite set of indiscernibles, we use the Erdős-Rado theorem of cardinal arithmetic (Lemma 4.10) to construct "coherent sets of k-variable indiscernibles" and the Weak Completeness Theorem to piece them together to get a model with a set of indiscernibles.

We use $[X]^n$ to denote the set

$$\{x\subseteq X \mid \operatorname{card}(x)=n\}.$$

4.10 Lemma (Erdős-Rado Theorem). *Let κ be an infinite cardinal and let $0<n<\omega$. Let X be a set with $\operatorname{card}(X)>\beth_{n-1}(\kappa)$ and suppose $[X]^n$ is partitioned into $\leqslant\kappa$ subsets, say $[X]^n=\bigcup_{i\in I}C_i$ where $\operatorname{card}(I)\leqslant\kappa$. There is an $X_0\subseteq X$ and an $i_0\in I$ such that*

$$\operatorname{card}(X_0)>\kappa \quad \textit{and} \quad [X_0]^n\subseteq C_{i_0}.$$

Proof. If the reader is not familiar with this result, he can find a proof in most advanced books on set theory, in Keisler [1971] or in Chang-Keisler [1973]. □

Let $\mathfrak{M}$ be a structure for L and let $\langle X, < \rangle$ be linearly ordered with $X \subseteq \mathfrak{M}$. Let $k < \omega$ be fixed. We say that $\langle X, < \rangle$ is a set of *k-variable indiscernibles* for $\mathsf{L}_{\mathbb{A}}$ in $\mathfrak{M}$ if, for all increasing k-tuples

$$x_1 < \cdots < x_k, \qquad y_1 < \cdots < y_k$$

in $\langle X, < \rangle$, we have

$$(\mathfrak{M}, x_1, \ldots, x_k) \equiv (\mathfrak{M}, y_1, \ldots, y_k).$$

Thus $\langle X, < \rangle$ is a set of indiscernibles iff it is a set of k-variable indiscernibles for each $k < \omega$. Also note that if $\langle X, < \rangle$ is a set of k-variable indiscernibles then $\langle X, < \rangle$ is a set of l-variable indiscernibles for all $l < k$. Any linearly ordered $\langle X, < \rangle$ with $X \subseteq \mathfrak{M}$ is a set of 0-variable indiscernibles. The notion of *set of k-variable indiscernibles over U* (when $\mathfrak{M} = \langle M, U, \ldots \rangle$) is defined in the same way.

As a first simple use of the Erdős-Rado Theorem we can prove a result which is useful when $h_\Sigma(\mathbb{A}) = \omega$.

4.11 Proposition. *Let $\mathsf{L}_{\mathbb{A}}$ be a fragment of $\mathsf{L}_{\infty\omega}$ with $\mathrm{card}(\mathsf{L}_{\mathbb{A}}) = \kappa$. Let $0 < k < \omega$ be fixed and let $\mathfrak{M}$ be a structure for L.*

(i) *If $\mathrm{card}(\mathfrak{M}) > \beth_k(\kappa)$ then there is an infinite set $\langle X, < \rangle$ of k-variable indiscernibles for $\mathsf{L}_{\mathbb{A}}$ in $\mathfrak{M}$.*

(ii) *If $\mathfrak{M} = \langle M, U, \ldots \rangle$ where $\mathrm{card}(U) \geqslant \kappa$ and $\mathrm{card}(M) > \beth_k(\mathrm{card}(U))$, then there is an infinite set $\langle X, < \rangle$ of k-variable indiscernibles over U for $\mathsf{L}_{\mathbb{A}}$ in $\mathfrak{M}$.*

Proof. (i) Let $<$ be a linear ordering of M and, for each k-tuple $\vec{x} = x_1 < \cdots < x_k$ from M, let

$$T_{\vec{x}} = \{\varphi(v_1, \ldots, v_k) \mid \mathfrak{M} \models \varphi[x_1, \ldots, x_k]\}.$$

This partitions $[M]^k$ up into $\leqslant 2^\kappa$ distinct sets, since there are $\leqslant 2^\kappa$ different sets of formulas of $\mathsf{L}_{\mathbb{A}}$. Since

$$\mathrm{card}(M) > \beth_k(\kappa) = \beth_{k-1}(2^\kappa),$$

the Erdős-Rado Theorem tells us that there is an $X \subseteq M$ (of power $> 2^\kappa > \aleph_0$) such that every element of $[X]^k$ is in one fixed member of the partition. That is, $T_{\vec{x}} = T_{\vec{y}}$ whenever $\vec{x} = x_1 < \cdots < x_k$, $\vec{y} = y_1 < \cdots < y_k$ and $x_1, \ldots, x_k, y_1, \ldots, y_k \in X$. Thus $\langle X, < \restriction X \rangle$ is a set of k-variable indiscernibles in $\mathfrak{M}$. To prove (ii), let $C = \{\mathsf{c}_u \mid u \in U\}$ and apply (i) to $\mathsf{L}_{\mathbb{A}}(C)$ and $\mathfrak{M}' = (\mathfrak{M}, u)_{u \in U}$ with κ replaced by $\mathrm{card}(\mathsf{L}_{\mathbb{A}}(C))$. □

Theorem 4.1 follows easily from Lemma 4.9 (i) and the following theorem.

4.12 Theorem. *Let* $L_{\mathbb{A}}$ *be an admissible fragment, let* card$(\mathbb{A})$, $\alpha = h_\Sigma(\mathbb{A})$. *Let* T *be a* Σ_1 *theory of* $L_{\mathbb{A}}$. *If for each* $\beta < \alpha$, T *has model of power* $\geqslant \beth_\beta(\kappa)$, *then* T *has a model with an infinite set of indiscernibles.*

Proof. We may assume by 2.4 that $L_{\mathbb{A}}$ is a Skolem fragment, that $T_{\text{Skolem}} \subseteq T$ and that $L_{\mathbb{A}}$ is Δ_1 on $\mathbb{A}$. We assume that T has models but no model with a set of indiscernibles for $L_{\mathbb{A}}$ and prove that, for some $\beta < \alpha$, T has no models of power $\geqslant \beth_\beta(\kappa)$. Let $L'_{\mathbb{A}}$ be a Skolem fragment containing $L_{\mathbb{A}}$ and two new symbols X, $<$. Let

$$C = \{c_n \mid 0 < n < \omega\}$$

be a set of new constant symbols. We will be concerned with all the languages

$$L_{\mathbb{A}}(c_1, \ldots, c_n), \qquad L'_{\mathbb{A}}(c_1, \ldots, c_n),$$
$$L_{\mathbb{A}}(C), \qquad L'_{\mathbb{A}}(C).$$

These are all Skolem fragments with constants. For $n \geqslant 0$ define $\mathfrak{S}_n$ to be the set of all supervalidity properties $\mathcal{D}$ for $L'_{\mathbb{A}}(c_1, \ldots, c_n)$ with the following properties:

(a) $T \subseteq \mathcal{D}$;

(b) "X is linearly ordered by $<$ and has no last element" $\in \mathcal{D}$;

(c) "$c_i \in X \wedge c_i < c_{i+1}$" $\in \mathcal{D}$ for $0 < i < n$;

(d) $\forall x_1, \ldots, x_n \in X\, [x_1 < \cdots < x_n \rightarrow (\varphi(x_1, \ldots, x_n) \leftrightarrow \varphi(c_1, \ldots, c_n))] \in \mathcal{D}$ for each $\varphi(v_1, \ldots, v_n) \in L_{\mathbb{A}}$, when $n > 0$.

It follows immediately from the Weak Completeness Theorem that

(1) *$\mathcal{D} \in \mathfrak{S}_n$ iff $\mathcal{D}$ is an s.v.p. for $L'_{\mathbb{A}}(c_1, \ldots, c_n)$ given by some structure*

$$(\mathfrak{M}, X, <, a_1, \ldots, a_n)$$

where $\mathfrak{M} \models T$, $\langle X, < \rangle$ is an infinite set of n-variable indiscernibles for $L_{\mathbb{A}}$ in $\mathfrak{M}$ and $a_1 < \cdots < a_n$ in $\langle X, < \rangle$.

Let $\mathfrak{S} = \bigcup_n \mathfrak{S}_n$. Note that each $\mathcal{D} \in \mathfrak{S}$ is in exactly one $\mathfrak{S}_n$ for $n \geqslant 0$; this n is called the *level of* $\mathcal{D}$ and we can determine the level n of $\mathcal{D}$ by seeing whether $(c_n = c_n) \in \mathcal{D}$ but $(c_{n+1} = c_{n+1}) \notin \mathcal{D}$. Let $l(\mathcal{D})$ be the level of $\mathcal{D}$. We define an order $\prec$ on $\mathfrak{S}$ by

$$\mathcal{D}' \prec \mathcal{D} \quad \text{iff} \quad l(\mathcal{D}') > l(\mathcal{D}) \quad \text{and} \quad \mathcal{D} \cap L_{\mathbb{A}}(c_1, \ldots, c_{l(\mathcal{D})}) \subseteq \mathcal{D}'.$$

Thus, if $\mathcal{D}' \prec \mathcal{D}$ then $\mathcal{D}$ and $\mathcal{D}'$ contain exactly the same formulas from the language $L_{\mathbb{A}}(c_1, \ldots, c_n)$, $n = l(\mathcal{D})$, but not necessarily from $L'_{\mathbb{A}}(c_1, \ldots, c_n)$.

The crucial step in the proof is to realize that

(2) *$\langle \mathfrak{S}, \prec \rangle$ is well founded.*

Suppose that it were not well founded and let

$$\cdots \prec \mathscr{D}_{n+1} \prec \mathscr{D}_n \prec \cdots \prec \mathscr{D}_1 \prec \mathscr{D}_0$$

be an infinite descending chain. If $\mathscr{D} \in \mathfrak{S}_n$ and $n > m$ then $\mathscr{D} \cap \mathsf{L}'_{\mathbb{A}}(\mathsf{c}_1, \ldots, \mathsf{c}_m) \in \mathfrak{S}_m$ so we may suppose that the level of $\mathscr{D}_n$ is n. Let $\mathscr{D}_n^0 = \mathscr{D}_n \cap \mathsf{L}_{\mathbb{A}}(\mathsf{c}_1, \ldots, \mathsf{c}_n)$ and let $\mathscr{D}_\infty^0 = \bigcup_n \mathscr{D}_n^0$. By the union of chain lemma, $\mathscr{D}_\infty^0$ is an s.v.p. for $\mathsf{L}_{\mathbb{A}}(C)$. Let $(\mathfrak{M}, a_1, \ldots, a_n, \ldots)$ be a model of $\mathscr{D}_\infty^0$, by the Weak Completeness Theorem. Then $\mathfrak{M} \models T$ and $X = \{a_1, a_2, \ldots\}$ is an infinite set of indiscernibles for $\mathsf{L}_{\mathbb{A}}$ in $\mathfrak{M}$ when ordered by $a_i < a_j$ if $i < j$. This proves (2).

Using (2), we can define the usual rank function on $\mathfrak{S}$:

$$\rho(\mathscr{D}) = \sup\{\rho(\mathscr{D}') + 1 \mid \mathscr{D}' \prec \mathscr{D}\},$$
$$\rho(\mathfrak{S}) = \sup\{\rho(\mathscr{D}) + 1 \mid \mathscr{D} \in \mathfrak{S}\}.$$

Since $\mathfrak{S}_0 \neq \emptyset$, $\rho(\mathfrak{S}) > 0$. We will prove later that $\rho(\mathfrak{S}) < h_\Sigma(\mathbb{A})$.

(3) *Assume* $\rho(\mathfrak{S}) = n < \omega$. *Then no model* $\mathfrak{M}$ *of* T *has an infinite set of* n*-variable indiscernibles.*

For suppose $\mathfrak{M} \models T$ and $\langle X, < \rangle$ is an infinite set of n-variable indiscernibles. Let, for $0 \leqslant m \leqslant n$,

$$\mathfrak{M}_m = (\mathfrak{M}, X, <, a_1, \ldots, a_m)$$

and let $\mathscr{D}_m$ be the s.v.p. for $\mathsf{L}'_{\mathbb{A}}(\mathsf{c}_1, \ldots, \mathsf{c}_m)$ given by $\mathfrak{M}_m$. Then $\mathscr{D}_m \in \mathfrak{S}_m$ and

$$\rho(\mathscr{D}_0) > \rho(\mathscr{D}_1) > \rho(\mathscr{D}_2) > \cdots > \rho(\mathscr{D}_n) \geqslant 0$$

so $\rho(\mathscr{D}_0) \geqslant n$ and hence $\rho(\mathfrak{S}) > n$, contrary to hypothesis.

From (3) and Proposition 4.11 (i), we immediately obtain

(4) *If* $\rho(\mathfrak{S}) = n < \omega$ *then* T *has no model of power* $> \beth_n(\kappa)$.

If $\rho(\mathfrak{S}) \geqslant \omega$ then we cannot put such an *a priori* upper bound on the "size" n of a set $\langle X, < \rangle$ of n-variables indiscernibles, but we can put a bound on $\operatorname{card}(X)$.

(5) *Suppose* $\mathfrak{M} \models T$, $\langle X, < \rangle$ *is a set of* n*-variable indiscernibles for* $\mathsf{L}_{\mathbb{A}}$ *in* $\mathfrak{M}$ *and that* $a_1 < \cdots < a_n$ *in* $\langle X, < \rangle$. *Let* $\mathscr{D}$ *be the s.v.p. in* $\mathfrak{S}_n$ *given by* $(\mathfrak{M}, X, <, a_1, \ldots, a_n)$. *If* $\beta = \rho(\mathscr{D})$ *then* $\operatorname{card}(X) < \beth_{\omega(\beta+1)}(\kappa)$.

We prove (5) by induction on β using the Erdős-Rado Theorem as in 4.11 (i). So suppose we know the result for ordinals $\gamma < \beta$ ($\beta > 0$) and suppose

$\operatorname{card}(X) \geqslant \beth_{\omega(\beta+1)}(\kappa)$. For each increasing $n+1$ tuple $\vec{x} = x_1 < \cdots < x_n < x_{n+1}$ from $\langle X, < \rangle$, let

$$T_{\vec{x}} = \{\varphi(v_1, \ldots, v_{n+1}) \in \mathsf{L}_{\mathbb{A}} \mid \mathfrak{M} \models \varphi[x_1, \ldots, x_{n+1}]\}.$$

This partitions $[X]^{n+1}$ into $\leqslant 2^\kappa$ sets. Since $2^\kappa \leqslant \beth_{\omega\beta+1}(\kappa)$ and

$$\begin{aligned} \operatorname{card}(X) &\geqslant \beth_{\omega(\beta+1)}(\kappa) \\ &= \beth_{\omega(\beta+1)}(2^\kappa) \\ &> \beth_{\omega\beta+n}(2^\kappa) \\ &= \beth_n(\beth_{\omega\beta}(\kappa)) \end{aligned}$$

we can apply the Erdős-Rado Theorem to find an $X_0 \subseteq X$ with $\operatorname{card}(X_0) > \beth_{\omega\beta}(\kappa)$ such that every member of $[X_0]^{n+1}$ lies in one member of the partition. That is, for $n+1$-tuples $x_1 < \cdots < x_{n+1}$ from X_0,

$$T_{\vec{x}} = T_{\vec{y}}$$

so that $\langle X_0, < \rangle$ forms a set of $(n+1)$-variable indiscernibles in $\mathfrak{M}$. Let $a_1 < \cdots < a_{n+1}$ be chosen from X_0 and let $\mathscr{D}_0$ be the s.v.p. given by

$$\mathfrak{M}_0 = (\mathfrak{M}, X_0, < \restriction X_0, a_1, \ldots, a_{n+1}).$$

Then $\mathscr{D}_0 \prec \mathscr{D}$ so $\rho(\mathscr{D}_0) < \beta$. But then $\mathfrak{M}_0$ contradicts the inductive hypothesis since $\operatorname{card}(X_0) > \beth_{\omega\beta}(\kappa) \geqslant \beth_{\omega(\gamma+1)}(\kappa)$ where $\gamma = \rho(\mathscr{D}_0)$. This contradiction proves (5) for $\beta > 0$. The case for $\beta = 0$ is easier and is left to the ideal student.

From (5) we get at once:

(6) *Every model* $\mathfrak{M}$ *of* T *has power* $< \beth_{\omega\beta}(\kappa)$, *where* $\beta = \rho(\mathfrak{S})$.

For let $X = M$ and let $<$ be any linear ordering of X. Recall that $\langle X, < \rangle$ is a set of 0-ary indiscernibles for $\mathfrak{M}$. Then, if $\mathscr{D}$ is the s.v.p. for $\mathsf{L}'_{\mathbb{A}}$ given by

$$(\mathfrak{M}, X, <)$$

then $\rho(\mathscr{D}) < \beta$ and $\operatorname{card}(\mathfrak{M}) = \operatorname{card}(X) < \beth_{\omega(\rho(\mathscr{D})+1)}(\kappa)$ which is $\leqslant \beth_{\omega\beta}(\kappa)$.

Finally, we claim that

(7) $\rho(\mathfrak{S}) < h_\Sigma(\mathbb{A})$.

To see that this concludes the proof, we see that if $h_\Sigma(\mathbb{A}) = \omega$ then the result follows from (4). If $\rho(\mathfrak{S}) = \beta$ and $h_\Sigma(\mathbb{A}) > \omega$ then $\omega\beta < h_\Sigma(\mathbb{A})$ by Corollary 3.5, so the conclusion follows from (6). (This is the only use of anything remotely approaching admissibility in the entire proof.)

It remains only to prove (7). We will see in § VIII.6 that $\langle \mathfrak{S}, \prec \rangle$ is a Π definable well-founded tree of subsets of A and that every such tree has rank less than $h_\Sigma(\mathbb{A})$. That is probably the simplest proof of (7). It's good for the soul, though, and gives added appreciation of the machinery developed in § VIII.6, to give a direct proof. We present a sketch to be filled in by the student.

Our goal then is to write down a Σ_1 theory $T'(<)$ of $\mathsf{L}_\mathbb{A}$ which pins down $\beta=\rho(\mathfrak{S})$. As is our custom, we first describe the intended model $\mathfrak{M}$ of $T'(<)$, the one where $<^{\mathfrak{M}}$ has order type β. Let $\mathfrak{M}$ be the following structure:

$$\langle M; \beta, <; \mathbb{A}; \mathrm{Power}(\mathbb{A}), E; \mathfrak{S}, \prec, F, G, x\rangle_{x\in\mathbb{A}}$$

where

$$\begin{aligned}
M &= \beta\cup\mathbb{A}\cup\mathrm{Power}(\mathbb{A}),\\
< &= \in\restriction\beta,\\
G(\mathscr{D}) &= \text{level of } \mathscr{D} \text{ for } \mathscr{D}\in\mathfrak{S}\\
&\quad = \text{some constant} \notin\omega, \text{ otherwise},\\
F(\mathscr{D}) &= \rho(\mathscr{D}) \text{ if } \mathscr{D}\in\mathfrak{S}\\
&\quad = \text{some constant} \notin\beta \text{ otherwise},\\
E &= \in\cap(\mathbb{A}\times\mathrm{Power}(\mathbb{A})).
\end{aligned}$$

Now suppose that

$$\mathfrak{M}' = \langle M'; B, <', \mathfrak{A}'; P, E', \mathfrak{S}', \prec', F', G', x\rangle_{x\in A_N}$$

satisfies all the finitary first order sentences true in M and that

$$\mathbb{A}\subseteq_{\mathrm{end}}\mathfrak{A}'.$$

We will show that $\langle B, <'\rangle$ is well ordered. The proof will show that the set of finitary sentences we actually use is Σ_1 on $\mathbb{A}$ so that will conclude the proof.

By the axiom of Extensionality for $\mathrm{Power}(\mathbb{A})$, we may assume that

$$P\subseteq\mathrm{Power}(\mathfrak{A}'),\quad E'=\in\cap(\mathfrak{A}'\times P),\quad \text{and}\quad \mathfrak{S}'\subseteq P.$$

Now suppose that the linear ordering $\langle B, <'\rangle$ is not well ordered, so that there is a subset $B_0\subseteq B$ with no $<'$-minimal element. Let

$$\mathfrak{S}_0' = \{\mathscr{D}\in\mathfrak{S}' \mid F'(\mathscr{D})\in B_0\}$$

and let

$$\mathfrak{S}_0'' = \{\mathscr{D}\cap\mathsf{L}'_\mathbb{A}(\mathsf{c}_1,\dots,\mathsf{c}_m) \mid \mathscr{D}\in\mathfrak{S}_0', m<\omega, m\leqslant G'(\mathscr{D})\},$$

where we must remember that $G'(\mathscr{D})$ might be a nonstandard integer. It is not difficult, though tedious, to see that $\mathfrak{S}_0''\subseteq\mathfrak{S}$, since each $\mathscr{D}\in\mathfrak{S}'$ claims to be an s.v.p. for $\mathsf{L}'_\mathbb{A}(\mathsf{c}_1,\dots,\mathsf{c}_{G(\mathscr{D})})$ of the appropriate kind, and the relevant quantifiers are all universal. So $\mathfrak{S}_0''$ must have a minimal element $\mathscr{D}$. By chasing $\mathscr{D}$ back into B, a contradiction easily results by considering the cases $G'(\mathscr{D})$ standard and $G'(\mathscr{D})$ nonstandard separately. □

4.13 Proof of Theorem 4.1. Again, using 2.4 we may assume $\mathsf{L}_{\mathbb{A}}$ is a Skolem fragment and that $T_{\text{Skolem}} \subseteq T$. Then 4.1 follows from 4.12 and 4.9 (i). □

4.14 Corollary. *Let $\mathsf{L}_{\mathbb{A}}$ be an admissible Skolem fragment with $h_\Sigma(\mathbb{A}) = \omega$. Let T be a Σ_1 theory of $\mathsf{L}_{\mathbb{A}}$. If for each $k < \omega$, T has a Skolem model with an infinite set of k-variable indiscernibles, then T has a Skolem model with an infinite set of indiscernibles for $\mathsf{L}_{\mathbb{A}}$.*

Proof. See line (3) of the proof of Theorem 4.12. □

We next turn to the analogous theorem for two cardinal models.

4.15 Theorem. *Let $\mathsf{L}_{\mathbb{A}}$ be an admissible fragment with $\kappa = \text{card}(\mathbb{A})$, $\alpha = h_\Sigma(\mathbb{A})$. Let $T = T(\mathsf{U}, \ldots)$ be a Σ_1 theory of $\mathsf{L}_{\mathbb{A}}$. If for each $\beta < \alpha$, there is a $\lambda \geqslant \kappa$ such that T admits $(\beth_\beta(\lambda), \lambda)$, then T has a model $\mathfrak{M} = \langle M, U, \ldots \rangle$ with an infinite set of indiscernibles over U for $\mathsf{L}_{\mathbb{A}}$.*

Proof. We indicate the changes necessary in the proof of Theorem 4.12. We may again assume that $\mathsf{L}_{\mathbb{A}}$ is a Skolem fragment and that $T_{\text{Skolem}} \subseteq T$. We may also assume (by adding κ new constant symbols and some axioms of the form $\mathsf{U}(\mathsf{c}_x)$, $\mathsf{c}_x \neq \mathsf{c}_y$ to T) that every model $\mathfrak{M}$ of T has $\text{card}(U) \geqslant \kappa$.

Let $\mathsf{L}'_{\mathbb{A}}(\mathsf{c}_1, \ldots, \mathsf{c}_n)$ be as before and let $\mathscr{D} \in \mathfrak{S}_n$ iff $\mathscr{D}$ is an s.v.p. for $\mathsf{L}'_{\mathbb{A}}(\mathsf{c}_1, \ldots, \mathsf{c}_n)$ with properties (a), (b), (c), (d) as before plus

(e) $\mathsf{U}(t(\mathsf{c}_1, \ldots, \mathsf{c}_n)) \to \forall x_1, \ldots, x_n \in X\, [x_1 < \cdots < x_n \to t(x_1, \ldots, x_n) = t(\mathsf{c}_1, \ldots, \mathsf{c}_n)]$ for all terms $t(v_1, \ldots, v_n)$ of $\mathsf{L}_{\mathbb{A}}$.

The analogue of (1) is the one way result:

(1′) *$\mathscr{D} \in \mathfrak{S}_n$ if $\mathscr{D}$ is the s.v.p. for $\mathsf{L}'_{\mathbb{A}}(\mathsf{c}_1, \ldots, \mathsf{c}_n)$ given by some $(\mathfrak{M}, X, <, a_1, \ldots, a_n)$ where $\langle X, < \rangle$ is a set of n-variable indiscernibles over $\mathsf{U}^{\mathfrak{M}}$ for $\mathsf{L}_{\mathbb{A}}$ and $a_1 < \cdots < a_n$ in $\langle X, < \rangle$.*

Luckily, we never really used the other half of (1).

The relation $\prec$ on $\mathfrak{S} = \bigcup_n \mathfrak{S}_n$ is defined just as before. Again we have, assuming T does not have a model $\mathfrak{M}$ with a set of indiscernibles over $\mathsf{U}^{\mathfrak{M}}$,

(2′) *$\langle \mathfrak{S}, \prec \rangle$ is well founded.*

This is just a bit trickier than (2). Suppose

$$\cdots \prec \mathscr{D}_{n+1} \prec \mathscr{D}_n \prec \cdots \prec \mathscr{D}_1 \prec \mathscr{D}_0.$$

Again, we may assume that each $\mathscr{D}_n$ has level n. Let $\mathscr{D}_n^0 = \mathscr{D}_n \cap \mathsf{L}_{\mathbb{A}}(\mathsf{c}_1,\ldots,\mathsf{c}_n)$ and let $\mathscr{D}_\infty^0 = \bigcup_n \mathscr{D}_n^0$. By the Union of Chain Lemma, $\mathscr{D}_\infty^0$ is an s.v.p. for $\mathsf{L}_{\mathbb{A}}(C)$. Let $(\mathfrak{M}_0, a_1, a_2, \ldots, a_n, \ldots)$ be a model for $\mathscr{D}_\infty^0$ let $X = \{a_1, a_2, \ldots\}$, $a_i < a_j$ iff $i < j$. Let $\mathfrak{M} = \mathrm{Hull}_{\mathfrak{M}_0}(X)$. By Exercise 2.16, $\mathfrak{M} \models \mathscr{D}_\infty^0$. Thus, $\mathfrak{M}$ is a model for T and $\langle X, < \rangle$ is a set of indiscernibles for $\mathsf{L}_{\mathbb{A}}$ in $\mathfrak{M}$. We need to see that $\langle X, < \rangle$ is a set of indiscernibles *over* $\mathsf{U}^{\mathfrak{M}}$. Thus suppose $u \in \mathsf{U}^{\mathfrak{M}}$. We need to see that increasing n-tuples from $\langle X, < \rangle$ satisfy the same formulas in $(\mathfrak{M}, u)$. (The case with more that one u is similar.) Since $\mathfrak{M} = \mathrm{Hull}(X)$, there is a term $t(v_1, \ldots, v_m)$ such that

$$\mathfrak{M} \models u = t(a_1, \ldots, a_m).$$

Then, by (e)

$$\mathfrak{M} \models u = t(x_1, \ldots, x_m)$$

whenever $x_1 < \cdots < x_m$ in $\langle X, < \rangle$. Now suppose $n < \omega$, $x_1 < \cdots < x_n$, $y_1 < \cdots < y_n$ in $\langle X, < \rangle$. We need to see that for all formulas $\varphi(v_1, \ldots, v_n, v_{n+1})$, if $\mathfrak{M} \models \varphi[x_1, \ldots, x_n, u]$ then $\mathfrak{M} \models \varphi[y_1, \ldots, y_n, u]$. Pick an increasing m-tuple $w_1 < \cdots < w_m$ such that $w_1 > x_n$, $w_1 > y_n$. Now consider the formula $\psi(v_1, \ldots, v_n, v_{n+1}, \ldots, v_{n+m})$ given by

$$\varphi(v_1, \ldots, v_n, t(v_{n+1}, \ldots, v_{n+m})/v_{n+1}).$$

Then, since $u = t(w_1, \ldots, w_m)$,

$$\mathfrak{M} \models \psi[x_1, \ldots, x_n, w_1, \ldots, w_m]$$

and hence,

$$\mathfrak{M} \models \psi[y_1, \ldots, y_n, w_1, \ldots, w_m]$$

by the indiscernibility of $\langle X, < \rangle$ in $\mathfrak{M}$. Thus

$$\mathfrak{M} \models \varphi[y_1, \ldots, y_n, u].$$

Thus $\langle X, < \rangle$ is indiscernible over U, proving (2).

Define $\rho(\mathscr{D})$, $\rho(\mathfrak{S})$ as before.

(3′) *Assume* $\rho(\mathfrak{S}) = n < \omega$. *Then no model* $\mathfrak{M}$ *of* T *has an infinite set of* n*-variable indiscernibles over* $\mathsf{U}^{\mathfrak{M}}$.

The proof of (3′) is just like the proof of (3).

Using (3′) and 4.11(ii), we get

(4′) *If* $\rho(\mathfrak{S}) = n < \omega$ *then* T *has no models of type* $(\beth_{n+1}(\lambda), \lambda)$ *for any* λ.

Corresponding to (5) we have

(5′) $\begin{cases} \textit{Suppose } M \models T, \langle X, < \rangle \textit{ is a set of n-variable indiscernibles over } U(=\mathsf{U}^{\mathfrak{M}}) \\ \textit{and that } a_1 < \cdots < a_n \textit{ in } \langle X, < \rangle. \textit{ Let } \mathscr{D} \textit{ be the s.v.p. in } \mathfrak{S}_n \textit{ given by} \\ (\mathfrak{M}, X, <, a_1, \ldots, a_n) \textit{ and let } \beta = \rho(\mathscr{D}). \textit{ Then } \operatorname{card}(X) < \beth_{\omega(\beta+1)}(\operatorname{card}(U)). \end{cases}$

The proof is by induction on β and uses the Erdős-Rado Theorem. The proof it too similar to the proof of (5) to present. From (5′) we get

(6′) *If* $\mathfrak{M} \models T$ *then* $\operatorname{card}(\mathfrak{M}) < \beth_{\omega\beta}(\operatorname{card}(\mathsf{U}^{\mathfrak{M}}))$ *where* $\beta = \rho(\mathfrak{S})$.

The proof is concluded by showing that

(7′) $\rho(\mathfrak{S}) < h_\Sigma(\mathbb{A})$.

The proof of (7′) is just like the proof of (7). □

Theorem 4.3 follows from 4.15 just as Theorem 4.1 followed from 4.12.

4.16—4.20 Exercises

4.16. Prove that if α is admissible then

$$\begin{aligned} \beth_\alpha(\aleph_0) &= \beth_\alpha \quad \text{if } \alpha > \omega, \\ \beth_\alpha(\aleph_0) &= \beth_\omega(\aleph_0) \\ &= \beth_{\omega+\omega} \quad \text{if } \alpha = \omega. \end{aligned}$$

4.17. Let $\mathbb{A}_{\mathfrak{N}}$ be admissible above $\mathfrak{N}$, $\kappa_0 = \operatorname{card}(\mathfrak{N})$, $\alpha = o(\mathbb{A})$. Prove that

$$\operatorname{card}(\mathbb{A}_{\mathfrak{N}}) \leqslant \beth_\alpha(\kappa_0).$$

Let $\kappa_1 = \operatorname{card}(\mathbb{A}_{\mathfrak{N}})$. Prove that if $h_\Sigma(\mathbb{A}) = \beta > \alpha$ then

$$\beth_\beta(\kappa_1) = \beth_\beta(\kappa_0).$$

4.18. Let $\mathbb{A}$ be an admissible set, $\alpha = h(\mathbb{A})$. Prove that the Hanf number for single sentences of $\mathsf{L}_{\mathbb{A}}$ is at least

$$\lambda = \sup\{\beth_\alpha(\kappa) \mid \kappa = \operatorname{card}(X) \text{ for some } X \in A\}.$$

That is, show that for $\lambda_0 < \lambda$ there is a sentence φ of $\mathsf{L}_{\mathbb{A}}$ which has models of power $\geqslant \lambda_0$ but none of power $\geqslant \lambda$. [Given $X \in A$, $\beta < h(\mathbb{A})$, formalize $\mathsf{V}_X(\beta)$.] Prove that the Hanf number is always of the form $\beth_\lambda$ for some limit ordinal λ.

4.19. Let $\mathbb{A}$ be an admissible set with $o(\mathbb{A})>\omega$.

(i) Prove that each $\varphi\in\mathbb{A}$ can be put in a Skolem fragment $\mathsf{L}_B\in\mathbb{A}$ in such a way that every model of φ (not just those in $\mathbb{A}$) can be expanded to a model of T_{Skolem}. [Use Infinity to carry out the proof of 2.4 inside $\mathbb{A}$.]

(ii) Prove that the Hanf number for single sentences of $\mathsf{L}_{\mathbb{A}}$ is

$$\lambda=\sup\{\beth_\alpha(\kappa)\,|\,\kappa=\operatorname{card}(X),\ X\in A\}$$

where $\alpha=h(\mathbb{A})$. That is, prove that if $\varphi\in\mathsf{L}_{\mathbb{A}}$ does not have a model of every power $\geqslant\operatorname{card}(\mathbb{A})$ then there is an $X\in\mathbb{A}$ and a $\beta<h(\mathbb{A})$ such that φ has no model of power $\geqslant\beth_\beta(\operatorname{card}(X))$. [The set X will be the L_B of (i). Modify the proof of 4.12.]

(iii) Prove that if $\mathbb{A}$ is a pure admissible set then the Hanf number for single sentences of $\mathsf{L}_{\mathbb{A}}$ is $\beth_{h(\mathbb{A})}$, even if $o(\mathbb{A})=\omega$.

4.20. Let $\mathbb{A}$ be an admissible set, let $\alpha=h_\Sigma(\mathbb{A})$ and let

$$\lambda_0=\sup\{\beth_\alpha(\kappa)\,|\,\kappa=\operatorname{card}(X)\text{ some }X\in\mathbb{A}\},$$

$$\lambda_1=\beth_\alpha(\operatorname{card}(\mathbb{A})).$$

Theorem 4.12 states that the Hanf number for Σ_1 theories of $\mathsf{L}_{\mathbb{A}}$ is $\leqslant\lambda_1$.

(i) Prove that this Hanf number is $\geqslant\lambda_0$.

(ii) Prove that if $\mathbb{A}$ is countable and $\neq\mathbb{HF}_{\mathfrak{M}}$, or if $h_\Sigma(\mathbb{A})>o(\mathbb{A})$, then $\lambda_0=\lambda_1$. It is an open problem to describe this Hanf number in general. Is it λ_0 or λ_1 or something in between?

4.21 Notes. Morley [1965] shows that the Hanf number for single sentences of $\mathsf{L}_{\omega_1\omega}$ was $\beth_{\omega_1}$. (This follows from 4.2.) Morley [1967] showed that the Hanf number for single sentences of ω-logic was $\beth_\alpha$ where $\alpha=\omega_1^c$. (The hard half of this follows from 4.2 with $\mathbb{A}=L(\alpha)$.) Barwise [1967] generalized this to obtain the Hanf number for any countable, admissible fragment. This was generalized in Barwise-Kunen [1971] to obtain 4.19(iii). The theorems of this section are a reworking of the ideas from Barwise-Kunen [1971] so that they apply to theories, not just single sentences. Theorem 4.3 is a generalization of Morely's Two Cardinal Theorem of Morley [1965]. The student should consult lectures 16 and 17 of Keisler [1971] for a different proof of the countable versions of these results.

The student should be aware of a difference between the results of this section and those in Chapter III. The use of admissible sets was absolutely essential in Chapter III to obtain our results. Here they provide a convenient setting but weaker assumptions would do. Of course we need to know that the countable set $\mathbb{A}$ is admissible to know that $h_\Sigma(\mathbb{A})=o(\mathbb{A})$.

5. Partially Isomorphic Structures

Having seen in the previous sections that the model theory of uncountable fragments is not completely beyond our control, even if it is less tractable than for countable fragments, we now investigate some uses of uncountable sentences.

One way to appreciate $\mathsf{L}_{\infty\omega}$ is to see the role it plays in algebra, but this is not the book to discuss such topics. We can only give a few exercises. The topics we discuss are of a more logical nature. These final sections are completely independent of the first half of the chapter. Admissible sets will not appear in an essential way until § 7.

A *partial isomorphism* f from $\mathfrak{M}$ into $\mathfrak{N}$ is simply an isomorphism

$$f: \mathfrak{M}_0 \cong \mathfrak{N}_0$$

where $\mathfrak{M}_0$, $\mathfrak{N}_0$ are substructures of $\mathfrak{M}$ and $\mathfrak{N}$ respectively. A set I of partial isomorphisms from $\mathfrak{M}$ into $\mathfrak{N}$ has the *back and forth property* if

(1) for every $f \in I$ and every $x \in \mathfrak{M}$ (or $y \in \mathfrak{N}$) there is a $g \in I$ with $f \subseteq g$ and $x \in \mathrm{dom}(g)$ (or $y \in \mathrm{rng}(g)$, resp.).

We write

$$I: \mathfrak{M} \cong_p \mathfrak{N}$$

if I is a nonempty set of partial isomorphisms and I has the back and forth property. If there is an I such that $I: \mathfrak{M} \cong_p \mathfrak{N}$ then we say that $\mathfrak{M}$, $\mathfrak{N}$ are *partially isomorphic* and write $\mathfrak{M} \cong_p \mathfrak{N}$. (Some authors prefer the more picturesque terminology *potentially isomorphic*, to suggest that $\mathfrak{M}$ and $\mathfrak{N}$ would become isomorphic if only they were to become countable, say in some larger universe of set theory.) Note that if $f: \mathfrak{M} \cong \mathfrak{N}$, then $\{f\}: \mathfrak{M} \cong_p \mathfrak{N}$.

5.1 Examples. (i) The canonical example is given by two dense linear orderings $\mathfrak{M} = \langle M, < \rangle$ nd $\mathfrak{N} = \langle N, < \rangle$ without end-points. Let I be the set of all *finite* partial isomorphisms from $\mathfrak{M}$ into $\mathfrak{N}$. Then

$$I: \mathfrak{M} \cong_p \mathfrak{N}$$

regardless of the cardinalities of $\mathfrak{M}$ and $\mathfrak{N}$. This is quite easy to verify. Combined with Theorem 5.2, this shows that the theory of dense linear orderings without end points is $\aleph_0$-categorical, i. e., that all its countable models are isomorphic.

(ii) If $\mathfrak{M}$, $\mathfrak{N}$ are dense linear orderings with first elements x_0, y_0 respectively, but without last elements, then $\mathfrak{M} \cong_p \mathfrak{N}$ but the set I used in (i) no longer has the back and forth property. Let

$$I_0 = \{f \in I \mid x_0 \in \mathrm{dom}(f), f(x_0) = y_0\}.$$

Then $I_0: \mathfrak{M} \cong_p \mathfrak{N}$.

(iii) We can generalize (i), (ii) as follows. Let $\mathsf{L}_{\mathbb{A}}$ be a countable fragment and let T be an $\aleph_0$-categorical theory of $\mathsf{L}_{\mathbb{A}}$. Then for any two infinite models $\mathfrak{M}$, $\mathfrak{N}$ of T,

$$\mathfrak{M} \cong_p \mathfrak{N}.$$

A proof of this will be given in 5.5 below.

(iv) We can get a different generalization of (i) and (ii) by looking at $\aleph_0$-saturated structures $\mathfrak{M}$ and $\mathfrak{N}$. If $\mathfrak{M} \equiv \mathfrak{N}$ $(\mathsf{L}_{\omega\omega})$ then $\mathfrak{M} \cong_p \mathfrak{N}$. The set I is defined as follows: Consider those partial isomorphisms

$$f: \mathfrak{M}_0 \cong \mathfrak{N}_0$$

where $\mathfrak{M}_0$ is finitely generated by some $a_1, \ldots, a_n$. We will let $f \in I$ iff

$$(\mathfrak{M}, a_1, \ldots, a_n) \equiv (\mathfrak{N}, f(a_1), \ldots, f(a_n)) \quad (\mathsf{L}_{\omega\omega}).$$

A simple use of $\aleph_0$-saturation shows that I has the back and forth property.

Traditionally, the back and forth property has been used for constructing isomorphisms of countable structures.

5.2 Theorem. *Let $\mathfrak{M}$, $\mathfrak{N}$ be countable structures for the same language and let $I: \mathfrak{M} \cong_p \mathfrak{N}$. For every $f_0 \in I$ there is an isomorphism*

$$f: \mathfrak{M} \cong \mathfrak{N}$$

with $f_0 \subseteq f$.

Proof. Enumerate $\mathfrak{M} = \{x_1, x_2, \ldots\}$, $\mathfrak{N} = \{y_1, y_2, \ldots\}$. Define

$$f_{2n+1} = \text{some } g \in I \quad \text{with} \quad f_{2n} \subseteq g,\ x_n \in \operatorname{dom}(g),$$
$$f_{2n+2} = \text{some } g \in I \quad \text{with} \quad f_{2n+1} \subseteq g,\ y_n \in \operatorname{rng}(g)$$

by using the back and forth property (1). Let $f = \bigcup_n f_n$. Then f maps $\mathfrak{M}$ onto $\mathfrak{N}$ and preserves atomic and negated atomic formulas so $f: \mathfrak{M} \cong \mathfrak{N}$.

The examples and Theorem 5.2 should suggest to the student of the previous chapter that $\cong_p$ could be the absolute version of $\cong$. After all, they agree on countable structures and $\cong_p$ does not seem to depend on cardinality. At first glance, though, it is not obvious that $\cong_p$ is absolute, but merely that it is Σ_1:

$$\mathfrak{M} \cong_p \mathfrak{N} \quad \text{iff} \quad \exists I\, [I: \mathfrak{M} \cong_p \mathfrak{N}]$$

where the part within brackets is Δ_0. This is no better that $\cong$, itself a Σ_1 notion. The Π_1 equivalent of $\cong_p$ is given by the next result. There is, of course, no Π_1 equivalent of $\cong$. This result as well as 5.7 appear in Karp [1965].

5.3 Karp's Theorem. *If* $\mathfrak{M}$, $\mathfrak{N}$ *are structures for the language* L, *then* $\mathfrak{M}\cong_p\mathfrak{N}$ *iff* $\mathfrak{M}\equiv\mathfrak{N}$ $(\mathsf{L}_{\infty\omega})$.

Proof. We first prove ($\Rightarrow$). Let $I:\mathfrak{M}\cong_p\mathfrak{N}$. We prove, by induction on formulas $\varphi(v_1,\dots,v_n)$ of $\mathsf{L}_{\infty\omega}$ that if $f\in I$, $x_1,\dots,x_n\in\mathrm{dom}(f)$ then

$$\mathfrak{M}\models\varphi[x_1,\dots,x_n] \quad \text{iff} \quad \mathfrak{N}\models\varphi[f(x_1),\dots,f(x_n)].$$

(The theorem follows by considering those $\varphi\in\mathsf{L}_{\infty\omega}$ which are sentences.) If φ is atomic, the result follows from the fact that each $f\in I$ is a partial isomorphism and so preserves atomic and negated atomic formulas. The case where φ is a propositional combination of simpler formulas is immediate by the induction hypothesis. The back and forth property (1) comes into play only in getting past quantifiers. Suppose φ is $\exists v_{n+1}\,\psi(v_1,\dots,v_{n+1})$. Let $f,x_1,\dots,x_n$ be given. We assume $\mathfrak{M}\models\varphi[x_1,\dots,x_n]$ and prove $\mathfrak{N}\models\varphi[f(x_1),\dots,f(x_n)]$, the other half being similar. Thus, there is a $y\in\mathfrak{M}$ so that

$$\mathfrak{M}\models\psi[x_1,\dots,x_n,y].$$

Use (1) to get a $g\in I$ with $f\subseteq g$, $y\in\mathrm{dom}(g)$. Then, by the induction hypothesis,

$$\mathfrak{N}\models\psi[g(x_1),\dots,g(x_n),g(y)]$$

so

$$\mathfrak{N}\models\exists v_{n+1}\,\psi[g(x_1),\dots,g(x_n),v_{n+1})]$$

and $g(x_i)=f(x_i)$ so

$$\mathfrak{N}\models\varphi[f(x_1),\dots,f(x_n)],$$

as desired. Since $\forall v_n\,\psi\leftrightarrow\neg\exists v_n\neg\psi$, we need not treat $\forall$ separately.

Now assume $\mathfrak{M}\equiv\mathfrak{N}$ $(\mathsf{L}_{\infty\omega})$. What should our set I be? The proof of the first half of the theorem tells use. Let $f\in I$ iff

$$f:\mathfrak{M}_0\cong\mathfrak{N}_0, \qquad \mathfrak{M}_0\subseteq\mathfrak{M}, \qquad \mathfrak{N}_0\subseteq\mathfrak{N}$$

where $\mathfrak{M}_0$ is finitely generated by some $x_1,\dots,x_n$ and

$$(\mathfrak{M},x_1,\dots,x_n)\equiv_{\infty\omega}(\mathfrak{N},f(x_1),\dots,f(x_n))$$

by which we mean that $x_1,\dots,x_n$ satisfies the same formula of $\mathsf{L}_{\infty\omega}$ in M that $f(x_1),\dots,f(x_n)$ satisfy in $\mathfrak{N}$. (Note that we need Π_1 Separation to define I so that we cannot carry out this proof in KPU.) Since $\mathfrak{M}\equiv\mathfrak{N}$ $(\mathsf{L}_{\infty\omega})$, the trivial partial isomorphism is in I. We claim that I has the back and forth property. Let $f\in I$ be as above and let x_{n+1} be a new element which we need to add to the domain of f. It suffices to find a $y\in\mathfrak{N}$ so that

$$(\mathfrak{M},x_1,\dots,x_n,x_{n+1})\equiv_{\infty\omega}(\mathfrak{N},f(x_1),\dots,f(x_n),y)$$

for then we may set $g(x_{n+1})=y$ and extend to the substructure generated by $x_1, \ldots, x_{n+1}$ in the canonical fashion. So suppose there is no such y. Then, for every $y \in \mathfrak{N}$ there is a formula $\varphi_y(v_1, \ldots, v_{n+1})$ such that

$$\mathfrak{M} \models \varphi_y[x_1, \ldots, x_n, x_{n+1}],$$
$$\mathfrak{N} \models \neg\varphi_y[f(x_1), \ldots, f(x_n), y]$$

Let $\psi(v_1, \ldots, v_n)$ be

$$\exists v_{n+1} \bigwedge_{y \in N} \varphi_y(v_1, \ldots, v_n, v_{n+1}).$$

Then $\mathfrak{M} \models \psi[x_1, \ldots, x_n]$ by letting $v_{n+1} = x_{n+1}$ but

$$\mathfrak{N} \models \neg\psi[f(x_1), \ldots, f(x_n)].$$

This contradicts $f \in I$. ▯

This theorem has a number of important uses. Here we state those having to do with absoluteness.

5.4 Corollary. *$\cong_p$ is the absolute version of $\cong$.*

Proof. $\mathfrak{M} \equiv \mathfrak{N}(\mathsf{L}_{\infty\omega})$ is a Π_1 predicate of $\mathfrak{M}, \mathfrak{N}$, by the results of § III.1, so $\cong_p$ is Δ_1. It agrees with $\cong$ on countable structures by Theorem 5.2. ▯

5.5 Corollary. *Example 5.1 (iii) is true.*

Proof. Let $T, \mathsf{L}_{\mathbb{A}}$ be as in 5.1 (iii). We need to show that

$$\forall \mathfrak{M} \forall \mathfrak{N} [\mathfrak{M}, \mathfrak{N} \text{ infinite} \wedge \mathfrak{M} \models T \wedge \mathfrak{N} \models T \rightarrow \mathfrak{M} \cong_p \mathfrak{N}].$$

By 5.4, the part within brackets is absolute (in the countable parameter T), so we need only verify the result for $\mathfrak{M}$, $\mathfrak{N}$ countable. But for such $\mathfrak{M}$, $\mathfrak{N}$, the result follows from the hypothesis that T is $\aleph_0$-categorical. ▯

This result (5.5) shows us that if a countable theory T is $\aleph_0$-categorical, then we should be able to prove this by a back and forth argument.

5.6 Corollary. *Let $\mathfrak{M}, \mathfrak{N}$ be partially isomorphic structures for a finite language L.*
(i) *For all α, $\mathrm{L}(\alpha)_{\mathfrak{M}} \cong_p \mathrm{L}(\alpha)_{\mathfrak{N}}$.*
(ii) *For all α, α is $\mathfrak{M}$-admissible iff α is $\mathfrak{N}$-admissible.*
(iii) *$o(\mathbb{HYP}_{\mathfrak{M}}) = o(\mathbb{HYP}_{\mathfrak{N}})$.*

Proof. (i) This is a Π_1 condition on $\mathfrak{M}, \mathfrak{N}$ which clearly holds when $\mathfrak{M}$, $\mathfrak{N}$ are countable since then they are isomorphic. Part (ii) follows immediately from (i) since α is $\mathfrak{M}$-admissible iff $\mathrm{L}(\alpha)_{\mathfrak{M}} \models \mathrm{KPU}^+$. Part (iii) follows from (ii). ▯

One of the advantages of Theorem 5.3 is that it allows us to approximate the relation $\mathfrak{M} \cong_p \mathfrak{N}$ by approximating

$$\mathfrak{M} \equiv \mathfrak{N}(\mathsf{L}_{\infty\omega}).$$

Define the *quantifier rank* of a formula φ of $\mathsf{L}_{\infty\omega}$ recursively as follows:

$$\mathrm{qr}(\varphi) = 0 \text{ if } \varphi \text{ is atomic},$$
$$\mathrm{qr}(\exists v \varphi) = \mathrm{qr}(\forall v \varphi) = \mathrm{qr}(\varphi) + 1,$$
$$\mathrm{qr}(\neg \varphi) = \mathrm{qr}(\varphi),$$
$$\mathrm{qr}(\bigwedge \Phi) = \mathrm{qr}(\bigvee \Phi) = \sup\{\mathrm{qr}(\varphi) \mid \varphi \in \Phi\}.$$

Thus $\mathrm{qr}(\varphi)$ is an ordinal number. Since qr is defined by Σ Recursion in KPU, we have $\mathrm{qr}(\varphi) < o(\mathbb{A})$ whenever φ is in the admissible fragment $\mathsf{L}_{\mathbb{A}}$.

We write

$$\mathfrak{M} \equiv^{\alpha} \mathfrak{N}$$

if for all sentences φ of $\mathsf{L}_{\infty\omega}$ with $\mathrm{qr}(\varphi) \leqslant \alpha$,

$$\mathfrak{M} \models \varphi \quad \text{iff} \quad \mathfrak{N} \models \varphi.$$

Thus $\mathfrak{M} \equiv \mathfrak{N}(\mathsf{L}_{\infty\omega})$ iff for all α, $\mathfrak{M} \equiv^{\alpha} \mathfrak{N}$.

The following is a refinement of Karp's Theorem also due to Karp [1965].

5.7 Theorem. *Given structures $\mathfrak{M}$, $\mathfrak{N}$ for L, $\mathfrak{M} \equiv^{\alpha} \mathfrak{N}$ iff the following condition holds: There is a sequence*

$$I_0 \supseteq I_1 \supseteq \cdots \supseteq I_\beta \supseteq \cdots \supseteq I_\alpha \qquad (\beta \leqslant \alpha)$$

where each I_β is a nonempty set of partial isomorphisms from $\mathfrak{M}$ into $\mathfrak{N}$ and such that whenever $\beta + 1 \leqslant \alpha$, $f \in I_{\beta+1}$ and $x \in \mathfrak{M}$ (or *$y \in \mathfrak{N}$*) *there is a $g \in I_\beta$ such that $f \subseteq g$ and $x \in \mathrm{dom}(g)$* (*resp., $y \in \mathrm{rng}(g)$*).

Proof. The proof is a routine refinement of the proof of Karp's Theorem. To prove ($\Leftarrow$), one shows that if

$$\mathrm{qr}(\varphi(v_1, \ldots, v_n)) \leqslant \beta, \; f \in I_\beta, \; x_1, \ldots, x_n \in \mathrm{dom}(f)$$

then

$$\mathfrak{M} \models \varphi[x_1, \ldots, x_n] \quad \text{iff} \quad \mathfrak{N} \models \varphi[f(x_1), \ldots, f(x_n)].$$

To prove ($\Rightarrow$), let I_β be the set of those finitely generated partial isomorphisms f which preserved satisfaction of formulas φ with $\mathrm{qr}(\varphi) \leqslant \beta$. $\square$

5.8—5.12 Exercises

5.8. Prove that if a theory T of $\mathsf{L}_{\omega\omega}$ is $\aleph_0$-categorical then every model of T is $\aleph_0$-saturated. [Use 5.1(iii), Theorem 5.3 and the fact that $\aleph_0$-saturation can be defined by a conjunction of sentences from $\mathsf{L}_{\omega_1\omega}$.]

5.9. Let $\mathfrak{M}$, $\mathfrak{N}$ be partially isomorphic structures for a finite language. Show that for every α, the pure sets in $\mathrm{L}(\alpha)_{\mathfrak{M}}$ and $\mathrm{L}(\alpha)_{\mathfrak{N}}$ are the same.

5.10. Let λ be a limit ordinal. Prove that if $\mathfrak{M} \equiv^{\beta} \mathfrak{N}$ for all $\beta < \lambda$ then $\mathfrak{M} \equiv^{\lambda} \mathfrak{N}$. [Each sentence of quantifier rank λ is a propositional combination of sentences of smaller quantifier rank.]

5.11. Show that the following notions are definable by a single sentence of $\mathsf{L}_{\infty\omega}$.
(i) G is an $\aleph_1$-free group.
(ii) G is an abelian p-group of length $\leqslant \alpha$ (for any ordinal α).

5.12. (i) Show that if G is a reduced abelian p-group and $G \equiv H$ $(\mathsf{L}_{\infty\omega})$ then H is a reduced abelian p-group.

(ii) Show that the notion of a reduced abelian p-group is not definable by a single sentence of $\mathsf{L}_{\infty\omega}$. [Hint: There are reduced p-groups of every ordinal length. Show that if the notion were definable then there would be a sentence which pinned down all ordinals, contrary to Theorem 4.1.]

6. Scott Sentences and their Approximations

One of the tasks the mathematician sets for himself is the discovery of invariants which classify a structure $\mathfrak{M}$ up to isomorphism (homomorphism, homeomorphism, etc.) among similar structures. In this section we consider the problem of characterizing arbitrary structures up to $\cong_p$. We will associate with each structure $\mathfrak{M}$, in a reasonably effective manner, a canonical object $\sigma_{\mathfrak{M}}$ such that

$$\mathfrak{M} \cong_p \mathfrak{N} \quad \text{iff} \quad \sigma_{\mathfrak{M}} = \sigma_{\mathfrak{N}} .$$

Hence, if $\mathfrak{M}, \mathfrak{N}$ are countable we will have $\mathfrak{M} \cong \mathfrak{N}$ iff $\sigma_{\mathfrak{M}} = \sigma_{\mathfrak{N}}$. Our invariants will not be cardinal or ordinal numbers, though, as is often the case. Rather, they will be sentences of $\mathsf{L}_{\infty\omega}$ with the additional properties:

$$\mathfrak{M} \models \sigma_{\mathfrak{M}} , \quad \text{and}$$

$$\mathfrak{N} \models \sigma_{\mathfrak{M}} \quad \text{implies} \quad \mathfrak{M} \cong_p \mathfrak{N} .$$

The sentence $\sigma_{\mathfrak{M}}$ is called the *canonical Scott sentence* of $\mathfrak{M}$.

The canonical Scott sentence is built up from its approximations defined below. We use s to range over finite sequences $\langle x_1, \ldots, x_n \rangle$ from $\mathfrak{M}$ and $s^\wedge x$ to denote the extension $\langle x_1, \ldots, x_n, x \rangle$ of s by x.

6.1 Definition. Let $\mathfrak{M}$ be a structure for a language L. For each ordinal α and each sequence $s=\langle x_1,\dots,x_n\rangle$ we define a formula $\sigma_s^\alpha(v_1,\dots,v_n)$, the *$\alpha$-characteristic of s in $\mathfrak{M}$*, by recursion on α:

(i) $\sigma_s^0(v_1,\dots,v_n)$ is

$$\bigwedge\{\varphi(v_1,\dots,v_n)\mid\varphi \text{ is atomic or negated atomic and } \mathfrak{M}\models\varphi[s]\}.$$

(ii) $\sigma_s^{\beta+1}(v_1,\dots,v_n)$ is the conjunction of the following three formulas

$$\sigma_s^\beta(v_1,\dots,v_n); \tag{1}$$

$$\forall v_{n+1}\bigvee_{x\in\mathfrak{M}}\sigma_{s^\frown x}^\beta(v_1,\dots,v_n); \tag{2}$$

$$\bigwedge_{x\in\mathfrak{M}}\exists v_{n+1}\,\sigma_{s^\frown x}^\beta(v_1,\dots,v_n). \tag{3}$$

(iii) If $\lambda>0$ is a limit ordinal then $\sigma_s^\lambda(v_1,\dots,v_n)$ is

$$\bigwedge_{\beta<\lambda}\sigma_s(v_1,\dots,v_n).$$

If we need to indicate the dependence on $\mathfrak{M}$ we write $\sigma_{(\mathfrak{M},s)}^\alpha$ for σ_s^α. If s is the empty sequence we write σ^α or $\sigma_{\mathfrak{M}}^\alpha$.

6.2 Lemma. *Fix $\mathfrak{M},\alpha$ and $s=\langle x_1,\dots,x_n\rangle$.*

(i) $\mathrm{qr}(\sigma_s^\alpha)=\alpha$.

(ii) $\mathfrak{M}\models\sigma_s^\alpha[s]$.

(iii) *If $\alpha\geqslant\beta$ then*

$$\models\forall v_1,\dots,v_n(\sigma_s^\alpha(v_1,\dots,v_n)\to\sigma_s^\beta(v_1,\dots,v_n)).$$

(iv) *If κ is an infinite cardinal and* $\mathrm{card}(\mathfrak{M})<\kappa$, $\mathrm{card}(\mathsf{L})<\kappa$ *and* $\alpha<\kappa$ *then* $\mathrm{card}(\mathrm{sub}(\sigma_s^\alpha))<\kappa$.

Proof. A simple induction on α proves all these facts. □

The crucial properties of the α-characteristics are given by the next result. In this section we write

$$(\mathfrak{M},x_1,\dots,x_n)\equiv_{\infty\omega}(\mathfrak{N},y_1,\dots,y_n)$$

(and

$$(\mathfrak{M},x_1,\dots,x_n)\equiv^\alpha(\mathfrak{N},y_1,\dots,y_n))$$

to indicate that all $\langle x_1,\dots,x_n\rangle$ satisfies the same formulas $\varphi(v_1,\dots,v_n)$ (of quantifier rank at most α) in $\mathfrak{M}$ that $\langle y_1,\dots,y_n\rangle$ satisfies in $\mathfrak{N}$.

6.3 Theorem. *Let $\mathfrak{M},\mathfrak{N}$ be L-structures, $s=\langle x_1,\dots,x_n\rangle$ a sequence from $\mathfrak{M}$, $t=\langle y_1,\dots,y_n\rangle$ a sequence from $\mathfrak{N}$. The following are equivalent:*

(i) $(\mathfrak{M},x_1,\dots,x_n)\equiv^\alpha(\mathfrak{N},y_1,\dots,y_n)$.

(ii) $\mathfrak{N}\models\sigma_{(\mathfrak{M},s)}^\alpha[t]$.

(iii) *The α-characteristic of s in $\mathfrak{M}$ is identical with the α-characteristic of t in $\mathfrak{N}$.*

Proof. The proofs of (i)⇒(ii) and (iii)⇒(ii) and both trivial. The first follows immediately form 6.2(i), (ii). The second implication also follows from 6.2(ii), since $\mathfrak{N} \models \sigma^{\alpha}_{(\mathfrak{N},t)}[t]$, so if $\sigma^{\alpha}_{(\mathfrak{M},s)} = \sigma^{\alpha}_{(\mathfrak{N},t)}$, then

$$\mathfrak{N} \models \sigma^{\alpha}_{(\mathfrak{M},s)}.$$

We are left with task of proving (ii)⇒(i) and (ii)⇒(iii). To prove (ii)⇒(i), we use Theorem 5.7. Assume

$$\mathfrak{N} \models \sigma^{\alpha}_{(\mathfrak{M},s)}[t]$$

and define, for $\beta \leqslant \alpha$, a set I_β as follows: $f \in I_\beta$ iff

$$f: \mathfrak{M}_0 \cong \mathfrak{N}_0, \quad \mathfrak{M}_0 \subseteq \mathfrak{M}, \quad \mathfrak{N}_0 \subseteq \mathfrak{N}, \quad \text{where}$$

$\mathfrak{M}_0$ is generated by some $z_1, \ldots, z_k$, and

$$\mathfrak{N} \models \sigma^{\beta}_{(\mathfrak{M},z_1,\ldots,z_k)}[f(z_1), \ldots, f(z_k)].$$

The map f_0 generated by sending x_i to y_i $(i=1,\ldots,n)$ is in I_α by hypothesis. By 6.2(iii), we have

$$I_0 \supseteq I_1 \supseteq \cdots \supseteq I_\beta \supseteq \cdots \supseteq I_\alpha \quad (\beta \leqslant \alpha).$$

The final condition on this sequence, the one demanded by 5.7, follows immediately from the definition of $\sigma^{\beta+1}_{(\mathfrak{M},z_1,\ldots,z_k)}$.

Finally, we prove (ii)⇒(iii) by induction on α. The cases for $\alpha=0$ and α a limit ordinal are trivial. So suppose

$$\mathfrak{N} \models \sigma^{\beta+1}_{(\mathfrak{M},s)}[t].$$

By 6.1(ii), we need to prove that

(4) $\quad \sigma^{\beta}_{(\mathfrak{M},s)}$ is $\sigma^{\beta}_{(\mathfrak{N},t)}$,

(5) for each $x \in \mathfrak{M}$ there is a $y \in \mathfrak{N}$ such that

$$\sigma^{\beta}_{(\mathfrak{M},s^\frown x)} \text{ is } \sigma^{\beta}_{(\mathfrak{N},t^\frown y)},$$

and

(6) for each $y \in \mathfrak{N}$ there is an $x \in \mathfrak{M}$ such that

$$\sigma^{\beta}_{(\mathfrak{M},s^\frown x)} \text{ is } \sigma^{\beta}_{(\mathfrak{N},t^\frown y)}.$$

Now, by the induction hypothesis, (4) is true, (5) reduces to

(5′) $\quad \mathfrak{N} \models \bigwedge_{x \in \mathfrak{M}} \exists v_{n+1}\, \sigma^{\beta}_{(\mathfrak{M},s^\frown x)}(v_{n+1})[t]$

and (6) reduces to

(6′) $$\mathfrak{N} \models \forall v_{n+1} \bigvee_{x \in \mathfrak{M}} \sigma^{\beta}_{(\mathfrak{M}, s^\frown x)}(v_{n+1})[t].$$

But (5′), (6′) are immediate consequences of

$$\mathfrak{N} \models \sigma^{\beta+1}_{(\mathfrak{M}, s)}$$

by lines (3), (2) respectively. □

If we apply 6.3 to the empty sequence, we obtain the following result.

6.4 Corollary. For all $\mathfrak{M}, \mathfrak{N}$, the following are equivalent:
(i) $\mathfrak{M} \equiv^{\alpha} \mathfrak{N}$;
(ii) $\mathfrak{N} \models \sigma^{\alpha}_{\mathfrak{M}}$;
(iii) $\sigma^{\alpha}_{\mathfrak{M}} = \sigma^{\alpha}_{\mathfrak{N}}$. □

6.5 Definition. The *Scott rank of a structure* $\mathfrak{M}$, $\mathrm{sr}(\mathfrak{M})$, is the least ordinal α such that for all finite sequences $x_1, \ldots, x_n$, $y_1, \ldots, y_n$ from $\mathfrak{M}$,

$$(\mathfrak{M}, x_1, \ldots, x_n) \equiv^{\alpha} (\mathfrak{M}, y_1, \ldots, y_n)$$

implies

$$(\mathfrak{M}, x_1, \ldots, x_n) \equiv^{\alpha+1} (\mathfrak{M}, y_1, \ldots, y_n).$$

We will see, quite soon, that if $\alpha = \mathrm{sr}(\mathfrak{M})$ then

$$(\mathfrak{M}, x_1, \ldots, x_n) \equiv^{\alpha} (\mathfrak{M}, y_1, \ldots, y_n)$$

actually implies

$$(\mathfrak{M}, x_1, \ldots, x_n) \equiv_{\infty\omega} (\mathfrak{M}, y_1, \ldots, y_n).$$

It is more convenient to use 6.5 as the definition, though, since then the next lemma becomes obvious.

6.6 Lemma. *If* κ *is an infinite cardinal and* $\mathrm{card}(\mathfrak{M}) < \kappa$ *then* $\mathrm{sr}(\mathfrak{M}) < \kappa$.

Proof. The proof is easy and we will get a much better bound in the next section, so we leave the proof to the student. □

6.7 Definition. Let $\mathfrak{M}$ be a structure for L, let $\mu = \mathrm{sr}(\mathfrak{M})$. The *canonical Scott theory* of $\mathfrak{M}$, $S_{\mathfrak{M}}$ consists of the sentences below:

$$\sigma^{\mu}_{\mathfrak{M}},$$

$$\forall v_1, \ldots, v_k \left[\sigma^{\mu}_{(\mathfrak{M}, s)}(v_1, \ldots, v_k) \to \sigma^{\mu+1}_{(\mathfrak{M}, s)}(v_1, \ldots, v_k)\right]$$

for all finite sequences $s=\langle x_1,\ldots,x_n\rangle$ from $\mathfrak{M}$. The *canonical Scott sentence* of $\mathfrak{M}$, $\sigma_{\mathfrak{M}}$, is the conjunction of the canonical Scott theory of $\mathfrak{M}$:

$$\sigma_{\mathfrak{M}}=\bigwedge S_{\mathfrak{M}}.$$

Note that $\mathrm{qr}(\sigma_{\mathfrak{M}})=\mathrm{sr}(\mathfrak{M})+\omega$. Also, from the definition of $\mathrm{sr}(\mathfrak{M})$ we see that

$$\mathfrak{M}\models\sigma_{\mathfrak{M}}.$$

We now come to the main theorem on Scott sentences.

6.8 Theorem. *Given structures $\mathfrak{M},\mathfrak{N}$ for a language L, the following are equivalent:*
(i) $\mathfrak{M}\cong_p\mathfrak{N}$;
(ii) $\mathfrak{N}\models\sigma_{\mathfrak{M}}$;
(iii) $\sigma_{\mathfrak{M}}=\sigma_{\mathfrak{N}}$.

Proof. We already know that $\mathfrak{M}\cong_p\mathfrak{N}$ iff $\mathfrak{M}\equiv_{\infty\omega}\mathfrak{N}$. Since $\mathfrak{M}\models\sigma_{\mathfrak{M}}$ we see that (i)⇒(ii) is immediate. Similarly, since $\mathfrak{N}\models\sigma_{\mathfrak{N}}$, (iii)⇒(ii) is immediate. To prove (ii)⇒(i) define I_β, for all β, just as in the proof of 6.3. The hypothesis that $\mathfrak{N}\models\sigma_{\mathfrak{M}}$ insures that $I_{\mu+1}=I_\mu$ so

$$I_\mu\colon\mathfrak{M}\cong_p\mathfrak{N}.$$

Finally, we prove that (i)⇒(iii). Assume that $\mathfrak{M}\equiv_{\infty\omega}\mathfrak{N}$. Then $\mathrm{sr}(\mathfrak{M})=\mathrm{sr}(\mathfrak{N})$. Let $\mu=\mathrm{sr}(\mathfrak{M})$. For each $x_1,\ldots,x_n\in\mathfrak{M}$ there is a sequence $y_1,\ldots,y_n\in\mathfrak{N}$ such that

$$\mathfrak{N}\models\sigma^{\mu}_{(\mathfrak{M},x_1,\ldots,x_n)}[y_1,\ldots,y_n]$$

and vice versa. Then, by 6.3, every $\sigma^{\mu}_{(\mathfrak{M},s)}$ is some $\sigma^{\mu}_{(\mathfrak{M},t)}$ and vice versa. Thus $S_{\mathfrak{M}}=S_{\mathfrak{N}}$ and $\sigma_{\mathfrak{M}}=\sigma_{\mathfrak{N}}$. □

The remainder of this section is devoted to corollaries of Theorem 6.8. First we have Scott's original result.

6.9 Corollary (Scott's Theorem). *Let L be a countable language and let $\mathfrak{M}$ be a countable structure for L. The Scott sentence $\sigma_{\mathfrak{M}}$ is a sentence of $\mathsf{L}_{\omega_1\omega}$ with the property that*

$$\mathfrak{M}\cong\mathfrak{N}\quad \textit{iff}\quad \mathfrak{N}\models\sigma_{\mathfrak{M}}$$

for all countable L-structures $\mathfrak{N}$.

Proof. $\sigma_{\mathfrak{M}}$ is in $\mathsf{L}_{\omega_1\omega}$ by Lemma 6.2(iv). The result is then an immediate consequence of Theorem 5.2 and 6.8. □

An n-ary relation P on $\mathfrak{M}=\langle M,R_1,\ldots,R_l\rangle$ is *invariant* if for every automorphism f of $\mathfrak{M}$ and every $x_1,\ldots,x_n\in\mathfrak{M}$,

$$P(x_1,\ldots,x_n)\quad\text{iff}\quad P(f(x_1),\ldots,f(x_n)).$$

From now on (in this section) we assume L is countable. Whenever we refer to $\mathbb{HYP}_{\mathfrak{M}}$ we assume L is finite.

6.10 Corollary. *If $\mathfrak{M}$ is a countable structure for L and P is an n-ary relation on $\mathfrak{M}$, then P is invariant iff P is definable by some formula $\varphi(v_1, \ldots, v_n)$ of $\mathsf{L}_{\omega_1\omega}$ (without additional parameters):*

$$P(x_1, \ldots, x_n) \quad \textit{iff} \quad \mathfrak{M} \models \varphi[x_1, \ldots, x_n].$$

Proof. If P is defined by φ then P must be invariant since $f: \mathfrak{M} \cong \mathfrak{M}$ and $\mathfrak{M} \models \varphi[x_1, \ldots, x_n]$ implies $\mathfrak{M} \models \varphi[f(x_1), \ldots, f(x_n)]$. Now assume P is invariant. Let $\varphi(v_1, \ldots, v_n)$ be

$$\bigvee \{\sigma^{\mu}_{\langle x_1, \ldots, x_n \rangle}(v_1, \ldots, v_n) \mid P(x_1, \ldots, x_n)\},$$

where $\mu = \mathrm{sr}(\mathfrak{M})$. It is clear that $P(x_1, \ldots, x_n)$ implies $\mathfrak{M} \models \varphi[x_1, \ldots, x_n]$.

To prove the converse, suppose that $\mathfrak{M} \models \varphi[y_1, \ldots, y_n]$, so that $\mathfrak{M} \models \sigma^{\mu}_{\langle x_1, \ldots, x_n \rangle}[y_1, \ldots, y_n]$ for some $x_1, \ldots, x_n$ with $P(x_1, \ldots, x_n)$. Then

$$(\mathfrak{M}, x_1, \ldots, x_n) \equiv_{\infty\omega} (\mathfrak{M}, y_1, \ldots, y_n),$$

so that there is an automorphism f of $\mathfrak{M}$ with $f(x_i) = y_i$ by 5.2. Since P is invariant, $P(y_1, \ldots, y_n)$ holds. ☐

6.11 Corollary. *Let $\mathfrak{M}$ be a countable structure for L and let $x \in \mathfrak{M}$ be an element fixed by every automorphism of $\mathfrak{M}$. Then x is definable by a formula $\varphi(v)$ of $\mathsf{L}_{\omega_1\omega}$:*

$$\mathfrak{M} \models \exists! v \varphi(v),$$
$$\mathfrak{M} \models \varphi[x].$$

Conversely, a definable element of $\mathfrak{M}$ is fixed by every automorphism.

Proof. Apply 6.10 with $P = \{x\}$. ☐

A *rigid* structure is one with only one automorphism, the identity map.

6.12 Corollary. *If $\mathfrak{M}$ is a countable structure for L then $\mathfrak{M}$ is rigid iff every element x of $\mathfrak{M}$ is definable by a formula $\varphi(v)$ of $\mathsf{L}_{\omega_1\omega}$:*

$$\mathfrak{M} \models \exists! v \varphi(v),$$
$$\mathfrak{M} \models \varphi[x]. \quad ☐$$

These results will be improved in the next section.

6.13—6.14 Exercises

6.13. Let $\mathfrak{M}$ be a countable structure with $x_1, \ldots, x_n \in \mathfrak{M}$ such that $(\mathfrak{M}, x_1, \ldots, x_n)$ is rigid; i.e., no nontrivial automorphisms of $\mathfrak{M}$ fix, $x_1, \ldots, x_n$. Show that $\mathfrak{M}$ has $\leqslant \aleph_0$ automorphisms.

6.14. Let $\mathfrak{M}$ be a countable L-structure with $< 2^{\aleph_0}$ automorphisms.

(i) Prove that there is a finite sequence $x_1, \ldots, x_n$ from $\mathfrak{M}$ such that $(\mathfrak{M}, x_1, \ldots, x_n)$ is rigid. [Hint (P. M. Cohn): Let σ_n fix $x_1, \ldots, x_n$ but move, say, x_{n+1}. Let $\sigma = \ldots \sigma_n^{\varepsilon_n} \ldots \sigma_2^{\varepsilon_2} \sigma_1^{\varepsilon_1}$ where $\varepsilon_i = 0$ or $= 1$. Show that this gives $2^{\aleph_0}$ automorphisms.]

(ii) Show that for all $\mathfrak{N}$, $\mathfrak{N} \models \sigma_{\mathfrak{M}}$ implies $\mathfrak{M} \cong \mathfrak{N}$; i.e., that there are no uncountable $\mathfrak{N}$ with $\mathfrak{M} \equiv \mathfrak{N}$ $(\mathsf{L}_{\omega_1\omega})$.

6.15. Show that if G is an $\aleph_1$-free abelian group then $G \cong_p H$ iff H is $\aleph_1$-free. Thus the notion of free group is not definable in $\mathsf{L}_{\infty\omega}$.

6.15 Notes. Scott's Theorem and Corollary 6.10 were announced in Scott [1965]. A proof, in the context of invariant Borel sets, appears in Scott [1964]. The Scott sentences used here are derived from Chang's proof of Scott's Theorem in Chang [1968]. The presentation follows that used in the survey article Barwise [1973]. Exercises 6.13, 6.14, 6.15 are due to Kueker. They are proved in Barwise [1973].

7. *Scott Sentences and Admissible Sets*

The first systematic study of the relationship between α-characteristics, canonical Scott sentences and admissible sets was undertaken by Nadel in his doctoral dissertation. His idea was to use α-characteristics and Scott sentences as approximations of models, asking to which admissible sets the formulas $\sigma_{\mathfrak{M}}, \sigma_{\mathfrak{M}}^{\alpha}$ belong as an alternative to asking to which admissible sets $\mathfrak{M}$ itself belongs. This has proven to be a fruitful idea. In this section we delve into the more elementary parts of the theory.

To simplify matters we assume the underlying language L of $\mathsf{L}_{\infty\omega}$ has no function symbols. Since function symbols can always be replaced by relation symbols, this is no essential loss. (The sole point in this restriction is that if L is an element of an admissible set $\mathbb{A}$ then the set of atomic and negated atomic formulas of the form

$$\varphi(v_1, \ldots, v_n)$$

(for fixed $n < \omega$) is a set in $\mathbb{A}$ *if* L has no function symbols, or *if* $o(\mathbb{A}) > \omega$, but *not* if L has a function symbol and $o(\mathbb{A}) = \omega$).

7.1 Proposition. *The formula*

$$\sigma^{\alpha}_{\mathfrak{M},s}(v_1, \ldots, v_n)$$

is definable in KPU *as a* Σ_1 *operation of* $\mathfrak{M}, s, \alpha$.

Proof. Consider sequences s as functions with dom(s) some $n < \omega$ and range $\subseteq \mathfrak{M}$. Let

$$F(\mathfrak{M}, s, \alpha) = \sigma^{\alpha}_{\mathfrak{M},s}(v_1, \ldots, v_n)\,.$$

If we write out the definition of F as given in 6.1 it takes the following form:

$$F(\mathfrak{M}, s, \alpha) = y \quad \text{iff} \quad \text{(i)} \vee \text{(ii)} \vee \text{(iii)}$$

where

(i) $\alpha = 0 \wedge \Delta_0(\mathfrak{M}, s, y)$ (a Δ_0 predicate of $\mathfrak{M}, s$ and y);

(ii) $\alpha = \beta + 1$ for some $\beta < \alpha$ and $y = \bigwedge\{\theta_1, \theta_2, \theta_3\}$ where

$$\theta_1 = F(\mathfrak{M}, s, \beta)\,,$$

θ_2 is $\forall v_{n+1} \bigvee \Phi$ where

$$\forall x \in \mathfrak{M}\, \exists z \in \Phi\, F(\mathfrak{M}, s^\wedge x, \beta) = z\,,$$

$$\forall z \in \Phi\, \exists x \in \mathfrak{M}\, F(M, s^\wedge x, \beta) = z\,, \quad \text{and}$$

θ_3 is similar to θ_2.

(iii) $\mathrm{Lim}(\alpha) \wedge y = \bigwedge\{F(\mathfrak{M}, s, \beta) \mid \beta < \alpha\}$.

This definition clearly falls under the second recursion theorem. □

7.2 Corollary. *If* $\mathsf{L}_{\mathbb{A}}$ *is an admissible fragment and* $\mathfrak{M}$ *is an* L*-structure in the admissible set* $\mathbb{A}$ *then, for any* L*-structure* $\mathfrak{N}$,

$$\mathfrak{M} \equiv \mathfrak{N}\ (\mathsf{L}_{\mathbb{A}}) \quad \textit{implies} \quad \mathfrak{M} \equiv^{\alpha} \mathfrak{N}$$

where $\alpha = o(\mathbb{A})$.

Proof. By Exercise 5.10 it suffices to prove that

$$\mathfrak{M} \equiv^{\beta} \mathfrak{N}$$

for all $\beta < \alpha$. But for $\beta < \alpha$, $\sigma^{\beta}_{\mathfrak{M}} \in \mathsf{L}_{\mathbb{A}}$ by 7.1 and $\mathfrak{M} \models \sigma^{\beta}_{\mathfrak{M}}$ so $\mathfrak{N} \models \sigma^{\beta}_{\mathfrak{M}}$. But then $\mathfrak{M} \equiv^{\beta} \mathfrak{N}$ by Corollary 6.4. □

If $\sigma_{\mathfrak{M}}$ were definable as a Σ_1 operation of $\mathfrak{M}$ in KPU then we could extend 7.2 to read

$$\mathfrak{M} \equiv \mathfrak{N}\ (\mathsf{L}_{\mathbb{A}}) \quad \text{implies} \quad \mathfrak{M} \equiv_{\infty\omega} \mathfrak{N}\,,$$

since then $\sigma_{\mathfrak{M}}$ would be in $\mathsf{L}_{\mathbb{A}}$. This, however, is not true. Unlike its approximations, *the canonical Scott sentence $\sigma_{\mathfrak{M}}$ is not definable in* KPU *as a Σ_1 operation of $\mathfrak{M}$*. The problem is that $\mathrm{sr}(\mathfrak{M})$ may be just a bit too big; that is, $\mathrm{sr}(\mathfrak{M})$ may equal $o(\mathbb{HYP}_{\mathfrak{M}})$. (See Exercise 7.13, 7.14.) This is as big as it can get, though, as we see in Corollary 7.4.

7.3 Theorem. *Let $\mathsf{L}_{\mathbb{A}}$ be an admissible fragment of $\mathsf{L}_{\infty\omega}$ and let $\mathfrak{M}, \mathfrak{N}$ be* L*-structures which are both elements of the admissible set $\mathbb{A}$. Then*

$$\mathfrak{M} \equiv \mathfrak{N}\ (\mathsf{L}_{\mathbb{A}}) \quad \textit{implies} \quad \mathfrak{M} \equiv_{\infty\omega} \mathfrak{N}.$$

Proof. By 7.2 we see that $\mathfrak{M} \equiv^{\alpha} \mathfrak{N}$ where $\alpha = o(\mathbb{A})$. Let I be the set of finite partial isomorphisms $f = \{\langle x_1, y_1\rangle, \ldots, \langle x_n, y_n\rangle\}$ $(0 \leqslant n < \omega)$ such that

$$(1) \qquad (\mathfrak{M}, x_1, \ldots, x_n) \equiv^{\alpha} (\mathfrak{N}, y_1, \ldots, y_n).$$

Since $\mathfrak{M} \equiv^{\alpha} \mathfrak{N}$, the trivial map is in I so $I \neq \emptyset$. We will prove that

$$I : \mathfrak{M} \cong_p \mathfrak{N}.$$

Suppose (1) holds and that a new $x_{n+1} \in \mathfrak{M}$ is given. We need to find a $y_{n+1} \in \mathfrak{N}$ such that

$$(\mathfrak{M}, x_1, \ldots, x_n, x_{n+1}) \equiv^{\alpha} (\mathfrak{N}, y_1, \ldots, y_n, y_{n+1}).$$

By Exercise 5.10 it suffices to insure that

$$(\mathfrak{M}, x_1, \ldots, x_n, x_{n+1}) \equiv^{\beta} (\mathfrak{N}, y_1, \ldots, y_n, y_{n+1})$$

for each $\beta < \alpha$. Suppose that no such y_{n+1} exists. Then

$$\forall y_{n+1} \in \mathfrak{N}\, \exists \beta < \alpha (\mathfrak{N} \models \neg \sigma^{\beta}_{(\mathfrak{M}, s)}[y_1, \ldots, y_n, y_{n+1}])$$

where $s = \langle x_1, \ldots, x_{n+1}\rangle$. By Σ Reflection in $\mathbb{A}$, there is a $\gamma < \alpha$ such that

$$\forall y_{n+1} \in \mathfrak{N}\, \exists \beta < \gamma (\mathfrak{N} \models \neg \sigma^{\beta}_{\mathfrak{M}, s}[y_1, \ldots, y_n, y_{n+1}])$$

and hence

$$\mathfrak{N} \models \forall v_{n+1} \neg \sigma^{\gamma}_{\mathfrak{M}, s}(v_{n+1})[y_1, \ldots, y_n]$$

so

$$\mathfrak{N} \models \neg \sigma^{\gamma+1}_{(\mathfrak{M}, s)}[y_1, \ldots, y_n]$$

contradicting

$$(\mathfrak{M}, x_1, \ldots, x_n) \equiv^{\alpha} (\mathfrak{N}, y_1, \ldots, y_n).$$

This establishes the "forth" half of the back and forth property; the "back" half follows from the symmetry of $\mathfrak{M}$ and $\mathfrak{N}$ in the theorem. $\square$

Theorem 7.3 is sometimes called Nadel's Basis Theorem. The reason for calling it a basis theorem is seen by stating the converse of its conclusion: If there is a sentence φ of $\mathsf{L}_{\infty\omega}$ true in $\mathfrak{M}$ and false in $\mathfrak{N}$, then there is such a sentence in $\mathsf{L}_{\mathbb{A}}$.

Our first application of 7.3 is to get the best possible bound on sr($\mathfrak{M}$). Another proof of this can be given by means of inductive definitions.

7.4 Corollary. *Let $\mathfrak{M}$ be a structure in an admissible set $\mathbb{A}$. Then*

$$\mathrm{sr}(\mathfrak{M}) \leqslant o(\mathbb{A}).$$

Proof. Let $\alpha = o(\mathbb{A})$. Let $x_1, \ldots, x_n, y_1, \ldots, y_n \in \mathfrak{M}$ be such that

$$(\mathfrak{M}, x_1, \ldots, x_n) \equiv^{\alpha} (\mathfrak{M}, y_1, \ldots, y_n).$$

But then

$$(\mathfrak{M}, x_1, \ldots, x_n) \equiv (\mathfrak{M}, y_1, \ldots, y_n)\,(\mathsf{L}_{\mathbb{A}})$$

so, by 7.3,

$$(\mathfrak{M}, x_1, \ldots, x_n) \equiv_{\infty\omega} (\mathfrak{M}, y_1, \ldots, y_n). \quad \square$$

The remainder of this section deals with uses of Nadel's Basis Theorem to improve the results of the previous section.

7.5 Theorem. *Let $\mathfrak{M}$ be an L-structure and let P be a relation on $\mathfrak{M}$ which is definable by some formula of $\mathsf{L}_{\infty\omega}$ without parameters. Let $\mathbb{A}$ be any admissible set with $(\mathfrak{M}, P) \in \mathbb{A}$. Then P is definable by a formula of $\mathsf{L}_{\mathbb{A}}$ without parameters.*

Proof. Let us suppose, for convenience, that P is unary. We assume that P is not definable by any formula of $\mathsf{L}_{\mathbb{A}}$. If we can find an x, y such that

$$P(x),\ \neg P(y), \quad \text{and} \quad (\mathfrak{M}, x) \equiv (\mathfrak{M}, y)\,(\mathsf{L}_{\mathbb{A}})$$

then, by 7.3,

$$(\mathfrak{M}, x) \equiv_{\infty\omega} (\mathfrak{M}, y)$$

so P is not definable by any formula of $\mathsf{L}_{\infty\omega}$. To find such an x, y we proceed as follows. Define, for $\beta < \alpha$, $\varphi_\beta(v)$ to be the formula

$$\bigvee \{\sigma_x^\beta(v) \mid x \in P\}.$$

Then $\varphi_\beta(v) \in \mathsf{L}_{\mathbb{A}}$ by 7.1 and

(2) $$\models \varphi_\beta(v) \rightarrow \varphi_\gamma(v)$$

for $\beta \geqslant \gamma$. Since $\mathfrak{M} \models \varphi_\beta[x]$ for all $x \in P$, and φ_β does not define P (nothing in $\mathsf{L}_{\mathbb{A}}$ does) there must be some $y \in M - P$ such that $\mathfrak{M} \models \varphi_\beta[y]$.

We claim that there is a fixed $y \in M - P$ which works for all $\beta < \alpha$:

$$(3) \qquad \exists y \in M - P \,\forall \beta < \alpha (\mathfrak{M} \models \varphi_\beta[y]) .$$

For otherwise we would have

$$\forall y \in M - P \,\exists \beta < \alpha \,(\mathfrak{M} \models \neg \varphi_\beta[y]) .$$

But then by Σ Reflection there is a $\gamma < \alpha$ such that for all $y \in M - P$

$$\mathfrak{M} \models \bigvee_{\beta < \gamma} \neg \varphi_\beta[y]$$

and hence by (2),

$$\forall y \in M - P \,(\mathfrak{M} \models \neg \varphi_\gamma[y]) ,$$

a contradiction. Thus (3) is established. Let y be as in (3). For each β there is an $x \in P$ such that

$$\mathfrak{M} \models \sigma_x^\beta[y]$$

by the definition of φ_β. By an argument entirely analogous to the proof of (3), we see that

$$\exists x \in P \,\forall \beta < \alpha \,(\mathfrak{M} \models \sigma_x^\beta[y]) .$$

For any such x we have $(\mathfrak{M}, x) \equiv^\alpha (\mathfrak{M}, y)$ and hence $(\mathfrak{M}, x) \equiv (\mathfrak{M}, y)(\mathsf{L}_{\mathbb{A}})$, as desired. ☐

7.6 Corollary. *Let $\mathfrak{M} = \langle M, R_1, \ldots, R_l \rangle$ be a countable structure for L. A relation P on $\mathfrak{M}$ is invariant on $\mathfrak{M}$ iff it is definable by a formula in $\mathsf{L}_{\infty\omega} \cap \mathbb{HYP}_{(\mathfrak{M}, P)}$.*

Proof. Combine 6.10 with 7.5. ☐

7.7 Corollary. *Let $\mathsf{L}_{\mathbb{A}}$ be an admissible fragment of $\mathsf{L}_{\infty\omega}$. If $\mathfrak{M}$ is an L-structure, $\mathfrak{M} \in \mathbb{A}$, then every element of $\mathfrak{M}$ definable by some formula of $\mathsf{L}_{\infty\omega}$ is definable by a formula of $\mathsf{L}_{\mathbb{A}}$.*

Proof. Apply 7.5 with $P = \{x\}$. ☐

7.8 Corollary. *Let $\mathfrak{M} = \langle M, R_1, \ldots, R_l \rangle$ be a countable L-structure. Then $\mathfrak{M}$ is rigid iff every element of $\mathfrak{M}$ is definable by a formula of $\mathsf{L}_{\infty\omega} \cap \mathbb{HYP}_{\mathfrak{M}}$.*

Proof. Combine 6.12 with 7.7. ☐

7.9 Corollary. *If* $\mathfrak{M}=\langle M, R_1, \ldots, R_l\rangle$ *is a countable rigid structure then* $\mathrm{sr}(\mathfrak{M})<O(\mathfrak{M})$.

Proof. By 7.8 we know that

$$\forall x\in M\,\exists\beta\,\mathfrak{M}\models\exists!\,v\,\sigma_x^\beta(v)\,.$$

Let $\beta(x)$ be the least such β. Then $\sigma_x^{\beta(x)}(v)$ is a $\mathbb{HYP}_{\mathfrak{M}}$-recursive function of x so, by Σ Replacement,

$$\Phi(v)=\{\sigma_x^{\beta(x)}(v)\mid x\in M\}$$

is in $\mathbb{HYP}_{\mathfrak{M}}$ and every element of M is definable by some member of it. Let $\gamma=\sup\{\beta(x)\mid x\in M\}$. We claim that $\mathrm{sr}(\mathfrak{M})\leqslant\gamma$. For suppose

$$(\mathfrak{M}, x_1, \ldots, x_n)\equiv^{\gamma}(\mathfrak{M}, y_1, \ldots, y_n)\,.$$

Then

$$\mathfrak{M}\models\varphi_{x_i}^{\beta(x_i)}[y_i]$$

so $x_i=y_i$ for $i=1,\ldots,n$, and hence

$$(\mathfrak{M}, x_1, \ldots, x_n)\equiv_{\infty\omega}(\mathfrak{M}, y_1, \ldots, y_n)\,. \quad \square$$

We can improve 7.9 by replacing the requirement that $\mathfrak{M}$ is rigid by the requirement that $\mathfrak{M}$ have $<2^{\aleph_0}$ automorphisms. See Exercise 7.15.

We end this section by returning to our old favorite, recursively saturated structures, to see what some of our results say in this case.

7.10 Corollary. *Let* $\mathfrak{M}=\langle M, R_1, \ldots, R_l\rangle$ *be a recursively saturated* L*-structure and let* P *be a relation on* $\mathfrak{M}$ *definable by some formula of* $\mathsf{L}_{\infty\omega}$. *Then* $(\mathfrak{M}, P)$ *is recursively saturated iff* P *is definable by a finitary formula of* $\mathsf{L}_{\omega\omega}$.

Proof. The ($\Rightarrow$) half follows from 7.5 with $\mathbb{A}=\mathbb{HYP}_{(\mathfrak{M},P)}$. To prove the ($\Leftarrow$) half, note that if P is definable by a formula $\varphi\in\mathbb{HYP}_{\mathfrak{M}}$ then $P\in\mathbb{HYP}_{\mathfrak{M}}$ by Δ_1 Separation so $o(\mathbb{HYP}_{(\mathfrak{M},P)})=\omega$. $\square$

Note that if $\mathfrak{M}$ is recursively saturated then so is $(\mathfrak{M},\vec{x})$ for any $\vec{x}\in\mathfrak{M}$ so 7.10 also applies to relations definable by a fixed finite number of parameters. The same remark applies to the next result.

7.11 Corollary. *Let* $\mathfrak{M}=\langle M, R_1, \ldots, R_l\rangle$ *be an infinite recursively saturated* L*-structure and let*

$$\mathscr{Df}(\mathfrak{M})=\{y\in M\mid y \text{ is definable by some formula } \varphi(v) \text{ of } \mathsf{L}_{\infty\omega} \text{ without parameters}\}\,.$$

Then we have the following:

(i) *Every element of* $\mathscr{D}\!f(\mathfrak{M})$ *is definable by a finitary formula of* $\mathsf{L}_{\omega\omega}$.

(ii) $\mathscr{D}\!f(\mathfrak{M})$ *is* Σ_1 *on* $\mathbb{HYP}_{\mathfrak{M}}$, *hence inductive* on* $\mathfrak{M}$.

(iii) *If* $\mathscr{D}\!f(\mathfrak{M})$ *is hyperelementary* on* $\mathfrak{M}$ *(i.e., if it is in* $\mathbb{HYP}_{\mathfrak{M}}$*) then* $\mathscr{D}\!f(\mathfrak{M})$ *is finite.*

(iv) $\mathfrak{M}-\mathscr{D}\!f(\mathfrak{M})$ *is infinite.*

Proof. (i) follows from 7.7 and (i)$\Rightarrow$(ii). To prove (iii) suppose that $\mathscr{D}\!f(\mathfrak{M})\in\mathbb{HYP}_{\mathfrak{M}}$. Let

$$\Phi=\{\sigma_x^n(v)\,|\,x\in\mathscr{D}\!f(\mathfrak{M}),\ \mathfrak{M}\models\exists!v\,\sigma_x^n(v) \text{ and } \mathfrak{M}\models\bigwedge_{m<n}\neg\exists!v\,\varphi_x^m(v)\}.$$

Then, exactly as in the proof of 7.9, Φ is an element of $\mathbb{HYP}_{\mathfrak{M}}$. But Φ is a pure set and $o(\mathbb{HYP}_{\mathfrak{M}})=\omega$ so Φ is finite. Thus $\mathscr{D}\!f(\mathfrak{M})$ must also be finite, since every member is defined by a formula in Φ. Part (iv) is immediate from (iii), for if $\mathfrak{M}-\mathscr{D}\!f(\mathfrak{M})$ is finite then $\mathscr{D}\!f(\mathfrak{M})\in\mathbb{HYP}_{\mathfrak{M}}$. □

7.12. Example. Let $\mathcal{N}'$ be a nonstandard model of Peano Arithmetic and let $x\in N'$ be a nonstandard integer. Let $\mathcal{N}[x]$ be the submodel of $\mathcal{N}'$ with universe

$$\mathscr{D}\!f((\mathcal{N}',x)).$$

The axiom of induction insures that

$$\mathcal{N}[x]\prec\mathcal{N}'.$$

Corollary 7.11(iv) (applied to $(\mathcal{N}[x],x)$) shows that models of the form $\mathcal{N}[x]$ can never be recursively saturated. Hence, the standard integers of $\mathcal{N}[x]$ form a hyperelementary subset of $\mathcal{N}[x]$ by VI.5.1(ii). From this it follows that such models can never be expanded to a model of second order arithmetic, by Exercise IV.5.13.

7.13—7.18 Exercises

7.13. Let M be countable, α a countable admissible ordinal, $\alpha>\omega$, and let η be the order type of the rationals.

(i) Prove that if $<_1$ is a linear ordering of M of order type $\alpha(1+\eta)$ than, setting $\mathfrak{M}_1=\langle M,<_1\rangle$,

$$\mathbb{HYP}_{\mathfrak{M}_1}\models\text{“}<_1 \text{ is well founded”},$$

$$\alpha=o(\mathbb{HYP}_{\mathfrak{M}_1}).$$

[See the proof of IV.6.1.]

(ii) Let $\mathfrak{M}_0=\mathscr{W}\!f(\mathfrak{M}_1)$. Let $\mathsf{L}_{\mathbb{A}}$ be the admissible fragment of $\mathsf{L}_{\infty\omega}$ given by $\mathbb{HYP}_{\mathfrak{M}_1}$, where $\mathsf{L}=\{<\}$. Prove that

$$\mathfrak{M}_0\prec\mathfrak{M}_1\quad[\mathsf{L}_{\mathbb{A}}].$$

[Use the Tarski Criterion for $\mathsf{L}_{\mathbb{A}}$ (Exercise 2.13) and the fact that any x in the non-wellfounded part of $<_1$ can be moved by an automorphism of $\mathfrak{M}_1$.]

(iii) Prove that

$$\mathfrak{M}_0 \equiv^{\alpha} \mathfrak{M}_1 .$$

(iv) Prove that $\mathrm{sr}(\mathfrak{M}_1)=\alpha$.

(v) Conclude that $\sigma_{\mathfrak{M}}$ is not definable in KPU as a Σ_1 operation of $\mathfrak{M}$.

7.14. Prove that $\mathrm{sr}(\mathfrak{M})$ and σ_M are Σ_1 definable in KPU + Infinity + Σ_1 Separation, as operations of $\mathfrak{M}$.

7.15. Use 6.14(i) to improve 7.9 to the case where $\mathfrak{M}$ has $<2^{\aleph_0}$ automorphisms.

7.16. Prove that if $o(\mathbb{HYP}_{\mathfrak{M}})>\omega$ and $\mathrm{sr}(\mathfrak{M})<o(\mathbb{HYP}_{\mathfrak{M}})$ then $\sigma_{\mathfrak{M}} \in \mathbb{HYP}_{\mathfrak{M}}$.

7.17. Prove that the absolute version of

"P is invariant on $\mathfrak{M}$"

is

"P is definable by a formula of $\mathsf{L}_{\infty\omega} \cap \mathbb{HYP}_{(\mathfrak{M},P)}$".

7.18. Prove that the absolute version of "$\mathfrak{M}$ is rigid" is "Every element of $\mathfrak{M}$ is definable by a formula of $\mathsf{L}_{\infty\omega} \cap \mathbb{HYP}_{\mathfrak{M}}$".

7.19 Notes. There are a number of interesting and important results which could be gone into at this point, but they would take us too far afield. The student is urged to read Makkai [1975] and Nadel [1974].

Theorem 7.3 is from Nadel [1971] (and Nadel [1974]) as are Collaries 7.7 and 7.8. Theorem 7.5 is new here but it is a fairly routine generalization of Nadel's 7.7. The important example 7.13 is also taken from Nadel [1971]. The last sentence of Example 7.12 is a theorem of Ehrenfeucht and Kreisel. [Added in proof: A recent paper by Nadel and Stari called "The pure part of $\mathbb{HYP}$" (to appear in the Journal of Symbolic Logic) has a number of interesting and highly relevant results. In particular, they characterize the pure part of $\mathbb{HYP}_{\mathfrak{M}}$ in terms of the sentences $\sigma^{\beta}_{\mathfrak{M}}$ for $\beta<O(\mathfrak{M})$.]

Chapter VIII
Strict Π_1^1 Predicates and König Principles

1. The König Infinity Lemma

In this section we discuss some of the uses of the Infinity Lemma in ordinary recursion theory. The applications chosen for discussion are those which become important new "axioms" or *König Principles*, when stated in the abstract.

Let $T = \langle T, \prec \rangle$ be the *full binary tree*, as pictured below.

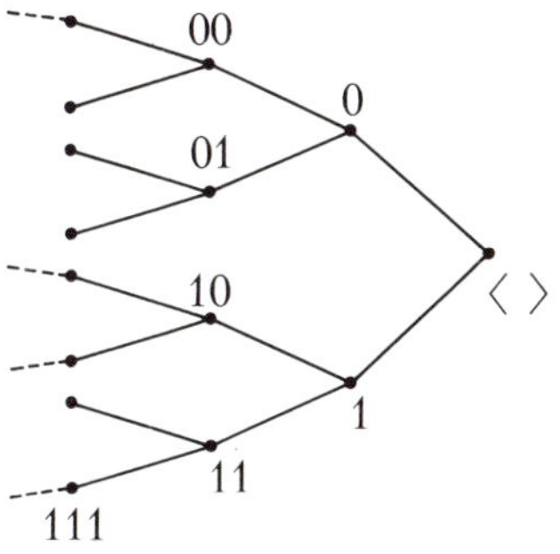

The set T is the set of nodes (finite sequences of 0's and 1's) ordered by

$$d' \prec d$$

if the sequence d' properly extends the sequence d. If $S \subseteq T$ is such that $d_0 \in S$ and $d_0 \prec d_1$ implies $d_1 \in S$, then $S = \langle S, \prec \upharpoonright S \rangle$ is called a *subtree* of T. If S is a subtree then any maximal $\prec$-linearly ordered subset b of S is called a *branch through* S.

1.1 König Infinity Lemma. *Let* $S = \langle S, \prec \upharpoonright S \rangle$ *be any subtree of the full binary tree. The following are equivalent:*

(i) *S has no infinite branch,*

(ii) *S is well founded,*

(iii) *S is well founded and has finite rank,*

(iv) *S is finite.*

Proof. Each of the implications (iv)$\Rightarrow$(iii)$\Rightarrow$(ii)$\Rightarrow$(i) is completely trivial so we need only prove (i)$\Rightarrow$(iv), or equivalently, $\neg$(iv)$\Rightarrow\neg$(i). Suppose S is infinite. Let $d_0 = \langle\ \rangle \in S$. Either 0 or 1 has infinitely many predecessors in S so let d_1 be the least of these which has infinitely many predecessors in S. Let $d_n \in S$ have infinitely many predecessors in S and let d_{n+1} be the first of $d_n\frown 0$, $d_n\frown 1$ which has infinitely many predecessors in S. One of them must. Then

$$b = \{d_0, d_1, d_2, \ldots\}$$

is an infinite branch through S. $\square$

One can generalize 1.1 trivially by allowing each node to have any finite number of immediate predecessors, instead of exactly two, but once you allow infinitely many, the theorem becomes false, as the following tree shows.

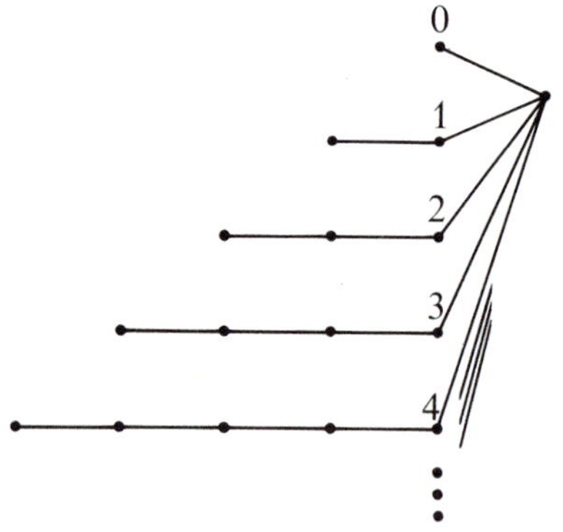

Indeed, the Infinity Lemma is so tied to the notion of finiteness and the integers that it is difficult to generalize in a really useful way. So, rather than generalize the Infinity Lemma itself, we go back and look for useful consequences of the Infinity Lemma. Three of these consequences have turned out to play important roles when generalized to other admissible sets. In this section we prove these three results.

A predicate $P(x, f)$ of integers x and number theoretic functions f is r.e. iff there is a recursive predicate $R(x, y)$ of integers such that

$$P(x, f) \leftrightarrow \exists n\, R(x, \bar{f}(n))$$

where $\bar{f}(n)$ is (a code for) the finite sequence $\langle f(0), \ldots, f(n-1)\rangle$ and $R(x, \bar{f}(n))$ implies $R(x, \bar{f}(m))$ for all $m \geqslant n$. (This may be taken as the definition or verified easily from any other reasonable definition. This is the natural extension of r.e. to predicates $P(x, f)$ of numbers and functions.)

1.2 Definition. A predicate $S(\vec{x})$ on the integers is *strict*-Π_1^1 (or *s*-Π_1^1 for short) if it can be written in the form

$$S(\vec{x}) \leftrightarrow \forall f \in 2^\omega\, P(\vec{x}, f)$$

where P is r.e.

Here we use 2^ω to denote the set of characteristic functions, i.e., those functions mapping ω into $2 = \{0,1\}$. The word "strict" refers to the fact that f ranges only over 2^ω, not over all number theoretic functions.

Our first application of the Infinity Lemma is to prove the following result.

1.3 Theorem (s-Π_1^1 = r.e., on ω). *A predicate P on ω is strict-Π_1^1 iff P is r.e.*

Proof. To prove ($\Leftarrow$) just add a superfluous function quantifier. To prove ($\Rightarrow$) write

$$(1) \qquad P(x) \leftrightarrow \forall f \in 2^\omega\, \exists n\, R(x, \bar{f}(n))$$

where R is recursive and satisfies

$$R(x, \bar{f}(n)) \wedge m \geqslant n \Rightarrow R(x, \bar{f}(m)).$$

For $f \in 2^\omega$, each $\bar{f}(n)$ is a sequence of 0's and 1's and so is really just a node on the full binary tree T. The condition on R above asserts that

$$R(x,d) \wedge d' \prec d \Rightarrow R(x,d')$$

or, turning it around,

$$\neg R(x,d') \wedge d' \prec d \Rightarrow \neg R(x,d).$$

Thus, $S_x = \{d \mid \neg R(x,d)\}$ is a subtree of T. If we restate (1) in terms of trees, it becomes

$$(2) \qquad P(x) \text{ iff } S_x \text{ has no infinite path,}$$

which becomes, by the Infinity Lemma,

$$\begin{aligned} P(x) &\leftrightarrow S_x \text{ is finite} \\ &\leftrightarrow \exists N\, \forall d \text{ of length } N,\ d \notin S_x \\ &\leftrightarrow \exists N\, \forall d \text{ of length } N,\ R(x,d) \\ &\leftrightarrow \exists N\, R'(x,N), \end{aligned}$$

where R' is recursive. More informally, $P(x)$ holds iff you can find a finite subtree such that $R(x,d)$ holds for every end-node d on S. ☐

In proving 1.3 we also proved the next theorem. This (or a relativized form of it) is what Shoenfield [1967] refers to as the Brouwer-König Infinity Lemma.

1.4 Theorem (s-Π_1^1 Reflection for ω). *Let*

$$P(x) \leftrightarrow \forall f \in 2^\omega \, \exists n \, R(x, \bar{f}(n))$$

define a strict-Π_1^1 predicate. Then for any x

$$P(x) \leftrightarrow \exists N \, \forall f \in 2^\omega \, \exists n \leqslant N \, R(x, \bar{f}(n)).$$

Proof. This is contained in the proof of 1.3. ▯

We will see in § 4 that the equation "s-$\Pi_1^1 = \text{r.e.}$" can be viewed as abstract formulation of the completeness theorem for $\mathsf{L}_{\omega\omega}$ and that "s-Π_1^1 Reflection for ω" corresponds to the compactness theorem for $\mathsf{L}_{\omega\omega}$. The fact that our proof of 1.3 also gives 1.4 corresponds to the fact that most proofs of the completeness theorem yield compactness, but not vice versa.

Our final application of the Infinity Lemma is to the notion of implicit ordinal. We state the definition in general to save repeating the definition in § 5.

1.5 Definition. Let $\mathfrak{M}$ be a structure for some language L, let R, S be two new n-ary relation symbols and let $\varphi(\mathsf{R},\mathsf{S})$ be a sentence of $\mathsf{L}(\mathsf{R},\mathsf{S})$, possibly containing parameters from $\mathfrak{M}$. Let α be an ordinal. The sentence $\varphi(\mathsf{R},\mathsf{S})$ *implicitly defines* α *over* $\mathfrak{M}$ if the relation $\prec_\varphi$ defined by

$$R \prec_\varphi S \quad \text{iff} \quad (\mathfrak{M}, R, S) \models \varphi(\mathsf{R},\mathsf{S})$$

is well founded and α is its rank, i. e., $\alpha = \rho(\prec_\varphi)$.

Our final application of the Infinity Lemma shows that if a Π_1^0 relation $\varphi(\mathsf{R},\mathsf{S})$ on ω implicitly defines an ordinal, then that ordinal is finite. In § 6 we will learn that any α implicitly defined by even a Σ_1^1 sentence on ω is just the order type of a recursive (explicit) well-ordering of ω. These two facts explain why the notion of implicitly defined ordinal does not arise explicitly in ordinary recursion theory.

A predicate $\varphi(R,S)$ on ω is Π_1^0 iff $\neg\varphi(R,S)$ is r.e. (To fit this into our definition of r.e. replace R, S by their characteristic functions.)

1.6 Theorem. *Let $\varphi(R,S)$ be a Π_1^0 predicate of n-ary relations on ω. If the relation $\prec_\varphi$ defined by*

$$R \prec_\varphi S \quad \textit{iff} \quad \varphi(R,S)$$

is well founded, then its rank is finite.

Proof. By use of pairing functions we can assume $n=1$, i.e., that R, S range over subsets of ω. Assume $\prec_\varphi$ is well founded so that

(2) $$\forall R_1, R_2 \ldots \exists n \neg\varphi(R_{n+1}, R_n).$$

For any $f \in 2^\omega$ let $(f)_n = \{x \mid f(2^x 3^n) = 1\}$. We can restate (2) as

(3) $$\forall f \in 2^\omega \, \exists n \neg\varphi((f)_{n+1}, (f)_n).$$

Since φ is Π_1^0, the predicate $\neg\varphi((f)_{n+1}, (f)_n)$ is an r.e. predicate of f, n. By s-Π_1^1 reflection there is an $N < \omega$ such that

$$\forall f \, \exists n \leqslant N \neg\varphi((f)_{n+1}, (f)_n)$$

which says that there is no sequence

$$R_{N+1} \prec_\varphi R_N \prec_\varphi \cdots \prec_\varphi R_1 \prec_\varphi R_0.$$

Thus $\rho(\prec_\varphi) \leqslant N+1$. □

1.7—1.10 Exercises

1.7. Prove the relativized version of the theorems of this section. (The fact that s-Π_1^1 Reflection holds relativized to any relation R on ω is expressed by saying that ω is *strict*-Π_1^1 *indescribable*.)

1.8. Let $R \prec S$ iff $R, S \subseteq \omega$ and the least member of R is less than the least member of S. Show that this is an r.e. predicate of R, S and that it implicitly defines ω.

1.9. Let $R \prec S$ iff $R, S \subseteq \omega \times \omega$, R, S are well-orderings and R is a proper initial segment of S. Show that this is a Π_1^1 relation which implicitly defines the first uncountable ordinal.

1.10. Let $\prec$ be a well-founded relation on subsets of ω defined by a Σ_1^1 sentence φ. Show that the rank of $\prec$ is $< \omega_1^c$. [Hint: Show that $\rho(\prec)$ can be pinned down by a sentence of $\mathsf{L}_{\omega_1^c}$.]

1.11 Notes. The equation "s-Π_1^1 = r.e., on ω" was first observed by Kreisel in the proof of the Kreisel Basis Theorem (cf. p. 187 of Shoenfield [1967]).

2. Strict Π_1^1 Predicates: Preliminaries

Over ω, or $\mathbb{HF}$, the strict-Π_1^1 predicates coincide with the r.e. predicates (by 1.3) so it is difficult to see the exact role that the notion of strict-Π_1^1 plays in traditional model theory and recursion theory. In general, however, strict-Π_1^1 does not coincide with Σ_1. By studying the s-Π_1^1 predicates in the general case, then, we see more clearly the role they play over ω.

Let $\mathsf{L}^* = \mathsf{L}(\in, \ldots)$ be the language for KPU. We assume that there are only relation and constant symbols in L^*, no function symbols. (This is not an essential restriction—see Exercise V.1.8.) Let $\mathsf{R}_1, \mathsf{R}_2, \ldots$ be an infinite list of new relation symbols, an infinite number of arity n for each $n < \omega$. Let $\mathsf{L}^*(\vec{\mathsf{R}})$ be the expanded language.

2.1 Definition. i) The *strict-Π_1^1 formulas* (s-Π_1^1 for short) of $\mathsf{L}^*(\vec{\mathsf{R}})$ form the smallest class containing the Δ_0 formulas of $\mathsf{L}^*(\vec{\mathsf{R}})$ closed under $\wedge$, $\vee$, $\forall u \in v$, $\exists u \in v$, $\exists u$ and the clause

$$\text{if } \Phi(\mathsf{R}_i) \text{ is strict-}\Pi_1^1 \text{ so is } \forall \mathsf{R}_i\, \Phi(\mathsf{R}_i).$$

The strict-Π_1^1 formulas of L^* consist of those s-Π_1^1 formulas of $\mathsf{L}^*(\vec{\mathsf{R}})$ which have only quantified occurrences of the new relation symbols $\mathsf{R}_1, \mathsf{R}_2, \ldots$.

ii) The *strict-Σ_1^1 formulas* form the dual class; that is, they form the smallest class containing the Δ_0 formulas closed under $\wedge$, $\vee$, $\forall u \in v$, $\exists u \in v$, $\forall u$, $\exists \mathsf{R}_i$.

There are two essential restrictions in the definition of strict-Π_1^1 formula. First, only existential quantifiers over individuals are permitted. Second, only universal second order quantifiers are allowed, and then only over relations, not over functions. If we were to allow universal second order quantification over functions, then we could build in first order universal quantification (by the manipulations discussed in § IV.2). These observations are summarized by the diagram:

strict-Π_1^1 formulas

$\cup$ $\qquad\qquad\qquad$ $\cap$

Σ formulas $\qquad\qquad\qquad$ Π_1^1 formulas

$\cap$ $\qquad\qquad\qquad$ $\cup$

first order formulas

All inclusions are proper.

Don't forget that L^* may have extra relation symbols (like a symbol for the power set relation) which are allowed to occur in Δ_0, hence in s-Π_1^1, formulas.

Satisfaction of s-Π_1^1 and s-Σ_1^1 formulas is defined in the classical second order manner. Thus

$$\mathfrak{M} \models \forall \mathsf{R}\, \Phi(\mathsf{R})$$

means that for every relation R on $\mathfrak{M}$ (of the correct number of places)

$$(\mathfrak{M}, R) \models \Phi(\mathsf{R}).$$

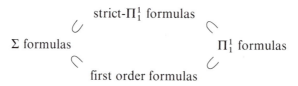

The study of s-Π_1^1 predicates is one of the few places in logic where the difference between relation symbols and function symbols really matters. In § 1 we defined s-Π_1^1 over ω in terms of quantification over *characteristic* functions, rather than the relations they describe, just to fit with standard practice in ordinary recursion theory. Here the approach with relation symbols is more natural.

The following simple lemma expresses one of the most crucial properties of strict-Π_1^1 formulas.

2.2 Lemma. *Strict-Π_1^1 formulas persist upwards under end extensions. That is, if $\mathfrak{A}_{\mathfrak{M}}, \mathfrak{B}_{\mathfrak{N}}$ are L^*-structures with $\mathfrak{A}_{\mathfrak{M}} \subseteq_{\mathrm{end}} \mathfrak{B}_{\mathfrak{N}}$, and if $\Phi(v_1,\ldots,v_n)$ is a s-Π_1^1 formula of L^* then*

$$\mathfrak{A}_{\mathfrak{M}} \models \Phi[x_1,\ldots,x_n] \quad \textit{implies} \quad \mathfrak{B}_{\mathfrak{N}} \models \Phi[x_1,\ldots,x_n]$$

for all $x_1,\ldots,x_n \in \mathfrak{A}_{\mathfrak{M}}$.

Proof. We need to prove a bit more to keep the induction on s-Π_1^1 formulas going. Let $\mathfrak{A}_{\mathfrak{M}} \subseteq_{\mathrm{end}} \mathfrak{B}_{\mathfrak{N}}$ be given. We prove by induction on s-Π_1^1 formulas $\Phi(\mathsf{R}_1,\ldots,\mathsf{R}_m, v_1,\ldots,v_n)$ of $\mathsf{L}^*(\vec{\mathsf{R}})$ that for all relations $R_1,\ldots,R_m$ on $\mathfrak{B}_{\mathfrak{N}}$ and all $x_1,\ldots,x_n \in \mathfrak{A}_{\mathfrak{M}}$

$$(\mathfrak{A}_{\mathfrak{M}}, R_1 {\restriction} \mathfrak{A}_{\mathfrak{M}},\ldots,R_m {\restriction} \mathfrak{A}_{\mathfrak{M}}) \models \Phi[x_1,\ldots,x_n]$$

implies

$$(\mathfrak{B}_{\mathfrak{N}}, R_1,\ldots,R_m) \models \Phi[x_1,\ldots,x_n].$$

The proof is just the proof of persistence of Σ formulas with a new case for $\forall \mathsf{S}$ thrown in. Suppose

$$(1) \qquad (\mathfrak{A}_{\mathfrak{M}}, R_1 {\restriction} \mathfrak{A}_{\mathfrak{M}},\ldots,R_m {\restriction} \mathfrak{A}_{\mathfrak{M}}) \models \forall \mathsf{S}\ \Phi(\vec{x},\vec{\mathsf{R}},\mathsf{S}).$$

Let S be any relation on $\mathfrak{B}_{\mathfrak{M}}$ of the correct number of places. By (1)

$$(2) \qquad (\mathfrak{A}_{\mathfrak{M}}, R_1 {\restriction} \mathfrak{A}_{\mathfrak{M}},\ldots,R_m {\restriction} \mathfrak{A}_{\mathfrak{M}}, S {\restriction} \mathfrak{A}_{\mathfrak{M}}) \models \Phi(\vec{x},\vec{\mathsf{R}},\mathsf{S})$$

so, by the induction hypothesis,

$$(\mathfrak{B}_{\mathfrak{N}}, R_1,\ldots,R_m, S) \models \Phi(\vec{x},\vec{\mathsf{R}},\mathsf{S}),$$

as desired. Notice that if we had allowed quantification over function symbols step (2) would fail; just because S is a total function on $\mathfrak{B}_{\mathfrak{N}}$ is no reason to suppose that $S {\restriction} \mathfrak{A}_{\mathfrak{M}}$ is a total function. ☐

Let $\mathfrak{A}_{\mathfrak{M}}$ be a structure for L^*. A relation P on $\mathfrak{A}_{\mathfrak{M}}$ is s-Π_1^1 if it can be defined by a s-Π_1^1 formula of L^* with parameters from $\mathfrak{A}_{\mathfrak{M}}$. P is s-Σ_1^1 if P can be defined by a s-Σ_1^1 formula with parameters. P is strict-Δ_1^1 if P is both s-Π_1^1 and s-Σ_1^1.

A function is s-Π_1^1, s-Σ_1^1 on s-Δ_1^1 iff its graph is s-Π_1^1, s-Σ_1^1 or s-Δ_1^1 respectively.

2.3 Lemma. *If a total function f on $\mathfrak{A}_{\mathfrak{M}}$ is s-Π_1^1 then it is s-Δ_1^1.*

Proof. Since f is total,

$$f(x_1,\dots,x_n)\neq y \quad \text{iff} \quad \exists z\,[f(x_1,\dots,x_n)=z \wedge z\neq y].$$

Replacing $f(\vec{x})=z$ by its s-Π_1^1 definition gives us a s-Π_1^1 definition of $f(\vec{x})\neq y$; or equivalently, a s-Σ_1^1 definition of the graph of f. □

2.4 Examples. Let $\mathbb{A}$ be an admissible set. We give three examples of strict-Π_1^1 predicates which are not, in general, Σ_1. Note, however, that if A is countable then these relations are Σ_1 (for rather trivial reasons).

(i) Define $P(a,b)$ iff $\operatorname{card}(a)<\operatorname{card}(b)$. Then P is s-Π_1^1 on $\mathbb{A}$.

(ii) Define $P(a)$ iff $\operatorname{card}(a)<\operatorname{card}(A)$. Then P is s-Π_1^1 on $\mathbb{A}$.

(iii) Define $P(a,b)$ iff $b=\operatorname{Power}(a)$, the real power set of a. P is s-Π_1^1 on $\mathbb{A}$. If $\mathbb{A}$ is closed under the power set operation then P is s-Δ_1^1 on $\mathbb{A}$.

Proof. (i) We can write $\operatorname{card}(a)<\operatorname{card}(b)$ as

$$\forall \mathsf{R}\,[\mathsf{R}\subseteq b\times a \wedge \forall x\in b\,\exists y\in a\,\mathsf{R}(x,y) \\ \rightarrow \exists x,x'\in b\,\exists y\in a\,(x\neq x' \wedge \mathsf{R}(x,y)\wedge \mathsf{R}(x',y))]$$

which asserts that no relation on $b\times a$ can be a one-one map of b into a. Schematically, we can rewrite this as

$$\forall \mathsf{R}\,[\Pi_1(\mathsf{R})\wedge\Delta_0(\mathsf{R})\rightarrow\Delta_0(\mathsf{R})].$$

Replacing $\rightarrow$ by $\vee$ we get

$$\forall \mathsf{R}\,[\Sigma_1(\mathsf{R})\vee\Delta_0(\mathsf{R})\vee\Delta_0(\mathsf{R})]$$

which is s-Π_1^1. The proof of (ii) is much the same. To prove that $b=\operatorname{Power}(a)$ is s-Π_1^1, note that $b=\operatorname{Power}(a)$ iff

$$\forall x\in b\,(x\subseteq a)\wedge\forall \mathsf{R}\,\exists y\in b\,\forall x\in a\,(x\in y\leftrightarrow \mathsf{R}(x)).$$

The second sentence of (iii) follows from 2.3. □

A formula is in s-Π_1^1 *normal form* if it is of the form

$$\forall \mathsf{R}\,\exists y_1,\dots,y_m\,\varphi(v_1,\dots,v_n,y_1,\dots,y_m,\mathsf{R})$$

where φ is Δ_0. The next lemma states that every s-Π_1^1 formula is logically equivalent to one in normal form.

2.5 s-Π_1^1 Normal Form Lemma. *Assume that* L^* *has a constant symbol* 0. *For every* s-Π_1^1 *formula* $\Phi(\vec{x},\vec{\mathsf{R}})$ *of* $\mathsf{L}^*(\vec{\mathsf{R}})$ *there is a* s-Π_1^1 *formula* Φ' *in normal form, with exactly the same free variables and free relation symbols, such that for all* L^* *structures* $\mathfrak{A}_{\mathfrak{M}}$:

$$\mathfrak{A}_{\mathfrak{M}}\models\forall\vec{\mathsf{R}}\,\forall\vec{x}\,[\Phi(\vec{x},\vec{\mathsf{R}})\leftrightarrow\Phi'(\vec{x},\vec{\mathsf{R}})].$$

Proof. We describe five quantifier-pushing manipulations which allow us to put any s-Π_1^1 formula in normal form.

$$\text{(i)}\ \forall \mathsf{R}_1\, \forall \mathsf{R}_2\, \Phi(\mathsf{R}_1,\mathsf{R}_2) \leftrightarrow \forall \mathsf{S}\, \Phi'(\mathsf{S})$$

where S is $n+m$-ary, n being the arity of R_1, m the arity of R_2, and where $\Phi'(\mathsf{S})$ results from Φ by replacing

$$\mathsf{R}_1(t_1,\ldots,t_n) \quad \text{by} \quad \mathsf{S}(t_1,\ldots,t_n,\mathbf{0},\ldots,\mathbf{0}),$$
$$\mathsf{R}_2(t_1,\ldots,t_m) \quad \text{by} \quad \mathsf{S}(\mathbf{0},\ldots,\mathbf{0},t_1,\ldots,t_m).$$

$$\text{(ii)}\ \forall \mathsf{R}_1\, \exists \vec{x}\, \varphi \wedge \forall \mathsf{R}_2\, \exists \vec{y}\, \psi \leftrightarrow \forall \mathsf{R}_1\, \forall \mathsf{R}_2\, \exists \vec{x}\, \exists \vec{y}\, (\varphi \wedge \psi)$$

as long as the various symbols are distinct. Similarly for $\vee$.

$$\text{(iii)}\ \forall x\, \exists \mathsf{R}\, \Phi(x,\mathsf{R}) \leftrightarrow \exists \mathsf{R}'\, \forall x\, \Phi'(x,\mathsf{R}')$$

where Φ' results from Φ by replacing $\mathsf{R}(t_1,\ldots,t_n)$ by $\mathsf{R}'(x,t_1,\ldots,t_n)$. Taking negations on both sides of (iii) we get

$$\text{(iv)}\ \exists x\, \forall \mathsf{R}\, \Phi(x,\mathsf{R}) \leftrightarrow \forall \mathsf{R}'\, \exists x\, \Phi'(x,\mathsf{R}')$$

which lets us pull $\forall \mathsf{R}$ out in front of $\exists x$. The bounded existential quantifier step follows from (ii) $\wedge$ (iv). The only remaining step is the bounded universal quantifier.

$$\text{(v)}\ \forall x \in a\, \forall \mathsf{R}\, \exists y\, \Phi(x,\mathsf{R},y)$$
$$\leftrightarrow \forall \mathsf{U}\, \forall \mathsf{R}\, \exists y\, [\mathrm{card}(a \cap \mathsf{U}) \leqslant 1 \rightarrow \forall x \in a\, (\mathsf{U}(x) \rightarrow \Phi(x,\mathsf{R},y))].$$

The part in brackets is Δ_0 since it can be written

$$\forall x \in a\, \forall z \in a\, (\mathsf{U}(x) \wedge \mathsf{U}(z) \rightarrow x=z) \rightarrow \forall x \in a\, (\mathsf{U}(x) \rightarrow \Phi(x,\mathsf{R},y)).$$

It is now clear, by induction on s-Π_1^1 formulas, that we can put every s-Π_1^1 formula in normal form. ☐

The Normal Form Lemma is quite useful. We use it in proving the next theorem, and repeatedly in this sections to come.

Recall, from § IV.3, that for countable structures $\mathfrak{M}$

$$\Pi_1^1 \text{ on } \mathfrak{M} = \Sigma_1 \text{ on } \mathbb{HYP}_{\mathfrak{M}}.$$

We proved an absolute version of this theorem in § VI.5, by showing that

$$\text{inductive* on } \mathfrak{M} = \Sigma_1 \text{ on } \mathbb{HYP}_{\mathfrak{M}}.$$

We close this section by proving a different generalization.

2.6 Theorem. *Let $\mathfrak{M}=\langle M,R_1,\ldots,R_l\rangle$ be an infinite structure. A relation S on $\mathfrak{M}$ is Π_1^1 on $\mathfrak{M}$ iff S is strict-Π_1^1 on $\mathbb{HYP}_{\mathfrak{M}}$.*

To see that this is a generalization of the countable result we need to know that, for $\mathfrak{M}$ countable,

$$s\text{-}\Pi_1^1 \text{ on } \mathbb{HYP}_{\mathfrak{M}} = \Sigma_1 \text{ on } \mathbb{HYP}_{\mathfrak{M}}.$$

This follows from Theorem 3.1 of the next section.

Proof of Theorem 2.6. We first prove the easy half ($\Rightarrow$). Suppose

$$S(x)\leftrightarrow\mathfrak{M}\models\forall \mathsf{R}\ \varphi(x,\mathsf{R}).$$

Then

$$S(x)\leftrightarrow\mathbb{HYP}_{\mathfrak{M}}\models\forall \mathsf{R}\left[\mathsf{R}\subseteq M^n\rightarrow\varphi^{(M)}(x,\mathsf{R})\right].$$

The part within brackets is Σ_1 since all quantifiers in $\varphi^{(M)}$ are bounded by M, an element of $\mathbb{HYP}_{\mathfrak{M}}$. To prove ($\Leftarrow$) we must reexamine the proof of Theorem IV.3.3, the result that

$$\Sigma_1 \text{ on } \mathbb{HYP}_{\mathfrak{M}} \Rightarrow \Pi_1^1 \text{ on } \mathfrak{M}$$

regardless of $\mathfrak{M}$'s cardinality. Suppose S is s-Π_1^1 on $\mathbb{HYP}_{\mathfrak{M}}$, $S\subseteq\mathfrak{M}$. By the Normal Form Lemma we can write

$$S(p) \quad \text{iff} \quad \mathbb{HYP}_{\mathfrak{M}}\models\forall \mathsf{P}\ \exists\vec{y}\ \varphi(p,\vec{y},\mathsf{P},\vec{z})$$

for some $\vec{z}=z_1,\ldots,z_k\in\mathbb{HYP}_{\mathfrak{M}}$. As in the proof mentioned above, we can replace all parameters z_i by good Σ_1 definitions and so assume all parameters are from $M\cup\{M\}$. Let's say

$$S(p) \quad \text{iff} \quad \mathbb{HYP}_{\mathfrak{M}}\models\forall P\ \exists\vec{y}\ \varphi(p,\vec{y},\mathsf{P},q,M).$$

By the persistence of s-Π_1^1 formulas under end extensions, and by the truncation lemma, $S(p)$ is equivalent to

(1′) $(\mathfrak{A}_{\mathfrak{M}},P)\models\exists\vec{y}\ \varphi(p,\vec{y},\mathsf{P},q,M)$ for all P and all models $\mathfrak{A}_{\mathfrak{M}}$ of KPU^+ (of cardinality $\mathrm{card}(\mathfrak{M})$).

From here the proof proceeds exactly like the proof of IV.3.3, by coding up (1′) on M, with the extra $\forall P$ riding along for free. □

2.7 Exercise. Let $\mathfrak{A}_{\mathfrak{M}}$ be any structure for L^* and let Γ be a s-Π_1^1 inductive definition on $\mathfrak{A}_{\mathfrak{M}}$; i. e.

$$\vec{x} \in \Gamma(R) \text{ iff } (\mathfrak{A}_{\mathfrak{M}}, R) \models \Phi(x, \mathsf{R}_+)$$

where Φ is s-Π_1^1. Show that the fixed point I_Γ is s-Π_1^1 on $\mathfrak{A}_{\mathfrak{M}}$.

2.8 Notes. The only theorem of this section comes from Barwise-Gandy-Moschovakis [1971].

3. *König Principles on Countable Admissible Sets*

Strict-Π_1^1 formulas give us a language for expressing important new principles, or axioms, for admissible sets; principles that isolate important aspects of the Infinity Lemma.

In this section we discuss three König principles and show that they hold on all countable admissible sets. Their role in the general case is discussed in the remaining sections of this chapter.

K_1: *An admissible set* $\mathbb{A}$ *satisfies*

$$s\text{-}\Pi_1^1 = \Sigma_1$$

if every strict-Π_1^1 *relation on* $\mathbb{A}$ *is already a* Σ_1 *relation on* $\mathbb{A}$.

It is important to remember that this equation (s-$\Pi_1^1 = \Sigma_1$) depends very much on just what extra relations may be part of our admissible set $\mathbb{A} = (\mathfrak{M}; A, \in, \ldots)$ in those three little dots. Add a new relation to $\mathbb{A}$ and you increase both the number of s-Π_1^1 formulas and the number of Σ_1 formulas. It should also be kept in mind that the Σ_1 formula defining a s-Π_1^1 predicate P may have parameters not appearing in a given s-Π_1^1 definition of P.

3.1 Theorem. *Every countable admissible set satisfies* s-$\Pi_1^1 = \Sigma_1$.

Proof. We will prove more; namely, that every Σ_1 complete admissible set $\mathbb{A}$ satisfies s-$\Pi_1^1 = \Sigma_1$. Let P be s-Π_1^1 on $\mathbb{A}$. By the Normal Form Lemma we can write P in the form

$$P(x) \quad \text{iff} \quad \mathbb{A} \models \forall \mathsf{R}\, \varphi(x, \mathsf{R})$$

for some Σ_1 formula φ. Let T be the usual infinitary diagram of $\mathbb{A}$:

$$\text{diagram}(\mathbb{A}),$$
$$\forall v\, [v \in \bar{\mathsf{a}} \to \textstyle\bigvee_{x \in a} v = \bar{\mathsf{x}}].$$

By the persistence of s-Π_1^1 formulas under end extensions (Lemma 2.2) the following are equivalent:

$$P(x),$$

$$\mathbb{A} \models \forall \mathsf{R}\, \varphi(\mathsf{R}, x),$$

$$\mathfrak{B} \models \forall \mathsf{R}\, \varphi(\mathsf{R}, x) \quad \text{for all} \quad \mathfrak{B} \supseteq_{\text{end}} \mathbb{A},$$

$$(\mathfrak{B}, R) \models \varphi(\mathsf{R}, \mathsf{x}) \quad \text{for all} \quad \mathfrak{B} \supseteq_{\text{end}} \mathbb{A} \quad \text{and all} \quad R \subseteq \mathfrak{B}^n,$$

$$T \models \varphi(\mathsf{R}, \mathsf{x}).$$

If $\mathbb{A}$ is Σ_1 complete then the set of x such that $T \models \varphi(\mathsf{R}, \mathsf{x})$ is a Σ_1 set. □

Before stating the second König Principle, K_2, we need to define the notation $\Phi^{(a)}$ for second order formulas Φ. To obtain $\Phi^{(a)}$ one relativizes all unbounded first order quantifiers to a (replace $\exists u$ by $\exists u \in a$, $\forall u$ by $\forall u \in a$) and replaces

$$\exists \mathsf{R}(\ldots) \quad \text{by} \quad \exists \mathsf{R}\,[\mathsf{R} \subseteq a^n \wedge (\ldots)],$$

$$\forall \mathsf{R}(\ldots) \quad \text{by} \quad \forall \mathsf{R}\,[\mathsf{R} \subseteq a^n \rightarrow (\ldots)].$$

Note that if Φ is strict-Π_1^1, or even Π_1^1, then $\Phi^{(a)}$ is strict-Π_1^1 with free variables those of Φ and the new variable a.

3.2 Lemma. *For every structure $\mathfrak{A}_{\mathfrak{M}}$ for L^* and every s-Π_1^1 formula $\Phi(v_1, \ldots, v_n)$ of L^*, the following are true in $\mathfrak{A}_{\mathfrak{M}}$:*

(i) $\forall a\, \forall v_1, \ldots, v_n \in a\, [\mathrm{Tran}(a) \wedge \Phi^{(a)}(\vec{v}) \rightarrow \Phi(\vec{v})]$;

(ii) $\forall a, b, \forall v_1, \ldots, v_n \in a\, [\mathrm{Tran}(a) \wedge a \subseteq b \wedge \Phi^{(a)}(\vec{v}) \rightarrow \Phi^{(b)}(\vec{v})]$.

Proof. This is just another version of the persistence of s-Π_1^1 formulas under end extensions. It can be proved directly or deduced from Lemma 2.2. □

K_2: *An admissible set $\mathbb{A}$ satisfies strict-Π_1^1 reflection if for every s-Π_1^1 formula $\Phi(v_1, \ldots, v_n)$ and every $x_1, \ldots, x_n \in \mathbb{A}$, $\mathbb{A}$ satisfies*

$$\Phi(\vec{x}) \rightarrow \exists a\, [\mathrm{Tran}(a) \wedge x_1, \ldots, x_n \in a \wedge \Phi^{(a)}(\vec{x})].$$

We will see in §§ 4, 6 and 7 that s-Π_1^1 reflection is a strong assumption. For now we prove that it holds in all countable admissible sets.

3.3 Theorem. *Every countable admissible set satisfies s-Π_1^1 reflection.*

Proof. Again we prove more with an eye toward the next section. This time we prove that if $\mathbb{A}$ is Σ_1 compact then $\mathbb{A}$ satisfies s-Π_1^1 reflection. Let $\Phi(v_1, \ldots, v_n)$ be s-Π_1^1. By the Normal Form Lemma there is a formula $\Psi(v_1, \ldots, v_n)$ in s-Π_1^1 normal form logically equivalent to Φ. It follows that $\Psi^{(a)}(v_1, \ldots v_n)$ is logically equivalent to $\Phi^{(a)}(v_1,\ldots,v_n)$ so it suffices to prove reflection for formulas in s-Π_1^1 normal form. So suppose $\Phi(\vec{v})$ is $\forall \mathsf{R}\, \varphi(\vec{v},\mathsf{R})$ and that $\mathbb{A}=\mathbb{A}_{\mathfrak{M}}$ and

$$\mathbb{A}\models\forall \mathsf{R}\, \varphi(x_1,\ldots,x_n,\mathsf{R})$$

where φ is a Σ_1 formula. Let T be the infinitary diagram of $\mathbb{A}$, as in 3.2. As we saw in that proof,

$$T\models\varphi(\mathsf{x}_1,\ldots,\mathsf{x}_n,\mathsf{R}).$$

By Σ_1 compactness there is a $T_0\subseteq T$, $T_0\in\mathbb{A}$ such that

$$T_0\models\varphi(\mathsf{x}_1,\ldots,\mathsf{x}_n,\mathsf{R}).$$

Let $a_0=\{y\,|\,\mathsf{y} \text{ occurs in } T_0\}\cup\{x_1,\ldots,x_n\}$ and let $a=\mathrm{TC}(a_0)$ so that $a\in\mathbb{A}$. Then

$$(\mathfrak{M}\cap a;\, a,\in,\ldots)\models T_0$$

so

$$(\mathfrak{M}\cap a;\, a,\in,\ldots)\models\forall \mathsf{R}\, \varphi(x_1,\ldots,x_n,\mathsf{R})$$

which is another way of saying that $\Phi^{(a)}[x_1,\ldots,x_n]$ holds. $\square$

The third König principle concerns the notion of implicit ordinal introduced in 1.5 and is suggested by Theorem 1.6.

An ordinal α is a Π *implicit ordinal* over $\mathbb{A}$ if there is a Π sentence $\varphi(\mathsf{R},\mathsf{S})$, possibly containing parameters from $\mathbb{A}$, which implicitly defines α over $\mathbb{A}$ (in the sense of 1.5). The notion of a s-Σ_1^1 implicit ordinal is defined in a parallel fashion. (It will turn out that every s-Σ_1^1 implicit ordinal is less than some Π implicit ordinal; see 3.11 or 6.3). It is easy to see that every $\beta<o(\mathbb{A})$ is a Π implicit ordinal over $\mathbb{A}$.

K_3: *The third König principle asserts that every Π implicit ordinal over $\mathbb{A}$ is an element of $\mathbb{A}$.*

One might paraphrase K_3 by saying that the Π implicit ordinals over $\mathbb{A}$ are explicitly in $\mathbb{A}$.

3.4. Theorem. *Every countable admissible set satisfies the third König principle.*

Proof. Since $h_\Sigma(\mathbb{A})=o(\mathbb{A})$ for countable $\mathbb{A}$, 3.4 follows from 3.5. $\square$

Admissible sets do not, in general, satisfy K_3. In general, the Π implicit ordinals know new bounds.

3.5 Theorem. *Let $\mathbb{A}$ be admissible and let α be a s-Σ_1^1 implicit ordinal over $\mathbb{A}$. If $\beta=h_\Sigma(\mathbb{A})$ then $\alpha<\beta$.*

Proof. Let $\mathbb{A}=\mathbb{A}_{\mathfrak{M}}$ be admissible and let α be the rank of the well-founded relation $\prec_\Phi$ where Φ is s-Σ_1^1; say

$$R \prec_\Phi S \quad \text{iff} \quad (\mathbb{A}, R, S) \models \exists \mathsf{Q}\, \varphi(\mathsf{Q}, \mathsf{R}, \mathsf{S})$$

and φ is a Π sentence. We can assume that $\mathsf{Q}, \mathsf{R}, \mathsf{S}$ are all unary (i. e. range over subsets of $\mathbb{A}$) since the pairing function $\langle x, y\rangle$ is $\mathbb{A}$-recursise. We need to find a Σ_1 theory T of $\mathsf{L}_{\mathbb{A}}$ which pins down α. The crucial observation is contained in (1).

$$(1)\begin{cases}\text{If } \mathbb{A}_{\mathfrak{M}} \subseteq_{\text{end}} \mathfrak{B}_{\mathfrak{N}} \text{ and } (\mathfrak{B}_{\mathfrak{N}}, R, S) \models \exists \mathsf{Q}\, \varphi(\mathsf{Q}, \mathsf{R}, \mathsf{S}), \text{ and if } R_0 = R \cap \mathbb{A}_{\mathfrak{M}},\\ S_0 = S \cap \mathbb{A}_{\mathfrak{M}} \text{ then } R_0 \prec_\Phi S_0.\end{cases}$$

This follows from $(\mathbb{A}_{\mathfrak{M}}, R_0, S_0) \subseteq_{\text{end}} (\mathfrak{B}_{\mathfrak{N}}, R, S)$ by the s-Σ_1^1 version of Lemma 2.2. From (1) we get (2).

$$(2)\begin{cases}\text{Let } \mathbb{A}_{\mathfrak{M}} \subseteq_{\text{end}} \mathfrak{B}_{\mathfrak{N}}. \text{ The relation } \prec' \text{ on subsets of } \mathfrak{B}_{\mathfrak{N}} \text{ defined by}\\ \qquad R \prec' S \quad \text{iff} \quad (\mathfrak{B}_{\mathfrak{N}}, R, S) \models \exists \mathsf{Q}\, \varphi(\mathsf{Q}, \mathsf{R}, \mathsf{S})\\ \text{is well founded. Hence any subrelation } \prec'' \text{ of } \prec' \text{ is well founded.}\end{cases}$$

For, by (1), any infinite descending sequence in $\prec'$ would give rise to an infinite descending sequence in $\prec_\Phi$.

It is pretty obvious how to use (2) to pin down the ordinal α by building the hypothesis of (2) into a Σ_1 theory $T = T(<, \ldots)$. The language for T will contain the symbols of $\mathsf{L}^* = \mathsf{L}(\in, \ldots)$; a constant symbol x for each $x \in A$; unary symbols A (for A), P (for Power(A)), U (for α); binary symbols E (for $\in \cap (A \times P)$), $\prec$ (for $\prec_\Phi$), $<$ (for $\in \restriction \alpha$); and a function symbol F. The intended model for T, the one with $<^{\mathfrak{M}}$ of order type α, is:

$$\mathfrak{M} = \langle A \cup \text{Power}(A) \cup \alpha;\ A, \ldots;\ \text{Power}(A), \in \cap (A \times \text{Power}(A)), \prec_\Phi;\ \alpha, <, F\rangle$$

where $F(x)=0$ for $x \notin \text{field}(\prec_\Phi)$, $F(R) = \prec_\Phi$-rank of R for R in field of $\prec_\Phi$. Hence $\text{rng}(F) = \alpha$ and $R \prec_\Phi S$ implies $F(R) < F(S)$. The theory T contains:

$\forall x\, [\mathsf{A}(x) \vee \mathsf{P}(x) \vee \mathsf{U}(x)]$,

Infinitary diagram of $\mathbb{A}$,

Extensionality for E,

"$\mathsf{E} \subseteq \mathsf{A} \times \mathsf{P}$".

(3) $\forall r, s\, [r \prec s \leftrightarrow \mathsf{P}(r) \wedge \mathsf{P}(s) \wedge \exists y\, (\mathsf{P}(q) \wedge \varphi(q, r, s))]$.

(4) "$<$ linearly orders U, $\text{rng}(\mathsf{F}) = \mathsf{U}$, $\mathsf{F}(x) = 0$ for $x \notin \text{field}(\prec)$, and $\mathsf{F}(s) = <\text{-sup}\{\mathsf{F}(r)+1 : r \prec s\}$ for $s \in \text{field}(\prec)$".

In line (3), $\varphi(q,r,s)$ denotes the result of replacing $\mathsf{R}(x)$ by $x\mathsf{E}r$, $\neg\mathsf{R}(x)$ by $\neg(x\mathsf{E}r)$ and similar for Q, S. (We are abusing notation since q does not range over urelements here.) To see that T pins down α it remains only to prove that for any other model $\mathfrak{M}$ of T, $<^{\mathfrak{M}}$ is well ordered. Let $\mathfrak{M}$ be any model of T. We can obviously assume $\mathfrak{M}$ has the form

$$\mathfrak{M}=\langle \mathfrak{B}_{\mathfrak{N}}\cup P_1\cup U;\mathfrak{B}_{\mathfrak{N}},\ldots;P,\in\cap(\mathfrak{B}_{\mathfrak{N}}\times P),\prec'',U,<,F\rangle$$

where $\mathbb{A}_{\mathfrak{M}}\subseteq_{\text{end}}\mathfrak{B}_{\mathfrak{N}}$ and $P\subseteq \text{Power}(\mathfrak{B}_{\mathfrak{N}})$. To see that $<$ is well ordered it suffices, by (4), to prove that $\prec''$ is well founded. But this is immediate from (2) and (3). $\square$

3.6 Corollary. *If* $\mathfrak{M}=\langle M,R_1,\ldots,R_l\rangle$ *is countable and* α *is a first order, or even* Σ_1^1, *implicit ordinal over* $\mathfrak{M}$ *then* $\alpha<O(\mathfrak{M})$. *In particular,* α *is countable.*

Proof. If α is Σ_1^1 over $\mathfrak{M}$ then it is s-Σ_1^1 over $\mathbb{HYP}_{\mathfrak{M}}$ so the result follows from 3.5. $\square$

3.7 Corollary. *Every* Σ_1^1 *implicit ordinal over* ω *is less than* ω_1^c.

Proof. Immediate from 3.6 since $\omega_1^c=O(\mathcal{N})$. $\square$

As we mentioned in § 1, Theorem 1.6 and Corollary 3.7 together account for the fact that implicit ordinals seldom appear in ordinary recursion theory on ω. They *do* appear in parts of mathematics far removed from the theory of admissible sets. We present one example suggestive of others.

3.8 Example. Let $\mathfrak{M}$ be a Noetherian module (over a ring with identity), that is, a module with no infinite chain

$$M_0\subset M_1\subset M_2\subset\cdots$$

of submodules. Then

$$M'\prec M'' \quad\text{iff}\quad M',M'' \text{ are submodules,}\quad M''\subset M'$$

defines a first order implicit well-founded relation. Its rank $\alpha=\rho(\prec)$ is called the length of $\mathfrak{M}$, $\alpha=l(\mathfrak{M})$. Thus $l(\mathfrak{M})$ is a first order implicit ordinal over M. This ordinal plays an important role in the structure theory of Noetherian modules.

3.9—3.12 Exercises

3.9. Prove a uniform version of 3.1. That is, show that for every s-Π_1^1 formula $\Phi(v_1,\ldots,v_n)$ there is a Σ_1 formula $\varphi(v_1,\ldots,v_n)$ such that for every countable admissible set $\mathbb{A}$,

$$\mathbb{A}\models\forall\vec{v}\,[\Phi(\vec{v})\leftrightarrow\varphi(\vec{v})].$$

3.10. Prove directly that every s-Σ_1^1 implicit ordinal is $\leqslant$ some Π implicit ordinal over $\mathbb{A}$.

3.11. Let $\mathfrak{M}=\langle M, R_1,\ldots,R_l\rangle$ and let α be a Σ_1^1 implicit ordinal over $\mathfrak{M}$. Improve 3.5 to show that $\alpha\leqslant h(\mathbb{HYP}_{\mathfrak{M}})$.

3.12. Prove that on the class of admissible sets $\mathbb{A}$ and relations P on $\mathbb{A}$,

"P is Σ_1 on $\mathbb{A}$"

is the absolute version of

"P is s-Π_1^1 on $\mathbb{A}$".

4. König Principles K_1 *and* K_2 *on Arbitrary Admissible Sets*

To summarize, the three König Principles introduced in § 3 are:

K_1: strict-$\Pi_1^1=\Sigma_1$;
K_2: strict-Π_1^1 reflection;
K_3: Every Π implicit ordinal over $\mathbb{A}$ is an element of $\mathbb{A}$.

These three principles are generalized recursion theoretic statements which attempt to capture different aspects of the Infinity Lemma. Each of them has a model-theoretic counterpart for the infinitary logic $\mathsf{L}_{\mathbb{A}}$. In this section we discuss the counterparts of K_1 and K_2.

The basic tool for the study of all three of these principles is the Weak Completeness Theorem of § VII.2. Our first theorem explains the reason for referring to that result as a completeness theorem.

4.1 Theorem. *Let* $\mathbb{A}$ *be admissible and let* T *be a set of sentence of* $\mathsf{L}_{\mathbb{A}}$ *which is strict-*Π_1^1 *definable on* $\mathbb{A}$. *The set*

$$\mathrm{Cn}(T)=\{\varphi\in\mathsf{L}_{\mathbb{A}}: T\models\varphi\}$$

is also strict-Π_1^1 *on* $\mathbb{A}$.

Theorem 4.1 will follow from the Weak Completeness Theorem and the next Lemma.

4.2 Lemma. *Let* $\mathbb{A}$ *be admissible and let* $\mathsf{L}_{\mathbb{A}}$ *be a Skolem fragment which is* Δ_1 *on* $\mathbb{A}$ *(in the sense of Lemma* VII.2.4*). There is a* Π *sentence* $\varphi(\mathsf{D})$ *such that for any* $\mathscr{D}\subseteq\mathbb{A}$:

$$\mathscr{D} \text{ is an s.v.p. for } \mathsf{L}_{\mathbb{A}} \quad \text{iff} \quad (\mathbb{A},\mathscr{D})\models\varphi(\mathsf{D}).$$

Proof. Since $\mathsf{L}_{\mathbb{A}}$ is Δ_1 on $\mathbb{A}$, T_{Skolem} is a Δ_1 subset of $\mathbb{A}$. $\mathscr{D}$ is an s.v.p. for $\mathsf{L}_{\mathbb{A}}$ iff $(\mathbb{A}, \mathscr{D})$ satisfies all the following conditions:

$$\mathscr{D} \subseteq \mathsf{L}_{\mathbb{A}},$$

$$\forall \varphi\,[(\varphi \text{ an axiom (A 1)—(A 7) of } \mathsf{L}_{\mathbb{A}}) \rightarrow (\varphi \in \mathscr{D})],$$

$$\mathscr{D} \text{ is closed under (R 1)—(R 3)},$$

$$\forall \varphi\,[\varphi \in \mathscr{D} \rightarrow (\neg \varphi) \notin \mathscr{D}],$$

$$T_{\text{Skolem}} \subseteq \mathscr{D},$$

$$\forall \Phi\,[\bigvee \Phi \in \mathscr{D} \wedge (\bigvee \Phi \text{ a sentence}) \rightarrow \exists \varphi \in \Phi (\varphi \in \mathscr{D})].$$

Each of these conditions is naturally expressed as a Π condition on $\mathscr{D}$, so the lemma is proved. Note the important role played here by Skolem fragments. If we had to do without "$T_{\text{Skolem}} \subseteq \mathscr{D}$", we would have to add the clause

$$\forall x\,[x = \varphi(v) \wedge (\exists v\, \varphi(v)) \in \mathscr{D} \rightarrow \exists t\,[\varphi(t/v) \in \mathscr{D}]]$$

which is not Π due to the unbounded $\exists t$. ◻

Proof of Theorem 4.1. We may assume, by VII.2.4, that $\mathsf{L}_{\mathbb{A}}$ is a Skolem fragment Δ_1 on $\mathbb{A}$ and that every model of T can be expanded to a Skolem model. By the Weak Completeness Theorem we have $T \models \varphi$ iff

$$\forall \mathscr{D}\,[\mathscr{D} \text{ an s.v.p. for } \mathsf{L}_{\mathbb{A}} \wedge T \subseteq \mathscr{D} \rightarrow \varphi \in \mathscr{D}].$$

By 4.2 this takes the form

$$\forall \mathscr{D}\,[\Pi(\mathscr{D}) \wedge \forall x\,(\Phi(x) \rightarrow x \in \mathscr{D}) \rightarrow \varphi \in \mathscr{D}]$$

where $\Phi(v)$ defines the s-Π_1^1 theory T. The hypothesis of the outer implication is s-Σ_1^1 so the whole becomes a s-Π_1^1 predicate of φ. ◻

4.3 Corollary. *An admissible set $\mathbb{A}$ satisfies s-$\Pi_1^1 = \Sigma_1$ iff $\mathbb{A}$ is Σ_1 complete.*

Proof. The implication ($\Rightarrow$) follows from 4.1. The other direction was proved explicitly in the proof of Theorem 3.1. ◻

4.4 Corollary. *The set of valid sentences of the admissible fragment $\mathsf{L}_{\mathbb{A}}$ is always s-Π_1^1 on $\mathbb{A}$.*

Proof. Let $T = 0$ in 4.1. ◻

At various places in the book we have referred to Σ_1 as a syntactic generalization of r.e. on ω and to strict-Π_1^1 as a semantic version of r.e. on ω. The next corollary of 4.1 makes this precise.

A subset $X \subseteq \mathbb{A}$ is a *complete* Σ_1 *set (or complete strict-*Π_1^1 *set)* for the admissible set $\mathbb{A}$ if X is Σ_1 (resp. s-Π_1^1), and for any other Σ_1 set (resp. s-Π_1^1 set) Y on $\mathbb{A}$ there is a one-one total $\mathbb{A}$-recursive function F such that

$$y \in Y \quad \text{iff} \quad F(y) \in X$$

for all $y \in \mathbb{A}$.

Recall that $T \vdash \varphi$ means that φ is provable from T in the sense of $\mathsf{L}_{\mathbb{A}}$. (This notation occurs in § III.5.)

4.5 Corollary. *Let* $\mathbb{A}$ *be admissible. Let* L' *contain* $\mathsf{L}^*(\vec{\mathsf{R}})$ *and a symbol* x *for each* $x \in \mathbb{A}$ *and let* $\mathsf{L}'_{\mathbb{A}}$ *be the admissible fragment given by* $\mathbb{A}$. *Let* T *be the infinitary diagram of* $\mathbb{A}$.

(i) *The set* $X_0 = \{\varphi \in \mathsf{L}'_{\mathbb{A}} \mid T \vdash \varphi\}$ *is complete* Σ_1 *for* $\mathbb{A}$.
(ii) *The set* $X_1 = \{\varphi \in \mathsf{L}'_{\mathbb{A}} \mid T \models \varphi\}$ *is complete* s-Π_1^1 *for* $\mathbb{A}$.

Proof. (i) is implicit in much of Chapters V and VI. It can also be obtained simply as the absolute version of (ii). To prove (ii) note that X_1 is s-Π_1^1 by 4.1 and that every s-Π_1^1 set is one-one reducible to X_1 by the proof of 3.1. □

An analogous proof shows that on "bad" admissible sets, s-Π_1^1 can be as far from Σ_1 as is conceivable.

4.6 Corollary. *Let* $\mathbb{A}$ *be a self-definable admissible set. Then* $\Pi_1^1 = $ *strict-*Π_1^1 *on* $\mathbb{A}$. *That is, every* Π_1^1 *relation on* $\mathbb{A}$ *can be defined by a strict-*Π_1^1 *formula.*

Proof. Let T be a Σ_1 theory of $\mathsf{L}_{\mathbb{A}}$ which self-defines $\mathbb{A}$. By 4.1, $\mathrm{Cn}(T)$ is s-Π_1^1. But by Lemma VII.1.9, every Π_1^1 relation on $\mathbb{A}$ is one-one reducible to $\mathrm{Cn}(T)$ so every Π_1^1 relation is s-Π_1^1. □

For example, $\Pi_1^1 = \text{strict-}\Pi_1^1$ on $H(\aleph_{\alpha+1})$ for all $\alpha \geqslant 0$, by VII.1.4.

We now turn to consider the logical role of strict-Π_1^1 reflection.

4.7 Theorem. *An admissible set is* Σ_1 *compact iff it satisfies strict-*Π_1^1 *reflection.*

Proof. The implication ($\Rightarrow$) was proved explicitly in the proof of Theorem 3.3. To prove the converse, suppose that $\mathbb{A}$ is admissible and satisfies s-Π_1^1 reflection and that T is a Σ_1 theory of $\mathsf{L}_{\mathbb{A}}$. Assume further that every $T_0 \subseteq T$, $T_0 \in \mathbb{A}$ has a model. By Lemma VII.2.4 we may assume that $\mathsf{L}_{\mathbb{A}}$ is a Skolem fragment and that every $T_0 \subseteq T$, $T_0 \in \mathbb{A}$ has a Skolem model. We will prove that T has a Skolem model. Suppose, aiming at a contradiction, that T has no Skolem model. By the Weak Completeness Theorem, no s.v.p. $\mathscr{D}$ for $\mathsf{L}_{\mathbb{A}}$ can contain T as a subset. Hence $(\mathbb{A}, T)$ satisfies the s-Π_1^1 sentence $\Psi(\mathsf{T})$ expressing:

$$\forall \mathscr{D}\,[\mathscr{D} \text{ an s.v.p. for } \mathsf{L}_{\mathbb{A}} \rightarrow \exists x\,(x \in \mathsf{T} \wedge x \notin \mathscr{D})].$$

Let $\theta(v)$ be the Σ_1 formula defining T. The line above becomes

$$\mathbb{A} \models \forall \mathscr{D}\,[\mathscr{D} \text{ an s.v.p. for } \mathsf{L}_{\mathbb{A}} \rightarrow \exists x\,(\theta(x) \wedge x \notin \mathscr{D})]$$

which is a $s\text{-}\Pi_1^1$ sentence $\Phi(\vec{y})$ with parameters $\vec{y}$ those of $\theta(v)=\theta(v, y_1,\dots, y_k)$. By $s\text{-}\Pi_1^1$ reflection there is a transitive set $a\in\mathbb{A}$ with $\vec{y}\in a$ such that $\mathbb{A}\models\Phi^{(a)}[\vec{y}]$. Let $\mathbb{A}_0=(\mathfrak{M}\cap a; a,\in,\dots)$ and let

$$T_0=\{x\in a\mid \theta^{(a)}(x)\}$$

so that $T_0\in\mathbb{A}$ by Δ_0 Separation and $T_0\subseteq T$ since θ is Σ_1. Since $\Phi^{(a)}[\vec{y}]$ holds we have

$$(\mathbb{A}_0, T_0)\models\Psi(\mathsf{T}).$$

We don't really know what $\Psi(\mathsf{T})$ says on $\mathbb{A}_0$, but $(\mathbb{A}_0, T_0)\subseteq_{\text{end}}(\mathbb{A}, T_0)$ so, by the persistence of $s\text{-}\Pi_1^1$ formulas

$$(\mathbb{A}, T_0)\models\Psi(\mathsf{T}).$$

But this says that T_0 is not a subset of any s.v.p. $\mathscr{D}$ for $\mathsf{L}_{\mathbb{A}}$. Hence T_0 has no Skolem model, a contradiction. ☐

Thus we see that two different aspects of the König Infinity Lemma, those expressed by K_1 and K_2, reflect themselves in related but apparently distinct aspects of first order logic. K_1 is responsible for the Completeness Theorem, K_2 for the Compactness Theorem.

One usually thinks of the Completeness Theorem as implying the Compactness Theorem. The corollary to the next result shows this to be the case for *resolvable* admissible sets. The general case is still open.

4.8 Proposition. *The resolvable admissible sets are divided into two disjoint classes: those that are Σ_1 compact and those that are self-definable.*

Proof. Proposition VII.1.3 shows that no self-definable $\mathbb{A}$ can be Σ_1 compact. Now let $\mathbb{A}$ be a resolvable, admissible set which is not Σ_1 compact. We must show that it is self-definable. Since $\mathbb{A}$ is resolvable there is a total $\mathbb{A}$-recursive function $J: o(\mathbb{A})\to\mathbb{A}$ such that

$$\alpha<\beta\Rightarrow J(\alpha)\in J(\beta),$$

$$J(\alpha) \text{ is transitive, for all } \alpha,$$

$$\mathbb{A}=\bigcup_{\alpha<o(\mathbb{A})} J(\alpha).$$

Since $\mathbb{A}$ is not Σ_1 compact, $\mathbb{A}$ does not satisfy $s\text{-}\Pi_1^1$ reflection, by 4.7. Thus there is a $s\text{-}\Pi_1^1$ formula $\Phi(v)$, and an $x\in\mathbb{A}$ such that $\mathbb{A}$ satisfies:

$$\Phi(x),$$

$$\neg\exists b\,[\mathrm{Tran}(b)\wedge x\in b\wedge\Phi^{(b)}(x)].$$

The second formula is s-Σ_1^1 and so is logically equivalent to a s-Σ_1^1 formula

$$\exists \mathsf{R}\; \varphi(x, \mathsf{R})$$

where φ is Π_1. Let $\sigma(u,v)$ be a Σ_1 formula defining J:

$$J(\alpha)=y \quad \text{iff} \quad \mathbb{A}\models\sigma(\alpha, y).$$

The Σ_1 theory used to self-define $\mathbb{A}$ consists of:

The infinitary diagram of $\mathbb{A}$,

$\forall u, v\,[\mathsf{J}(u)=v\leftrightarrow\sigma(u,v)]$,

$\forall u,u'\,[\mathrm{Ord}(u)\wedge \mathrm{Ord}(u')\wedge u<u'\rightarrow \mathsf{J}(u)\in \mathsf{J}(u')]$,

$\forall x\,\exists u\,[\mathrm{Ord}(u)\wedge x\in \mathsf{J}(u)]$,

$\forall u\,[\mathrm{Tran}(\mathsf{J}(u))]$,

$\varphi(\mathsf{x}, \mathsf{R})$.

Let $(\mathfrak{B}_{\mathfrak{N}}, J)$ be any model of T. We can assume $\mathbb{A}_{\mathfrak{M}}\subseteq_{\mathrm{end}}\mathfrak{B}_{\mathfrak{N}}$. We need to show that $\mathbb{A}_{\mathfrak{M}}=\mathfrak{B}_{\mathfrak{N}}$. If not, let $x\in\mathfrak{B}_{\mathfrak{N}}-\mathbb{A}_{\mathfrak{M}}$. Then by the axioms on σ,

$$\mathfrak{B}_{\mathfrak{N}}\models\exists a\,[\mathrm{Ord}(a)\wedge x\in \mathsf{J}(a)].$$

Pick such an a. Then a is an ordinal of $\mathfrak{B}_{\mathfrak{N}}$ but $a\notin\mathbb{A}_{\mathfrak{M}}$, for if $a\in\mathbb{A}_{\mathfrak{M}}$ then $J(a)\in\mathbb{A}$ which implies $x\in\mathbb{A}$. Thus $\alpha>\beta$ for all $\beta\in\mathbb{A}$. But then $J(\beta)\subseteq J(a)$ holds in $\mathfrak{B}_{\mathfrak{N}}$, for each $\beta\in\mathbb{A}$. Thus $\mathbb{A}\subseteq_{\mathrm{end}}\langle J(a), E\rangle$ and so $\Phi^{(J(a))}(x)$ holds. This contradicts

$$\mathfrak{B}_{\mathfrak{N}}\models\exists \mathsf{R}\; \varphi(\mathsf{x}, \mathsf{R})$$

since this asserts that $\Phi^{(a)}$ fails for all transitive b, in particular for $b=J(a)$. □

4.9 Corollary. *Every Σ_1 complete resolvable admissible set is Σ_1 compact. In other words, K_1 implies K_2 on resolvable admissible sets.*

Proof. If s-$\Pi_1^1=\Sigma_1$ then s-$\Pi_1^1\neq\Pi_1^1$ and hence $\mathbb{A}$ cannot be self-definable, by 4.6. Then by 4.8, $\mathbb{A}$ must be Σ_1 compact. □

What is wrong with the following argument? If s-$\Pi_1^1=\Sigma_1$ then (since Σ reflection holds in all admissible sets) we must have s-Π_1^1 reflection. If you try to fill in the steps in this argument you see that one is missing a certain uniformity in the equation s-$\Pi_1^1=\Sigma_1$. This uniformity is captured by the next definition.

Let $\mathbb{A}=\mathbb{A}_{\mathfrak{M}}$ be transitive, let $\Phi(v_1,\ldots,v_n)$ be strict-Π_1^1 and let $\varphi(v_1,\ldots,v_n)$ be Σ_1, where extra parameters from $\mathbb{A}$ are permitted in φ. We say that φ is *uniformly equivalent* to Φ on $\mathbb{A}$ if

(1) $\quad \mathbb{A}\models\forall v_1,\ldots,v_n\,[\Phi(\vec{v})\rightarrow\varphi(v)]$,

(2) $\quad \mathbb{A}\models\forall a\,\forall v_1,\ldots,v_n\in a\,[\mathrm{Tran}(a)\wedge\varphi^{(a)}(\vec{v})\rightarrow\Phi^{(a)}(\vec{v})]$.

4.10 Lemma. *Let $\mathbb{A}=\mathbb{A}_{\mathfrak{M}}$ be transitive, closed under pairs and* TC. *Let $\varphi(v_1,\ldots,v_n)$ be a Σ_1 formula which is uniformly equivalent to the s-Π_1^1 formula $\Phi(v_1,\ldots,v_n)$ on $\mathbb{A}$. For all $x_1,\ldots,x_n\in\mathbb{A}$, $\mathbb{A}$ satisfies:*

$$\begin{aligned}\Phi(\vec{x})&\leftrightarrow\varphi(\vec{x})\\ &\leftrightarrow\exists a\,[\mathrm{Tran}(a)\wedge\vec{x}\in a\wedge\varphi^{(a)}(\vec{x})]\\ &\leftrightarrow\exists a\,[\mathrm{Tran}(a)\wedge\vec{x}\in a\wedge\Phi^{(a)}(\vec{x})].\end{aligned}$$

Proof. Write $\varphi(v_1,\ldots,v_n)$ as $\exists y\,\psi(v_1,\ldots,v_n,y)$ where y is Δ_0. By (1),

$$\mathbb{A}\models(\Phi(\vec{x})\to\exists y\,\psi(\vec{x},y)).$$

If $\mathbb{A}\models\psi(\vec{x},y)$ then let $a=\mathrm{TC}(\{y,x_1,\ldots,x_n\})$. Then $a\in\mathbb{A}$ and $\varphi^{(a)}(\vec{x})$ holds so

$$\varphi(\vec{x})\to\exists a\,[\mathrm{Tran}(a)\wedge\vec{x}\in\varphi^{(a)}(\vec{x})].$$

By (2), the right hand side of the above line implies

$$\exists a\,[\mathrm{Tran}(a)\wedge\vec{x}\in a\wedge\Phi^{(a)}(\vec{x})].$$

By Lemma 3.2 (i), the above implies $\Phi(\vec{x})$. □

4.11 Definition. An admissible set $\mathbb{A}$ satisfies

$$s\text{-}\Pi_1^1=\Sigma_1 \textit{ uniformly}$$

if for each s-Π_1^1 formula $\Phi(v_1,\ldots,v_n)$ of L^* there is a Σ_1 formula $\varphi(v_1,\ldots,v_n)$, possibly with additional parameters from $\mathbb{A}$, such that φ is uniformly equivalent to Φ on $\mathbb{A}$.

4.12 Theorem. *An admissible set $\mathbb{A}$ satisfies s-$\Pi_1^1=\Sigma_1$ uniformly iff $\mathbb{A}$ satisfies s-$\Pi_1^1=\Sigma_1$ and s-Π_1^1 Reflection.*

Proof. The implication ($\Rightarrow$) is immediate from Lemma 4.10. To prove that converse, assume that $\mathbb{A}$ satisfies K_1 and K_2 and that $\Phi(v_1,\ldots,v_n)$ is s-Π_1^1. We must find a Σ_1 formula $\varphi(v_1,\ldots,v_n)$ uniformly equivalent to $\Phi(v_1,\ldots,v_n)$ on $\mathbb{A}$. Let $\Psi(v_1,\ldots,v_n,b)$ be the s-Π_1^1 formula

$$[\mathrm{Tran}(b)\wedge v_1,\ldots,v_n\in b\wedge\Phi^{(b)}(v_1,\ldots,v_n)].$$

Since s-$\Pi_1^1=\Sigma_1$ there is a Σ_1 formula $\psi(v_1,\ldots,v_n,b)$ equivalent to $\Psi(v_1,\ldots,v_n,b)$ on $\mathbb{A}$. Let $\varphi(v_1,\ldots,v_n)$ be

$$\exists b\,\psi(v_1,\ldots,v_n,b).$$

To prove $\varphi(\vec{v})$ uniformly equivalent to $\Phi(\vec{v})$ first suppose that $\Phi(\vec{x})$ holds in $\mathbb{A}$. By s-Π_1^1 Reflection there is a $b\in\mathbb{A}$ such that $\Psi(\vec{x},b)$ holds in $\mathbb{A}$. But then $\psi(\vec{x},b)$

holds so $\varphi(\vec{x})$ holds. To prove (2), suppose that $a \in \mathbb{A}$ is transitive, that $x_1, \ldots, x_n \in a$ and that $\varphi^{(a)}(\vec{x})$ holds. Then there is a $b \in a$ such that $\psi(\vec{x}, b)^{(a)}$ holds in $\mathbb{A}$. Hence $\psi(\vec{x}, b)$ holds in $\mathbb{A}$ and so

$$\operatorname{Tran}(b) \wedge x_1, \ldots, x_n \in b \wedge \Phi^{(b)}(\vec{x}).$$

Since $b \in a$ and a is transitive, $b \subseteq a$ so, by 3.2 (ii), $\Phi^{(a)}(\vec{x})$ holds, as desired. □

4.13 Corollary. *An admissible set $\mathbb{A}$ satisfies s-$\Pi_1^1 = \Sigma_1$ uniformly iff $\mathbb{A}$ is Σ_1 complete and Σ_1 compact.* □

4.14 Corollary. *On resolvable admissible sets the condition s-$\Pi_1^1 = \Sigma_1$ is equivalent to the condition s-$\Pi_1^1 = \Sigma_1$ uniformly.*

Proof. By 4.9 and 4.12. □

The condition s-$\Pi_1^1 = \Sigma_1$ uniformly clearly captures a great deal of the recursion theoretic and logical content of the Infinity Lemma. If you state it in the "s-$\Sigma_1^1 = \Pi_1$ *uniformly*" version, it even looks like the Infinity Lemma, at least from one point of view. We will use it in § 6 to help us find interesting uncountable Σ_1 complete and Σ_1 compact admissible sets.

4.15—4.21 Exercises

4.15. Let α be admissible but not recursively inaccessible, let $\mathbb{A} = L(\alpha)$. Prove that the valid sentences of $\mathsf{L}_{\mathbb{A}}$ form a complete s-Π_1^1 set. Show that for any admissible $\beta \leqslant \omega_1$, β is recursively inaccessible iff the set of valid sentences of L_β is β-recursive.

4.16. A subset X of an admissible set $\mathbb{A}$ in *bounded* if $X \subseteq a$ for some $a \in \mathbb{A}$. Let $\mathbb{A}$ be admissible and satisfy s-Π_1^1 Reflection. Let T be a set of sentences of $\mathsf{L}_{\mathbb{A}}$ which is s-Π_1^1 on $\mathbb{A}$. Prove that if every bounded subset $T_0 \subseteq T$ has a model then T has a model. (It is open whether one can improve this by replacing "bounded" by "$\mathbb{A}$-finite".)

4.17. Let $\mathbb{A}$ be admissible. Suppose that for every Δ_0 theory T of $\mathsf{L}_{\mathbb{A}}$, if every $T_0 \subseteq T$, $T_0 \in \mathbb{A}$ has a model, then T has a model. Show that $\mathbb{A}$ is Σ_1 compact. [Show that s-Π_1^1 Reflection holds.]

4.18. Let $\mathbb{A}$ be admissible, $\alpha = o(\mathbb{A})$. $\mathbb{A}$ is s-Δ_1^1 *resolvable* if there is a limit ordinal $\lambda \leqslant \alpha$ and a s-Δ_1^1 function $J : \lambda \to \mathbb{A}$ such that

$$\beta < \xi \Rightarrow J_\beta \in J_\xi \quad \text{for} \quad \beta, \xi < \lambda,$$

$$J_\beta \quad \text{is transitive for all} \quad \beta < \lambda,$$

$$\mathbb{A} = \textstyle\bigcup_{\beta < \lambda} J_\beta .$$

The ordinal λ is said to be the *length* of the hierarchy J on $\mathbb{A}$.

(i) Prove that if $\alpha = \beth_\alpha$ then $H(\beth_\alpha)$ is s-Δ_1^1 resolvable. [Hint: If $\alpha = \beth_\alpha$ then $H(\beth_\alpha) = \mathrm{V}(\alpha)$. Let $J_\beta = \mathrm{V}(\beta)$.]

(ii) Prove that if $\mathbb{A}$ is s-Δ_1^1 resolvable and if J is as above with $\lambda < o(\mathbb{A})$ then $\mathbb{A}$ is self-definable.

(iii) Strengthen Proposition 4.8 to: The class of s-Δ_1^1 resolvable admissible sets are divided into disjoint two classes, the Σ_1 compact and the self-definable.

(iv) Let $\kappa = \beth_\kappa$. Show that $\langle H(\kappa), \in, \mathscr{P} \rangle$ satisfies K_1 iff it satisfies K_2.

4.19. Kunen [1968] introduced an invariant definability approach to generalized recursion theory on admissible sets by introducing the notions of a.i.d., i.i.d., and s.i.i.d. (see below) as generalizations of the concepts of finite, recursive and r.e. In Barwise [1968], [1969b] we showed that s-$\Pi_1^1 =$ s.i.i.d. (see (ii)). (This leads to the formulation of s-Π_1^1 Reflection and the results of this section in Barwise [1968], [1969b].) Let $\mathbb{A}$ be admissible and let P be an n-ary relation on $\mathbb{A}$. P is

(a) absolutely implicitly definable (a.i.d.) on $\mathbb{A}$,

(b) invariantly implicitly definable (i.i.d.) on $\mathbb{A}$,

(c) semi-invariantly implicitly definable (s.i.i.d.) on $\mathbb{A}$

if there is a finitary first order sentence $\theta(\mathsf{P}, \mathsf{S}_1, \ldots, \mathsf{S}_k)$ of $\mathsf{L}^*(\mathsf{P}, \ldots)$ such that

$$(\mathbb{A}, P) \models \exists \mathsf{S}_1, \ldots, \mathsf{S}_k, \theta(\mathsf{P}, \mathsf{S}_1, \ldots, \mathsf{S}_k)$$

and such that if $\mathbb{A} \subseteq_{\text{end}} \mathfrak{B}$ and $P' \subseteq \mathfrak{B}^n$ satisfies

$$(\mathfrak{B}, P') \models \exists \mathsf{S}_1, \ldots, \mathsf{S}_k, \theta(\mathsf{P}, \mathsf{S}_1, \ldots, \mathsf{S}_k)$$

then

(a) $P = P'$,

(b) $P = P' \cap \mathbb{A}^n$,

(c) $P \subseteq P' \cap \mathbb{A}^n$.

The sentence θ may contain parameters from $\mathbb{A}$.

(i) Prove that P is i.i.d. iff P, $\neg P$ are s.i.i.d.

(ii) Prove that P is s.i.i.d. iff P is s-Π_1^1. [One half of this uses 4.1.]

(iii) Prove that if $\mathbb{A}$ satisfies s-$\Pi_1^1 = \Sigma_1$ uniformly then

$$\text{s.i.i.d.} = \Sigma_1 \quad \text{on} \quad \mathbb{A},$$

$$\text{i.i.d.} = \Delta_1 \quad \text{on} \quad \mathbb{A},$$

$$\text{a.i.d.} = \text{element} \quad \text{of} \quad \mathbb{A}.$$

(iv) Prove that $\mathbb{A}$ is self-definable iff $\mathbb{A}$ is a.i.d. on $\mathbb{A}$.

(v) Prove that if $\mathbb{A}$ satisfies K_2 then every a.i.d. subset of $\mathbb{A}$ is bounded.

4.20. The notion of uniformity given by Definition 4.11 is really suggested by the notion of proof. Prove directly, using the Extended Completeness Theorem that if $\mathbb{A}$ is countable and admissible then $\mathbb{A}$ satisfies $s\text{-}\Pi_1^1 = \Sigma_1$ uniformly.

4.21. A more recursion theoretic version of the uniformity discussed in 4.11 and 4.20 was discovered by Nyberg. Prove that the admissible set $\mathbb{A}$ satisfies $s\text{-}\Pi_1^1 = \Sigma_1$ uniformly iff for every $s\text{-}\Pi_1^1$ formula $\Phi(x, \mathsf{T}_+)$ in an extra relation symbol T there is a Σ_1 formula $\varphi(x, \mathsf{T}_+)$ such that for all Σ_1 relations T on $\mathbb{A}$, $(\mathbb{A}, T)$ satisfies

$$\forall x\,[\Phi(x, \mathsf{T}_+) \leftrightarrow \varphi(x, \mathsf{T}_+)].$$

[Show that this condition is equivalent to $K_1 \wedge K_2$. Note that in the proof of 4.7, T occurs positively in $\Psi(\mathsf{T})$.]

4.22 Notes. See Exercise 4.19 for the way $s\text{-}\Pi_1^1$ predicates found their way into the subject. Corollary 4.9 was observed by Nyberg. For the record, it is still open whether every Σ_1 complete admissible set is Σ_1 compact. (Surely not!) It follows from Theorem 8.3 (applied to $\mathrm{L}(\alpha)$) that there are lots of resolvable Σ_1 compact sets which are not Σ_1 complete.

5. König's Lemma and Nerode's Theorem: a Digression

In this section we interupt our study to apply the condition

$$s\text{-}\Pi_1^1 = \Sigma_1 \quad \text{uniformly}$$

to notions of relative definability.

One of the starkest applications of the Infinity Lemma in ordinary recursion theory is the proof of Nerode's Theorem:

B is truth table reducible to C iff there is a total general recursive operator $\mathfrak{F}$ with $\mathfrak{F}(K_C) = K_B$.

Here $B, C \subseteq \omega$ and K_B is the characteristic function of B. This says, in effect, that the total general recursive operators rather trivial.

Since Nerode's Theorem uses so little about ω, other than the Infinity Lemma, it becomes a good test case for abstract versions of the Infinity Lemma, the matter which concerns us in this chapter.

Turing reducibility breaks up into many non-equivalent notions over an arbitrary set. We discuss generalizations of Nerode's Theorem for three of these:

$\leqslant_{\mathrm{d}}$ is "Δ definable from",

$\leqslant_{\mathrm{w}}$ is "weakly meta-recursive in",

$\leqslant_{\mathrm{mr}}$ is "meta-recursive in".

5.1 Definition. Let $\mathbb{A}$ be admissible and let $\varphi(x,\mathsf{C})$, $\psi(x,\mathsf{C})$ be Σ formulas, possibly containing parameters from $\mathbb{A}$. Let B, C be subsets of $\mathbb{A}$. We say that $B \leqslant_{\mathrm{d}} C$ (via $\langle\varphi,\psi\rangle$) if for all $x \in \mathbb{A}$:

$$x \in B \quad \text{iff} \quad (\mathbb{A}, C) \models \varphi(x,\mathsf{C}),$$

$$x \notin B \quad \text{iff} \quad (\mathbb{A}, C) \models \psi(x,\mathsf{C}).$$

If, for every C there is a B such that $B \leqslant_{\mathrm{d}} C$ via $\langle\varphi,\psi\rangle$ then the pair $\langle\varphi,\psi\rangle$ is called a *general Δ definability operator* $\mathfrak{F}$ over $\mathbb{A}$, and we write $\mathfrak{F}(C) = B$.

By the relativized version of Theorem II.2.3, if $\mathbb{A} = \mathbb{HF}$ then $B \leqslant_{\mathrm{d}} C$ iff B is recursive in C, so that $\leqslant_{\mathrm{d}}$ coincides with Turing reducibility.

What is the most obvious way to define a general Δ definability operator? It seems to be captured by the following definition. If $\mathbb{A}$ is admissible then $\Delta_0(\mathbb{A})$ denotes the Δ_0 formulas when all total $\mathbb{A}$-recursive functions are denoted by terms of the language.

5.2 Definition. Let $\mathbb{A}$ be admissible and let $\varphi(x,\mathsf{C})$ be a $\Delta_0(\mathbb{A})$ formula. Then $B \leqslant_{\mathrm{d}}^{\mathrm{tt}} C$ via φ if, for all $x \in \mathbb{A}$,

$$x \in B \leftrightarrow (\mathbb{A}, C) \models \varphi(x,\mathsf{C}).$$

5.3 Lemma. *Let $\mathbb{A}$ be admissible.*

(i) *Every $\Delta_0(\mathbb{A})$ formula $\varphi(x,\mathsf{C})$ defines a general Δ definability operator.*

(ii) *If F is $\mathbb{A}$-recursive and*

$$x \in B \quad \textit{iff} \quad F(x) \in C$$

then $B \leqslant_{\mathrm{d}}^{\mathrm{tt}} C$.

(iii) $\mathbb{A} - B \leqslant_{\mathrm{d}}^{\mathrm{tt}} B$.

(iv) $\leqslant_{\mathrm{d}}^{\mathrm{tt}}$ *is transitive.*

Proof. They are all trivial. For example, to prove (i) you simply replace any function symbol in φ by its definition as in § I.4. Note that

$$x \notin B \quad \text{iff} \quad (\mathbb{A}, C) \models \neg\varphi(x,\mathsf{C})$$

and $\neg\varphi$ is also a $\Delta_0(\mathbb{A})$ formula. □

5.4 Theorem. *Let $\mathbb{A}$ be a resolvable, admissible set satisfying s-$\Pi_1^1 = \Sigma_1$ uniformly. Let $\mathfrak{F}$ be any general Δ definability operator over $\mathbb{A}$. There is a $\Delta_0(\mathbb{A})$ formula $\varphi(\mathsf{C})$ such that for all $C \subseteq \mathbb{A}$*

$$\mathfrak{F}(C) \leqslant_{\mathrm{d}}^{\mathrm{tt}} C \quad \textit{via} \quad \varphi.$$

Proof. Let $\theta(x,\mathsf{C})$, $\psi(x,\mathsf{C})$ be Σ formulas such that $\mathfrak{F}$ is defined by

$$x \in \mathfrak{F}(C) \quad \text{iff} \quad (\mathbb{A}, C) \models \theta(x,\mathsf{C}),$$

$$x \notin \mathfrak{F}(C) \quad \text{iff} \quad (\mathbb{A}, C) \models \psi(x,\mathsf{C}).$$

Then, for every x, the following s-Π_1^1 formula $\Phi(x)$ holds on $\mathbb{A}$:

$$\forall \mathsf{C}\,[\theta(x,\mathsf{C}) \vee \psi(x,\mathsf{C})].$$

Let $\varphi(x)$ be a Σ_1 formula uniformly equivalent to $\Phi(x)$ on $\mathbb{A}$. Since $\mathbb{A}$ is resolvable there is an $\mathbb{A}$-recursive function $J: o(\mathbb{A}) \to \mathbb{A}$ such that $\mathbb{A} = \bigcup_{\alpha < o(\mathbb{A})} J(\alpha)$ and $J(\alpha)$ is transitive for all α. Now for each x, $\Phi(x)$ holds so there is an $\alpha \in \mathbb{A}$ such that $\varphi(x)^{(J(\alpha))}$. Define

$$G(x) = J(\text{least}\,\alpha\,[\varphi(x)^{(J(\alpha))}]).$$

Then G is $\mathbb{A}$-recursive and total, $G(x)$ is always transitive and

$$\varphi(x)^{(G(x))}.$$

But then for every x, $\Phi^{(G(x))}(x)$, by the uniform equivalence of φ and Φ. Thus, for every $C \subseteq \mathbb{A}$, either $\theta(x,C)^{(G(x))}$ or $\psi(x,C)^{(G(x))}$. We claim that

(1) $\qquad x \in \mathfrak{F}(C) \quad \text{iff} \quad \theta(x,C)^{(G(x))}.$

For if $x \in \mathfrak{F}(C)$ then $(\mathbb{A},C) \models \theta(x,\mathsf{C})$ so $\psi(x,C)^{(G(x))}$ cannot hold so $\varphi(x,C)^{(G(x))}$ must hold. Similarly, if $x \notin \mathfrak{F}(C)$ then $\theta(x,C)^{(G(x))}$ cannot hold. Let $\sigma(v,\mathsf{C})$ be $\theta(v,\mathsf{C})^{(F(v))}$. Then $\sigma(v,\mathsf{C})$ is $\Delta_0(\mathbb{A})$ and $\mathfrak{F}(C) \leqslant_d^{tt} C$ via σ. □

Since "$s-\Pi_1^1 = \Sigma_1$" implies "$s-\Pi_1^1 = \Sigma_1$ uniformly" on resolvable admissible sets, we could have used the weaker condition in the statement of the theorem. This seems to conceal the main point of the theorem, though, since it is the uniformity which really matters in the above proof. Since the above proof is virtually identical (in outline) to the proof of Nerode's Theorem (in, say Rogers [1967]) this gives further support to the feeling that "$s-\Pi_1^1 = \Sigma_1$ uniformly" captures a great deal of the recursion theoretic content of the Infinity Lemma.

The relation $\leqslant_d$ is quite sensible from a definability point of view. It has been studied very little, however, because one does not have the tools from ordinary recursion theory available. Put another way, the relation $B \leqslant_d C$ is not sensible in terms of computation if the expanded structure $(\mathbb{A}, C)$ fails to be admissible, for then in checking whether or not $x \in B$ one may have to use all of C, not just an $\mathbb{A}$-finite amount of information about C. This never comes up for $\mathbb{HF}$, or for any other $H(\kappa)$, κ-regular, since every expansion of $H(\kappa)$ is still admissible.

These observations prompt one to define a new notion of reducibility, one where an answer to "$x \in B$?" is determined by an $\mathbb{A}$-finite amount of information about C. Let K_C be the characteristic function of C:

$$K_C(x) = \begin{cases} 0 & \text{if } x \in C, \\ 1 & \text{if } x \notin C \end{cases}$$

and let $\mathrm{Ch}_{\mathbb{A}}(C) = \{f \in A \mid f \subseteq K_C\}$. Thus $\mathrm{Ch}_{\mathbb{A}}(C)$ is the set of all $\mathbb{A}$-finite bits of information about membership in C.

5.5 Definition. Let $\mathbb{A}$ be admissible and let $\varphi(x,f)$, $\psi(x,f)$ be Σ_1 formulas with parameters from $\mathbb{A}$. We say that B is *weakly metarecursive in C via* $\langle\varphi,\psi\rangle$, written $B\leqslant_{\mathrm{w}}C$ via $\langle\varphi,\psi\rangle$, if for all $x\in\mathbb{A}$

$$x\in B \quad \text{iff} \quad \exists f\in \mathrm{Ch}_{\mathbb{A}}(C)[\mathbb{A}\models\varphi(x,f)],$$

$$x\notin B \quad \text{iff} \quad \exists f\in \mathrm{Ch}_{\mathbb{A}}(C)[\mathbb{A}\models\psi(x,f)].$$

If for every C there is a B such that $B\leqslant_{\mathrm{w}}C$ via $\langle\varphi,\psi\rangle$ then the pair $\langle\varphi,\psi\rangle$ is called a *general weak metarecursive operator* $\mathfrak{F}$ and we write $\mathfrak{F}(C)=B$.

The notion of tt-reducibility corresponding to $\leqslant_{\mathrm{w}}$ is complicated by the following observations. On $\mathbb{HF}$ one can define a recursive function H by

$$H(x)=\{f\,|\,f \text{ is a characteristic function with } \mathrm{dom}(f)=x\}.$$

Then given any recursive predicate P of finite functions one can "split" it by

$$F(x)=\{f\in H(x)\,|\,P(f)\},$$

$$G(x)=\{f\in H(x)\,|\,\neg P(f)\}.$$

Then F,G are recursive and, for each $x\in\mathbb{HF}$ and each $C\subseteq\mathbb{HF}$, $\mathrm{Ch}_{\mathbb{HF}}(C)$ meets (has nonempty intersection with) exactly one of the sets $F(x)$, $G(x)$ (depending on whether or not $K_C\restriction x$ satisfies P or not). This triviality simplifies a lot of the recursion theory on $\mathbb{HF}$, especially when contrasted with a general admissible set $\mathbb{A}$ where $H(x)$ need not be a subset of $\mathbb{A}$, let alone an element of $\mathbb{A}$.

5.6 Definition. Let $\mathbb{A}$ be admissible. An $\mathbb{A}$-*recursive splitting* is a pair F,G of total $\mathbb{A}$-recursive functions such that

(i) for each $x\in\mathbb{A}$, $F(x)$, $G(x)$ are sets of $\mathbb{A}$-finite characteristic functions,

(ii) for each $x\in\mathbb{A}$ and each $C\subseteq\mathbb{A}$, $\mathrm{Ch}_{\mathbb{A}}(C)$ meets exactly one of $F(x)$, $G(x)$.

5.7 Lemma. *Let $\mathbb{A}$ be admissible and let F,G be an $\mathbb{A}$-recursive splitting. Define*

$$\mathfrak{F}(C)=\{x\,|\,\mathrm{Ch}_{\mathbb{A}}(C)\cap F(x)\neq 0\}.$$

Then $\mathfrak{F}$ is a general weak metarecursive operator on $\mathbb{A}$.

Proof. Let $\varphi(x,f)$ be $f\in F(x)$, $\psi(x,f)$ be $f\in G(x)$. Then

$$x\in\mathfrak{F}(C) \quad \text{iff} \quad \exists f\in\mathrm{Ch}_{\mathbb{A}}(C)\,\varphi(x,f),$$

$$x\notin\mathfrak{F}(C) \quad \text{iff} \quad \exists f\in\mathrm{Ch}_{\mathbb{A}}(C)\,\psi(x,f)$$

so $\mathfrak{F}(C)\leqslant_{\mathrm{w}}C$ via $\langle\varphi,\psi\rangle$, for all $C\subseteq\mathbb{A}$. $\square$

The next theorem shows that for some admissible sets, every general weak metarecursive operator arises as in the above lemma.

5.8 Theorem. *Let* $\mathbb{A}$ *be a resolvable admissible set satisfying* $s-\Pi_1^1=\Sigma_1$ *uniformly. Let* $\mathfrak{F}$ *be any general weak metarecursive operator. There is an* $\mathbb{A}$*-recursive splitting* F, G *such that for all* C,

$$\mathfrak{F}(C)=\{x \mid \mathrm{Ch}_{\mathbb{A}}(C)\cap F(x)\neq 0\}.$$

Proof. The proof is very much like the proof of Theorem 5.4. Let $\theta(x,f)$, $\psi(x,f)$ be Σ_1 formulas such that $\mathfrak{F}$ is defined by

$$x\in\mathfrak{F}(C) \quad \text{iff} \quad \exists f\in \mathrm{Ch}_{\mathbb{A}}(C)[A\models\theta(x,f)],$$

$$x\notin\mathfrak{F}(C) \quad \text{iff} \quad \exists f\in \mathrm{Ch}_{\mathbb{A}}(C)[A\models\psi(x,f)].$$

Then for each $x\in\mathbb{A}$ the following $s-\Pi_1^1$ formula $\Phi(x)$ holds:

$$\forall C\,\exists f[f\in\mathrm{Ch}_{\mathbb{A}}(C)\wedge[\theta(x,f)\vee\psi(x,f)]].$$

Let $\varphi(x)$ be uniformly equivalent to $\Phi(x)$ on $\mathbb{A}$. Let $J\colon o(\mathbb{A})\to\mathbb{A}$ be as in the proof of 5.4 and define

$$H(x)=J(\text{least}\,\alpha\,\varphi(x)^{(J(\alpha))}).$$

As in the proof of 5.4 we see that H is a total $\mathbb{A}$-recursive function, that $H(x)$ is always transitive, and $\Phi(x)^{(H(x))}$; i.e.,

$$(2) \qquad \forall C\,\exists f\in H(x)[f\in\mathrm{Ch}_{\mathbb{A}}(C)\wedge\theta(x,f)\vee\psi(x,f)]^{(H(x))}.$$

Let

$$F(x)=\{f\in H(x)\mid f \text{ is a characteristic function } \wedge\theta(x,f)^{(H(x))}\},$$

$$G(x)=\{f\in H(x)\mid f \text{ is a characteristic function } \wedge\psi(x,f)^{(H(x))}\}.$$

We claim that for all x, C

$$x\in\mathfrak{F}(C) \quad \text{iff} \quad F(x)\cap\mathrm{Ch}_{\mathbb{A}}(C)\neq 0,$$

$$x\notin\mathfrak{F}(C) \quad \text{iff} \quad G(x)\cap\mathrm{Ch}_{\mathbb{A}}(C)\neq 0.$$

This will prove that F, G is an $\mathbb{A}$-recursive splitting and the conclusion of the theorem. First suppose $x\in\mathfrak{F}(C)$. From line (2) we see that $F(x)\cap\mathrm{Ch}_{\mathbb{A}}(C)\neq 0$. But line (2) also implies that $G(x)\cap\mathrm{Ch}_{\mathbb{A}}(C)=0$ for if $f\in\mathrm{Ch}_{\mathbb{A}}(C)\wedge\psi(x,f)^{(H(x))}$ then $\psi(x,f)$ holds in $\mathbb{A}$, since ψ is Σ_1, so $x\notin\mathfrak{F}(C)$. The other half is similar. □

The relation $\leqslant_w$ has been studied a fair amount by the Sacks school (on admissible sets of the form $L(\alpha)$). In particular, it has shown that $\leqslant_w$ is not transitive. This is not too surprising given the disparity between the amount of information used about C (namely $f\in\mathrm{Ch}_{\mathbb{A}}(C)$) and the amount of information received ($x\in B$ or $x\notin B$). Thus Sacks defines

$$B\leqslant_{mr}C \quad \text{iff} \quad \mathrm{Ch}_{\mathbb{A}}(B)\leqslant_w C.$$

This is equivalent to the existence of a single Σ_1 formula $\varphi(x,f)$ such that

$$(3) \qquad g\in \mathrm{Ch}_{\mathbb{A}}(B) \quad \text{iff} \quad \exists f\in \mathrm{Ch}_{\mathbb{A}}(C)[\mathbb{A}\models \psi(x,f)].$$

There doesn't seem to be a very elegant notion of tt-reducibility to go along with $\leqslant_{\mathrm{mr}}$, but we do get one out of Theorem 5.8.

5.9 Corollary. *Let $\mathbb{A}$ be resolvable and satisfy $s-\Pi_1^1=\Sigma_1$ uniformly. Let ψ be such that for every C there is a B satisfying line* (3) *above. There is an $\mathbb{A}$-recursive splitting F,G such that for all C,B as in* (3)

$$g\in \mathrm{Ch}_{\mathbb{A}}(B) \quad \textit{iff} \quad F(g)\cap \mathrm{Ch}_{\mathbb{A}}(C)\neq 0\,. \quad \square$$

5.10 Exercise (R. Shore). Show that if $\mathrm{V}=\mathrm{L}$ then the conclusions of 5.4 and 5.8 fail for $\mathbb{A}=\mathrm{L}(\omega_1)$. [For 5.8 define $\mathfrak{F}(C)=\mathbb{A}$ if $C\cap\omega$ is infinite, $=0$ otherwise. For 5.4 let $R\subset\mathscr{P}(\omega)$ by Δ_1 on $\mathbb{A}$ but not Δ_0 and define $\mathfrak{F}(C)=\mathbb{A}$ if $R(C\cap\omega)$, $=0$ otherwise.]

5.11 Notes. The reader should consult Simpson's forthcoming book in this series for more about reducibilities on admissible sets.

6. Implicit Ordinals on Arbitrary Admissible Sets

For the model theory of an admissible fragment $\mathsf{L}_{\mathbb{A}}$, the ordinal $h_\Sigma(\mathbb{A})$ plays a more important role than $o(\mathbb{A})$. For countable $\mathbb{A}$ we have $h_\Sigma(\mathbb{A})=o(\mathbb{A})$. In general, we will see that this condition again goes back to the König Infinity Lemma.

6.1 Theorem. *An admissible set satisfies the third König principle iff $h_\Sigma(\mathbb{A})=o(\mathbb{A})$.*

Proof. This is an immediate consequence of the next theorem. $\square$

The ordinal $h_\Sigma(\mathbb{A})$ is not an absolute notion. That is, the size (cardinality) of $h_\Sigma(\mathbb{A})$ may vary drastically from one model of set theory to another (cf. Theorem 4.2 in Barwise-Kunen [1971]). The important point for application, though, is that $h_\Sigma(\mathbb{A})$ has a precise description in terms of the generalized recursion theory of $\mathbb{A}$.

6.2 Theorem. *Let $\mathbb{A}$ be admissible:*

$$h_\Sigma(\mathbb{A})=\sup\{\xi\colon \xi \textit{ is a } \Pi \textit{ implicit ordinal over } \mathbb{A}\}\,.$$

Proof. The inequality $\geqslant$ follows from Theorem 3.5. To prove the theorem it suffices to prove that every ordinal $\beta<h_\Sigma(\mathbb{A})$ is less than some Π implicit ordinal ξ.

We can read this off the proof of Theorem VII.3.1. Since $\beta < h_\Sigma(\mathbb{A})$, β can be pinned down by some Σ_1 theory T of $\mathsf{L}_{\mathbb{A}}$. Now consider the proof of VII.3.1 for this particular theory T. In particular, consider the well-founded relation $\langle \mathfrak{S}, \prec \rangle$ constructed there. Since every ordinal pinned down by T is less than the rank $\xi = \rho(\prec)$ of this well-founded relation, it suffices to prove that this ξ is Π implicit over $\mathbb{A}$ or, at least less than or equal to some Π implicit ordinal.

Case 1. If $o(\mathbb{A}) > \omega$ *then* ξ *is a* Π *implicit ordinal.*
For if $o(\mathbb{A}) > \omega$. then we can write

(1) $$\mathscr{D} \in \mathfrak{S} \wedge \mathscr{D}' \in \mathfrak{S} \wedge \mathscr{D} \prec \mathscr{D}$$

out as a Π sentence $\varphi(\mathscr{D}', \mathscr{D})$ using 4.2:

$$\begin{aligned} \exists n < \omega\, \exists m < \omega [n > m \wedge T \subseteq \mathscr{D} \subseteq \mathscr{D}' \wedge \mathscr{D}' &\text{ is an s.v.p. for } \mathsf{L}_{\mathbb{A}}(\mathsf{c}_1, \ldots, \mathsf{c}_n), \\ \mathscr{D} &\text{ is an s.v.p. for } \mathsf{L}_{\mathbb{A}}(\mathsf{c}_1, \ldots, \mathsf{c}_m) \\ &\wedge \forall i < n [0 < i \rightarrow (\mathsf{c}_{i+1} < \mathsf{c}_i) \in \mathscr{D}']. \end{aligned}$$

All these clauses are Π and the others follows from these. Thus $\varphi(\mathscr{D}', \mathscr{D})$ is a Π sentence which implicitly defines $\xi = \rho(\prec)$.

Case 2. If $o(\mathbb{A}) = \omega$ *then* $\xi \leqslant \xi'$ *for some* Π *implicit ordinal* ξ'.
Let $\psi(\mathscr{D}', \mathscr{D})$ be the Π sentence expressing the following:

$\mathscr{D}, \mathscr{D}'$ are sets of sentences of $\mathsf{L}_{\mathbb{A}}(C)$,
$T \subseteq \mathscr{D} \subseteq \mathscr{D}'$,
$\mathscr{D} \cap \mathsf{L}_{\mathbb{A}}$ is an s.v.p. for $\mathsf{L}_{\mathbb{A}}$,
$\forall m [(\mathsf{c}_m = \mathsf{c}_m) \in \mathscr{D} \rightarrow \mathscr{D} \cap \mathsf{L}_{\mathbb{A}}(\mathsf{c}_1, \ldots, \mathsf{c}_m)$ is an s.v.p.
for $L_{\mathbb{A}}(\mathsf{c}_1, \ldots, \mathsf{c}_m)$ and $\forall i < m [0 < i \rightarrow (\mathsf{c}_{i+1} < \mathsf{c}_i) \in \mathscr{D}$,
the same sentence for $\mathscr{D}'$,
$(\mathsf{c}_1 = \mathsf{c}_1) \in \mathscr{D}'$,
$\forall m [(\mathsf{c}_m = \mathsf{c}_m) \in \mathscr{D} \rightarrow (\mathsf{c}_{m+1} = \mathsf{c}_{m+1}) \in \mathscr{D}']$.

If $\mathscr{D}, \mathscr{D}' \in S$ and $\mathscr{D}' \prec \mathscr{D}$ then $\psi(\mathscr{D}', \mathscr{D})$. If

$$\mathscr{D}' \prec' \mathscr{D} \quad \text{iff} \quad (\mathbb{A}, \mathscr{D}, \mathscr{D}') \models \psi(\mathscr{D}', \mathscr{D})$$

defines a well-founded relation then its rank is $\geqslant \xi = \rho(\prec)$ since $\prec$ is a subrelation. So we need only prove that $\prec'$ is well founded. Suppose not. That is, suppose

$$\ldots \prec' \mathscr{D}_{n+1} \prec' \mathscr{D}_n \prec' \cdots \prec' \mathscr{D}_1 \prec' \mathscr{D}_0 .$$

Let $\overline{\mathscr{D}}_n = \mathscr{D}_n \cap \mathsf{L}_{\mathbb{A}}(\mathsf{c}_1, \ldots, \mathsf{c}_n)$. Since $\psi(\mathscr{D}_1, \mathscr{D}_0)$ holds it follows that $\overline{\mathscr{D}}_0$ is an s.v.p. for $\mathsf{L}_{\mathbb{A}}$, that $(\mathsf{c}_1 = \mathsf{c}_1) \in \overline{\mathscr{D}}_2$ and hence that $\overline{\mathscr{D}}_1$ is an s.v.p. for $\mathsf{L}_{\mathbb{A}}(\mathsf{c}_1)$. That is, $\overline{\mathscr{D}}_0 \in \mathfrak{S}_0$, $\mathscr{D}_1 \in \mathfrak{S}_1$ and $\overline{\mathscr{D}}_1 \prec \overline{\mathscr{D}}_0$. By induction on n we see that $\overline{\mathscr{D}}_n \in \mathfrak{S}_n$ and $\overline{\mathscr{D}}_{n+1} \prec \overline{\mathscr{D}}_n$.

This contradicts the well-foundedness of $\prec$. Thus $\prec'$ is well founded. Since ψ is a Π formula, the rank $\xi'=\rho(\prec')$ is a Π implicit ordinal and $\beta<\xi\leqslant\xi'$. □

Theorem 6.2 would have simplified the proofs of the theorems in § VII.4 since it is usually easier to show that a given well-founded relation $\prec$ is definable by a Π sentence than to prove $\rho(\prec)<h_\Sigma(\mathbb{A})$.

6.3 Corollary. *Every $s-\Sigma_1^1$ implicit ordinal over the admissible set $\mathbb{A}$ is less than some Π implicit ordinal over $\mathbb{A}$.*

Proof. Immediate from 3.5 and 6.2. □

6.4 Corollary. *If $\mathbb{A}$ is a resolvable admissible set and $h_\Sigma(\mathbb{A})=o(\mathbb{A})$ then $\mathbb{A}$ is Σ_1 compact; i.e., K_3 implies K_2 on resolvable admissible sets.*

Proof. By 4.8, if $\mathbb{A}$ fails to be Σ_1 compact then $\mathbb{A}$ is self-definable, and hence $h_\Sigma(\mathbb{A})>o(\mathbb{A})$ by Proposition VII.1.5. □

6.5 Corollary. *Let $\mathfrak{M}=\langle M,R_1,\ldots,R_l\rangle$ and let $\mathbb{A}=\mathbb{HYP}_{\mathfrak{M}}$. Then $\mathbb{A}$ is Σ_1 compact iff $h_\Sigma(\mathbb{A})=o(\mathbb{A})$.*

Proof. One half follows from 6.4, since $\mathbb{HYP}_{\mathfrak{M}}$ is resolvable; the other half ($\Rightarrow$) from VII.3.8. □

We conclude this section with a theorem that explains why Π implicit ordinals are so important from a theoretical, not just a practical, point of view.

Let $\varphi(v,\mathsf{R},\mathsf{S})$ be a formula with R,S n-ary, v a free variable, which may contain parameters from $\mathbb{A}$. For $x\in\mathbb{A}$ we write $\prec_\varphi^x$ for the relation defined by

$$R\prec_\varphi^x S \quad\text{iff}\quad (\mathbb{A},R,S)\models\varphi(x,\mathsf{R},\mathsf{S}).$$

6.6 Lemma. *If $\varphi(v,\mathsf{R},\mathsf{S})$ is a Π (or even $\mathrm{s}-\Sigma_1^1$) formula then*

$$P(x) \text{ iff } \prec_\varphi^x \text{ is well founded}$$

defines a $s-\Pi_1^1$ predicate P over the admissible set A.

Proof. $P(x)$ holds iff

$$\forall \mathsf{Q}\,\exists m\,\neg\varphi(x,(\mathsf{Q})_{m+1},(\mathsf{Q})_m)$$

where Q is $n+1$ ary and $\varphi(x,(\mathsf{Q})_{m+1},(\mathsf{Q})_m)$ denotes the result of replacing $\mathsf{R}(x_1,\ldots,x_n)$ by $\mathsf{Q}(x_1,\ldots,x_n,m+1)$, $\mathsf{S}(x_1,\ldots,x_n)$ by $\mathsf{Q}(x_1,\ldots,x_m,m)$. □

6.7 Theorem. *Let $\mathbb{A}$ be admissible. There is a Π formula $\varphi(v,\mathsf{R},\mathsf{S})$ such that*

$$\{x\colon \prec_\varphi^x \text{ is well founded}\}$$

is a complete strict $-\Pi_1^1$ set for $\mathbb{A}$.

Proof. The set in question is always $s-\Pi_1^1$ by 6.6. Let X_1 be the complete $s-\Pi_1^1$ set defined in Corollary 4.5: $X_1=\{\varphi\in \mathsf{L}'_{\mathbb{A}} \mid T\models\varphi\}$. We can assume $\mathsf{L}'_{\mathbb{A}}$ (of 4.5) is a Skolem fragment which is Δ_1 on $\mathbb{A}$ and that every model of T can be expanded to a Skolem model. We will show how to write "Is $x\in X_1$?" in terms of asking whether or not a certain tree of theories of $\mathsf{L}'_{\mathbb{A}}$ is well founded. Let $\varphi(x,\mathsf{T}'',\mathsf{T}')$ express:

$$
\begin{aligned}
&T', T'' \text{ are sets of sentences of } \mathsf{L}'_{\mathbb{A}},\\
&T\cup\{\neg x\}\subseteq T'\subseteq T'',\\
&\forall y[y\in T''\to(\neg y)\notin T''],\\
&\forall y[y=\textstyle\bigwedge\Phi\in T'\to\forall\psi\in\Phi(\psi\in T'')],\\
&\forall y[y=\textstyle\bigvee\Phi\in T'\to\exists\psi\in\Phi(\psi\in T'')],\\
&\forall y[y=\forall v\varphi(v)\in T'\to\forall t(t \text{ a closed term of } \mathsf{L}'\to\varphi(t/v)\in T'')],\\
&\forall y[y=\exists v\varphi(v)\in T'\to\varphi(\mathsf{F}_{\exists v\varphi}/v)\in T''],\\
&\forall y,z[y,z \text{ closed terms of } \mathsf{L}'\wedge(y=z)\in T'\to(z=y)\in T''],\\
&\forall y,z,w[w=\varphi(v)\in \mathsf{L}'_{\mathbb{A}}\wedge y,z \text{ closed terms of } \mathsf{L}'\wedge\varphi(y/v)\in T'\\
&\qquad\wedge(z=y)\in T'\to\varphi(z/v)\in T''].
\end{aligned}
$$

If $x\notin X_1$ (i.e. $T\not\models x$) then $T\cup\{\neg x\}$ has a Skolem model $\mathfrak{M}$. Let T' be the set of sentences true in $\mathfrak{M}$. Then $\psi(x,T',T')$ holds so $\prec_\varphi^x$ is not well founded. Now suppose $\prec_\varphi^x$ is not well founded, so there is a sequence

$$\cdots\prec_\varphi^x T_{n+1}\prec_\varphi^x T_n\prec\cdots\prec_\varphi^x T_1\,.$$

Let $T_\omega=\bigcup_n T_n$. Then T satisfies all the conditions of Lemma VII.2.9, so T_ω has a model. But $T\cup\{\neg x\}\subseteq T_\omega$ so $T\not\models x$. Thus

$$x\in X_1 \quad\text{iff}\quad \prec_\varphi^x \text{ is well founded}\,. \quad \square$$

6.8—6.10 Exercises

6.8. Let $\mathfrak{M}$ be infinite. Show that if $P\subseteq\mathfrak{M}$ is Π_1^1 on $\mathfrak{M}$ then there is a first order formula $\varphi(v,\mathsf{R},\mathsf{S})$ such that

$$P(x) \text{ iff } \prec_\varphi^x \text{ is well founded}\,.$$

This is analogous to the normal form for Π_1^1 relations on $\mathcal{N}$.

6.9. Show that if α is the rank of some well-founded relation on Power($\mathbb{A}$) then $\alpha<(2^{\mathrm{card}(\mathbb{A})})^+$. Conclude that $h_\Sigma(\mathbb{A})<(2^{\mathrm{card}(\mathbb{A})})^+$.

6.10 (Open). Prove that $h_\Sigma(\mathbb{HYP}_{\mathfrak{M}})=\sup\{\xi:\xi$ is a first order implicit ordinal over $\mathfrak{M}\}$.

7. Trees and Σ_1 Compact Sets of Cofinality ω

The results of this chapter would be vacuous if there were no uncountable admissible sets sytisfying the König principles K_1—K_3. We exhibit such admissible sets in this and the next section.

A set $\mathbb{A}$ is *essentially uncountable* if every countable subset of $\mathbb{A}$ is an element of $\mathbb{A}$. All of the Σ_1 compact sets exhibited in this section have *cofinality* ω in the sense that

$$\mathbb{A} = \bigcup_{n<\omega} A_n$$

where each $A_n \in \mathbb{A}$. Hence none of them is essentially uncountable. We give a proof of the existence of essentially uncountable Σ_1 compact sets in the next section, though no explicit such sets are known. An explanation for this phenomenon will be found in § 9.

Let us return to our discussion of trees from § 1, and attempt to give a generalization of the Infinity Lemma soley in terms of trees.

In this section we turn the full binary tree around and think of it as pictured in Fig. 7A.

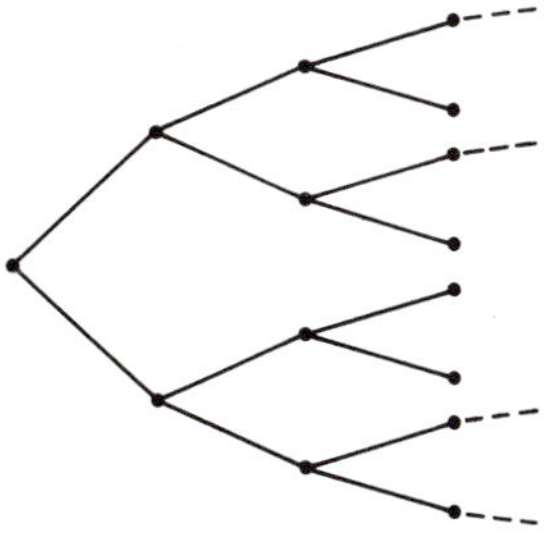

Fig. 7A. Another view of the full binary tree

Another tree, one with paths of length ω^2, is pictured in Fig. 7B.

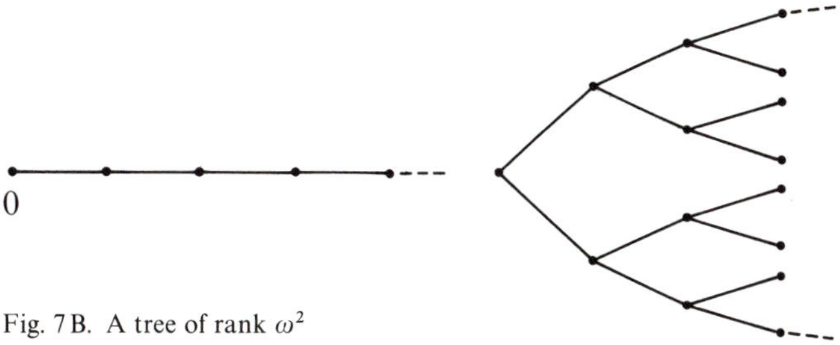

Fig. 7B. A tree of rank ω^2

In general, a *tree* is a well-founded partial ordering $\mathcal{T} = \langle T, \prec \rangle$, with a least element (usually denoted by 0), such that for each $x \in T$, the set $\{y \in T : y \prec x\}$

of predecessors of x is well ordered by $\prec$. A subset $C \subseteq T$ is a *chain* in $\mathcal{T}$ if for each $x, y \in C$,

$$x \prec y \quad \text{or} \quad x = y \quad \text{or} \quad y \prec x .$$

A *path thru* $\mathcal{T}$ is a maximal chain. Thus every path is well ordered by $\prec$.

Let $\mathcal{T} = \langle T, \prec \rangle$ be a tree. Since $\prec$ is well founded we have the usual rank function $\rho = \rho_\prec$ associated with $\mathcal{T}$:

$$\rho(x) = \sup\{\rho(y) + 1 : y \prec x\}$$

and $\mathcal{T}$ has a *rank* $\rho(\mathcal{T})$:

$$\begin{aligned}\rho(\mathcal{T}) &= \rho(\prec) \\ &= \sup\{\rho(x) + 1 : x \in T\} .\end{aligned}$$

A *branch* thru the tree $\mathcal{T}$ is a path of length $\rho(\mathcal{T})$. Not every tree has a branch, as Fig. 7C demonstrates.

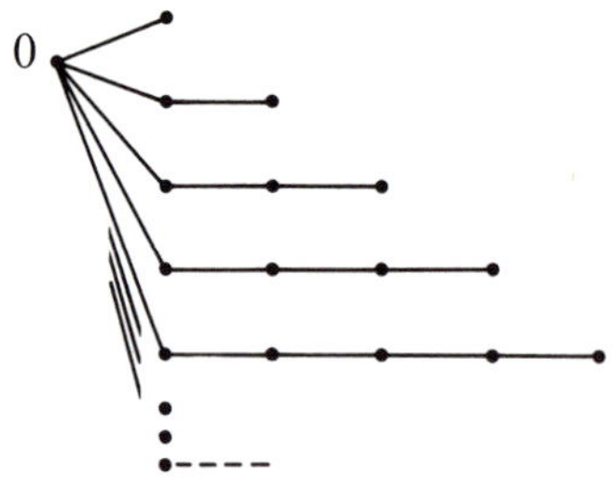

Fig. 7C. A tree with no branch

This tree has rank $\rho(\mathcal{T}) = \omega$ but every path is finite. Thus $\mathcal{T}$ has no branch. We call the elements x of a tree $\mathcal{T}$ with $\rho(x) = \beta$ the *nodes of level* β. Thus $\mathcal{T}$ has nodes of every level $\beta < \rho(\mathcal{T})$. Let $\text{lev} = \text{lev}_{\mathcal{T}}$ be the function with domain $\rho(\mathcal{T})$ defined by

$$\text{lev}(\beta) = \{x \in T : \rho(x) = \beta\} .$$

Let $\mathbb{A}$ be an admissible set. A tree $\mathcal{T} = \langle T, \prec \rangle$ is an $\mathbb{A}$-*tree* if $T \subseteq \mathbb{A}$, $T, \prec, \text{lev}_{\mathcal{T}}$ are $\mathbb{A}$-recursive and the rank $\rho(\mathcal{T})$ of $\mathcal{T}$ is $o(\mathbb{A})$. In particular, for each $\beta < o(\mathbb{A})$, T has nodes of level β (since $\rho(\mathcal{T}) = o(\mathbb{A})$) but the set of all nodes of level β is $\mathbb{A}$-finite (since $\text{lev}(\beta) \in \mathbb{A}$).

The König Infinity Lemma can be restated as follows. Let $\mathbb{A} = \langle \mathbb{HF}, \in, R \rangle$. Then every $\mathbb{A}$-tree has a branch.

7.1 Theorem. *If $\mathbb{A}$ is a Σ_1 compact admissible set then every $\mathbb{A}$-tree has a branch.*

Proof. It is easy to prove this by means of Σ_1 compactness, by constructing a nonstandard extension of the tree, picking a node d of nonstandard length and letting the branch be defined by

$$B = \{x \in T \mid x \prec d\}.$$

An even easier proof, though, is by means of $s-\Pi_1^1$ Reflection. Let $\mathscr{T} = \langle T, \prec \rangle$ be an $\mathbb{A}$-tree and suppose $\mathscr{T}$ has no branch, i.e. no path of length $\rho(\mathscr{T}) = o(\mathbb{A})$. Then $\mathbb{A}$ satisfies the $s-\Pi_1^1$ sentence:

$$\forall C[C \text{ is a chain} \rightarrow \exists \beta \, \forall x \in \text{lev}(\beta) \, (x \notin C)].$$

By $s-\Pi_1^1$ Reflection there is a $\gamma \in \mathbb{A}$ such that

(1) $$\forall C[C \text{ is a chain} \rightarrow \exists \beta < \gamma \, \forall x \in \text{lev}(\beta) \, (x \notin C)].$$

But then lev(γ) must be empty, for if $y \in \text{lev}(\gamma)$ then

$$C = \{x \in T \mid x \prec y\}$$

would violate (1). But then $\rho(\mathscr{T}) \leqslant \gamma < o(\mathbb{A})$, contradicting the definition of $\mathbb{A}$-tree. ☐

The hypothesis "every $\mathbb{A}$-tree has a branch" looks like it ought to be called a König principle. The next theorem shows that it is in fact too weak to be of general interest.

7.2 Theorem. *Let $\mathbb{A}$ be an admissible set whose ordinal $\alpha = o(\mathbb{A})$ has cofinality ω. Then every $\mathbb{A}$-tree has a branch.*

Proof. Since this is a direct generalization of the Infinity Lemma it is not surprising that the proof is an amplification of the proof of that lemma. Let $\alpha = \sup\{\alpha_n : n < \omega\}$ where $\alpha_0 < \alpha_1 < \cdots < \alpha_n < \cdots < \alpha$. Let $\mathscr{T} = \langle T, \prec \rangle$ be an $\mathbb{A}$-tree. We claim that we can find $x_0, x_1, \ldots$ such that $x_n \in \text{lev}(\alpha_n)$ and $x_0 \prec x_1 \prec \ldots$. If we can do this, then

$$B = \{y \in T : y \prec x_n \text{ for some } n\}$$

will be a branch thru $\mathscr{T}$. To find the x's, let $x_0 \in \text{lev}(\alpha_0)$ be such that

$$\forall \beta > \alpha_0 \, \exists z \in \text{lev}(\beta) \quad (x_0 \prec z).$$

(We must see that there is such an x_0.) Given x_0, let $x_1 \succ x_0$ be choosen so that $x_1 \in \text{lev}(\alpha_1)$ and

$$\forall \beta > \alpha_1 \, \exists z \in \text{lev}(\beta) \, [x_1 \prec z].$$

Continuing in this way gives the desired sequence of x_n's. Let us now prove that x_0 exists. (The proof that given x_n we can find x_{n+1} as above is almost identical.) Suppose there were no such x_0. Then

$$\forall x \in \mathrm{lev}(\alpha_0)\, \exists \beta > \alpha_0\, \forall z \in \mathrm{lev}(\beta)\, [x \not\prec z].$$

By Σ_1 Reflection there is a $\gamma \in \mathbb{A}$ such that

(2) $$\forall x \in \mathrm{lev}(\alpha_0)\, \exists \beta < \gamma\, (\alpha_0 < \beta \wedge \forall z \in \mathrm{lev}(\beta)\, [x \not\prec z].$$

Let $w \in \mathrm{lev}(\gamma)$. Now w has a predecessor x of level α_0 and a predecessor z_β for each $\beta < \gamma$. But then $x \prec z_\beta$, contradicting (2). □

Thus, e. g., $\mathbb{A} = H(\beth_{\omega+\omega})$ satisfies "Every $\mathbb{A}$-tree has a branch" but it does not satisfy s-Π_1^1 Reflection. Still, we did use the Infinity Lemma to prove s-Π_1^1 Reflection in § 1, so there should be some context in which the tree proof generalizes. If you analyze that proof you see that we also used two other facts: every subset of an $a \in \mathbb{HF}$ is in $\mathbb{HF}$ and, moreover, we can effectively find the set of all subsets of a. It easy to see that a pure admissible set $\mathbb{A}$ such that

$$b \subseteq a \in \mathbb{A} \Rightarrow b \in \mathbb{A}$$

(sometimes called *supertransitive*) must be of the form $H(\kappa)$ for some κ, so we restrict attention to $H(\kappa)$'s for the time being. In order for $H(\kappa)$ to be closed under $\mathscr{P}$, the power set, it is necessary and sufficient that κ be a strong limit cardinal ($\lambda < \kappa \Rightarrow 2^\lambda < \kappa$). Note that $H(\beth_{\omega+\omega})$ is closed under the power set but that

$$\langle H(\beth_{\omega+\omega}), \in, \mathscr{P} \rangle$$

is *not* admissible (for the same reason that $L(\omega+\omega)$ is not admissible). We write $\langle H(\kappa), \in, \mathscr{P} \rangle$ rather than the correct $\langle H(\kappa), \in, \mathscr{P} \cap H(\kappa)^2 \rangle$.

7.3 Theorem. *Let κ be a strong limit cardinal and suppose that $\mathbb{A} = \langle H(\kappa), \in, \mathscr{P}, R \rangle$ is admissible. Then $\mathbb{A}$ is Σ_1 compact iff every $\mathbb{A}$-tree has a branch.*

Proof. We have ($\Rightarrow$) by 7.1. To prove the converse we assume that every $\mathbb{A}$-tree has a branch and prove s-Π_1^1 Reflection. Since $H(\kappa)$ is closed under the power set and since $\langle H(\kappa), \in, \mathscr{P} \rangle$ is admissible, the usual definition by recursion of $\mathrm{V}(\alpha)$ shows that $\alpha \mapsto \mathrm{V}(\alpha)$ is an $\mathbb{A}$-recursive function of α. The usual "ZF-proof" that every set is in some $\mathrm{V}(\alpha)$ shows that $\mathrm{V}(\kappa) = H(\kappa)$. Let

(3) $$\forall \mathsf{S}\, \exists y\, \varphi(\mathsf{S}, x, y)$$

be a typical s-Π_1^1 formula *true in* $\mathbb{A}$, where φ is Δ_0 and we assume S is unary, for simplicity. Let $\xi = \mathrm{rk}(x)$ so that $x \in \mathrm{V}(\alpha)$ for all $\alpha > \xi$. We suppose that for each α, $\xi < \alpha < \kappa$,

$$\neg \forall S \subseteq \mathrm{V}(\alpha)\, \exists y \in \mathrm{V}(\alpha)\, \varphi(S, x, y)$$

and get a contradiction, thus establishing s-Π_1^1 Reflection. Thus we are assuming that for each $\alpha, \xi<\alpha<\kappa$

(4) $$\exists S \subseteq \mathrm{V}(\alpha)\, \forall y \in \mathrm{V}(\alpha) \neg\varphi(S,x,y).$$

We define a tree $\mathscr{T}=\langle T, \prec\rangle$ by

$$T=\{\langle\alpha,S\rangle : S\subseteq \mathrm{V}(\alpha) \wedge (\xi<\alpha \rightarrow \forall y \in \mathrm{V}(\alpha) \neg\varphi(S,x,y))\}$$

$$\langle\alpha,S\rangle \prec \langle\beta,S'\rangle \quad \text{iff} \quad \alpha<\beta \quad \text{and} \quad S=S'\cap \mathrm{V}(\alpha).$$

Each such S is an element of $H(\kappa)$, by supertransitivity. The least member of T is $\langle 0,0\rangle$. The predecessors of some $\langle\alpha,S\rangle\in T$ are just the pairs of the form $\langle\beta, S\cap \mathrm{V}(\beta)\rangle$ for $\beta<\alpha$ and hence have order type α under $\prec$. Thus the level of a pair $\langle\alpha,S\rangle$ is just α. Furthermore, since

$$\mathrm{lev}(\alpha)\subseteq \{\alpha\}\times\mathscr{P}(\mathrm{V}(\alpha)),$$

$\mathrm{lev}(\alpha)\in\mathbb{A}$ and, by the above comments, lev is $\mathbb{A}$-recursive. Line (4) says that $\mathrm{lev}(\alpha)\neq 0$ for all $\alpha<\kappa$. Thus $\mathscr{T}$ is an $\mathbb{A}$-tree. Let B be a branch thru $\mathscr{T}$, that is, a path of order type κ. B is a set of pairs

$$\langle\alpha,S_\alpha\rangle\in T,$$

exactly one pair for each $\alpha<\kappa$, linearly ordered by $\prec$. Furthermore, $\alpha<\beta$ implies $S_\alpha=S_\beta\cap \mathrm{V}(\alpha)$. Let $S=\bigcup_{\alpha<\kappa} S_\alpha$. Then

$$S_\alpha=S\cap \mathrm{V}(\alpha)$$

for each α. We claim that $(\mathbb{A},S)$ satisfies

$$\neg\exists y\, \varphi(S,x,y)$$

contradicting (3). For let $y\in\mathbb{A}$ be arbitrary. Pick $\alpha<\kappa$ such that $\xi<\alpha$ and $y\in \mathrm{V}(\alpha)$. Since $\langle\alpha,S_\alpha\rangle\in T$

$$\langle \mathrm{V}(\alpha),\in,\mathscr{P}\restriction \mathrm{V}(\alpha), R\restriction \mathrm{V}(\alpha), S_\alpha\rangle \models \neg\varphi(\mathsf{S},x,y).$$

But $(\mathbb{A},S)$ is an end extension of this structure so it also satsfies the Δ_0 formula $\neg\varphi(\mathsf{S},x,y)$, establishing our contradiction to (3). $\square$

A cardinal is said to be *(strongly) inaccessible* if κ is a regular strong limit cardinal. It follows from Theorem II.3.2 that if κ is inaccessible then

$$\langle H(\kappa),\in,\mathscr{P},R\rangle$$

is admissible for all $R \subseteq H(\kappa)$, so that Theorem 7.3 applies. A simple Löwenheim-Skolem argument shows that one can find $\lambda < \kappa$ such that

$$\langle H(\lambda), \in, \mathscr{P}, R \upharpoonright H(\lambda) \rangle$$

is admissible and $\mathrm{cf}(\lambda) = \omega$. Alternatively, one can drop all talk of inaccessibles and prove directly (using the reflection theorem of Lévy) that for any definable class R there are cardinals λ with $\mathrm{cf}(\lambda) = \omega$ such that $\langle H(\lambda), \in, \mathscr{P}, R \upharpoonright H(\lambda) \rangle$ is admissible. Thus the hypothesis of the next theorem is not vacuous. It is this theorem which has been the aim of the first part of this section.

7.4 Theorem. *Let κ be a strong limit cardinal of cofinality ω and assume that $\mathbb{A} = \langle H(\kappa), \in, \mathscr{P}, R \rangle$ is admissible. Then $\mathbb{A}$ is Σ_1 compact.*

Proof. This is immediate from 7.2 and 7.3. ▯

Exercise 7.10 shows that $\mathbb{A}$ is also Σ_1 complete. Exercise 7.11 shows that it satisfies K_3.

The urelement versions of 7.3 and 7.4 are not very interesting since 7.3 only goes through for $\langle \mathfrak{M}; H(\kappa)_{\mathfrak{M}}, \in, \mathscr{P} \rangle$ when $\mathrm{card}(\mathfrak{M}) < \kappa$, in which case $\mathfrak{M}$ is already contained in $H(\kappa)$, up to isomorphism.

We will return briefly to the notion of tree in § 9. Now we go on to discuss two rather different examples of Σ_1 compact admissible sets.

The following theorem of Nyberg gives quite concrete examples of Σ_1 compact and Σ_1 complete admissible sets.

7.5 Theorem. *Let α be a limit ordinal of cofinality ω, let $\mathbb{A}$ be of the form $\langle H(\beth_\alpha), \in, R \rangle$ and let $\mathbb{B}$ be admissible with $\mathbb{A} \in \mathbb{B}$, $\mathbb{B}$ projectible into $\mathbb{A}$. Then $\mathbb{B}$ satisfies s-$\Pi_1^1 = \Sigma_1$ uniformly and hence is Σ_1 complete and Σ_1 compact. ($\mathbb{A}$ is not necessarily admissible.)*

The proof of this theorem is sketched in Exercise 7.16. Note that it applies to $\mathbb{HYP}(H(\beth_\alpha))$ whenever $\mathrm{cf}(\alpha) = \omega$. This is a resolvable admissible set satisfying s-$\Pi_1^1 = \Sigma_1$ uniformly (and hence the theorems of the previous sections). On the other hand, if $\mathrm{cf}(\alpha) > \omega$ then $\mathbb{HYP}(H(\beth_\alpha))$ is strongly self-definable, hence not Σ_1 complete or Σ_1 compact.

We conclude this section with a different kind of example.

A structure $\mathfrak{M} = \langle M, R_1, \ldots, R_l \rangle$ is *recursively Σ_1^1 saturated* iff for every finite expansion $\mathsf{L}' = \mathsf{L}(\mathsf{S}_1, \ldots, \mathsf{S}_k)$ of L and every recursive (equivalently, r.e.) set $\Phi(v_1, \ldots, v_n, \mathsf{S}_1, \ldots, \mathsf{S}_k)$ of formulas of $\mathsf{L}'_{\omega\omega}$, $\mathfrak{M}$ is a model of:

$$\forall \vec{v} \left[\bigwedge\nolimits_{\Phi_0 \in S_\omega(\Phi)} \exists \mathsf{S}_1, \ldots, \mathsf{S}_k \bigwedge \Phi_0(\vec{v}, \vec{\mathsf{S}}) \rightarrow \exists \mathsf{S}_1, \ldots, \mathsf{S}_k \bigwedge \Phi(\vec{v}, \vec{\mathsf{S}}) \right].$$

It is easy to see that every recursively Σ_1^1 saturated structure is recursively saturated. Theorem IV.5.7 shows that if $\mathfrak{M}$ is countable and recursively saturated then it is recursively Σ_1^1 saturated. The following theorem characterizes the recursively Σ_1^1 saturated structures among the class of recursively saturated structures.

7.6 Theorem. *Let $\mathfrak{M}=\langle M,R_1,\ldots,R_l\rangle$ be an infinite recursively saturated structure of any cardinality. Then $\mathfrak{M}$ is recursively Σ_1^1 saturated iff $\mathbb{HYP}_{\mathfrak{M}}$ is Σ_1 compact.*

Proof. The "if" part of the theorem is established by the very proof of Theorem IV.5.7. To prove the converse, we assume that $\mathfrak{M}$ is recursively Σ_1^1 saturated and prove that $\mathbb{HYP}_{\mathfrak{M}}$ satisfies s-Π_1^1 Reflection. Suppose

(5) $\qquad \mathbb{HYP}_{\mathfrak{M}}\models\forall \mathsf{R}\ \varphi(\mathsf{R},\vec{x})$

where φ is a Σ_1 formula. Since $\mathbb{HYP}_{\mathfrak{M}}=\mathrm{L}(\mathfrak{M},\omega)$, we need to exclude the possibility that for every $n<\omega$

$(6)_n$ $\qquad \mathrm{L}(\mathfrak{M},n)\models\exists \mathsf{R}\,\neg\varphi(\mathsf{R},\vec{x})$.

Since each $x\in\mathbb{HYP}_{\mathfrak{M}}$ has a good Σ_1 definition in terms of parameters from $M\cup\{M\}$, we may assume that each x_i in the sequence x is either in M or is M itself. Let us rewrite (5) as

(7) $\qquad \mathrm{L}(\mathfrak{M},\omega)\models\forall \mathsf{R}\ \exists y\ \psi(\mathsf{R},p,y,M)$

where ψ is a Δ_0 formula with no other parameters. We can rewrite (6) as: for every $n<\omega$

$(8)_n$ $\qquad \mathrm{L}(\mathfrak{M},\omega)\models\exists \mathsf{R}\ \forall y\in \mathrm{L}(M,n)\,\neg\psi(\mathsf{R},p,y,M)$.

Let $\Phi(p)$ be the set of formulas in $\mathsf{L}(\mathsf{U},\mathsf{A},\mathsf{E},\mathsf{F},\mathsf{R},\mathsf{R}_1',\ldots,\mathsf{R}_l')$ which express the following about $\mathfrak{M}$, p:

KPU$^+$ relativized to U (for urelements), S (for sets), E (for $\in$),

"$\mathsf{F}:\langle M,\mathsf{R}_1,\ldots,\mathsf{R}_l\rangle\cong\langle\mathsf{U},\mathsf{R}_1',\ldots,\mathsf{R}_l'\rangle$",

$\forall y\in \mathrm{L}(\mathsf{U},n)\,\neg\psi(\mathsf{R},\mathsf{F}(p),y,\mathsf{U})$ (for all $n<\omega$).

Every finite subset $\Phi_0(p)$ of $\Phi(p)$ is satisfiable on $\mathfrak{M}$ by choosing relations which code up $\mathbb{HYP}_{\mathfrak{M}}$ on $\mathfrak{M}$ itself and using $(8)_n$ to satisfy the last sentence in $\Phi_0(p)$. Since $\mathfrak{M}$ is recursively Σ_1^1 saturated there are relations on $\mathfrak{M}$ which make the whole set $\Phi(p)$ true:

(9) $\qquad \langle M,R_1,\ldots,R_l,\,U,R_1',\ldots,R_l',\,F,A,E,R\rangle\models\Phi(p)$.

But then $(\langle U,R_1',\ldots,R'\rangle;\,A,E,R)$ is isomorphic to $(\mathfrak{A}_{\mathfrak{M}},R)$ for some $\mathfrak{A}_{\mathfrak{M}}\supseteq_{\mathrm{end}}\mathbb{HYP}_{\mathfrak{M}}$.

Let $R^* = R \upharpoonright \mathbb{HYP}_{\mathfrak{M}}$. By (7) there is a $y \in \mathbb{HYP}_{\mathfrak{M}}$ such that

$$(\mathrm{L}(\mathfrak{M}, \omega), R^*) \models \psi(R, p, y, M).$$

But $y \in \mathrm{L}(M, n)$ for some $n < \omega$. Since ψ is Δ_0 and

$$(\mathbb{HYP}_{\mathfrak{M}}, R^*) \subseteq_{\text{end}} (\mathfrak{A}_{\mathfrak{M}}, R),$$

we have

$$(\mathfrak{A}_{\mathfrak{M}}, R) \models y \in \mathrm{L}(M, n) \wedge \psi(\mathsf{R}, p, y, M)$$

contradicting (9), since (9) asserts, among other things, that

$$(\mathfrak{A}_{\mathfrak{M}}, R) \models \forall y \in \mathrm{L}(M, n) \neg \psi(\mathsf{R}, p, y, M). \quad \square$$

To see that this result gives us lots of uncountable Σ_1 compact sets, we must know that there are lots of recursively Σ_1^1 saturated models. We assume the reader is familiar with saturated or special models, referring him to Chang-Keisler [1973] for the relevant definitions and properties.

7.7 Proposition. *Every saturated (or even every special) model* $\mathfrak{M} = \langle M, R_1, \ldots, R_l \rangle$ *is recursively* Σ_1^1 *saturated.*

Proof. If we assume the GCH we can get rid of the requirement that the set of formulas is recursive; the proof not involving the GCH is sketched in Exercise 7.17. Let $\mathfrak{M}$ be a special model and let $\Phi(\overline{p}, S)$ be a set of sentences such that for each finite $\Phi_0 \subseteq \Phi$,

$$(\mathfrak{M}, p) \models \exists \mathsf{S} \bigwedge \Phi_0(\overline{\mathsf{p}}, \mathsf{S}).$$

Then the first order theory $Th(\mathfrak{M}, p) \cup \Phi(\overline{\mathsf{p}}, \mathsf{S})$ is consistent and so has a special model $(\mathfrak{M}', p', S')$ of power $\mathrm{card}(\mathfrak{M})$, by the GCH.

But then $(\mathfrak{M}, p) \equiv (\mathfrak{M}', p')$ $(\mathsf{L}_{\omega\omega})$, and both models are special so

$$(\mathfrak{M}, p) \cong (\mathfrak{M}', p').$$

Hence

$$(\mathfrak{M}, p) \models \exists \mathsf{S} \bigwedge \Phi(\overline{p}, \mathsf{S}). \quad \square$$

7.8—7.19 Exercises

7.8. Prove that the pure admissible set $\mathbb{A}$ is supertransitive iff $\mathbb{A} = H(\kappa)$ for some cardinal κ.

7.9. Prove the following: *Let* $\mathbb{A}$ *be pure, admissible, supertransitive and* Σ_1 *compact. There is a cardinal* $\kappa = \beth_\kappa$ *such that* $\mathbb{A} = H(\kappa)$. *Let* $\mathbb{A}' = (\mathbb{A}, \mathscr{P})$. *Then* $\mathbb{A}'$ *is admissible and satisfies* s-$\Pi_1^1 = \Sigma_1$ *uniformly.* More slowly, prove:

(i) $\mathbb{A}$ is closed under $\mathscr{P}$, using s-Π_1^1 Reflection.
(ii) $\mathbb{A}'$ satisfies s-Π_1^1 Reflection ($\mathscr{P}$ is s-Δ_1^1 on $\mathbb{A}$).
(iii) $\mathbb{A}'$ is admissible (using (ii)).
(iv) $\mathbb{A}'$ satisfies s-$\Pi_1^1 = \Sigma_1$.

7.10. Prove that the following are equivalent, where κ is a strong limit cardinal and $\mathbb{A} = \langle H(\kappa), \in, \mathscr{P}, R\rangle$ is admissible:
(i) $\mathbb{A}$ is Σ_1 compact (s-Π_1^1 Reflection),
(ii) $\mathbb{A}$ is Σ_1 complete (s-$\Pi_1^1 = \Sigma_1$),
(iii) Every $\mathbb{A}$-tree has a branch.

7.11. Let $\mathbb{A} = \langle H(\kappa), \in, R\rangle$ be Σ_1 compact. Prove that $h_\Sigma(\mathbb{A}) = \kappa$. [Use 7.9 and s-Π_1^1 Reflection plus trivial cardinality considerations.]

7.12. Let $\lambda = \mathrm{card}(\mathfrak{M})$ and let κ be a limit ordinal. Prove that the following are equivalent:
(i) $(\mathfrak{M}; \mathrm{V}_{\mathfrak{M}}(\kappa), \in)$ is admissible,
(ii) $\kappa = \beth_\kappa(\lambda)$,
(iii) κ is a cardinal and $\mathrm{V}_{\mathfrak{M}}(\kappa) = H(\kappa)_{\mathfrak{M}}$.

7.13. Prove in ZFC that there are arbitrarily large cardinals $\kappa = \beth_\kappa$ of cofinality ω such that $\langle H(\kappa), \in, \mathscr{P}\rangle$ is admissible.

7.14. Let κ be the Hanf number of second order logic. Show that $\langle H(\kappa), \in, \mathscr{P}\rangle$ satisfies the hypothesis of 7.4.

7.15. Let α be a limit ordinal, let $\mathbb{A}$ be admissible and let $\mathrm{V}(\alpha) \in \mathbb{A}$. Prove that $H(\beth_\alpha) \in \mathbb{A}$. [Consider the set $X = \{E \in \mathrm{V}(\alpha)\}$: E is well-founded$\}$.]

7.16. Theorem 7.5 follows from the following result of Nyberg. Prove that if $\mathfrak{M}$ is a uniform Kleene structure and $\mathbb{A}_{\mathfrak{M}}$ is admissible above $\mathfrak{M}$ and projectible into $\mathfrak{M}$ then $\mathbb{A}_{\mathfrak{M}}$ satisfies s-$\Pi_1^1 = \Sigma_1$ uniformly. [Use the alternate form of "s-$\Pi_1^1 = \Sigma_1$ uniformly" given in Exercise 4.21.]

7.17. A structure $\mathfrak{M} = \langle M, R_1, \ldots, R_l\rangle$ is resplendent if for every finitary Σ_1^1 sentence $\exists \mathsf{S}\, \varphi(\mathsf{S})$ with constants from $\mathfrak{M}$, if $\mathfrak{N} \models \exists \mathsf{S}\, \varphi(\mathsf{S})$ for some $\mathfrak{N} \succ \mathfrak{M}$, then $\mathfrak{M} \models \exists \mathsf{S}\, \varphi(\mathsf{S})$.

(i) Prove that every special model is ω-resplendent (Kueker [1971]).

(ii) Prove that every resplendent model is recursively Σ_1^1 saturated. [Use the techniques of IV.2.]

(iii) Associate with any finitary Σ_1^1 formula $\Phi(x)$ a recursive closed game formula $\mathscr{G}_\Phi(x)$ such that

$$\mathfrak{M} \models \forall x\, (\Phi(x) \rightarrow \mathscr{G}_\Phi(x))$$

for all $\mathfrak{M}$ and, for $\mathfrak{M}$ countable,

$$\mathfrak{M} \models \forall x\,(\mathscr{G}_\Phi(x) \leftrightarrow \Phi(x)). \tag{10}$$

Such a $\mathscr{G}_\Phi$ is given by (the proof of) Svenonius's Theorem. Prove that if $\mathfrak{M}$ is recursively saturated then $\mathfrak{M}$ is resplendent iff

$$\mathfrak{M} \models \forall x\,(\mathscr{G}_\Phi(x) \leftrightarrow \Phi(x))$$

for all Σ_1^1 formulas Φ.

(iv) Prove that if $\mathfrak{M}$ is resplendent then Π_1^1 on $\mathfrak{M} = \Sigma_1$ on $\mathbb{HYP}_{\mathfrak{M}}$.

(v) (Schlipf). Improve (iv) by showing that if $\mathfrak{M}$ is resplendent then $\mathbb{HYP}_{\mathfrak{M}}$ satisfies K_1.

7.18. (Open). Characterize those $\mathfrak{M}$ such that $\mathbb{HYP}_{\mathfrak{M}}$ is Σ_1 compact.

7.19. (Open). Characterize those $\mathfrak{M}$ such that $\mathbb{HYP}_{\mathfrak{M}}$ is Σ_1 complete.

7.20 Notes. Theorem 7.4 is due to Barwise [1968] and, independently, and by a completely different proof, to Karp [1968]. Theorem 7.3 is a refinement of a classical result about weakly compact cardinals, contained in Theorem 9.10.

8. Σ_1 *Compact Sets of Cofinality Greater than* ω

In this section we prove an existence theorem which shows that there are many Σ_1 compact admissible sets besides those exhibited in the previous section. In particular, we prove the existence of essentially uncountable Σ_1 compact admissible sets.

Let κ be an uncountable regular cardinal. A subset C of κ is *closed in* κ if for each initial segment C_0 of C,

$$(\sup C_0) < \kappa \quad \text{implies} \quad (\sup C_0) \in C.$$

This says that C is closed in the order topology on κ. C is *unbounded in* κ if

$$\forall \beta < \kappa\, \exists \gamma \in C\, (\beta < \gamma).$$

A set C is *c.u.b. in* κ if $C \subseteq \kappa$ and C is closed and unbounded in κ.

8.1 Lemma. *Let* $\kappa > \omega$ *be regular. If* C_0, C_1 *are c.u.b. in* κ *then so is* $C_0 \cap C_1$. *In particular,* $C_0 \cap C_1$ *is nonempty.*

Proof. The intersection $C_0 \cap C_1$ is closed since the intersection of closed sets is closed. To see that $C_0 \cap C_1$ is unbounded, let $\beta < \kappa$ be given. Let $\gamma_1 > \beta$ be in C_1. Let $\gamma_2 > \gamma_1$ be in C_0. Let $\gamma_3 > \gamma_2$ be in C_1 and so on for each $n < \omega$. Then $\gamma = \sup_{n<\omega} \gamma_n$ is less than κ since κ is regular. Since $\gamma = \sup_{n<\omega} \gamma_{2n}$ and C_0 is closed, $\gamma \in C_0$. Since $\gamma = \sup_{n<\omega} \gamma_{2n+1}$ and C_1 is closed, $\gamma \in C_1$. Thus $\beta < \gamma$ and $\gamma \in (C_0 \cap C_1)$. □

Thus, by 8.1, the collection

$$\mathfrak{F} = \{C \subseteq \kappa : C_0 \subseteq C \text{ for some } C_0 \text{ c.u.b. in } \kappa\}$$

defines a filter on the subsets of κ, called the c.u.b. filter on κ. We say that $P(\alpha)$ *holds for almost all* $\alpha < \kappa$ if

$$\{\alpha < \kappa : P(\alpha)\}$$

is a member of the c.u.b. filter on κ.

8.2 Lemma. *Let λ, κ be regular cardinals, $\omega \leqslant \lambda < \kappa$. If $P(\alpha)$ holds for almost all $\alpha < \kappa$ then $P(\alpha)$ holds for some α with $\mathrm{cf}(\alpha) = \lambda$.*

Proof. Let C be c.u.b. in κ be a subset of

$$\{\alpha < \kappa : P(\alpha)\}.$$

Let γ be the λth member of C, enumerated in the natural order. There is such a λth member since κ is regular and C is unbounded in κ. It is clear that $\mathrm{cf}(\gamma) = \mathrm{cf}(\lambda) = \lambda$ since λ is regular. □

In reading the next theorem, the student should think of J_α as $H(\aleph_\alpha)$ or $\mathrm{L}(\omega\alpha)$ or $\mathrm{L}(a, \omega\alpha)$, since these are the usual applications.

8.3 Theorem. *Let κ be an uncountable regular cardinal, let $R \subseteq H(\kappa)$ and let $J: \kappa \to H(\kappa)$ have the following properties:*

(i) *J_α is transitive and closed under pairs and union, for all $\alpha < \kappa$;*
(ii) *$\alpha < \beta < \kappa$ implies $J_\alpha \in J_\beta$;*
(iii) *if $\lambda < \kappa$ is a limit ordinal then $J_\lambda = \bigcup_{\alpha<\lambda} J_\alpha$; and*
(iv) *for each $\alpha < \kappa$, the structure*

$$\mathbb{J}_\alpha = \langle J_\alpha, \in, R \cap J_\alpha \rangle$$

satisfies Δ_0 Separation. Then, for almost all $\alpha < \kappa$, $\mathbb{J}_\alpha$ is a Σ_1 compact admissible set.

Proof. The idea for this proof goes back to the notion of stable ordinal. For the purposes of this proof we call an ordinal α *β-superstable* if $\alpha < \beta < \kappa$ and for every s-Π_1^1 formula $\Phi(v_1, \ldots, v_n)$ and every $a_1, \ldots, a_n \in J_\alpha$,

$$\text{if } \mathbb{J}_\beta \models \Phi(a_1, \ldots, a_n) \text{ then } \mathbb{J}_\alpha \models \Phi(a_1, \ldots, a_n).$$

We first prove:

(1) if α is β-superstable then $\mathbb{J}_\alpha$ is a Σ_1 compact admissible set.

So suppose α is β-superstable. Since Δ_0 Collection follows from s-Π_1^1 Reflection (in fact from Σ Reflection) it suffices to prove that $\mathbb{J}_\alpha$ satisfies s-Π_1^1 Reflection. Let $\Phi(a_1,\ldots,a_n)$ be a s-Π_1^1 formula true in J_α.

Then

$$\mathbb{J}_\beta \models \Phi^{(J_\alpha)}(a_1,\ldots,a_n)$$

and hence $\mathbb{J}_\beta$ is a model of the s-Π_1^1 formula $\Psi(a_1,\ldots,a_n)$

$$\exists b\,[\mathrm{Tran}(b) \wedge a_1,\ldots,a_n \in b \wedge \Phi^{(b)}(a_1,\ldots,a_n)]$$

since $J_\alpha \in J_\beta$. But then by superstability, $\mathbb{J}_\alpha \models \Psi(a_1,\ldots,a_n)$, so $\mathbb{J}_\alpha$ satisfies s-Π_1^1 Reflection, proving (1).

We will prove the theorem by proving that almost every $\alpha<\kappa$ is β-superstable for every $\beta, \alpha<\beta<\kappa$. To prove this we use normal functions. (A function $f:\kappa\to\kappa$ is *normal* if f is increasing ($\alpha<\beta<\kappa \Rightarrow f(\alpha)<f(\beta)$) and continuous ($\lambda$ a limit $<\kappa \Rightarrow f(\lambda)=\sup\{f(\alpha): \alpha<\lambda\}$). If $f:\kappa\to\kappa$ is normal then the set of fixed points of f,

$$\{\alpha<\kappa: f(\alpha)=\alpha\},$$

is always c.u.b. in κ, as is easily seen.) We define a normal function f such that $f(\alpha)=\alpha$ implies α is β-superstable for all β between α and κ. This will prove the theorem. Let $P(\alpha,\beta)$ be the following condition on $\alpha,\beta<\kappa$:

for all $\beta', \beta\leqslant\beta'<\kappa$, and for all s-Π_1^1 sentences $\Phi(a_1,\ldots,a_n)$ with constants from J_α, if $\mathbb{J}_{\beta'} \models \Phi(a_1,\ldots,a_n)$ then $\mathbb{J}_\beta \models \Phi(a_1,\ldots,a_n)$.

Note that $P(\alpha,\beta_0)$ implies $P(\alpha,\beta_1)$ for all β_1 between β_0 and κ. Since $\mathrm{card}(J_\alpha)<\kappa$ there are $<\kappa$ s-Π_1^1 formulas $\Phi(\vec{a})$ so a trivial cardinality argument proves that $\forall\alpha<\kappa\,\exists\beta<\kappa\,P(\alpha,\beta)$. Now define f by

$$f(\alpha)=\text{least}\,\beta\,[\beta>f(\gamma) \text{ for all } \gamma<\alpha, \text{ and } P(\alpha,\beta)].$$

Since κ is regular, $f(\alpha)$ is defined for all $\alpha<\kappa$. Thus $f:\kappa\to\kappa$ and f is increasing by definition. Let us prove that f is continuous. Let $\lambda<\kappa$ be a limit ordinal. Let $\beta=\sup\{f(\alpha): \alpha<\lambda\}$. We need to verify $P(\lambda,\beta)$. Thus let $\beta'\geqslant\beta$ and let Φ be a s-Π_1^1 sentence with parameters from J_λ which is true in $J_{\beta'}$. We must see that Φ is true in J_β. But Φ is defined in J_α for some $\alpha<\lambda$ so Φ is true in $J_{f(\alpha)}$ and hence in J_β by persistence of s-Π_1^1 formulas. Thus f is normal.

Now suppose $f(\alpha)=\alpha$. Then $P(\alpha,\alpha)$ holds so α is β'-stable for all $\beta'>\alpha$, $\beta'<\kappa$. By (1) this shows that almost every $\alpha<\kappa$ has $\mathbb{J}_\alpha$ Σ_1 compact. ☐

8.4 Corollary. *Let $\kappa > \omega$ be regular. Then for almost all $\alpha < \kappa$, $L(\alpha)$ is a Σ_1 compact admissible set.*

Proof. Apply 8.3 with $J_\alpha = L(\omega\alpha)$. Then for almost all $\alpha < \kappa$, $L(\omega\alpha)$ is Σ_1 compact. But $\omega\alpha = \alpha$ for almost all $\alpha < \kappa$ since $f(\alpha) = \omega\alpha$ is a normal function. ▯

8.5 Corollary. *Let $\kappa > \omega$ be regular and let $\mathfrak{M} = \langle M, R_1, \ldots, R_l \rangle$ be a structure of power less than κ. Then for almost all $\alpha < \kappa$, $L(\mathfrak{M}, \alpha)$ is Σ_1 compact.*

Proof. Similar to 8.4. Since there is isomorphic copy of $\mathfrak{M}$ in $H(\kappa)$. ▯

The next result gives us essentially uncountable Σ_1 compact admissible sets, when one applies Lemma 8.2 and the observation that $H(\kappa)$ is essentially uncountable iff $\mathrm{cf}(\kappa) > \omega$. $\mathscr{P}$ denotes the power set operation (restricted to $H(\lambda)$ in 8.6).

8.6 Theorem. *Let κ be inaccessible, $\kappa > \omega$. Let $R \subseteq H(\kappa)$. Then for almost all $\lambda < \kappa$, $\langle H(\lambda), \in, \mathscr{P}, R \cap H(\lambda) \rangle$ is Σ_1 compact.*

Proof. Let $J_\alpha = H(\beth_\alpha)$. Then $J: \kappa \to H(\kappa)$ since $\kappa = \beth_\kappa$, and $\mathrm{card}(H(\beth_\alpha) \leqslant \beth_{\alpha+1} < \kappa$. Thus, for almost all $\alpha < \kappa$, $\langle H(\beth_\alpha), \in, \mathscr{P}, R \cap H(\beth_\alpha) \rangle$ is Σ_1 compact. But $f(\alpha) = \beth_\alpha$ is a normal function so almost all $\alpha < \kappa$ have $\beth_\alpha = \alpha$. Thus almost all $\lambda < \kappa$ have

$$\langle H(\lambda), \in, \mathscr{P}, R \cap H(\lambda) \rangle$$

Σ_1 compact. ▯

We can reinterpret all of the above by thinking of the class of all ordinals as an inaccessible cardinal. We can restate Theorem 8.6 in this case as a result in ZFC.

8.7 Corollary. *Let R be any class. The class of λ such that $\langle H(\lambda), \in, \mathscr{P}, R \cap H(\lambda))$ is Σ_1 compact contains a closed proper class of cardinals. Hence for any regular κ there are arbitrarily large such λ's of cofinality κ.*

Proof. The last sentence follows from 8.2. ▯

A cardinal κ is a *Mahlo cardinal* if every c.u.b. set $C \subseteq \kappa$ contains an inaccessible cardinal (and hence contains κ such inaccessible cardinals $\lambda < \kappa$).

8.8 Corollary. *Let κ be a Mahlo cardinal and let $R \subseteq H(\kappa)$. There are κ inaccessible cardinals $\lambda < \kappa$ such that $\langle H(\lambda), \in, \mathscr{P}, R \cap H(\lambda) \rangle$ is Σ_1 compact.*

Proof. Immediate from 8.6. ▯

8.9 Exercise. Suppose $\langle H(\kappa), \in \rangle$ is Σ_1 compact. Prove that κ is not the first inaccessible. Prove, in fact, that if κ is inaccessible then κ is the κth inaccessible. [Use s-Π_1^1 Reflection.]

8.10 Notes. Theorem 8.3 is contained in Barwise [1969b].

9. Weakly Compact Cardinals

In this final section we consider weakly compact cardinals and their relationship to Σ_1 compact admissible sets.

Let L be a language with $\leqslant\kappa$ symbols coded as a Δ_1 subset of $H(\kappa)$. The language $\mathsf{L}_{\kappa\omega}$ consists of those $\varphi\in\mathsf{L}_{\infty\omega}$ with less than κ subformulas.

9.1 Definition. A cardinal $\kappa\geqslant\omega$ is *weakly compact* (*for* $\mathsf{L}_{\kappa\omega}$) if for every set $T\subseteq H(\kappa)$ of sentences of $\mathsf{L}_{\kappa\omega}$, if every subset $T_0\subseteq T$ of power $<\kappa$ has a model then T has a model.

This definition is usually expressed in terms of a stronger language $\mathsf{L}_{\kappa\kappa}$ (defined in Exercise 9.14) and it is usually assumed that κ is inaccessible in which case $H(\kappa)$ has power κ and hence T has power $\leqslant\kappa$. We will see that both of these apparent strengthenings follow from Definition 9.1. Note that ω is weakly compact.

9.2 Lemma. *Let* $\kappa\geqslant\omega$ *be a cardinal.*

(i) $\mathsf{L}_{\kappa\omega}=\mathsf{L}_{\infty\omega}\cap H(\kappa)$.

(ii) *If* κ *is regular then* $\mathsf{L}_{\kappa\omega}$ *is the least subset of* $\mathsf{L}_{\infty\omega}$ *containing* $\mathsf{L}_{\omega\omega}$ *closed under* $\neg$, $\forall$, $\exists$ *and*

$$\text{if}\quad \Phi\subseteq\mathsf{L}_{\kappa\omega}\quad\text{and}\quad \mathrm{card}(\Phi)<\kappa\quad\text{then}\quad \bigwedge\Phi\quad\text{and}\quad\bigvee\Phi\in\mathsf{L}_{\kappa\omega}.$$

(iii) *If* $\kappa>\omega$ *is a limit cardinal then*

$$\mathsf{L}_{\kappa\omega}=\bigcup\nolimits_{\lambda<\kappa}\mathsf{L}_{\lambda\omega}$$

where the union is over all infinite cardinals $\lambda<\kappa$.

(iv) κ *is weakly compact iff* $\langle H(\kappa),\in,R\rangle$ *is* Σ_1 *compact for every relation* $R\subseteq H(\kappa)$.

Proof. (i), (iii) and (iv) are immediate from the definitions. To prove (ii) let $\mathsf{L}'_{\kappa\omega}$ be the least class described. It is clear that $\mathsf{L}_{\kappa\omega}\subseteq\mathsf{L}'_{\kappa\omega}$. To prove $\mathsf{L}_{\kappa\omega}=\mathsf{L}'_{\kappa\omega}$ it suffices to prove that $\mathsf{L}_{\kappa\omega}$ is closed under $\neg$, $\forall$, $\exists$ and the clause

$$\text{if}\quad \Phi\subseteq\mathsf{L}_{\kappa\omega}\quad\text{and}\quad \mathrm{card}(\Phi)<\kappa\quad\text{then}\quad \bigwedge\Phi,\ \bigvee\Phi\in\mathsf{L}_{\kappa\omega}.$$

The first part is trivial. So suppose $\Phi\subseteq\mathsf{L}_{\kappa\omega}$ and $\mathrm{card}(\Phi)<\kappa$. We must verify that

$$\mathrm{card}(\mathrm{sub}(\textstyle\bigwedge\Phi))<\kappa.$$

But

$$\mathrm{sub}(\textstyle\bigwedge\Phi)=\{\bigwedge\Phi\}\cup\bigcup\{\mathrm{sub}(\varphi)\colon\varphi\in\Phi\}.$$

Since $\Phi\subseteq\mathsf{L}_{\kappa\omega}$ each $\mathrm{sub}(\varphi)$ has power $<\kappa$ for $\varphi\in\Phi$. But $\mathrm{card}(\Phi)<\kappa$ and κ is regular so $\mathrm{card}(\bigwedge\Phi)<\kappa$. Similarly, $\mathrm{card}(\bigvee\Phi)<\kappa$. □

Part (iv) of this lemma shows that the notion of weakly compact cardinal is just the relativization of the concept of Σ_1 compact admissible set to an arbitrary $R \subseteq H(\kappa)$.

Before we see just how strong the assumption that κ is weakly compact and uncountable is, let us stop to examine the plausibility of the existence of such cardinals. We want to show that the same kind of intuition which prompts one to admit ω, inaccessible cardinals and Mahlo cardinals as legitimate abstract objects also prompts one to admit weakly compact cardinals as legitimate objects in the hierarchy of sets.

There was a time when the existence of ω was considered problematic. One must accept each natural number, but it took years for the limit, the set of natural numbers, to be accepted as a legitimate abstract object, suitable for use in mathematics.

Once one accepts the basic principles of set theory, one sees how to generate many cardinal numbers, which must be accepted. Only fairly recently have inaccessible cardinals begun to be considered as the natural limit of the accessible cardinals and hence suitable for use in mathematics.

We saw in Corollary 8.7 that for any class R, almost all cardinals κ have the property that $\langle H(\kappa), \in, R \cap H(\kappa)\rangle$ is Σ_1 compact. Given any collection $\mathscr{R}$ of classes that can be coded by a single class, we see that almost all κ are such that $\langle H(\kappa), \in, R \cap H(\kappa)\rangle$ is Σ_1 compact for all $R \in \mathscr{R}$. A natural limiting assumption is that $\langle H(\kappa), \in, R\rangle$ should by Σ_1 compact for *all* $R \subseteq H(\kappa)$. This is the assumption that κ is weakly compact.

(Another argument that is often given for the existence of weakly compact cardinals, as well as measurable cardinals and strongly compact cardinals, cardinals we can see no argument for at all, is that they should exist "by analogy with ω". This seems like a very weak argument. The results of § 7 suggest that the crucial property of $\kappa = \omega$ for compactness is that $\mathrm{cf}(\kappa) = \omega$, whereas weakly compact cardinals are always inaccessible and hence regular. Of course ω is the only regular cardinal κ with $\mathrm{cf}(\kappa) = \omega$.)

Call κ a Σ_1 *compact cardinal* if $\langle H(\kappa), \in\rangle$ is Σ_1 compact. Call κ a $\Sigma_1(R)$ *compact cardinal* if $\langle H(\kappa), \in, R\rangle$ is Σ_1 compact. Thus κ is weakly compact iff κ is $\Sigma_1(R)$ compact for every $R \subseteq H(\kappa)$. We remind the reader once again that $\omega = \beth_\omega$.

9.3 Proposition. *Let* $\kappa \geqslant \omega$.

(i) *If* κ *is* Σ_1 *compact then* $\kappa = \beth_\kappa$.

(ii) *If* κ *is weakly compact then* κ *is inaccessible.*

Proof. Part (i) is a small part of Exercise 7.9 but we include its proof for completeness sake. Suppose κ is Σ_1 compact. We will first prove that

(1) $H(\kappa)$ is closed under the power set.

Suppose $a \in H(\kappa)$. Then $H(\kappa)$ satisfies the s-Π_1^1 formula

$$\forall U\, \exists b\, [b \subseteq a \wedge \forall x \in a\, (x \in b \leftrightarrow U(x))].$$

By s-Π_1^1 Reflection, $\mathscr{P}(a)\subseteq c$ for some $c\in H(\kappa)$ so $\mathscr{P}(a)\in H(\kappa)$. For (1) we see that $\kappa=\beth_\alpha$ for some limit ordinal α. Suppose $\alpha<\kappa$. Then $H(\kappa)$ satisfies the s-Π_1^1 formula expressing:

$$\begin{aligned}\forall\beta<\alpha\,\exists f\,[\mathrm{fun}(f)\wedge\mathrm{dom}(f)=\beta+1\\ f(0)=0\\ f(\gamma+1)=\mathscr{P}(f(\gamma))\quad&\text{for}\quad\gamma<\beta\\ f(\lambda)=\textstyle\bigcup_{\alpha<\lambda}f(\alpha)\quad&\text{for limit}\quad\lambda\leqslant\beta].\end{aligned}$$

Then s-Π_1^1 Reflection gives a contradiction since one would have an $a\in H(\kappa)$ such that $V(\alpha)\subseteq a$. This proves (i). To prove (ii) we need only see that κ is regular. Suppose $f:\alpha\to\kappa$ where $\alpha<\kappa$ and $\kappa=\sup\{f(\beta):\beta<\alpha\}$. We claim that $\langle H(\kappa),\in,f\rangle$ does not satisfy s-Π_1^1 Reflection. In fact it does not even satisfy Σ Reflection and hence is not admissible, since it satisfies the Σ formula

$$\forall\beta<\alpha\,\exists\gamma\,(f(\beta)=\gamma)$$

but there can be no bound $\xi<\kappa$ for the ordinals γ. $\square$

There are many characterizations of the class of weakly compact cardinals which fall out of our study. An admissible set $\mathbb{A}$ is *strict-Π_1^1 indescribable* if $(\mathbb{A},R)$ satisfies s-Π_1^1 Reflection for every $R\subseteq\mathbb{A}$. κ is *s-Π_1^1 indescribable* iff $\langle H(\kappa),\in\rangle$ is s-Π_1^1 indescribable.

9.4 Theorem. *An infinite cardinal κ is weakly compact iff it is strict-Π_1^1 indescribable.*

Proof. Immediate from Theorem 4.7. $\square$

An admissible set $\mathbb{A}$ satisfies Π_1^1 *Reflection* if for every Π_1^1 formula $\Phi(x_1,\ldots,x_n)$, $\mathbb{A}$ satisfies

$$\Phi(\vec{x})\to\exists a\,[\mathrm{Tran}(a)\wedge x_1,\ldots,x_n\in a\wedge\Phi^{(a)}(\vec{x})].$$

$\mathbb{A}$ is Π_1^1 *indescribable* if $(\mathbb{A},R)$ satisfies Π_1^1 Reflection for every $R\subseteq\mathbb{A}$. κ is Π_1^1 indescribable iff $\langle H(\kappa),\in\rangle$ is Π_1^1 indescribable. $\mathbb{HF}$ does not satisfy Π_1^1 Reflection or, for that matter, Π_2^0 Reflection since

$$\mathbb{HF}\models\forall x\,\exists y\quad(x\in y)$$

but no finite set can satisfy this sentence. Thus ω is certainly not Π_1^1 indescribable. We will see, however, that for κ with $\mathrm{cf}(\kappa)>\omega$, s-Π_1^1 Reflection implies Π_1^1 Reflection and s-Π_1^1 indescribability implies Π_1^1 indescribability. The secret to understanding this and a number of other facts is contained in the following surprising result.

Let the language L (of $\mathsf{L}_{\infty\omega}$) contain a distinguished binary relation symbol E. A *well-founded* L*-structure* is an L-structure $\mathfrak{M}$ with $\mathsf{E}^{\mathfrak{M}}$ well-founded.

9.5 Theorem. *Let $\mathbb{A}$ be an essentially uncountable Σ_1 compact admissible set. Let T be a Σ_1 theory of $\mathsf{L}_{\mathbb{A}}$. If every $\mathbb{A}$-finite $T_0 \subseteq T$ has a well-founded model then T has a well-founded model.*

Proof. Recall that $\mathbb{A}$ is essentially uncountable iff every countable subset of $\mathbb{A}$ is an element of $\mathbb{A}$. We know that $\mathbb{A}$ satisfies s-Π_1^1 Reflection since $\mathbb{A}$ is Σ_1 compact. The proof of this theorem is exactly like the proof that s-Π_1^1 Reflection implies Σ_1 compactness, once we have the following definitions and lemma. $\square$

We may assume that $\mathsf{L}_{\mathbb{A}}$ is a Skolem fragment which is Δ_1 on $\mathbb{A}$. Call an s.v.p. $\mathscr{D}$ for $\mathsf{L}_{\mathbb{A}}$ *well-founded* if there is no infinite sequence $\langle t_n : n<\omega\rangle$ of closed terms of L_A such that $(t_{n+1}\,\mathsf{E}\,t_n)\in\mathscr{D}$ for all $n<\omega$.

9.6 Lemma. *Let $\mathbb{A}$ be an essentially uncountable admissible set.*

(i) *There is a Π sentence $\varphi(\mathsf{D})$ such that for all $\mathscr{D}\subseteq\mathbb{A}$,*

$$(\mathbb{A},\mathscr{D})\models\varphi(\mathsf{D}) \text{ iff } \mathscr{D} \text{ is a well-founded s.v.p. for } \mathsf{L}_{\mathbb{A}}.$$

(ii) *If $\mathfrak{M}$ is a well-founded Skolem structure for $\mathsf{L}_{\mathbb{A}}$ then the* s.v.p. *$\mathscr{D}_{\mathfrak{M}}$ given by $\mathfrak{M}$ is well-founded.*

(iii) *If $\mathscr{D}$ is a well-founded* s.v.p. *for $\mathsf{L}_{\mathbb{A}}$ then $\mathscr{D}$ has a well-founded model.*

Proof. (i) Since $\mathbb{A}$ is essentially uncountable, every sequence $\langle t_n : n<\omega\rangle$ of terms of $\mathsf{L}_{\mathbb{A}}$ is actually an element of $\mathsf{L}_{\mathbb{A}}$. Thus the condition that $\mathscr{D}$ be well-founded is expressed by a universal quantifier over $\mathbb{A}$. The proof of (ii) is trivial. To prove (iii) let $\mathscr{D}$ be a well-founded s.v.p. By the Weak Completeness Theorem, $\mathscr{D}$ has a model $\mathfrak{M}_1$. Let $\mathfrak{M}$ be the smallest submodel of $\mathfrak{M}_1$. Then

$$\mathfrak{M}\prec\mathfrak{M}_1 \quad (\mathsf{L}_{\mathbb{A}}).$$

By Exercise VII.2.14 every element of $\mathfrak{M}$ is denoted by a closed term of $\mathsf{L}_{\mathbb{A}}$. Thus $\mathfrak{M}$ is well-founded and a model of the sentences in $\mathscr{D}$. $\square$

This lemma can also be used to prove a completeness theorem. See Exercise 9.11.

Theorem 9.5 explains why none of the explicitly described Σ_1 compact sets given in § 7 were essentially uncountable. The conclusion of Theorem 9.5 is so strong that it makes such sets very hard to find.

Our first use of Theorem 9.5 is to prove the results referred to above.

9.7 Theorem. *Let κ be a cardinal with $\mathrm{cf}(\kappa)>\omega$.*

(i) *If $\langle H(\kappa),\in,R\rangle$ satisfies s-Π_1^1 Reflection then it satisfies Π_1^1 Reflection.*

(ii) *If κ is s-Π_1^1 indescribable then κ is Π_1^1 indescribable.*

Proof. Part (ii) follows immediately from (i). To prove (i) let $\langle H(\kappa),\in,R\rangle$ satisfy s-Π_1^1 Reflection. By 9.3, $\kappa=\beth_\kappa$. Since $H(\kappa)$ is closed under $\mathscr{P}$, the graph of $\mathscr{P}$ is s-Π_1^1 on $H(\kappa)$ so $\mathbb{A}=\langle H(\kappa),\in,\mathscr{P},R\rangle$ also satisfies s-Π_1^1 Reflection and in particular, is admissible. Thus the definition of $\mathrm{V}(\alpha)$ is $\mathbb{A}$-recursive and $H(\kappa)=\mathrm{V}(\kappa)$. Suppose

$$\langle \mathrm{V}(\kappa),\in,R\rangle\models\forall \mathsf{S}\ \psi(\mathsf{S})$$

where ψ is first order but that for all α, $\alpha_0\leqslant\alpha<\kappa$,

$$\langle \mathrm{V}(\alpha),\in,R\cap\mathrm{V}(\alpha)\rangle\models\exists \mathsf{S}\ \neg\psi(\mathsf{S})$$

where α_0 is large enough so that all parameters in ψ are in $\mathrm{V}(\alpha_0)$. Let T be the following Σ_1 theory of $\mathsf{L}_{\mathbb{A}}$:

$$\mathrm{KP}+\mathrm{Power},$$

Infinitary diagram of $\langle\mathbb{A},\mathscr{P}\rangle$,

"c is an ordinal",

$(\mathsf{c}>\overline{\beta})$ for all $\beta<\kappa=o(\mathbb{A})$,

$$\forall\alpha\leqslant\mathsf{c}\ \exists\mathsf{S}\in\mathrm{V}(\alpha+1)\,\neg\psi(\mathsf{S})^{(\mathrm{V}(\alpha))}.$$

Every $\mathbb{A}$-finite subset of T has a well-founded model; one simply interprets c as some large $\alpha<\kappa$. By Theorem 9.5, T has a well-founded model $\mathfrak{M}$. Since it is well founded we can assume it is transitive. But then $\mathsf{c}^{\mathfrak{M}}$ is a real ordinal $\beta\geqslant\kappa$ and the last axiom of T implies that there is an $S\subseteq\mathrm{V}(\kappa)$ such that

$$\langle \mathrm{V}(\kappa),\in,R,S\rangle\models\neg\psi(\mathsf{S}).\quad\square$$

Theorem 9.7 is really rather remarkable since if κ is Σ_1 compact then s-$\Pi_1^1=\Sigma_1(\mathscr{P})$ and hence s-$\Pi_1^1\neq\Pi_1^1$.

9.8 Corollary. *If κ is weakly compact and greater than ω then κ is Mahlo.*

Proof. Since κ is weakly compact it is inaccessible. Since $\kappa>\omega$, 9.7 applies so κ is Π_1^1 indescribable. Let $C\subseteq\kappa$ be c.u.b. in κ. We must prove that there is a $\lambda<\kappa$ such that λ is inaccessible and $\lambda\in C$. Let $\mathbb{A}=\langle H(\kappa),\in,\mathscr{P},C\rangle$ and consider the Π_1^1 sentence Φ true in $\mathbb{A}$:

(2) $$\forall\mathsf{F}\,\forall\alpha\,[\mathsf{F}\text{ a function}\wedge\mathrm{dom}(\mathsf{F})=\alpha\wedge\forall\beta<\alpha\,(\mathsf{F}(\beta)\text{ is an ordinal})\rightarrow\exists\gamma\,\forall\beta<\alpha\,(\mathsf{F}(\beta)<\gamma)].$$

(3) $$\forall a\,\exists b\,\exists\beta\,\exists f\,[b=P(a)\wedge f:b\xrightarrow{1-1}\beta].$$

(4) $$\forall\alpha\,\exists\beta\,(\alpha<\beta\wedge\beta\in C).$$

The $\forall \mathsf{F}$ in (2) is the only second order quantifier; so Φ is Π_1^1 (but not s-Π_1^1). By Π_1^1 Reflection, there is transitive $B \in H(\kappa)$ such that $\Phi^{(B)}$ holds. Let $\lambda = o(B) = B \cap \mathrm{Ord}$. By (2), λ is a regular cardinal. By (3), λ is a strong limit cardinal. By (4), λ is the sup of elements of C. Since C is closed, $\lambda \in C$. ☐

We can connect weakly compact cardinals with trees as follows. A tree $\mathcal{T} = \langle T, \prec \rangle$ is a *κ-tree* if the rank of $\mathcal{T}$ is κ and for each $\alpha < \kappa$, $\mathcal{T}$ has less than κ nodes of level α. A cardinal κ has the *tree property* iff every κ-tree has a branch, that is, a path of length κ.

9.9 Theorem. *Let $\kappa \geqslant \omega$ be inaccessible. Then κ is weakly compact iff κ has the tree property.*

Proof. By Theorem 7.3 we see that, for κ inaccessible, κ is weakly compact iff for every $\mathbb{A}$ of the form $\langle H(\kappa), \in, \mathcal{P}, R \rangle$, every $\mathbb{A}$-tree has a branch. Clearly every such $\mathbb{A}$-tree is a κ-tree. Conversely, if $\mathcal{T}$ is a κ-tree then $\mathcal{T}$ is isomorphic to a tree on $H(\kappa)$. Thus T is isomorphic to an $\mathbb{A}$-tree for some expansion $\langle H(\kappa), \in, R \rangle$ of $H(\kappa)$. ☐

We summarize the characterizations of weakly compact cardinals obtained in the above by means of the following statement. We say that κ *satisfies* s-$\Pi_1^1(R) = \Sigma_1(R)$ *uniformly in R* if for every s-Π_1^1 formula $\Phi(v_1, \ldots, v_n, \mathsf{P}, \mathsf{R})$ there is a Σ_1 formula $\varphi(v_1, \ldots, v_n, \mathsf{P}, \mathsf{R})$ such that

$$\langle H(\kappa), \in, \mathcal{P}, R \rangle \models \forall \vec{v} \, [\Phi(\vec{v}, \mathsf{R}) \leftrightarrow \varphi(\vec{v}, \mathsf{R})]$$

for all $R \subseteq H(\kappa)$. (This is a different use of the word "uniformly".) We say that κ is *weakly compact* for $\mathsf{L}_{\kappa\omega}(\mathscr{W\!f})$ if for every $T \subseteq H(\kappa)$, if every subset of T_0 of power $< \kappa$ has a well-founded model, then T has a well-founded model.

9.10 Theorem (Summary). *Let κ be an infinite cardinal. The following are equivalent:*

(i) *κ is weakly compact for $\mathsf{L}_{\kappa\omega}$.*

(ii) *$\kappa = \omega$ or κ is weakly compact for $\mathsf{L}_{\kappa\omega}(\mathscr{W\!f})$.*

(iii) *κ is s-Π_1^1 indescribable.*

(iv) *$\kappa = \omega$ or κ is Π_1^1 indescribable.*

(v) *κ is inaccessible and has the tree property.*

(vi) *κ is inaccessible and for every $R \subseteq H(\kappa)$, $\langle H(\kappa), \in, R \rangle$ has a proper elementary end extension.*

(vii) *κ is inaccessible and $\kappa = \omega$ or else for every $R \subseteq H(\kappa)$, $\langle H(\kappa), \in, R \rangle$ has a proper well-founded elementary end extension.*

(viii) *κ is inaccessible and satisfies s-$\Pi_1^1(R) = \Sigma_1(R)$, uniformly in R.*

Proof. We list below the equivalences which have been already stated or else are immediate consequences of earlier results.

(i) $\Longleftrightarrow$ (ii) ($\Rightarrow$ by 9.5; $\Leftarrow$ by just adding E to a theory not mentioning it),
(i) $\Longleftrightarrow$ (iii) (by 9.4),
(iii) $\Longleftrightarrow$ (iv) (by 9.7 ii),
(i) $\Longleftrightarrow$ (v) (by 9.3 and 9.9).

The following implications are trivial:
(ii) $\Longrightarrow$ (vii) (trivial compactness argument),
(vii) $\Longrightarrow$ (vi) (trivial for $\kappa > \omega$, the case $\kappa = \omega$ follows from compactness of $\mathsf{L}_{\omega\omega}$).

The remaining implications (vi) $\Longrightarrow$ (v), and (iii) $\Longleftrightarrow$ (viii) are implicit in earlier results or proofs, but we will make them explicit. To prove (vi) $\Longrightarrow$ (v), let $\mathcal{T} = \langle T, \prec \rangle$ be a κ-tree. We may assume $T \subseteq \kappa$. Let $\mathbb{A} = \langle H(\kappa), \in, T, \prec, \text{lev} \rangle$. We can code up all of T, $\prec$, lev into one $R \subseteq H(\kappa)$ so, by assumption (vi), there is a proper elementary end extension $\mathfrak{B} = \langle B, E, T', \prec', \text{lev}' \rangle$ of $\mathbb{A}$. Let $b \in B$ be an ordinal, $b \notin A$. Let $x \in T$ satisfy

$$\mathfrak{B} \models x \in \text{lev}'(b).$$

Then $\{y \in A; y \prec' x\}$ is a branch through T. To prove (iii) $\Longrightarrow$ (viii), let $\Phi(x, \mathsf{R}) = \forall \mathsf{S}\, \varphi(x, \mathsf{R}, \mathsf{S})$ be a s-Π_1^1 formula involving an extra relation symbol R. For any R, $\langle H(\kappa), \in, \mathscr{P}, R \rangle$ satisfies one of the below iff it satisfies all:

$$\Phi(x, \mathsf{R}),$$

$$\forall S\, \varphi(x, \mathsf{R}, \mathsf{S}),$$

$$\exists a\, [\text{Tran}(a) \wedge x \in a \wedge \forall \mathsf{S} \subseteq a\, \varphi^{(a)}(x, \mathsf{R}, \mathsf{S})] \quad \text{(by (iii))},$$

$$\exists a\, \exists b\, [\text{Tran}(a) \wedge x \in a \wedge b = \mathscr{P}(a) \wedge \forall \mathsf{S} \in b\, \varphi^{(a)}(x, \mathsf{R}, \mathsf{S})].$$

The last line gives us a Σ_1 formula $\psi(x, \mathscr{P}, \mathsf{R})$ equivalent to $\Phi(x, \mathsf{R})$ for all R. To prove (viii) $\Longrightarrow$ (iii), notice that since κ is inaccessible, $H(\kappa) = \mathrm{V}(\kappa)$ and that $\mathbb{A} = \langle H(\kappa), \in, \mathscr{P}, R \rangle$ is resolvable, since $H(\kappa) = \bigcup_{\alpha < \kappa} \mathrm{V}(\alpha)$. Thus if $\mathbb{A}$ satisfies s-$\Pi_1^1 = \Sigma_1$ then $\mathbb{A}$ satisfies s-Π_1^1 Reflection by Corollary 4.9. $\square$

Some further equivalences are given in the Exercises.

Looking at this summary, one can hardly fail to be struck by the equivalence of notions coming to us from model theory, set theory and recursion theory. The summary is slightly misleading, however, in that it hides many important considerations which go into its proof, considerations including supervalidity properties, resolvability, essential uncountability, $\mathbb{A}$-trees, and so forth. It is only by understanding the earlier results involving these notions that one sees the various forces at work in Theorem 9.10.

9.11—9.16 Exercises

9.11. Let $\mathbb{A}$ be an essentially uncountable admissible set and let T be a s-Π_1^1 set of sentences of $\mathsf{L}_{\mathbb{A}}$. Let

$$\mathrm{Cn}_{\mathscr{W}\!f}(T) = \{\varphi \in \mathsf{L}_{\mathbb{A}} : \varphi \text{ is true in all } \textit{well-founded} \text{ models of } T\}.$$

Show that $\mathrm{Cn}_{\mathscr{W}\!f}(T)$ is s-Π_1^1.

9.12. Let κ be weakly compact, $\kappa > \omega$. Show that if $C \subseteq \kappa$ is c.u.b. then there is a Mahlo cardinal $\lambda < \kappa$, $\lambda \in C$.

9.13. Suppose that for every R, $\langle H(\kappa), \in, \mathscr{P}, R\rangle$ satisfies s-$\Pi_1^1 = \Sigma_1$. Show that $\langle H(\kappa), \in, \mathscr{P}\rangle$ satisfies s-$\Pi_1^1(R) = \Sigma_1(R)$, uniformly in R.

9.14. The definition of weakly compact cardinal is often given in terms of $\mathsf{L}_{\kappa\kappa}$. We sketch a proof that the two definitions are equivalent. We define $\mathsf{L}_{\infty\infty}$ to be the smallest collecting containing $\mathsf{L}_{\infty\omega}$ closed under $\neg$, $\bigwedge$, $\bigvee$ and

> if $\varphi \in \mathsf{L}_{\infty\omega}$ and V is a *set* of variables occurring in φ then $\exists V\varphi$ and $\forall V\varphi$ are in $\mathsf{L}_{\infty\infty}$.

For any κ, $\mathsf{L}_{\kappa\kappa} = L_{\infty\infty} \cap H(\kappa)$.

(i) Prove that $\mathsf{L}_{\kappa\kappa}$ consists of those $\varphi \in \mathsf{L}_{\infty\infty}$ with $<\kappa$ subformulas.

(ii) The following are sentences of $\mathsf{L}_{\omega_1\omega_1}$:

$$\forall \{v_1, \ldots, v_n, \ldots\} \bigvee_{n<\omega} \neg(v_{n+1} \mathsf{E} v_n),$$
$$\forall \{v_1, \ldots, v_n, \ldots\} \exists w \forall x [x \mathsf{E} w \leftrightarrow \bigvee_n x = v_n].$$

Give a formal definition of $\mathfrak{M} \models \varphi[s]$ for $\varphi \in \mathsf{L}_{\infty\infty}$ so that these sentences express well-foundedness and essential uncountability, respectively.

(iii) Show that every subformula of a sentence of $\mathsf{L}_{\kappa\kappa}$ has less than κ free variables.

(iv) Let κ be inaccessible and let $\varphi \in \mathsf{L}_{\kappa\kappa}$. Show that if φ has a model then it has one in $H(\kappa)$. Let $T \subseteq H(\kappa)$ be a set of sentences of $\mathsf{L}_{\kappa\kappa}$. Show that if T has a model then it has one of power κ. [Modify the usual Löwenheim-Skolem proof.]

(v) Let κ be weakly compact for $\mathsf{L}_{\kappa\omega}$. Show that κ is weakly compact for $\mathsf{L}_{\kappa\kappa}$. That is, let $T \subseteq \mathsf{L}_{\kappa\kappa}$ be a set of sentences such that every $T_0 \subseteq T$, $\mathrm{card}(T_0) < \kappa$, has a model. Show that T has a model. [For $\kappa = \omega$ this is trivial. For $\kappa > \omega$ apply 9.10 (vii) to $\langle H(\kappa), \in, \mathscr{P}, T\rangle$. Use the fact that (iv) holds in $H(\kappa)$ and hence in any elementary end extension. Also use the fact that $H(\kappa)$ is closed under sequences of length $<\kappa$.]

9.15. Show that κ is weakly compact iff $\kappa \to (\kappa)_2^2$; that is, iff for every partition

$$[\kappa]^2 = P_0 \cup P_1$$

of $[\kappa]^2 = \{\{\alpha, \beta\} : \alpha < \beta < \kappa\}$ into two sets, there is a subset $C \subseteq \kappa$ such that $[C]^2 \subseteq P_i$ for $i = 0$ or $i = 1$. [It is probably easiest to prove that 9.10 (vii) implies $\kappa \to (\kappa)_2^2$ and to prove $\omega \to (\omega)_2^2$ separately. To prove the other half show that $\kappa \to (\kappa)_2^2$ implies 9.10 (v).]

9.16. The parts (vi) and (vii) of Theorem 9.10 do not have significant lightface versions; that is, versions without the "for all R" clause, as the following example of Kunen shows. *Let κ be the least inaccessible cardinal such that $\langle H(\kappa), \in\rangle$ has* *an elementary end extension. Show that it has no well-founded elementary end extension.*

9.17 Notes. The "weakly" in weakly compact derives from the following. A cardinal κ is *strongly compact* if $\langle\mathfrak{M}; H(\kappa)_{\mathfrak{M}}, \in, R\rangle$ is Σ_1 compact for every structure $\mathfrak{M}=\langle M,S\rangle$ and every $R\subseteq H(\kappa)_{\mathfrak{M}}$, regardless of the size of $\mathfrak{M}$ as compared to κ. We see no convincing argument that strongly compact cardinals $>\omega$ are a natural limit of existing cardinals and so we do not study them here.

The equivalence, for $\kappa>\omega$, of weakly compact with Π_1^1 indescribability is due to Hanf and Scott [1961]. Some authors take Π_1^1 indescribability as the definition of weakly compact, thus ruling out ω. This seems not only silly (to rule out the one concrete example) but positively misleading since, as the proof of 9.7 shows, a number of considerations besides compactness are involved in the proof of Π_1^1 indescribability. The equivalences (in 9.10) (i)$\Longleftrightarrow$(ii)$\Longleftrightarrow$(v)$\Longleftrightarrow$(vi)$\Longleftrightarrow$(vii) are all well known. Similarly for the other equivalences given in the exercises. Corollary 9.8 and Exercise 9.12, which show that the first weakly compact $\kappa>\omega$ is much larger than the first inaccessible cardinal, are due to Hanf [1964]. The last equivalence ((i)$\Longleftrightarrow$(viii)) in 9.10 is a uniform version of a result in Kunen [1968].

The remarkable argument that strongly compact cardinals exist "by analogy with ω" always reminds me of the goofang, described in *The Book of Imaginary Beings*, by Jorge Luis Borges:

> The yarns and tall tales of the lumber camps of Wisconsin and Minnesota include some singular creatures, in which, surely, no one ever believed...
>
> There's another fish, the *Goofang*, that swims backward to keep the water out of its eyes. It's described as "about the size of a sunfish, only much bigger".

Appendix

Nonstandard Compactness Arguments and the Admissible Cover

One of the subjects we have not touched on in this book is applications of infinitary logic to constructing models of set theory and the relationship between compactness and forcing arguments. At one time we planned to include a chapter on these matters, but the book developed along other lines.

In this appendix we present one example of such a result because it leads very naturally to *the admissible cover* of a model $\mathfrak{M}$ of set theory. We want to treat this admissible set for two reasons. In the first place, it gives an example of an admissible set with urelements which has no counterpart in the theory without urelements, and it is as different from $\mathbb{HYP}_{\mathfrak{M}}$ as possible. Secondly, we promised (in Barwise [1974]) to present the details of the construction of this admissible set in this book.

1. Compactness Arguments over Standard Models of Set Theory

Let $\mathbb{A}=\langle A,\in\rangle$ be a countable transitive model of ZF. Then $\mathbb{A}$ is an admissible set and, moreover, $(\mathbb{A},R)$ is admissible for every definable relation R. We can therefore apply Completeness and Compactness to $\mathsf{L}_{\mathbb{A}}$ or $\mathsf{L}_{(\mathbb{A},R)}$, for any such R. There are many interesting results to be obtained in this way; we present one here and refer the reader to Barwise [1971], Barwise [1974], Friedman [1973], Krivine-MacAloon [1973], Suzuki-Wilmers [1973], and Wilmers [1973] for other examples. We also refer the reader to Keisler [1973] for connections with forcing.

The axiom $\mathrm{V}=\mathrm{L}$ asserts that every set is constructible.

1.1 Theorem. *Let* $\mathbb{A}$ *be a countable transitive model of* ZF. *There is an end extension* $\mathfrak{B}=\langle B,E\rangle$ *of* $\mathbb{A}$ *which is a model of* $\mathrm{ZF}+\mathrm{V}=\mathrm{L}$.

Proof. Let T be the theory of $\mathsf{L}_{\mathbb{A}}$ containing:

ZF.

The Infinitary diagram of $\mathbb{A}$.

We need to see that $T\cup\{\mathrm{V}=\mathrm{L}\}$ has a model. If not, then

$$T\models\mathrm{V}\neq\mathrm{L}$$

so

$$T \vdash \mathrm{V} \neq \mathrm{L}$$

by the Extended Completeness Theorem of § III.5. Thus $\mathbb{A}$ is a model of the Σ_1 sentence expressing:

(1) $\exists \Phi\, \exists p\, [p$ is a proof of $(\bigwedge \Phi) \to (\mathrm{V} \neq \mathrm{L})$ where $\forall x \in \Phi\, (x \in \mathrm{ZF}$ or x is a member of the infinitary diagram)$]$.

This Σ_1 sentence contains no parameters. Now let $\alpha = o(\mathbb{A})$ and let $\mathbb{A}_0 = \mathrm{L}(\alpha)$. Then $\mathbb{A}_0$ is a model of $\mathrm{ZF} + \mathrm{V} = \mathrm{L}$ (it is the constructible sets in the model $\mathbb{A}$ of ZF) and, interpreting Shoenfield's Lemma (Theorem V.8.1) in $\mathbb{A}$, we have: Any Σ_1 sentence true in $\mathbb{A}$ is true in $\mathbb{A}_0$.

Thus the sentence (1) is also true in $\mathbb{A}_0$. But this means that there is some subset T_0 of the infinitary diagram of $\mathbb{A}_0$ such that

$$T_0 + \mathrm{ZF} \vdash \mathrm{V} \neq \mathrm{L}$$

which is ridiculous since $\mathbb{A}_0$ itself is a model of $T_0 + \mathrm{ZF} + \mathrm{V} = \mathrm{L}$. ▯

There are a number of extensions of the above which will strike the reader; most of these are covered by the version contained in Theorem 3.1 of Barwise [1971]. What is not so obvious is how to extend the result from standard models of set theory to nonstandard models. For if $\mathfrak{A} = \langle A, E \rangle$ is a nonstandard model of ZF then we have no guarantee that a "proof" in the sense of $\mathfrak{A}$ proves anything at all. What we need is a new admissible set intimately related to $\mathfrak{A}$ which will allow us to carry out the above, and similar, proofs.

What is even less obvious is how to generalize results like Theorem 1.1 to the uncountable. There are uncountable models of ZFC with no end extension satisfying $\mathrm{V} = \mathrm{L}$, assuming of course that ZFC is consistent. Is there an uncountable generalization of Theorem 1.1, involving consideration like $h_\Sigma(\mathbb{A})$, which explains more satisfactorily why the result holds in the countable case? The same question applies to all the results in Barwise [1971] and Barwise [1974].

2. *The Admissible Cover and its Properties*

In this section we will be considering models of set theory as basic structures over which we build admissible sets. Thus we denote such structures by $\mathfrak{M} = \langle M, E \rangle$ where E is binary. Recall, for $x \in \mathfrak{M}$, the definition

$$x_E = \{y \in M \mid yEx\}.$$

Let L contain only the relation symbol E; let $\mathsf{L}^* = \mathsf{L}(\in, \mathsf{F})$ where F is a unary function symbol. Let (†) be the axiom of L^* given by

(†) $\qquad \forall p, x\, [x\, \mathsf{E} p \leftrightarrow x \in \mathsf{F}(p)] \wedge \forall a\, [\mathsf{F}(a) = 0]$.

An admissible set (for L^*), say $\mathbb{A}_{\mathfrak{M}}=(\mathfrak{M}; A,\in,F)$, is a *cover* of $\mathfrak{M}$ if $\mathbb{A}_{\mathfrak{M}}$ is a model of (†). That is, $\mathbb{A}_{\mathfrak{M}}$ is a cover of $\mathfrak{M}$ iff

$$F(x)=x_E \quad \text{for} \quad x\in M\,,$$

$$F(x)=0 \quad \text{for} \quad x\in A\,.$$

The point of the definition is pretty obvious, assuming that we are working in an admissible set $\mathbb{A}_{\mathfrak{M}}$ with $\mathfrak{M}\notin\mathbb{A}_{\mathfrak{M}}$. A quantifier like $\forall x\,(x\,\mathsf{E}\,y\rightarrow\ldots)$ is a bounded quantifier in the sense of L but it is not bounded, in general, in L^*. Using the axiom (†) however, it becomes equivalent to the bounded quantifier $\forall x\in\mathsf{F}(y)\,(\ldots)$.

In this way every formula φ of L translates into a formula $\hat{\varphi}$ of L^* with the properties:

if φ is Δ_0 (resp. Σ_1) is L then $\hat{\varphi}$ is Δ_0 (resp. Σ_1) in L^*

$$\text{KPU}+(\dagger)\vdash\forall p_1,\ldots,p_n\,[\varphi(p_1,\ldots,p_n)\leftrightarrow\hat{\varphi}(p_1,\ldots,p_n)]\,.$$

We use these remarks below without comment.

There are many admissible sets which cover a given structure $\mathfrak{M}$. For example, if $\mathbb{A}_{\mathfrak{M}}=(\mathfrak{M}; A,\in)$ is admissible *above* $\mathfrak{M}$ (in the sense of $\mathsf{L}(\in)$) then we can define an $\mathbb{A}_{\mathfrak{M}}$-recursive F by

$$F(x)=\{y\in M\mid yEx\}\,,\quad x\in M\,,$$

$$F(x)=0\,,\qquad x\notin M\,,$$

and then $(\mathbb{A}_{\mathfrak{M}},F)$ will be admissible in the sense of $\mathsf{L}(\in,\mathsf{F})$ and will cover $\mathfrak{M}$. These admissible sets are not tied closely enough to the intended interpretation of $\mathfrak{M}$ for the applications we have in mind; they are too big with too many subsets of $\mathfrak{M}$. What we would like would be an admissible set $\mathbb{A}_{\mathfrak{M}}$ which covers $\mathfrak{M}$ and whose only sets of urelements are the sets of the form p_E for $p\in\mathfrak{M}$.

2.1 Definition. Let $\mathfrak{M}=\langle M,E\rangle$ be an L-structure and let $\mathfrak{Cov}_{\mathfrak{M}}$ be the intersection of all admissible sets which cover $\mathfrak{M}$. More precisely,

$$\mathfrak{Cov}_{\mathfrak{M}}=(\mathfrak{M}; A,\in,F)$$

where:

$$A=\bigcap\{B\mid(\mathfrak{M}; B,\in,F)\text{ is admissible and covers }\mathfrak{M}\}.$$

$$F(p)=p_E \quad \text{for} \quad p\in M\,.$$

$$F(a)=0 \quad \text{for} \quad a\in A\,.$$

2.2 Theorem. *If $\mathfrak{M}$ is a model of* KP *then $\mathfrak{Cov}_{\mathfrak{M}}$ is admissible. $\mathfrak{Cov}_{\mathfrak{M}}$ is called the* admissible cover *of $\mathfrak{M}$.*

Proof. Deferred to § 3. □

If we proved this theorem right now, the proof would look complicated and *ad hoc*. What we shall do instead is to develop further properties of the admissible cover in this section until, by the end of the section, we will know almost exactly what $\mathfrak{Cov}_{\mathfrak{M}}$ looks like. This should make the proofs (in § 3) easier to follow.

The next property of the admissible cover suggests the main step in the proof of Theorem 2.2 and shows us that $\mathfrak{Cov}_{\mathfrak{M}}$ really lives in $\mathfrak{M}$. (The corollaries of Theorem 2.3 are easier to understand than 2.3 at a first reading.)

2.3 Theorem. *Let $\mathfrak{M}=\langle M,E\rangle$ be a model of* KP. *There is a single valued notation system p projecting $\mathfrak{Cov}_{\mathfrak{M}}$ into $\mathfrak{M}$ satisfying the following equations (where we use $\dot{x}$ for the unique y such that $p(x)=\{y\}$, where 0,1 denote the first two ordinals in the sense of $\mathfrak{M}$ and where $\langle,\rangle$ is the ordered pair operation as defined in $\mathfrak{M}$):*

(i) *For $x\in M$,*

$$\dot{x}=\langle 0,x\rangle$$

(ii) *for $a\in\mathfrak{Cov}_{\mathfrak{M}}$, there is a $y\in M$ such that*

$$\dot{a}=\langle 1,y\rangle$$

and $y_E=\{\dot{x}\,|\,x\in a\}$.

Proof. Deferred to § 3, 3.1—3.7. ☐

Call a set $a\subseteq\mathfrak{M}$ of urelements $\mathfrak{M}$-*finite* if $a=x_E$ for some $x\in\mathfrak{M}$.

2.4 Corollary. *Let $\mathfrak{M}\models$ KP and let $a\subseteq\mathfrak{M}$. Then a is $\mathfrak{M}$-finite iff $a\in\mathfrak{Cov}_{\mathfrak{M}}$. Hence for any $a\in\mathfrak{Cov}_{\mathfrak{M}}$, the support of a is $\mathfrak{M}$-finite. In particular, $M\notin\mathfrak{Cov}_{\mathfrak{M}}$.*

Proof. Let $a\subseteq\mathfrak{M}$, $a\in\mathfrak{Cov}_{\mathfrak{M}}$. Using the notation from 2.3,

$$\dot{a}=\langle 1,y\rangle$$

where $y_E=\{\dot{x}:x\in a\}$. But $a\subseteq\mathfrak{M}$ so $\dot{x}=\langle 0,x\rangle$ for all $x\in a$. Then we can define, inside the model $\mathfrak{M}$, the following set by Σ Replacement, remembering that $\mathfrak{M}\models$ KP:

$$z=\{x\,|\,\langle 0,x\rangle E y\}$$

and then $z_E=a$. The converse is trivial. ☐

Corollary 2.4 is very useful in compactness arguments involving $\mathfrak{Cov}_{\mathfrak{M}}$, for it tells us that if $T_0\in\mathfrak{Cov}_{\mathfrak{M}}$ is a set of infinitary sentences, then the set

$$\{x\in M:\mathsf{x}\text{ is mentioned in }T_0\}$$

is $\mathfrak{M}$-finite. Recall that x is the constant symbol used to denote x. ☐

We can use the projection from 2.3 to identify the pure sets in $\mathfrak{Cov}_{\mathfrak{M}}$ and the ordinals of $\mathfrak{Cov}_{\mathfrak{M}}$.

2.5 Corollary. *Let $\mathfrak{M} \models \mathrm{KP}$. Let $\mathbb{A}_0$ be the transitive set isomorphic to $\mathcal{Wf}(\mathfrak{M})$. The pure sets in $\mathfrak{Cov}_{\mathfrak{M}}$ are exactly the sets in $\mathbb{A}_0$. In particular, $o(\mathfrak{Cov}_{\mathfrak{M}}) = o(\mathbb{A}_0)$.*

Proof. Since $\mathbb{A}_0$ is admissible (by the Truncation Lemma) it is closed under TC so it suffices to prove that every transitive set $a \in \mathbb{A}_0$ is in $\mathfrak{Cov}_{\mathfrak{M}}$ in order to prove $\mathbb{A}_0 \subseteq \mathfrak{Cov}_{\mathfrak{M}}$, since $\mathfrak{Cov}_{\mathfrak{M}}$ is transitive. Let $a \in \mathbb{A}_0$ be transitive and let

$$\langle a, \in \rangle \cong \langle x, E \upharpoonright x_E \rangle$$

where $x \in \mathcal{Wf}(\mathfrak{M})$. Since $\mathfrak{Cov}_{\mathfrak{M}}$ is admissible, by 2.2, we can apply Theorem V.3.1 in $\mathfrak{Cov}_{\mathfrak{M}}$ to see that $a \in \mathfrak{Cov}_{\mathfrak{M}}$. To prove the other inclusion define the following function by recursion in $\mathfrak{M}$ (more precisely, define it by Σ Recursion in KP and interpret the result in $\mathfrak{M}$):

$$\langle 0, x \rangle' = x$$

$$\langle 1, x \rangle' = \{ y' \mid yEx \} .$$

(It is only the second clause which is relevant here but we'll use $'$ again later.) Let $\eta : \langle \mathcal{Wf}(\mathfrak{M}), E \rangle \cong \langle A_0, \in \rangle$ and consider the following diagram, where $D_0 = \{ \dot{a} \mid a \text{ a pure set in } \mathfrak{Cov}_{\mathfrak{M}} \}$:

$$\begin{array}{ccc} \text{Pure part}(\mathfrak{Cov}_{\mathfrak{M}}) & \stackrel{\cdot}{\longrightarrow} & D_0 \subseteq M \\ {\scriptstyle \mathrm{Id}} \downarrow & & \downarrow {\scriptstyle '} \\ \mathbb{A}_0 & \stackrel{\eta}{\longleftarrow} & \mathcal{Wf}(\mathfrak{M}) \end{array}$$

We claim that, for every pure set $a \in \mathfrak{Cov}_{\mathfrak{M}}$, $(\dot{a})' \in \mathcal{Wf}(\mathfrak{M})$ and $\eta((\dot{a})') = a$, which will conclude 2.5. The proof is by induction on $\in$. First, $\dot{a} = \langle 1, x \rangle$ where $x_E = \{ \dot{b} : b \in a \}$. But then $(\dot{a})' = z$ where

$$z_E = \{ y' \mid yEx \}$$
$$= \{ (\dot{b})' \mid b \in a \} .$$

Thus $(\dot{a})'_E \subseteq \mathcal{Wf}(\mathfrak{M})$ by part of the induction hypothesis, and hence $(\dot{a})' \in \mathcal{Wf}(\mathfrak{M})$. Computing $\eta((\dot{a})')$ we get

$$\eta((\dot{a})') = \{ \eta(y') \mid yEz \}$$
$$= \{ \eta((\dot{b})') \mid b \in a \} .$$

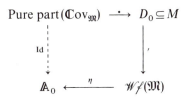

The other part of the induction hypothesis states that $\eta((\dot{b})')=b$ for $b\in a$ so we get

$$\eta((\dot{a})')=\{b\mid b\in a\}$$
$$=a. \quad \square$$

Using 2.4 and 2.5 we can give a picture of $\mathbb{C}\mathrm{ov}_{\mathfrak{M}}$. The dotted line in $\mathfrak{M}$ is the level at which it becomes nonstandard (if it is nonstandard).

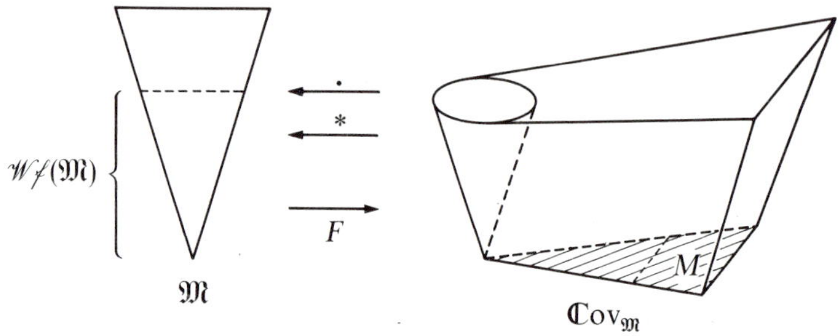

Fig. 2A. A model $\mathfrak{M}$ of set theory next to its admissible cover

The projection given in 2.3 is *ad hoc* in that we could have used others. The next function, by contrast, is canonical.

Let $\mathbb{A}_{\mathfrak{M}}=(\mathfrak{M};A,\in,F)$ be admissible and a cover of $\mathfrak{M}$. A function $*$ is an *$\in$-retraction of $\mathbb{A}_{\mathfrak{M}}$ onto $\mathfrak{M}$* if x^* is defined for every $x\in\mathbb{A}_{\mathfrak{M}}$ and satisfies the following equations:

$$(1)\begin{cases} p^*=p & \text{for} \quad p\in\mathfrak{M} \\ (a^*)_E=\{b^*\mid b\in a\} & \text{for all} \quad a\in\mathbb{A}_{\mathfrak{M}}. \end{cases}$$

We can use the projection given by Theorem 2.3 to prove the following characterization of $\mathbb{C}\mathrm{ov}_{\mathfrak{M}}$.

2.6 Corollary. *Let $\mathfrak{M}\models\mathrm{KP}$. $\mathbb{C}\mathrm{ov}_{\mathfrak{M}}$ has an $\in$-retraction into $\mathfrak{M}$ and it is the only admissible set covering $\mathfrak{M}$ which has such an $\in$-retraction.*

Proof. The proof is an elaboration of the proof of Theorem 2.5. It is clear that any admissible set $\mathbb{A}_{\mathfrak{M}}$ covering $\mathfrak{M}$ has a function $*$ satisfying (1), simply by the second recursion theorem for KPU:

$$\begin{aligned} x^*=y \quad \text{iff} \quad & (x \text{ is an urelement} \wedge y=x)\vee \\ & (x \text{ is a set and } F(y)=\{b^*\mid b\in x\}). \end{aligned}$$

The problem is that x^* won't usually be defined for all x. Let us first show that for $\mathbb{A}_{\mathfrak{M}}=\mathbb{C}\mathrm{ov}_{\mathfrak{M}}$, x^* *is* defined for all x. Define $'$ just as in the proof of 2.5. We claim that for all $x \in \mathbb{C}\mathrm{ov}_{\mathfrak{M}}$,

$$(\dot{x})' \text{ is defined}$$
$$(\dot{p})' = p \quad \text{for} \quad p \in M$$
$$((\dot{a})')_E = \{(\dot{x})' \mid x \in a\} \quad \text{for} \quad a \in M\,.$$

This is proved by induction just as in 2.5 and shows that x^* is defined for all x since $x^* = (\dot{x})'$. This proves that $\mathbb{C}\mathrm{ov}_{\mathfrak{M}}$ has an $\in$-retraction onto $\mathfrak{M}$. Let $\mathbb{A}_{\mathfrak{M}}$ be any other cover

$$(\mathfrak{M}; A, \in, F)$$

which has a totally defined $\in$-retraction $*$. Let D be the domain (in the peculiar sense of Definition V.5.1; that is $D = \mathrm{rng}(\cdot)$) of the notation system of Theorem 2.3 and let

$$|p| = \text{the unique } x \text{ such that } \quad \dot{x} = p$$

for $p \in D$. Thus $|\ |$ maps D onto $\mathbb{C}\mathrm{ov}_{\mathfrak{M}}$. Define an $\mathbb{A}_{\mathfrak{M}}$-recursive function f from $\mathbb{A}_{\mathfrak{M}}$ into $\mathfrak{M}$ using $*$:

$$f(p) = \langle 0, p\rangle$$
$$f(a) = \langle 1, \{f(b) : b \in a\}^*\rangle\,.$$

See Fig. 2B at this point.

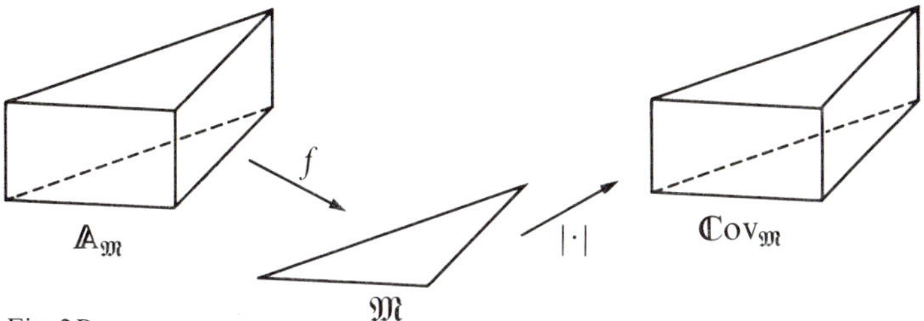

Fig. 2B.

A simple proof by induction on $\in$ shows that $f(x) \in D$ and $|f(x)| = x$, for all $x \in \mathbb{A}_{\mathfrak{M}}$. Thus $\mathbb{A}_{\mathfrak{M}} \subseteq \mathbb{C}\mathrm{ov}_{\mathfrak{M}}$ so $\mathbb{C}\mathrm{ov}_{\mathfrak{M}} = \mathbb{A}_{\mathfrak{M}}$ since $\mathbb{C}\mathrm{ov}_{\mathfrak{M}}$ is the smallest admissible set covering $\mathfrak{M}$. $\square$

The $\in$-retraction $*$ of $\mathbb{C}\mathrm{ov}_{\mathfrak{M}}$ onto $\mathfrak{M}$ is not one-one, of course, since $(a^*)^* = a^*$ but $a^* \neq a$, for any set $a \in \mathbb{C}\mathrm{ov}_{\mathfrak{M}}$. Otherwise, though, it is far more natural and less *ad hoc* than the projection of Theorem 2.3. We saw in the proof of 2.6 how to reconstruct the projection from $*$.

Also note that $*$ is $\mathbb{C}\mathrm{ov}_{\mathfrak{M}}$-recursive.

For applications of $\mathbb{C}\mathrm{ov}_{\mathfrak{M}}$ we need two more properties of $\mathbb{C}\mathrm{ov}_{\mathfrak{M}}$. The first tells us what Σ_1 on $\mathbb{C}\mathrm{ov}_{\mathfrak{M}}$ means in term of $\mathfrak{M}$.

2.7 Theorem. *Let* $\mathfrak{M} \models \mathrm{KP}$. *A relation* S *on* $\mathfrak{M}$ *is* Σ_1 *on* $\mathfrak{Cov}_{\mathfrak{M}}$ *iff* S *is* Σ_+ *inductive on* $\mathfrak{M}$; *that is, iff* S *is a section of* I_φ *where* $\varphi = \varphi(v_1, \ldots, v_n, \mathsf{R}_+)$ *is some* Σ *inductive definition (in the language* $\mathsf{L}(\mathsf{R})$*) interpreted over* $\mathfrak{M}$.

Proof. Deferred to 3.9. ▯

The last property we need relates the admissible covers of two different models $\mathfrak{M}, \mathfrak{N}$. Let $\mathfrak{M} = \langle M, E \rangle$, $\mathfrak{N} = \langle N, F \rangle$ where $\mathfrak{M} \subseteq \mathfrak{N}$. Note that $\mathfrak{M} \subseteq_{\mathrm{end}} \mathfrak{N}$ if $\mathfrak{Cov}_{\mathfrak{M}} \subseteq \mathfrak{Cov}_{\mathfrak{N}}$. If $\mathfrak{M}, \mathfrak{N} \models \mathrm{KP}$ and $\mathfrak{M} \subseteq_{\mathrm{end}} \mathfrak{N}$ then $\mathfrak{Cov}_{\mathfrak{M}} \subseteq \mathfrak{Cov}_{\mathfrak{N}}$, as the construction in § 3 makes translucent.

2.8 Theorem. *Let* $\mathfrak{M}, \mathfrak{N} \models \mathrm{KP}$, $\mathfrak{M} \subseteq_{\mathrm{end}} \mathfrak{N}$. *Then*

$$\mathfrak{M} \prec_1 \mathfrak{N}$$

if and only if

$$\mathfrak{Cov}_{\mathfrak{M}} \prec_1 \mathfrak{Cov}_{\mathfrak{N}}\,.$$

Proof. The translation $\varphi \to \hat{\varphi}$ defined at the beginning of this section makes the ($\Leftarrow$) half of this theorem immediate. The converse follows from the considerations of the next section. ▯

3. *An Interpretation of* KPU *in* KP

The proofs of the theorems of § 2 all involve interpreting the theory KPU of $\mathsf{L}(\in, \mathsf{F})$ in the theory KP of L, in the sense of § II.4, and then applying this interpretation to models $\mathfrak{M}$ of KP.

The interpretation is the one suggested by the projection of $\mathfrak{Cov}_{\mathfrak{M}}$ into $\mathfrak{M}$ which we want to construct to prove Theorem 2.3:

$$\dot{p} = \langle 0, p \rangle\,, \qquad \dot{a} = \langle 1, y \rangle$$

where

$$y_E = \{\dot{x} \mid xEa\}\,.$$

3.1 The Interpretation *I*. We are dealing with two separate set theories, KP formulated in L with E as a membership symbol and KPU+(†) formulated in $\mathsf{L}(\in, \mathsf{F})$ with $\in$ as the membership symbol, so this must make things a bit confusing no matter what we do. In this subsection we want to work axiomatically within KP so we use $\in$ for membership when we really ougth to use E, just because it seems the lesser of two evils. We use the usual notation for symbols defined in KP, symbols like $0, 1, \langle x, y \rangle, \mathsf{OP}$ (for ordered pair).

Define predicates within KP by the following:

$$\begin{aligned}
\mathsf{N}(x) &\leftrightarrow \exists y(x=\langle 0,y\rangle)\\
&\leftrightarrow \mathsf{OP}(x)\wedge \mathsf{1^{st}}(x)=0\\
x\,\mathsf{E}'y &\leftrightarrow \mathsf{N}(x)\wedge \mathsf{N}(y)\wedge(\mathsf{2^{nd}}(x)\in \mathsf{2^{nd}}(y))\\
\mathsf{Set}(x) &\leftrightarrow \exists y[x=\langle 1,y\rangle\wedge \forall z\in y(\mathsf{N}(z)\vee \mathsf{Set}(z))]\\
z\mathscr{E}x &\leftrightarrow \exists y[x=\langle 1,y\rangle\wedge z\in y]\\
&\leftrightarrow \mathsf{OP}(x)\wedge \mathsf{1^{st}}(x)=1\wedge z\in \mathsf{2^{nd}}(x)\\
\mathsf{F}'(x) &= \langle 1,\{\langle 0,y\rangle : y\in \mathsf{2^{nd}}(x)\}\rangle\,.
\end{aligned}$$

The predicates $\mathsf{N}, \mathsf{E}', \mathscr{E}$ and F' are defined by Δ_0 formulas. The predicate Set is defined, using the second recursion theorem, by a Σ_1 formula. We use these to define our interpretation as follows, where $\mathsf{L}^* = \mathsf{L}(\in, \mathsf{F})$ is considered as a one-sorted language with relation symbols U (for urelement), S (for set)

Symbol of L^*	*Interpretation in* KP *under I*
$\forall x$	$\forall x(\mathsf{N}(x)\vee \mathsf{Set}(x)\to\ldots)$
$=$	$=$
$\mathsf{U}(x)$	$\mathsf{N}(x)$
$\mathsf{S}(x)$	$\mathsf{Set}(x)$
$x\mathsf{E}y$	$x\mathsf{E}'y$
$x\in y$	$x\mathscr{E}y$
$\mathsf{F}(x)$	$\mathsf{F}'(x)$

3.2 Lemma. *I is an interpretation of* KPU+(†) *in* KP. *That is, for each axiom φ of* KPU+(†), *φ^I is a theorem of* KP.

Proof. We run quickly through the axioms, beginning with (†). The interpretation of (†) reads

$$\forall x\,\forall y[\mathsf{N}(x)\wedge \mathsf{N}(y)\to(x\mathsf{E}'y\leftrightarrow x\mathscr{E}\mathsf{F}'(y))]\,.$$

So suppose $\mathsf{N}(x)\wedge\mathsf{N}(y)$. Let $x=\langle 0,x_0\rangle$, $y=\langle 0,y_0\rangle$. Then the following are equivalent:

$$x\mathsf{E}'y\,,$$
$$x_0\in y_0\,,$$
$$x\mathscr{E}\mathsf{F}'(y)\,.$$

Extensionality: The interpretation of Extensionality asserts that if $\mathsf{Set}(x)$ and $\mathsf{Set}(y)$ and

$$\forall z[(\mathsf{N}(z)\vee\mathsf{Set}(z))\to(z\mathscr{E}x\leftrightarrow z\mathscr{E}y)]$$

then $x=y$. Assume the three hypotheses. Let $x=\langle 1,u\rangle$, $y=\langle 1,v\rangle$. Then $z\mathscr{E}x$ iff $z\in u$, $z\mathscr{E}y$ iff $z\in v$. Since every $z\in u\cup v$ satisfies $\mathsf{N}(z)\vee\mathsf{Set}(z)$, $u=v$ and hence $x=y$.

Foundation: Suppose there is an x such that

$$\mathsf{Set}(x)\wedge\varphi^I(x).$$

Choose such an x of least possible rank. Then since $y\mathscr{E}z\rightarrow rk(y)<rk(z)$, we have

$$\forall z[\mathsf{Set}(z)\wedge z\mathscr{E}x\rightarrow\neg\varphi^I(z)].$$

Pair: Suppose $\mathsf{N}(x)\vee\mathsf{Set}(x)$ and $\mathsf{N}(y)\vee\mathsf{Set}(y)$. Let

$$z=\langle 1,\{x,y\}\rangle.$$

Then $\mathsf{Set}(z)\wedge(u\mathscr{E}z\leftrightarrow(u=x\vee u=y))$.

Union: Suppose $\mathsf{Set}(x)$. Let

$$y_0=\{z\mid\exists u\mathscr{E}x(z\mathscr{E}u)\}$$

by Δ_0 Separation and let $y=\langle 1,y_0\rangle$.

Δ_0 *Separation:* Let φ be a Δ_0 formula of $\mathsf{L}(\in,\mathsf{F})$. The formula φ^I is a Δ_0 formula of L^* when L^* is expanded by the symbols $\mathsf{N},\mathsf{E}',\mathsf{E},\mathsf{F}'$. Suppose $\mathsf{Set}(x)$, say $x=\langle 1,x_0\rangle$. Let

$$y_0=\{z\in x_0\mid\varphi^I(z)\}$$

by Δ_0 Separation and let $y=\langle 1,y_0\rangle$. Then

$$z\mathscr{E}y \text{ iff } z\mathscr{E}x\wedge\varphi^I(z).$$

Δ_0 *Collection:* Suppose $\varphi(x,y)$ is Δ_0, suppose $\mathsf{Set}(a)$ and that

$$\forall x\mathscr{E}a\,\exists y[\mathsf{N}(y)\vee\mathsf{Set}(y))\wedge\varphi^I(x,y)].$$

Let $a=\langle 1,a_0\rangle$ so that the above becomes

$$\forall x\in a_0\,\exists y[(\mathsf{N}(y)\vee\mathsf{Set}(y))\wedge\varphi^I(x,y)].$$

By Σ Reflection there is a b such that

$$\forall x\in a_0\,\exists y\in b[(\mathsf{N}(y)\vee\mathsf{Set}(y))\wedge\varphi^I(x,y)]^{(b)}.$$

Let

$$b_0=\{y\in b\mid(\mathsf{N}(y)\vee\mathsf{Set}(y))^{(b)}\}$$

by Δ_0 Separation and let $b_1 = \langle 1, b_0 \rangle$. Then

$$\forall x \mathscr{E} a \, \exists y \mathscr{E} b_1 \, \varphi^I(x,y) . \quad \Box$$

3.3 The model $\mathfrak{M}^{-I}$. Let $\mathfrak{M} \models \mathrm{KP}$. Let $N, E', \mathrm{Set}, \mathscr{E}, F'$ be the predicates and function defined in $\mathfrak{M}$ by the corresponding symbols of KP. Then, letting $\mathfrak{N} = \langle N, E' \rangle$ we have

$$\begin{aligned} \mathfrak{M}^{-I} &= (\mathfrak{N}; \mathrm{Set}, \mathscr{E} \upharpoonright \mathrm{Set}, F') \\ &= \mathfrak{B}_{\mathfrak{N}}, \text{ say} . \end{aligned}$$

$\mathfrak{B}_{\mathfrak{N}}$ is a model of KPU + (†), by 3.2. The structure $\mathfrak{N}$ is isomorphic to $\mathfrak{M}$ via the map $x \mapsto \langle 0, x \rangle$. If $\mathbb{D}_{\mathfrak{M}}$ is any admissible set covering $\mathfrak{M}$ then

N, E', F' are $\mathbb{D}_{\mathfrak{M}}$-recursive, as is the isomorphism $x \mapsto \langle 0, x \rangle$.

Set, $\mathscr{E} \upharpoonright$ Set are $\mathbb{D}_{\mathfrak{M}}$-r.e.

by the remarks at the beginning of § 2.

3.4 The model $\mathscr{W}\!f(\mathfrak{M}^{-I})$. Let $\mathfrak{M} \models \mathrm{KP}$ and let $\mathfrak{B}_{\mathfrak{N}}$ be as defined in 3.3. $\mathscr{W}\!f(\mathfrak{B}_{\mathfrak{N}})$ is the largest well-founded substructure of $\mathfrak{B}_{\mathfrak{N}}$, *before being identified with a transitive set this time.* Notice that $\mathscr{W}\!f(\mathfrak{B}_{\mathfrak{N}})$ is closed under F' since $F'(x)$ is always a set of urelements. Thus by the Truncation Lemma, $\mathscr{W}\!f(\mathfrak{B}_{\mathfrak{N}})$ is a well-founded model of KPU + (†). If $\mathbb{D}_{\mathfrak{M}}$ is admissible and covers $\mathfrak{M}$ then

N, E', F' are $\mathbb{D}_{\mathfrak{M}}$-recursive, as is the isomorphism $x \mapsto \langle 0, x \rangle$ and $\mathrm{Set} \cap \mathscr{W}\!f(\mathfrak{B}_{\mathfrak{N}}), \mathscr{E} \upharpoonright (\mathrm{Set} \cap \mathscr{W}\!f(\mathfrak{B}_{\mathfrak{N}}))$ are $\mathbb{D}_{\mathfrak{M}}$-r.e.

The first follows from 3.2. The second line follows from Theorem V.3.1.

3.5 The admissible set isomorphic to $\mathscr{W}\!f(\mathfrak{M}^{-I})$. Let $\mathfrak{M} \models \mathrm{KP}$ and let

$$\mathscr{W}\!f(\mathfrak{B}_{\mathfrak{N}}) \cong (\mathfrak{N}; A, \in, F') = \mathbb{A}_{\mathfrak{N}}$$

where A is transitive (in $\mathbb{V}_{\mathfrak{N}}$). By 3.4, $\mathbb{A}_{\mathfrak{N}}$ is admissible and covers $\mathfrak{N}$. Let $\mathbb{D}_{\mathfrak{M}}$ be any admissible set which covers $\mathfrak{M}$. By 3.4 and Theorem V.3.1, there is a $\mathbb{D}_{\mathfrak{M}}$-recursive isomorphism of $\mathfrak{M}$ and $\mathfrak{N}$, and A is $\mathbb{D}_{\mathfrak{M}}$-r.e.

3.6 $\mathbb{C}\mathrm{ov}_{\mathfrak{M}}$ defined. Let $\mathfrak{M} \models \mathrm{KP}$ and let $\mathbb{A}_{\mathfrak{N}}$ be as in 3.5. The isomorphism $i: \mathfrak{N} \cong \mathfrak{M}$ extends to an isomorphism of $\mathbb{V}_{\mathfrak{N}}$ onto $\mathbb{V}_{\mathfrak{M}}$ by:

$$i(a) = \{ i(b) \mid b \in a \} ,$$

carrying every transitive set in $\mathbb{V}_{\mathfrak{N}}$ onto a transitive set of $\mathbb{V}_{\mathfrak{M}}$. In particular, $\mathbb{A}_{\mathfrak{N}}$ is carried over to an isomorphic admissible set over $\mathfrak{M}$, say $\mathbb{A}'_{\mathfrak{M}} =$

$$(\mathfrak{M}; A', \in, F)$$

where $A' = \{i(a) \mid a \in A\}$. We claim that this $\mathbb{A}'_{\mathfrak{M}}$ is the admissible cover of $\mathfrak{M}$. It clearly is admissible and covers $\mathfrak{M}$. Let $\mathbb{D}_{\mathfrak{M}}$ be admissible and cover $\mathfrak{M}$. The isomorphism i can be defined by $\in$-recursion in $\mathbb{D}_{\mathfrak{M}}$ and so $\mathbb{A}'_{\mathfrak{M}} \subseteq \mathbb{D}_{\mathfrak{M}}$. Thus $\mathbb{A}'_{\mathfrak{M}}$ is contained in every admissible set covering $\mathbb{M}$ so $\mathbb{A}'_{\mathfrak{M}} = \mathbb{C}\mathrm{ov}_{\mathfrak{M}}$. This proves Theorem 2.2.

3.7 The projection. It is clear from the above construction of $\mathbb{C}\mathrm{ov}_{\mathfrak{M}}$ that every $x \in M$ is "denoted by" $\langle 0, x\rangle$ and that every $a \in \mathbb{C}\mathrm{ov}_{\mathfrak{M}}$ is denoted by

$$\langle 1, y\rangle$$

where y_E is the set of "notations for" members of a. Turning this around gives the desired projection.

We saw, early in § 2, how to translate Σ_1 formulas of L into Σ_1 formulas of L^*, using the covering function. We now see how we can translate Σ_1 formulas of L^* into "formulas" about $\mathfrak{M}$.

3.8 Translation Lemma. *Let $\exists y\, \varphi(x,y)$ be a Σ_1 formula of L^*, where φ is Δ_0, and let $\psi(x,z)$ be the interpretation*

$$\exists y[\mathrm{rk}(y) = z \wedge \varphi(x,y)]^I,$$

a formula of L. Let $\mathfrak{M} \models \mathrm{KP}$, let $\alpha = o(\mathbb{C}\mathrm{ov}_{\mathfrak{M}})$ and let $x \in \mathbb{C}\mathrm{ov}_{\mathfrak{M}}$. Then

$$\mathbb{C}\mathrm{ov}_{\mathfrak{M}} \models \exists y\, \varphi(x,y)$$

iff there is a $\beta < \alpha$ such that

$$\mathfrak{M} \models \psi(\dot{x}, \dot{\beta}).$$

Proof. Suppose $\mathbb{C}\mathrm{ov}_{\mathfrak{M}} \models \varphi(x,y)$. Then

$$\mathfrak{M} \models \varphi^I(\dot{x}, \dot{y}) \wedge (\mathrm{rk}(\dot{y}) = z)^I$$

for some "standard ordinal" z of $\mathfrak{M}^{-I}$. Thus, by Corollary 2.5,

$$\mathfrak{M} \models \psi(\dot{x}, \dot{\beta})$$

for some $\beta < \alpha$. The other half follows from 3.3—3.7. □

3.9 Proof of Theorem 2.7. A complete proof of Theorem 2.7 would include a proof of the following fact. The Σ_+ inductive relations on $\mathfrak{M}$ contain all Σ relations and are closed under $\wedge$, $\vee$, $\exists$ and substitution by total Σ_1 functions. This is proved just as in Exercise VI.4.18. But, given this, we have an easy proof of Theorem 2.7

from 3.8. Suppose R is Σ_1 on $\mathfrak{Cov}_{\mathfrak{M}}$, say

$$R(p) \leftrightarrow \mathfrak{Cov}_{\mathfrak{M}} \models \exists y\, \varphi(p,y)$$

where φ is Δ_0. Let $\theta(x) = \mathrm{Ord}(x)^I$ and define

$$\Gamma(U) = \{x \mid M \models \theta(x) \wedge \forall y\, Ex\, U(y)\}\,.$$

Then Γ is a Σ_+ inductive definition over $\mathfrak{M}$ and I_Γ is the set of $\{\dot{\beta} \mid \beta < \alpha = o(\mathfrak{Cov}_{\mathfrak{M}})\}$. Furthermore

$$R(p) \text{ iff } \exists z \in I_\Gamma (\mathfrak{M} \models \psi(\langle 0, p\rangle, z))$$

so R is Σ_+ inductive. The other half is trivial since any Σ_+ inductive definition Γ over $\mathfrak{M}$ transforms into a Σ_+ inductive definition $\hat{\Gamma}$ over $\mathfrak{Cov}_{\mathfrak{M}}$, and then, by Gandy's Theorem, $I_{\hat{\Gamma}}$ is Σ_1 on $\mathfrak{Cov}_{\mathfrak{M}}$. □

3.10 Proof of Theorem 2.8. Suppose $\mathfrak{M} \subseteq_{\mathrm{end}} \mathfrak{N}$ and $\mathfrak{M} \prec_1 \mathfrak{N}$. Since $\mathfrak{M} \subseteq_{\mathrm{end}} \mathfrak{N}$, $\mathfrak{Cov}_{\mathfrak{M}} \subseteq_{\mathrm{end}} \mathfrak{Cov}_{\mathfrak{N}}$ so any Σ predicate true in $\mathfrak{Cov}_{\mathfrak{M}}$ is true in $\mathfrak{Cov}_{\mathfrak{N}}$. In particular, the projections for $\mathfrak{Cov}_{\mathfrak{M}}$ and $\mathfrak{Cov}_{\mathfrak{N}}$ agree on $a \in \mathfrak{Cov}_{\mathfrak{M}}$, so we may write $\dot{a}$ for this projection without fear of confusion. Suppose $a \in \mathfrak{Cov}_{\mathfrak{M}}$ and

$$\mathfrak{Cov}_{\mathfrak{N}} \models \exists y\, \varphi(a,y)$$

where φ is Δ_0. Then there is a $\beta < o(\mathfrak{Cov}_{\mathfrak{M}})$ such that $\mathfrak{N}$ is a model of

$$[\exists y(\mathrm{rk}(y) = \dot{\beta} \wedge \varphi(\dot{a}, y))]^I\,,$$

by 3.8. Hence $\mathfrak{N}$ is a model of

$$(1) \qquad \exists z\,[\mathrm{Ord}(z)^I \wedge [\exists y(\mathrm{rk}(y) = z \wedge \varphi(\dot{a}, y)]^I]\,.$$

Since $\mathfrak{M} \prec_1 \mathfrak{N}$, $\mathfrak{M}$ is also a model of (1). By Lemma 3.2, $\mathfrak{M}$ is a model of (Foundation)^I so $\mathfrak{M}$ is a model of

$$\exists z\,[\mathrm{Ord}(z)^1 \wedge [\exists y(\mathrm{rk}(y) = z \wedge \varphi(\dot{a}, y)]^I \wedge \\ [\forall w \in z \neg \exists y(\mathrm{rk}(y) = w \wedge \varphi(\dot{a}, y))]^I]\,.$$

Pick such a "least" z. Since $\mathfrak{M} \subseteq_{\mathrm{end}} \mathfrak{N}$, this least z must be $\leqslant \dot{\beta}$ in the sense of E, so it must be a standard ordinal. That is, there must be some $\gamma < o(\mathfrak{Cov}_{\mathfrak{M}})$ such that $\dot{\gamma} = z$. Thus $\mathfrak{M}$ is a model of

$$\exists y\,[\mathrm{rk}(y) = \dot{\gamma} \wedge \varphi(\dot{a}, y)]$$

so, by 3.8,

$$\mathfrak{Cov}_{\mathfrak{M}} \models \exists y\, \varphi(a,y)\,.$$

Thus $\mathfrak{Cov}_{\mathfrak{M}} \prec_1 \mathfrak{Cov}_{\mathfrak{N}}$. □

3.11—3.13 Exercises

3.11. Prove that a relation $S \subseteq \mathfrak{M}$ (a model of KP) is s-Π_1^1 over $\mathfrak{M}$ iff it is s-Π_1^1 over $\mathfrak{Cov}_{\mathfrak{M}}$.

3.12. Prove the following result of Aczel: Let $S \subseteq \mathfrak{M}$ (a countable model of KPU). Prove that S is s-Π_1^1 on $\mathfrak{M}$ iff S is Σ_+ inductive on $\mathfrak{M}$. [Combine 2.7, 3.9 and VII.3.1.]

3.13. Extend the construction above from models of KP to models of KPU.

4. *Compactness Arguments over Nonstandard Models of Set Theory*

In this final section we want to show how the admissible cover can be used to extend results from standard to nonstandard models. We give two simple examples.

We know from Theorem VII.1.3 that no countable admissible set $\mathbb{A}$ is self-definable. An equivalent statement (in view of Exercise VIII.4.19(iv)) is that if $\mathbb{A}$ is countable, admissible and

$$\mathbb{A} \models \exists \vec{\mathsf{R}}\, \varphi(\vec{\mathsf{R}})$$

for some first order sentence $\varphi(\vec{\mathsf{R}})$ (possibly involving constants from $\mathbb{A}$) then there is a *proper* end extension $\mathfrak{B}$ of $\mathbb{A}$ such that

$$\mathfrak{B} \models \exists \vec{\mathsf{R}}\, \varphi(\vec{\mathsf{R}}).$$

Phrased this way, the result holds for any countable model of KP, standard or nonstandard (or countable model of KPU by 3.13).

4.1 Theorem. *Let* $\mathfrak{M} = \langle M, E \rangle$ *be a countable model of* KP *such that*

$$\mathfrak{M} \models \exists \mathsf{R}\, \varphi(\mathsf{R})$$

for some sentence $\varphi(\mathsf{R})$. *There is a proper end extension* $\mathfrak{N}$ *of* $\mathfrak{M}$ *such that*

$$\mathfrak{N} \models \exists \mathsf{R}\, \varphi(\mathsf{R}).$$

Proof. Let $\mathbb{A} = \mathbb{A}_{\mathfrak{M}} = \mathfrak{Cov}_{\mathfrak{M}}$ and let $\mathsf{L}_{\mathbb{A}}$ be the admissible fragment given by $\mathbb{A}$. Let x be a constant symbol in $\mathbb{A}$ used to denote x, for each $x \in M$, and let T be the following Σ_1 theory of $\mathsf{L}_{\mathbb{A}}$:

$$\forall v [v \mathsf{E} \mathsf{x} \rightarrow \bigvee_{y \in x_E} v = \mathsf{y}]$$
$$\text{diagram}(\mathfrak{M})$$
$$\varphi(\mathsf{R})$$
$$c \neq \mathsf{x} \quad (\text{all } x \in M).$$

We can form the first sentences since $\mathbb{A}$ covers $\mathfrak{M}$. We must prove that T is consistent. Since $\mathbb{A}$ is a countable admissible set, the Compactness Theorem implies that if T is not consistent, then there is a $T_0 \subseteq T$, $T_0 \in \mathbb{A}$ such that T_0 is not consistent. By Corollary 2.4,

$$\{x \in M \mid \mathsf{x} \text{ occurs in } T_0\}$$

is $\mathfrak{M}$-finite. But then there is always some $y \in M$ left over to interpret c so T_0 is consistent. ▯

Our final result extends Theorem 1.1 from standard to nonstandard models of set theory.

4.2 Theorem. *Let* $\mathfrak{M} = \langle M, E\rangle$ *be any countable model of* ZF. *There is an end extension* $\mathfrak{N}$ *of* $\mathfrak{M}$ *which is a model of* ZF + V = L.

Proof. Let $\mathfrak{M}_0$ be the submodel of $\mathfrak{M}$ such that

$$M_0 = \{x \in M \mid \mathfrak{M} \models \text{"}\alpha \text{ is the first stable ordinal"} \wedge x \in L(\alpha)\}.$$

Then by Shoenfield's Absoluteness Lemma (see § V.8)

$$\mathfrak{M}_0 \prec_1 \mathfrak{M}.$$

Let $\mathbb{A} = \mathbb{C}\mathrm{ov}_{\mathfrak{M}}$, $\mathbb{A}_0 = \mathbb{C}\mathrm{ov}_{\mathfrak{M}_0}$, so that $\mathbb{A}_0 \prec_1 \mathbb{A}$ by Theorem 2.8. Let T be the theory of $\mathsf{L}_{\mathbb{A}}$ containing

$$\text{ZF}$$
$$\forall v[v \mathsf{E} \mathsf{x} \leftrightarrow \bigvee_{y \in x_E} v = \mathsf{y}], \quad \text{for all} \quad x \in M.$$

The proof now proceeds exactly like the proof of Theorem 1.1 except that the model of T_0 is not $\mathfrak{M}_0$ but the model $\mathfrak{M}_1$ where

$$M_1 = \{x \in M \mid \mathfrak{M} \models \text{"}x \text{ is constructible"}\}.$$

The reason for using $\mathfrak{M}_1$, rather than $\mathfrak{M}_0$, is that $\mathfrak{M}_1 \not\prec_1 \mathfrak{M}$ (parameters are not allowed in Shoenfield's Lemma) but the statement of Theorem 2.8 requires $\prec_1$. One could equally well improve 2.8. ▯

4.3—4.4 Exercises

4.3. Prove that both assumptions $\mathfrak{M} \models$ KP and $\mathfrak{M}$ is countable are needed for Theorem 4.1.

4.4. Show that if ZF is consistent then there is an uncountable model of ZFC which has no end extension satisfying ZF + V = L.

References

Aczel, P.
1970 Implicit and inductive definability (abstract). J. Symbolic Logic **35**, 599 (1970).

Aczel, P., Richter, W.
1973 Inductive definitions and reflecting properties of admissible ordinals. In: J. E. Fenstad and P. Hinman, eds., Generalized Recursion Theory, pp. 301—381. Amsterdam: North-Holland 1973.

Barwise, J.
1967 Infinitary Logic and Admissible Sets. Ph. D. Thesis. Stanford, CA: Stanford Univ. 1967.
1968 The Syntax and Semantics of Infinitary Logic, editor (Lecture Notes in Math., Vol. 72). Berlin-Heidelberg-New York: Springer 1968.
1969a Infinitary Logic and Admissible Sets. J. Symbolic Logic **34**, 226—252 (1969).
1969b Applications of Strict Π_1^1 predicates to Infinitary Logic, J. Symbolic Logic **34**, 409—423 (1969).
1971 Infinitary Methods in the Model Theory of Set Theory, Logic Colloquium 69, pp. 53—66. Amsterdam: North-Holland 1971.
1973a Back and Forth Through Infinitary Logic, Studies in Model Theory, pp. 5—34, edited by M. Morley. MAA Studies in Mathematics 1973.
1973b Abstract Logics and $L_{\infty\omega}$, Ann. of Math. Logic **4**, 309—340 (1973).
1973c A Preservation Theorem for Interpretations, Proceedings of the Cambridge Logic Conference, pp. 618—621 (Lecture Notes in Math., Vol. 337). Berlin-Heidelberg-New York: Springer 1973.
1974a Admissible Sets over Models of Set Theory, Generalized Recursion Theory, pp. 97—122. Amsterdam: North-Holland 1974.
1974b Mostowski's Collapsing Function and the Closed Unbounded Filter. Fund. Math. **82**, 95—103 (1974).

Barwise, J., Fisher, E.
1970 The Shoenfield Absoluteness Lemma. Israel J. Math. **8**, 329—339 (1970).

Barwise, J., Gandy, R., Moschovakis, Y.
1971 The next admissible set. J. Symbolic Logic **36**, 108—120 (1971).

Barwise, J., Kunen, K.
1971 Hanf numbers for fragments of $L_{\infty\omega}$. Israel J. Math. **10**, 306—320 (1971).

Chang, C. C.
1964 Some new results in definability. Bull. Amer. Math. Soc. **70**, 808—813 (1964).
1968 Some remarks on the model theory of infinitary languages. In: Barwise [1968], pp. 36—63.

Chang, C. C., Moschovakis, Y. N.
1968 On Σ_1^1-relations on special models. Notices Amer. Math. Soc. **15**, 934 (1968).
1970 The Suslin-Kleene theorem for V_κ with cofinality $(\kappa)=\omega$. Pacific J. Math. **35**, 565—569 (1970).

Church, A.
1938 The constructive second number class. Bull. Amer. Math. Soc. **44**, 224—232 (1938).

Church, A., Kleene, S. C.
1937 Formal definitions in the theory of ordinal numbers. Fund. Math. **28**, 11—21 (1937).

Cutland, N.
1972 Σ_1-compactness and ultraproducts. J. Symbolic Logic **37**, 668—672 (1972).

Devlin, K. V.
1973 Aspects of constructibility (Lecture Notes in Math., Vol. 354). Berlin-Heidelberg-New York: Springer 1973.

Dickmann, M. A.
1970 Model theory of infinitary languages (Aarhus University Lecture Notes, series No. 20) 1970.

Ehrenfeucht, A., Kreisel, G.
1966 Strong Models of Arithmetic. Bull. Acad. Polon. Sci. **14**, 107—110 (1966).

Ehrenfeucht, A., Mostowski, A.
1956 Models admitting automorphisms. Fund. Math. **43**, 50—63 (1956).

Enderton, H.
1972 A Mathematical Introduction to Logic. New York and London: Academic Press 1972.

Engeler, E.
1961 Unendliche Formeln in der Modell-Theorie. Z. Math. Logik Grundlagen Math. **7**, 154—160 (1961).

Feferman, S.
1965 Some applications of forcing and generic sets. Fund. Math. **56**, 325—345 (1965).
1968 Lectures on Proof Theory, Proceedings of the Summer School in Logic, Leeds 1967 (Lecture Notes in Math., Vol. 70). Berlin-Heidelberg-New York: Springer 1968.
1974 Some predicative set theories, Axiomatic Set Theory II. Amer. Math. Soc., Providence, R.I. (1974).

Feferman, S., Kreisel, G.
1966 Persistent and invariant formulas relative to theories of higher order. Bull. Amer. Math. Soc. **72**, 480—485 (1966).

Flum, J.
1972 Hanf numbers and well-ordering numbers. Arch. Math. Logik Grundlagenforsch. **15**, 164—178 (1972).

Friedman, H.
1973 Countable models of set theories, Cambridge Summer School in Math. Logic, pp. 539—573 (Lecture Notes in Math., Vol. 337). Berlin-Heidelberg-New York: Springer 1973.

Friedman, H., Jensen, R.
1968 Note on admissible ordinals. In: Barwise [1968], pp. 77—79.

Gaifman, H.
1970 On local arithmetical functions and their application for constructing models of Peano's arithmetic, Mathematical Logic and Foundations of Set Theory (Y. Bar-Hillel, ed.), pp. 105—121. Amsterdam: North-Holland 1970.

Gale, D., Stewart, F. M.
1953 Infinite games of perfect information. Ann. Math. Studies **28**, 245—266 (1953).

Gandy, R. O.
1974 Inductive definitions, Generalized Recursion Theory, pp. 265—300. Amsterdam: North-Holland 1974.
1975 Basic functions, Axiomatic Set Theory II. Amer. Math. Soc., Providence, R.I. (1974).

Gandy, R., Kriesel, G., Tait, W.
1960 Set existence. Bull. Acad. Polon. Ser. Sci. Math. Astron. Phys. **8**, 577—582 (1960).

Garland, S. J.
1972 Generalized interpolation theorems. J. Symbolic Logic **37**, 343—351 (1972).

Gödel, K.
1930 Die Vollständigkeit der Axiome des logischen Funktionenkalküls. Monatsh. Math. Phys. **37**, 349—360 (1930).
1931 Über formal unentscheidbare Sätze der Principia Mathematica und verwandter Systeme I. Monatsh. Math. Phys. **38**, 173—198 (1931).
1939 Consistency proof for the generalized continuum hypothesis. Proc. Natl. Acad. Sci. U.S.A. **25**, 220—224 (1939).
1940 The consistency of the Axiom of Choice and of the Generalized Continuum Hypothesis with the Axioms of Set Theory. Annals Math. Studies 3 (Princeton, N. J., Princeton Univ. Press).

Gordon, C.
1970 Comparisons between some generalizations of recursion theory. Compositio Math. **22**, 333—346 (1970).
1971 Finitistically computable functions and relations on an abstract structure. J. Symbolic Logic **36**, 704 (1971).

Grilliot, T.
1971 Inductive definitions and computability. Trans. Amer. Math. Soc. **158**, 309—317 (1971).
1972 Omitting types; applications to recursion theory. J. Symbolic Logic **37**, 81—89 (1972).

Grzegorczyk, A., Mostowski, A., Ryll-Nardzewski, C.
1958 The classical and ω-complete arithmetic. J. Symbolic Logic **23**, 188—206 (1958).
1961 Definability of sets in models of axiomatic theories. Bull. Acad. Polon. Sci. Ser. Sci. Math. Astronom. Phys. **9**, 163—167 (1961).

Hanf, W.
1964 Incompactness in languages with infinitely long expressions. Fund. Math. **LIII**, 309—324 (1964).

Hanf, W., Scott, D.
1961 Classifying inaccessible cardinals. Notices Amer. Math. Soc. **8**, 445 (1961).

Harrison, J.
1966 Recursive pseudo-wellorderings. Ph. D.Thesis. Stanford, CA: Stanford Univ.

Henkin, L.
1949 The completeness of the first-order predicate calculus. J. Symbolic Logic **14**, 159—166 (1949).
1954 A generalization of the concept of ω-consistency. J. Symbolic Logic **19**, 183—196 (1954).
1957 A generalization of the concept of ω-completeness. J. Symbolic Logic **22**, 1—14 (1957).

Jech, T.
1973 Some combinatorial problems concerning uncountable cardinals. Annals of Math. Logic **5**, 165—198 (1973).

Jensen, R. B.
1972 The fine structure of the constructible hierarchy. Annals of Math. Logic **4**, 229–308 (1972).

Jensen, R., Karp, C.
1971 Primitive recursive set functions, Axiomatic Set Theory. Proceedings of the set theory institute, UCLA, 1967, Amer. Math. Soc. (1971), Providence, R.I.

Karp, C.
1962 Independence proofs in predicate logic with infinitely long expressions. J. Symbolic Logic **27**, 171—188 (1962).
1964 Languages with Expressions of Infinite Length. Amsterdam: North-Holland 1964.
1965 Finite quantifier equivalence, The Theory of Models. (Edited by J. Addison, L. Henkin and A. Tarski) pp. 407—412. Amsterdam: North-Holland 1965.
1967 Nonaxiomatizability results for infinitary systems. J. Symbolic Logic **32**, 367—384 (1967).
1968 An algebraic proof of the Barwise compactness theorem. In: Barwise [1968] pp. 80—95.
1972 From countable to cofinality ω in infinitary model theory. J. Symbolic Logic **37**, 430 (1972).

Keisler, H. J.
1965 Finite approximations of infinitely long formulas, The Theory of Models (edited by J. Addison, L. Henkin and A. Tarski) pp. 158—169. Amsterdam: North-Holland 1965.
1968 Formulas with linearly ordered quantifiers. In: Barwise [1968], pp. 96—130.
1971 Model Theory for Infinitary Logic. Amsterdam: North-Holland.
1973 Forcing and the omitting types theorem, Studies in Model Theory, MAA (1973).

Kino, A., Takeuti, G.
1962 On Hierarchies of predicates of ordinal numbers. J. Math. Soc. Japan **14**, 199—232 (1962).

Kleene, S. C.
1938 On notations for ordinal numbers. J. Symbolic Logic **3**, 150—155 (1938).
1944 On the forms of the predicates in the theory of constructive ordinals. Amer. J. Math. **66**, 41—58 (1944).
1955a On the forms of the predicates in the theory of constructive ordinals [second paper]. Amer. J. Math. **77**, 405—428 (1955).
1955b Arithmetical predicates and function quantifiers. Trans. Amer. Math. Soc. **79**, 312—340 (1955).
1955c Hierarchies of number theoretic predicates. Bull. Amer. Math. Soc. **61**, 193—213 (1955).
1959a Quantification of number theoretic functions. Compositio Math. **14**, 23—40 (1959).
1959b Recursive functionals and quantifiers of finite types I. Trans. Amer. Math. Soc. **91**, 1—52 (1959).

Kreisel, G.
1961 Set theoretic problems suggested by the notion of potential totality. In: Infinitistic Methods, pp. 103—140. Oxford: Pergamon 1961.
1962 The axiom of choice and the class of hyperarithmetic functions. Indag. Math. **24**, 307—319 (1962).
1965 Model theoretic invariants: applications to recursive and hyperarithmetic operatons. In: J. Addison et al. (eds.) Theory of Models (Proceedings of the 1963 Berkeley Symposium) pp. 190—205. Amsterdam: North-Holland 1965.
1968 Choice of infinitary language by means of definability criteria. In: Barwise [1968] pp. 139—151.
1971 Some reasons for generalizing recursion theory. In: R. O. Gandy and C. E. M. Yates (eds.) Logic Colloquium 69, pp. 139—198. Amsterdam: North-Holland 1971.

Kreisel, G., Krivine, J. L.
1967 Elements of mathematical logic. Amsterdam: North-Holland 1967.

Kreisel, G., Sacks, G. E.
1965 Metarecursive sets. J. Symbolic Logic **30**, 318—338 (1965).

Kripke, S.
1964 Transfinite recursion on admissible ordinals, I, II (abstracts). J. Symbolic Logic **29**, 161—162 (1964).

Krivine, J. L., McAloon, K.
1973 Some true unprovable formulas for set theory. Proceedings of the Bertrand Russell Logic Conference, Denmark 1971, pp. 332—341. Leeds 1973.

Kueker, D.
1968 Definability, automorphisms, and infinitary languages. In: Barwise [1968] pp. 152—165.
1970 Generalized interpolation and definability. Annals Math. Logic **1**, 423—468 (1970).
1972 Löwenheim-Skolem and interpolation theorems in infinitary languages. Bull. Amer. Math. Soc. **78**, 211—215 (1972).

Kunen, K.
1968 Implicit definability and infinitary languages. J. Symbolic Logic **33**, 446—451 (1968).

Lévy, A.
1963 Transfinite computability, Notices Amer. Math. Soc. **10**, 286 (1963).
1965 A hierarchy of formulas in set theory. Memoir Amer. Math. Soc. **57** (1965).

Lopez-Escobar, E.
1965 An interpolation theorem for denumerably long sentences. Fund. Math. **LVII**, 253—272 (1965).
1966 On definable well-orderings. Fund. Math. **LVIX**, 13—21 and 299—300 (1966).

Machover, M.
1961 The theory of transfinite recursion. Bull. Amer. Math. Soc. **67**, 575—578 (1961).

Makkai, M.
1964 On a generalization of a theorem of E. W. Beth. Acta Math. Acad. Sci. Hungarica **15**, 227—235 (1964).
1969 An application of a method of Smullyan to logics on admissible sets. Bull. de l'Académie Polon. des Sci., Ser. Math. **17**, 341—346 (1969).
1973 Global definibility theory in $L_{\omega_1\omega}$. Bull. Amer. Math. Soc. **79**, 916—921 (1973).
1975 Applications of a result on weak definability theory in $L_{\omega_1\omega}$ (to appear).

Malitz, J.
1965 Problems in the model theory of infinite languages, Ph. D. Thesis. Berkeley: Univ. of California 1965.
1971 Infinitary analogues of theorems from first order model theory. J. Symbolic Logic **36**, 216—228 (1971).

Montague, R.
1968 Recursion theory as a branch of model theory. In: B. van Rootselaar et al. (eds.) Logic, Methodology and Philosophy of Science III (Proceedings of the 1967 Congress) pp. 63—86. Amsterdam: North-Holland 1968.

Morley, M.
1965 Omitting classes of elements, The Theory of Models (J. W. Addison, L. Henkin and A. Tarski, eds.) pp. 265—273. Amsterdam: North-Holland 1965.
1967 The Hanf number for ω-logic (abstract). J. Symbolic Logic **32**, 437 (1967).
1970 The number of countable models. J. Symbolic Logic **35**, 14—18 (1970).

Morley, M., Morley, V.
1967 The Hanf number for κ-logic (abstract). Notices Amer. Math. Soc. **14**, 556 (1967).

Moschovakis, Y. N.
1969a Abstract first order computability I. Trans. Amer. Math. Soc. **138**, 427—464 (1969).
1969b Abstract first order computability II. Trans. Amer. Math. Soc. **138**, 465—504 (1969).
1970 The Suslin-Kleene theorem for countable structures. Duke Math. J. **37**, 341—352 (1970).
1971 The game quantifier, Proc. Amer. Math. Soc. **31**, 245—250 (1971).
1974 Elementary Induction on Abstract Structures. Amsterdam: North-Holland 1974.

Mostowski, A.
1949 An undecidable arithmetical statement. Fund. Math. **36**, 143—164 (1949).
1961 Formal system of analysis based on an infinitistic rule of proof, Infinitistic Methods. Warsaw 1961.

Nadel, M.
1971 Model Theory in Admissible Sets, Ph. D. Thesis. Univ. of Wisconsin 1971.
1972a Some Löwenheim-Skolem results for admissible sets. Israel J. Math. **12**, 427 (1972).
1972b An application of set theory to model theory. Israel J. Math. **11**, 386 (1972).
1974 Scott sentences and admissible sets. Annals of Math. Logic **7**, 267—294 (1974).

Nerode, A.
1957 General topology and partial recursive functions, Summaries of talks presented at the Summer Institute for Symbolic Logic, pp. 247—251. Cornell Univ. 1957.

Orey, S.
1956 On ω-consistency and related properties. J. Symbolic Logic **21**, 246—252 (1956).

Platek, R.
1966 Foundations of recursion theory, Doctoral Dissertation and Supplement. Stanford, CA: Stanford Univ. 1966.

Reyes, G. E.
1970 Local definability theory. Annals of Math. Logic **1**, 95—137 (1970).

Richter, W.
1971 Recursively Mahlo ordinals and inductive definitions. In: R. O. Gandy and C. E. M. Yates (eds.) Logic Colloquium '69, pp. 273—288. Amsterdam: North-Holland 1971.

Rogers, H. Jr.
1967 Theory of recursive functions and effective computability. New York: McGraw-Hill 1967.

Scott, D.
1964 Invariant Borel Sets. Fund. Math. **56**, 117—128 (1964).
1965 Logic with denumerably long formulas and finite strings of quantifiers, The Theory of Models, pp. 329—341. Amsterdam: North-Holland 1965.

Shoenfield, J. R.
1961 The problem of predicativity, Essays on the Foundations of Mathematics, pp. 132—139. Amsterdam: North-Holland 1961.
1967 Mathematical logic. Reading, Mass.: Addison-Wesley 1967.

Simpson, S. G.
1974 Degree Theory on Admissible Ordinals, Generalized Recursion Theory, pp. 165—194. Amsterdam: North-Holland 1974.

Smullyan, R.
1963 A unifying principle in quantification theory. Proc. Nat. Acad. Sci. **49**, 828—832 (1963).
1965 A unifying principle in quantification theory, The Theory of Models, edited by J. Addison, L. Henkin and A. Tarski, pp. 443—434. Amsterdam: North-Holland 1965.

Spector, C.
1955 Recursive wellorderings. J. Symbolic Logic **20**, 151—163 (1955).
1960 Hyperarithmetical quantifiers. Fund. Math. **48**, 313—320 (1960).
1961 Inductively defined sets of natural numbers. In: Infinitistic methods, pp. 97—102. New York: Pergamon 1961.

Suzuki, Y., Wilmers, G.
1973 Nonstandard models for set theory, Proceedings of the Bertrand Russell Memorial Logic Conference, Denmark 1971, pp. 278—314. Leeds 1973.

Svenonius, L.
1965 On the denumerable models of theories with extra predicates, The Theory of Models, pp. 376—389. Amsterdam: North-Holland 1965.

Takeuti, G.
1960 On the recursive functions of ordinal numbers. J. Math. Soc. Japan **12**, 119—128 (1960).
1965 Recursive functions and arithmetic functions of ordinal numbers. In: Y. Bar-Hillel (ed.) Logic, Methodology and Philosophy of Science II (Proceedings of the 1964 Congress) pp. 179—196. Amsterdam: North-Holland 1965.

Tagué, T.
1959 Predicates recursive in a type-2 object and Kleene hierarchies. Comment. Math. Univ. St. Paul **8**, 97—117 (1959).
1964 On the partial recursive functions of ordinal numbers. J. Math. Soc. Japan **16**, 1—31 (1964).

Vaught, R.
1973 Descriptive set theory in $L_{\omega_1\omega}$, Cambridge Summer School in Math. Logic, pp. 574—598 (Lecture Notes in Math., Vol. 337). Berlin-Heidelberg-New York: Springer 1973.

Ville, F.
1974 More on set existence, Generalized Recursion Theory, pp. 195—208. Amsterdam: North-Holland 1974.

Wilmers, G.
1973 An $\aleph_1$-standard model for ZF which is an element of the minimal model, Proceedings of the Bertrand Russell Memorial Logic Conference, Denmark 1971, pp. 215—326. Leeds 1973.

Index of Notation

Chapter I

Chapter II

Chapter III

Chapter IV

Chapter V

Chapter VI

Chapter VII

Subject Index

Perspectives in Mathematical Logic

In recent years interconnections between different lines of research in mathematical logic and links with other branches of mathematics have proliferated. The subject is now both rich and varied. This series, organized by the Ω-Group, aims to provide, as it were, maps or guides to this complex terrain as seen from various angles. The group is not committed to any particular philosophical program. Nevertheless, the critical discussion which each planned book undergoes ensures that it will represent a coherent line of thought; and that, by developing certain themes, it will be of greater interest than a mere assemblage of results and techniques.

The books in the series differ in level: some are introductory, some highly specialized. They also differ in scope, some offering a wide view of an area while others present more specialized topics. Each book is, at its own level, reasonably self-contained. Although no book depends on another as prerequisite, authors are encouraged to fit their book in with other planned volumes—sometimes deliberately seeking coverage of the same material from different points of view.

Among the next volumes to appear will be:

P. Hinman, Inductive Definitions and Higher Types
D. S. Scott and P. Kraus, Languages and Structure
A. Levy, Basic Set Theory.

Some Lecture Notes in Logic

Lecture Notes in Mathematics

6. H. Hermes, Term Logic with Choice Operator
70. Proceedings of the Summer School in Logic, Leeds, 1967. Ed: M.H. Löb
72. The Syntax and Semantics of Infinitary Languages. Editor: J. Barwise
95. A.S. Troelstra, Principles of Intuitionism
120. O. Siefkes, Büchi's Monadic Second Order Successor Arithmetic
212. B. Scarpellini, Proof Theory and Intuitionistic Systems
217. T.J. Jech, Lectures in Set Theory. With Particular Emphasis on the Method of Forcing
223. U. Felgner, Models of ZF-Set Theory
255. Conference in Mathematical Logic—London '70. Editor: W. Hodges
306. H. Luckhardt, Extensional Gödel Functional Interpretation. A Consistency Proof of Classical Analysis
328. J.R. Büchi, D. Siefkes, The Monadic Second Order Theory of All Countable Ordinals. Decidable Theories II
337. Cambridge Summer School in Mathematical Logic. Eds: A.R.D. Mathias, H. Rogers
344. Metamathematical Investigation of Intuitionistic Arithmetic and Analysis. Editor: A.S. Troelstra
354. K.J. Devlin, Aspects of Constructibility
405. K.J. Devlin, H. Johnsbraten, The Souslin Problem
447. S.A. Toledo, Tableau Systems for First Order Number Theory and Certain Higher Order Theories
450. Algebra und Logic. Editor: J.N. Crossley
453. Logic Colloquium. Editor: R. Parikh
454. J. Hirschfeld, W.H. Wheeler, Forcing, Arithmetic, and Division Rings
492. D.W. Kueker, Infinitary Logic: In Memoriam Carol Karp
498. Model Theory and Algebra. A Memorial Tribute to Abraham Robinson. Edited by D.H. Saracino, V.B. Weispfenning
499. Logic Conference, Kiel 1974. Edited by G.H. Müller, A. Oberschelp, K. Potthoff
500. Proof Theory Symposium, Kiel 1974. Edited by J. Diller, G.H. Müller

Lecture Notes in Computer Science

37. C. Böhm, λ-Calculus and Computer Science Theory